Solved Problems in Classical M

Solved Problems in Classical Mechanics

Analytical and numerical solutions
with comments

O.L. de Lange and J. Pierrus
*School of Physics, University of KwaZulu-Natal,
Pietermaritzburg, South Africa*

OXFORD
UNIVERSITY PRESS

Great Clarendon Street, Oxford OX2 6DP

Oxford University Press is a department of the University of Oxford.
It furthers the University's objective of excellence in research, scholarship,
and education by publishing worldwide in

Oxford   New York

Auckland Cape Town Dar es Salaam Hong Kong Karachi
Kuala Lumpur Madrid Melbourne Mexico City Nairobi
New Delhi Shanghai Taipei Toronto

With offices in

Argentina Austria Brazil Chile Czech Republic France Greece
Guatemala Hungary Italy Japan Poland Portugal Singapore
South Korea Switzerland Thailand Turkey Ukraine Vietnam

Oxford is a registered trade mark of Oxford University Press
in the UK and in certain other countries

Published in the United States
by Oxford University Press Inc., New York

© O.L. de Lange and J. Pierrus 2010

The moral rights of the authors have been asserted
Database right Oxford University Press (maker)

First published 2010

All rights reserved. No part of this publication may be reproduced,
stored in a retrieval system, or transmitted, in any form or by any means,
without the prior permission in writing of Oxford University Press,
or as expressly permitted by law, or under terms agreed with the appropriate
reprographics rights organization. Enquiries concerning reproduction
outside the scope of the above should be sent to the Rights Department,
Oxford University Press, at the address above

You must not circulate this book in any other binding or cover
and you must impose the same condition on any acquirer

British Library Cataloguing in Publication Data

Data available

Library of Congress Cataloging in Publication Data

Data available

Typeset by SPI Publisher Services, Pondicherry, India
Printed in Great Britain
on acid-free paper by
CPI Antony Rowe, Chippenham, Wiltshire

ISBN 978–0–19–958252–5 (Hbk)
978–0–19–958251–8 (Pbk)

1 3 5 7 9 10 8 6 4 2

Preface

It is in the study of classical mechanics that we first encounter many of the basic ingredients that are essential to our understanding of the physical universe. The concepts include statements concerning space and time, velocity, acceleration, mass, momentum and force, and then an equation of motion and the indispensable law of action and reaction – all set (initially) in the background of an inertial frame of reference. Units for length, time and mass are introduced and the sanctity of the balance of units in any physical equation (dimensional analysis) is stressed. Reference is also made to the task of measuring these units – metrology, which has become such an astonishing science/art.

The rewards of this study are considerable. For example, one comes to appreciate Newton's great achievement – that the dynamics of the classical universe can be understood via the solutions of differential equations – and this leads on to questions regarding determinism and the effects of even small uncertainties or disturbances. One learns further that even when Newton's dynamics fails, many of the concepts remain indispensable and some of its conclusions retain their validity – such as the conservation laws for momentum, angular momentum and energy, and the connection between conservation and symmetry – and one discusses the domain of applicability of the theory. Along the way, a student encounters techniques – such as the use of vector calculus – that permeate much of physics from electromagnetism to quantum mechanics.

All this is familiar to lecturers who teach physics at universities; hence the emphasis on undergraduate and graduate courses in classical mechanics, and the variety of excellent textbooks on the subject. It has, furthermore, been recognized that training in this and related branches of physics is useful also to students whose careers will take them outside physics. It seems that here the problem-solving abilities that physics students develop stand them in good stead and make them desirable employees.

Our book is intended to assist students in acquiring such analytical and computational skills. It should be useful for self-study and also to lecturers and students in mechanics courses where the emphasis is on problem solving, and formal lectures are kept to a minimum. In our experience, students respond well to this approach. After all, the rudiments of the subject can be presented quite succinctly (as we have endeavoured to do in Chapter 1) and, where necessary, details can be filled in using a suitable text.

With regard to the format of this book: apart from the introductory chapter, it consists entirely of questions and solutions on various topics in classical mechanics that are usually encountered during the first few years of university study. It is suggested that a student first attempt a question with the solution covered, and only consult the solution for help where necessary. Both analytical and numerical (computer) techniques are used, as appropriate, in obtaining and analyzing solutions. Some of the numerical questions are suitable for project work in computational physics (see the Appendix). Most solutions are followed by a set of comments that are intended to stimulate inductive reasoning (additional analysis of the problem, its possible extensions and further significance), and sometimes to mention literature we have found helpful and interesting. We have included questions on bits of 'theory' for topics where students initially encounter difficulty – such as the harmonic oscillator and the theory of mechanical energy – because this can be useful, both in revising and cementing ideas and in building confidence.

The mathematical ability that the reader should have consists mainly of the following: an elementary knowledge of functions – their roots, turning points, asymptotic values and graphs – including the 'standard' functions of physics (polynomial, trigonometric, exponential, logarithmic, and rational); the differential and integral calculus (including partial differentiation); and elementary vector analysis. Also, some knowledge of elementary mechanics and general physics is desirable, although the extent to which this is necessary will depend on the proclivities of the reader.

For our computer calculations we use *Mathematica*®, version 7.0. In each instance the necessary code (referred to as a notebook) is provided in a shadebox in the text. Notebooks that include the interactive `Manipulate` function are given in Chapters 6, 10, 11 and 13 (and are listed in the Appendix). They enable the reader to observe motion on a computer screen, and to study the effects of changing relevant parameters. A reader without prior knowledge of *Mathematica* should consult the tutorial ('First Five Minutes with *Mathematica*') and the on-line Help. Also, various useful tutorials can be downloaded from the website www.Wolfram.com. All graphs of numerical results have been drawn to scale using Gnuplot.

In our analytical solutions we have tried to strike a balance between burdening the reader with too much detail and not heeding Littlewood's dictum that "two trivialities omitted can add up to an impasse". In this regard it is probably not possible to satisfy all readers, but we hope that even tentative ones will soon be able to discern footprints in the mist. After all, it is well worth the effort to learn that (on some level) the rules of the universe are simple, and to begin to enjoy "the unreasonable effectiveness of mathematics in the natural sciences" (Wigner).

Finally, we thank Robert Lindebaum and Allard Welter for their assistance with our computer queries and also Roger Raab for helpful discussions.

Pietermaritzburg, South Africa   O. L. de Lange
January 2010   J. Pierrus

Contents

1	Introduction	1
2	Miscellanea	11
3	One-dimensional motion	30
4	Linear oscillations	60
5	Energy and potentials	92
6	Momentum and angular momentum	127
7	Motion in two and three dimensions	157
8	Spherically symmetric potentials	216
9	The Coulomb and oscillator problems	263
10	Two-body problems	286
11	Multi-particle systems	325
12	Rigid bodies	399
13	Non-linear oscillations	454
14	Translation and rotation of the reference frame	518
15	The relativity principle and some of its consequences	557
	Appendix	588
	Index	590

1
Introduction

The following outline of the rudiments of classical mechanics provides the background that is necessary in order to use this book. For the reader who finds our presentation too brief, there are several excellent books that expound on these basics, such as those listed below.[1–4]

1.1 Kinematics and dynamics of a single particle

The goal of classical mechanics is to provide a quantitative description of the motion of physical objects. Like any physical theory, mechanics is a blend of definitions and postulates. In describing this theory it is convenient to first introduce the concept of a point object (a particle) and to start by considering the motion of a single particle.

To this end one must make an assumption concerning the geometry of space. In Newtonian dynamics it is assumed that space is three-dimensional and Euclidean. That is, space is spanned by the three coordinates of a Cartesian system; the distance between any two points is given in terms of their coordinates by Pythagoras's theorem, and the familiar geometric and algebraic rules of vector analysis apply. It is also assumed – at least in non-relativistic physics – that time is independent of space. Furthermore, it is supposed that space and time are 'sufficiently' continuous that the differential and integral calculus can be applied. A helpful discussion of these topics is given in Griffiths's book.[2]

With this background, one selects a coordinate system. Often, this is a rectangular or Cartesian system consisting of an arbitrarily chosen coordinate origin O and three orthogonal axes, but in practice any convenient system can be used (spherical, cylindrical, etc.). The position of a particle relative to this coordinate system is specified by a vector function of time – the position vector $\mathbf{r}(t)$. An equation for $\mathbf{r}(t)$ is known as the trajectory of the particle, and finding the trajectory is the goal mentioned above.

In terms of $\mathbf{r}(t)$ we define two indispensable kinematic quantities for the particle: the velocity $\mathbf{v}(t)$, which is the time rate of change of the position vector,

---

[1]   L. D. Landau, A. I. Akhiezer, and E. M. Lifshitz, *General physics: mechanics and molecular physics*. Oxford: Pergamon, 1967.
[2]   J. B. Griffiths, *The theory of classical dynamics*. Cambridge: Cambridge University Press, 1985.
[3]   T. W. B. Kibble and F. H. Berkshire, *Classical mechanics*. London: Imperial College Press, 5th edn, 2004.
[4]   R. Baierlein, *Newtonian dynamics*. New York: McGraw-Hill, 1983.

$$\mathbf{v}(t) = \frac{d\mathbf{r}(t)}{dt}, \tag{1}$$

and the acceleration $\mathbf{a}(t)$, which is the time rate of change of the velocity,

$$\mathbf{a}(t) = \frac{d\mathbf{v}(t)}{dt}. \tag{2}$$

It follows from (1) and (2) that the acceleration is also the second derivative

$$\mathbf{a} = \frac{d^2\mathbf{r}}{dt^2}. \tag{3}$$

Sometimes use is made of Newton's notation, where a dot denotes differentiation with respect to time, so that (1)–(3) can be abbreviated

$$\mathbf{v} = \dot{\mathbf{r}}, \qquad \mathbf{a} = \dot{\mathbf{v}} = \ddot{\mathbf{r}}. \tag{4}$$

The stage for mechanics – the frame of reference – consists of a coordinate system together with clocks for measuring time. Initially, we restrict ourselves to an inertial frame. This is a frame in which an isolated particle (one that is free of any applied forces) moves with constant velocity $\mathbf{v}$ – meaning that $\mathbf{v}$ is constant in both magnitude and direction (uniform rectilinear motion). This statement is the essence of Newton's first law of motion. In Newton's mechanics (and also in relativity) an inertial frame is not a unique construct: any frame moving with constant velocity with respect to it is also inertial (see Chapters 14 and 15). Consequently, if one inertial frame exists, then infinitely many exist. Sometimes mention is made of a primary inertial frame, which is at rest with respect to the 'fixed' stars.

Now comes a central postulate of the entire theory: in an inertial frame, if a particle of mass m is acted on by a force $\mathbf{F}$, then

$$\mathbf{F} = \frac{d\mathbf{p}}{dt}, \tag{5}$$

where

$$\mathbf{p} = m\mathbf{v} \tag{6}$$

is the momentum of the particle relative to the given inertial frame. Equation (5) is the content of Newton's second law of motion: it provides the means for determining the trajectory $\mathbf{r}(t)$, and is known as the equation of motion. If the mass of the particle is constant then (5) can also be written as

$$m\frac{d\mathbf{v}}{dt} = \mathbf{F}, \tag{7}$$

or, equivalently,

$$m\frac{d^2\mathbf{r}}{dt^2} = \mathbf{F}. \tag{8}$$

The theory is completed by postulating a restriction on the interaction between any two particles (Newton's third law of motion): if $\mathbf{F}_{12}$ is the force that particle 1 exerts on particle 2, and if $\mathbf{F}_{21}$ is the force that particle 2 exerts on particle 1, then

$$\mathbf{F}_{21} = -\mathbf{F}_{12}.\qquad(9)$$

That is, the mutual actions between particles are always equal in magnitude and opposite in direction. (See also Question 10.5.)

The realization that the dynamics of the physical world can be studied by solving differential equations is one of Newton's great achievements, and many of the problems discussed in this book deal with this topic. His theory shows that (on some level) it is possible to predict the future and to unravel the past.

The reader may be concerned that, from a logical point of view, two new quantities (mass and force) are introduced in the single statement (5). However, by using both the second and third laws, (5) and (9), one can obtain an operational definition of relative mass (see Question 2.6). Then (5) can be regarded as defining force.

Three ways in which the equation of motion can be applied are:

- ☞ Use a trajectory to determine the force. For example, elliptical planetary orbits – with the Sun at a focus – imply an attractive inverse-square force (see Question 8.13).
- ☞ Use a force to determine the trajectory. For example, parabolic motion in a uniform field (see Question 7.1).
- ☞ Use a force and a trajectory to determine particle properties. For example, the electric charge from rectilinear motion in a combined gravitational and electrostatic field, and the electric charge-to-mass ratio from motion in uniform electrostatic and magnetostatic fields (see Questions 3.11, 7.19 and 7.20).

1.2 Multi-particle systems

The above formulation is readily extended to multi-particle systems. We follow standard notation and let m_i and $\mathbf{r}_i$ denote the mass and position vector of the ith particle, where $i = 1, 2, \cdots, N$ for a system of N particles. The velocity and acceleration of the ith particle are denoted $\mathbf{v}_i$ and $\mathbf{a}_i$, respectively. The equations of motion are

$$\mathbf{F}_i = \frac{d\mathbf{p}_i}{dt} \qquad (i = 1, 2, \cdots, N),\qquad(10)$$

where $\mathbf{p}_i = m_i \mathbf{v}_i$ is the momentum of the ith particle relative to a given inertial frame, and $\mathbf{F}_i$ is the total force on this particle.

In writing down the $\mathbf{F}_i$ it is useful to distinguish between interparticle forces, due to interactions among the particles of the system, and external forces associated with sources outside the system. The total force on particle i is the vector sum of all interparticle and external forces. Thus, one writes

$$\mathbf{F}_i = \sum_{j \neq i} \mathbf{F}_{ji} + \mathbf{F}_i^{(e)} \qquad (i = 1, 2, \cdots, N),\qquad(11)$$

where $\mathbf{F}_{ji}$ is the force that particle j exerts on particle i, and $\mathbf{F}_i^{(e)}$ is the external force on particle i. In (11) the sum over j runs from 1 to N but excludes $j = i$. The interparticle forces are all assumed to obey the third law

$$\mathbf{F}_{ji} = -\mathbf{F}_{ij} \qquad (i, j = 1, 2, \cdots, N). \tag{12}$$

From (10) and (11) we have the equations of motion of a system of particles in terms of interparticle forces and external forces:

$$\frac{d\mathbf{p}_i}{dt} = \sum_{j \neq i} \mathbf{F}_{ji} + \mathbf{F}_i^{(e)} \qquad (i = 1, 2, \cdots, N). \tag{13}$$

If the masses m_i are all constant then (13) can be written as

$$m_i \frac{d^2 \mathbf{r}_i}{dt^2} = \sum_{j \neq i} \mathbf{F}_{ji} + \mathbf{F}_i^{(e)} \qquad (i = 1, 2, \cdots, N). \tag{14}$$

These are the equations of motion for the classical N-particle problem. In general, they are a set of N coupled differential equations, and they are usually intractable.

Two of the four presently known fundamental interactions are applicable in classical mechanics, namely the gravitational and electromagnetic forces. For the former, Newton's law of gravitation is usually a satisfactory approximation. For electromagnetic forces there are Coulomb's law of electrostatics, the Lorentz force, and multipole interactions. Often, it is impractical to deduce macroscopic forces (such as friction and viscous drag) from the electromagnetic interactions of particles, and instead one uses phenomenological expressions.

Another method of approximating forces is through the simple expedient of a spatial Taylor-series expansion, which opens the way to large areas of physics. Here, the first (constant) term represents a uniform field; the second (linear) term encompasses a 'Hooke's-law'-type force associated with linear (harmonic) oscillations; the higher-order (quadratic, cubic, ...) terms are non-linear (anharmonic) forces that produce a host of non-linear effects (see Chapter 13).

Also, there are many approximate representations of forces in terms of various potentials (Lennard-Jones, Morse, Yukawa, Pöschl–Teller, Hulthén, etc.), which are useful in molecular, solid-state and nuclear physics. The Newtonian concepts of force and potential have turned out to be widely applicable – even to the statics and dynamics of such esoteric yet important systems as flux quanta (Abrikosov vortices) in superconductors and line defects (dislocations) in crystals.

Some of the most impressive successes of classical mechanics have been in the field of astronomy. And so it seems ironic that one of the major unanswered questions in physics concerns observed dynamics – ranging from galactic motion to accelerating expansion of the universe – for which the source and nature of the force are uncertain (dark matter and dark energy, see Question 11.20).

1.3 Newton and Maxwell

The above outline of Newtonian dynamics relies on the notion of a particle. The theory can also be formulated in terms of an extended object (a 'body'). This is the form used originally by Newton, and subsequently by Maxwell and others. In his fascinating study of the *Principia Mathematica*, Chandrasekhar remarks that Maxwell's "is a rarely sensitive presentation of the basic concepts of Newtonian dynamics" and "is so completely in the spirit of the *Principia* and illuminating by itself ... ."[5]

Maxwell emphasized "that by the velocity of a body is meant the velocity of its centre of mass. The body may be rotating, or it may consist of parts, and be capable of changes of configuration, so that the motions of different parts may be different, but we can still assert the laws of motion in the following form:

Law I. – The centre of mass of the system perseveres in its state of rest, or of uniform motion in a straight line, except in so far as it is made to change that state by forces acting on the system from without.

Law II. – The change of momentum during any interval of time is measured by the sum of the impulses of the external forces during that interval."[5]

In Newtonian dynamics, the position of the centre of mass of any object is a unique point in space whose motion is governed by the two laws stated above. The concept of the centre of mass occurs in a straightforward manner[5] (see also Chapter 11) and it plays an important role in the theory and its applications.

Often, the trajectory of the centre of mass relative to an inertial frame is a simple curve, even though other parts of the body may move in a more complicated manner. This is nicely illustrated by the motion of a uniform rod thrown through the air: to a good approximation, the centre of mass describes a simple parabolic curve such as P in the figure, while other points in the rod may follow a more complicated three-dimensional trajectory, like Q. If the rod is 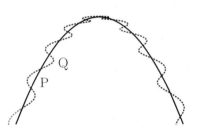 thrown in free space then its centre of mass will move with constant velocity (that is, in a straight line and with constant speed) while other parts of the rod may have more intricate trajectories. In general, the motion of a free rigid body in an inertial frame is more complicated than that of a free particle (see Question 12.22).

1.4 Newton and Lagrange

The first edition of the *Principia Mathematica* was published in July 1687, when Newton was 44 years old. Much of it was worked out and written between about August 1684 and May 1686, although he first obtained some of the results about twenty years earlier, especially during the plague years 1665 and 1666 "for in those days I was in the prime of my age for invention and minded Mathematicks and Philosophy more than at any time since."[5]

After Newton had laid the foundations of classical mechanics, the scene for many subsequent developments shifted to the Continent, and especially France, where

---

[5] S. Chandrasekhar, *Newton's Principia for the common reader*, Chaps. 1 and 2. Oxford: Clarendon Press, 1995.

important works were published by d'Alembert (1717–1783), Lagrange (1736–1813), de Laplace (1749–1827), Legendre (1725–1833), Fourier (1768–1830), Poisson (1781–1840), and others. In particular, an alternative formulation of classical particle dynamics was presented by Lagrange in his *Mécanique Analytique* (1788).

To describe this theory it is helpful to consider first a single particle of constant mass m moving in an inertial frame. We suppose that all the forces acting are conservative: then the particle possesses potential energy $V(\mathbf{r})$ in addition to its kinetic energy $K = \frac{1}{2}m\dot{\mathbf{r}}^2$, and the force is related to $V(\mathbf{r})$ by $\mathbf{F} = -\boldsymbol{\nabla}V$ (see Chapter 5). So, Newton's equation of motion in Cartesian coordinates x_1, x_2, x_3 has components

$$m\ddot{x}_i = F_i = -\partial V/\partial x_i \qquad (i = 1, 2, 3). \tag{15}$$

Also, $\partial K/\partial x_i = 0$, $\partial K/\partial \dot{x}_i = m\dot{x}_i$, and $\partial V/\partial \dot{x}_i = 0$. Therefore (15) can be recast in the form

$$\frac{d}{dt}\frac{\partial \mathsf{L}}{\partial \dot{x}_i} - \frac{\partial \mathsf{L}}{\partial x_i} = 0 \qquad (i = 1, 2, 3), \tag{16}$$

where $\mathsf{L} = K - V$. The quantity $\mathsf{L}(\mathbf{r},\dot{\mathbf{r}})$ is known as the Lagrangian of the particle. The Lagrange equations (16) imply that the action integral

$$I = \int_{t_1}^{t_2} \mathsf{L}\, dt \tag{17}$$

is stationary (has an extremum – usually a minimum) for any small variation of the coordinates x_i:

$$\delta I = 0. \tag{18}$$

Equations (16) hold even if V is a function of t, as long as $\mathbf{F} = -\boldsymbol{\nabla}V$.

This account can be generalized:

- ☞ It applies to systems containing an arbitrary number of particles N.
- ☞ The coordinates used need not be Cartesian; they are customarily denoted $q_1, q_2, \cdots, q_f$ ($f = 3N$) and are known as generalized coordinates. (In practice, the choice of these coordinates is largely a matter of convenience.) The corresponding time derivatives are the generalized velocities, and the Lagrangian is a function of these $6N$ coordinates and velocities:

$$\mathsf{L} = \mathsf{L}(q_1, q_2, \cdots, q_f\, ;\, \dot{q}_1, \dot{q}_2, \cdots, \dot{q}_f). \tag{19}$$

Often, we will abbreviate this to $\mathsf{L} = \mathsf{L}(q_i, \dot{q}_i)$.
- ☞ The Lagrangian is required to satisfy the action principle (18), and this implies the Lagrange equations

$$\frac{d}{dt}\frac{\partial \mathsf{L}}{\partial \dot{q}_i} - \frac{\partial \mathsf{L}}{\partial q_i} = 0 \qquad (i = 1, 2, \cdots, 3N), \tag{20}$$

where $\mathsf{L} = K - V$, and K and V are the total kinetic and potential energies of the system.[2]

☞ The Lagrangian formulation applies also to non-conservative systems such as charged particles in time-dependent electromagnetic fields and damped harmonic oscillators (see Question 4.16). Lagrangians can also be constructed for systems with variable mass. In these instances L is not of the form $K - V$.

☞ The Lagrange equations (20) can be expressed as

$$dp_i/dt = F_i, \tag{21}$$

where

$$p_i = \partial \mathsf{L}/\partial \dot{q}_i \quad \text{and} \quad F_i = \partial \mathsf{L}/\partial q_i \tag{22}$$

are known as the generalized momenta and generalized forces. In Cartesian coordinates, **p** is equal to mass × velocity.

☞ The action principle (18) is valid in any frame of reference, even a non-inertial frame (one that is accelerating relative to an inertial frame). However, in a non-inertial frame the Lagrangian is modified by the acceleration, and Lagrange's equations (16) yield the equation of motion (24) below – see Question 14.22.

Although the Newtonian formulation (based on force) and the Lagrangian formulation (based on a scalar L that often derives from kinetic and potential energies) look very different, they are completely equivalent and must yield the same results in practice. There are several reasons for the importance of the Lagrange approach, such as:

☞ It may be simpler to obtain the equation of motion by working with energy rather than by taking account of all the forces.

☞ Constrained motion is more easily treated.

☞ Conserved quantities can be readily identified.

☞ The action principle is a fundamental part of physics, and it provides a powerful formulation of classical mechanics. For example, the theory can be extended to continuous systems by introducing a Lagrangian density whose volume integral is the Lagrangian. In this version the Lagrangian formulation has important applications to field theory and quantum mechanics.

1.5 Non-inertial frames of reference

This section outlines a topic that is considered in more detail in Chapter 14 and is used occasionally in earlier chapters.

Often, the frame of reference that one uses is not inertial, either by circumstance (for example, a frame fixed on the Earth is non-inertial) or by choice (it may be convenient to solve a particular problem in a non-inertial frame). And so the question arises: what is the form of the equation of motion in a non-inertial frame (that is, a frame that is accelerating with respect to an inertial frame)?

This leads one to consider a frame S' that is translating and rotating with respect to an inertial frame S. These frames are depicted in the figure below, where **r** is the position vector of a particle of mass m relative to S and **r'** is its position vector relative to S'. The frame S' has origin O' and coordinate axes $x'y'z'$.

8 *Solved Problems in Classical Mechanics*

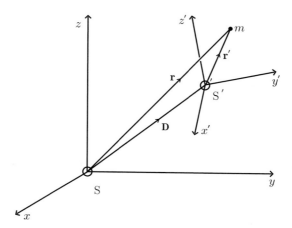

The motion of S' is described by two vectors: the position vector $\mathbf{D}(t)$ of the origin O' relative to S, and the angular velocity $\boldsymbol{\omega}(t)$ of S' relative to a third frame S'' that has origin at O' and axes $x''y''z''$, which are parallel to the corresponding axes xyz of S. This angular velocity is given in terms of a unit vector $\hat{\mathbf{n}}$ (that specifies the axis of rotation relative to S'') and the angle $d\theta$ rotated through in a time dt by

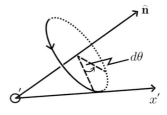

$$\boldsymbol{\omega} = \frac{d\theta}{dt}\hat{\mathbf{n}}, \qquad (23)$$

where the sense of rotation and the direction of $\hat{\mathbf{n}}$ are connected by the right-hand rule illustrated in the figure.

Starting from the equation of motion (8) for a single particle of constant mass m in an inertial frame S, it can be shown that the equation of motion in the translating and rotating frame S' can be expressed in the form (see Chapter 14)

$$m\frac{d^2\mathbf{r}'}{dt^2} = \mathbf{F}_{\mathrm{e}}. \qquad (24)$$

Here

$$\mathbf{F}_{\mathrm{e}} = \mathbf{F} + \mathbf{F}_{\mathrm{tr}} + \mathbf{F}_{\mathrm{Cor}} + \mathbf{F}_{\mathrm{cf}} + \mathbf{F}_{\mathrm{az}}, \qquad (25)$$

where

$$\mathbf{F}_{\mathrm{tr}} = -m\frac{d^2\mathbf{D}}{dt^2}, \qquad (26)$$

$$\mathbf{F}_{\mathrm{Cor}} = -2m\boldsymbol{\omega} \times \frac{d\mathbf{r}'}{dt}, \qquad (27)$$

$$\mathbf{F}_{\mathrm{cf}} = -m\boldsymbol{\omega} \times (\boldsymbol{\omega} \times \mathbf{r}'), \qquad (28)$$

$$\mathbf{F}_{\mathrm{az}} = -m\frac{d\boldsymbol{\omega}}{dt} \times \mathbf{r}'. \qquad (29)$$

We mention that (24) is not a separate postulate, but is a consequence of (8) and the assumptions that space is absolute (meaning $\mathbf{r} = \mathbf{r}' + \mathbf{D}$ in the first of the above figures), time is absolute (meaning $t' = t$), and mass is absolute (meaning $m' = m$). Note that the relation $\mathbf{r} = \mathbf{r}' + \mathbf{D}$ is not simply a consequence of the triangle law for addition of vectors, because $\mathbf{r}$ and $\mathbf{r}'$ are measured by observers who are moving relative to each other – see Chapter 15.

We can interpret the equation of motion (24) in the following way: if we wish to write Newton's second law in a non-inertial frame S' in the same way as in an inertial frame S (i.e. as force = mass × acceleration), then the force $\mathbf{F}$ due to physical interactions (such as electromagnetic interactions) must be replaced by an effective force $\mathbf{F}_e$ that includes the four additional contributions $\mathbf{F}_{tr}$, $\mathbf{F}_{Cor}$, $\mathbf{F}_{cf}$, and $\mathbf{F}_{az}$. Collectively, these contributions are variously referred to in the literature as:

- ☞ 'inertial forces' (because each involves the particle's inertial mass m);
- ☞ 'non-inertial forces' (because each is present only in a non-inertial frame);
- ☞ 'fictitious forces' (to emphasize that they are not due to physical interactions but to the acceleration of the frame S' relative to S).

Each of the forces (26)–(29) also has its own name: $\mathbf{F}_{tr}$ is known as the translational force (it occurs whenever the origin of the non-inertial system accelerates relative to an inertial frame); $\mathbf{F}_{Cor}$ is the Coriolis force (it acts on a moving particle unless the motion in S' is parallel or anti-parallel to $\boldsymbol{\omega}$); $\mathbf{F}_{cf}$ is the centrifugal force, and it acts even on a particle at rest in S'; $\mathbf{F}_{az}$ is the azimuthal force, and it occurs only if the non-inertial frame has an angular acceleration $d\boldsymbol{\omega}/dt$ relative to S.

1.6 Homogeneity and isotropy of space and time

In addition to the fact that the laws of motion assume their simplest forms in inertial frames, these frames also possess unique properties with respect to space and time. For a free particle in an inertial frame these are: First, all positions in inertial space are equivalent with regard to mechanics. This is known as the homogeneity of space in inertial frames. Secondly, all directions in space are equivalent. This is the isotropy of space. Thirdly, all instants of time are equivalent (homogeneity of time). Fourthly, there is invariance with respect to reversal of motion – the replacement $t \to -t$ (isotropy of time). These symmetries of space and time in inertial frames play a fundamental role in physics. For example, in the conservation laws for energy, momentum and angular momentum, and in the space-time transformation between inertial frames (see Chapters 14 and 15). In a non-inertial frame these properties do not hold. For example, if one stands on a rotating platform it is noticeable that positions on and off the axis of rotation are not equivalent: space is not homogeneous in such a frame.

Notwithstanding the fact that, in general, Newtonian dynamics is most simply formulated in inertial space, one should keep in mind the following proviso. Namely, that the solution to certain problems is facilitated by choosing a suitable non-inertial frame. Thus the trajectory of a particle at rest on a rotating turntable is simplest in the frame of the turntable, where the particle is in static equilibrium under the action of four forces (weight, normal reaction, friction and centrifugal force). Similarly, for a charged particle in a uniform magnetostatic field, one can transform away the magnetic force: relative to a specific rotating and translating frame the particle is in static equilibrium, whereas relative to inertial space the trajectory is a helix of constant pitch (see Question 14.25).

1.7 The importance of being irrelevant

There are several obvious questions one can ask concerning Newtonian dynamics, which can all be formulated: 'Does it matter if $\cdots$?' All are answered in the negative and have deep consequences for physics.

The first concerns the units in which mass, length, and time are measured. Humans (and probably also other life in the universe) have devised an abundance of different physical units. In principle, there are infinitely many and one can ask whether the validity of Newton's second law is affected by an arbitrary choice of units. The answer is 'no': the law is valid in any system of units because each side of the equation $\mathbf{F} = m\mathbf{a}$ must have the same units (see also Question 2.9). Thus, the unit of force in the MKS system (the newton) is, by definition, $1 \, \text{kg m s}^{-2}$.

This seemingly simple property is required of all physical laws: they do not depend on an arbitrary choice of units because each side of an equation expressing the law is required to have the same physical dimensions. The consequences of this are dimensional analysis (see Chapter 2), similarity and scaling.[6] The fact that physical laws are equally valid in all systems of units is an example of a 'relativity principle'.

Similarly, one can ask whether the mechanical properties of an isolated (closed) system depend in any way on its position or orientation in inertial space. The statement that they do not implies, respectively, the conservation of momentum and angular momentum of the system (see Questions 14.7, 14.18 and 14.19).

Furthermore, in Newtonian dynamics any choice of inertial frame (from among an infinite set of frames in uniform, rectilinear relative motion) is acceptable because the laws of motion are equally valid in all such frames. The extension of this property to all the laws of physics constitutes Einstein's relativity principle. A remarkable consequence of this principle is that there are just two possibilities for the space-time transformation between inertial frames: relative space-time (in a universe in which there is a finite universal speed) or Newton's absolute space-time (if this speed is infinite) – see Chapter 15.

Further extensions of this type of reasoning have led to a theory of elementary particles and their interactions.[7] So, this concept of irrelevance (or invariance, as it is known in physics) which emerged from Newton's mechanics, and was later emphasized particularly by Einstein, has turned out to be extremely fruitful. The reader may wonder what physics would be like if these invariances did not hold.

---

[6]   G. I. Barenblatt, *Scaling, self-similarity, and intermediate asymptotics*. Cambridge: Cambridge University Press, 1996.
[7]   See, for example, G. t' Hooft, "Gauge theories of the forces between elementary particles," Scientific American, vol. 242, pp. 90–116, June 1980.

2
Miscellanea

This chapter contains questions dealing with three disparate topics, namely sensitivity of trajectories to small changes in initial conditions; the reasons why we consider just one, rather than three types of mass; and the use of dimensional reasoning in the analysis of physical problems. The reader may wish to omit this chapter at first, and return to it at a later stage.

Question 2.1

A particle moves in one dimension along the x-axis, bouncing between two perfectly reflecting walls at $x = 0$ and $x = \ell$. In between collisions with the walls no forces act on the particle. Suppose there is an uncertainty Δv_0 in the initial velocity v_0. Determine the corresponding uncertainty Δx in the position of the particle after a time t.

Solution

In between the instants of reflection, the particle moves with constant velocity equal to the initial value. Thus, if the initial velocity is v_0 then the distance moved by the particle in a time t is $v_0 t$, whereas if the initial velocity is $v_0 + \Delta v_0$ the distance moved is $(v_0 + \Delta v_0)t$. Therefore, the uncertainty in position after a time t is

$$\begin{aligned} \Delta x &= (v_0 + \Delta v_0)t - v_0 t \\ &= (\Delta v_0)t\,. \end{aligned} \qquad (1)$$

Comments

(i) According to (1), after a time $t_c = \ell/\Delta v_0$ has elapsed, $\Delta x = \ell$, meaning that the position at time t_c is completely undetermined.
(ii) For times $t \ll t_c$, (1) shows that the uncertainty $\Delta x \ll \ell$, and one can still regard the motion as deterministic (in the sense mentioned in Question 3.1). However, if we wait long enough the particle can be found anywhere between the walls: determinism has changed into complete indeterminism.
(iii) It is only in the ideal (and unattainable) case $\Delta v_0 = 0$ (i.e. the initial velocity is known exactly) that deterministic motion persists indefinitely.
(iv) In non-linear systems the uncertainty can increase much faster with time (exponentially rather than linearly) due to chaotic motion (see Chapter 13).

Question 2.2

A ball moves freely on the surface of a round billiard table, and undergoes elastic reflections at the boundary of the table. The motion is frictionless, and once started it continues indefinitely. The initial conditions are that the ball starts at a point A on the boundary and that the chord AB drawn in the direction of the initial velocity subtends an angle α at the centre O of the table. Discuss the dependence of the trajectory of the ball on α.

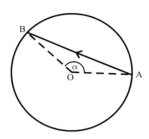

Solution

Because the collisions with the wall are elastic, the angles of incidence and reflection are equal (cf. the angles ϕ in the figure). Thus, the angular positions of successive points of impact with the boundary are each rotated through α (the chords AB, BC, ... in the figure all subtend an angle α at O). We may therefore distinguish between two types of trajectory:

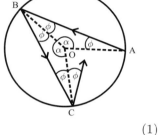

☞ α is equal to 2π times a rational number, that is

$$\alpha = 2\pi \frac{p}{q}, \tag{1}$$

where p and q are integers. Then, after q reflections at the wall the point of impact will have rotated through an angle

$$q\alpha = 2\pi p \tag{2}$$

from A. That is, the ball will have returned to A. The trajectory is a closed path of finite length, and the motion is periodic.

☞ α is equal to 2π times an irrational number. The angle of rotation of the point of impact with the wall ($q\alpha$ after q impacts) is not equal to 2π times an integer; the ball will never return to the starting position A – the trajectory is open and non-periodic.

Comments

(i) This question, like the previous one, shows that small causes can have big consequences. Here, the slightest change in the initial velocity can change a closed trajectory into an open one. Consequently, determinism over indefinitely long periods of time can be achieved only in the unphysical limit where the uncertainty in the initial velocity is precisely zero.

(ii) Other systems showing extreme sensitivity to initial conditions can readily be constructed (see Questions 3.3 and 4.2).

(iii) On the basis of these, questions were raised by Born and others concerning the deterministic nature of classical mechanics.[1,2] These examples show that "determinism is an idealization rather than a statement of fact, valid only under the assumption that unlimited accuracy is within our reach, an assumption which in view of the atomic structure of our measuring instruments is anything but realistic."[2] The examples depict "a curious half-way house, showing not so much the fall as the decline of causality – the point, that is, where the principle begins to lose its applicability."[2] (See also Chapter 13.) At the atomic level uncertainties of a more drastic sort were encountered that required the abandonment of deterministic laws in favour of the statistical approach of quantum mechanics.

Question 2.3

The active gravitational mass (m^A) of a particle is an attribute that enables it to establish a gravitational field in space, whereas the passive gravitational mass (m^P) is an attribute that enables the particle to respond to this field.

(a) Write Newton's law of universal gravitation in terms of the relevant active and passive gravitational masses.
(b) Show that the third law of motion makes it unnecessary to distinguish between active and passive gravitational mass.

Solution

(a) The gravitational force $\mathbf{F}_{12}$ that particle 1 exerts on particle 2 is proportional to the product of the active gravitational mass m_1^A of particle 1 and the passive gravitational mass m_2^P of particle 2. Thus, the inverse-square law of gravitation is

$$\mathbf{F}_{12} = -G \frac{m_1^A m_2^P}{r^2} \hat{\mathbf{r}}, \qquad (1)$$

where G is the universal constant of gravitation, r is the distance between the particles and $\hat{\mathbf{r}}$ is a unit vector directed from particle 1 to particle 2. By the same token, the force $\mathbf{F}_{21}$ which particle 2 exerts on particle 1 is

$$\mathbf{F}_{21} = G \frac{m_2^A m_1^P}{r^2} \hat{\mathbf{r}}. \qquad (2)$$

(b) According to Newton's third law, $\mathbf{F}_{12} = -\mathbf{F}_{21}$. It therefore follows from (1) and (2) that

$$\frac{m_2^A}{m_2^P} = \frac{m_1^A}{m_1^P}. \qquad (3)$$

We conclude from (3) that the ratio of the active to the passive gravitational mass of a particle is a universal constant. Furthermore, this constant can be

---
[1] M. Born, *Physics in my generation*, pp. 78–82. New York: Springer, 1969.
[2] F. Waismann, in *Turning points in physics*. Amsterdam: North-Holland, 1959. Chap. 5.

incorporated in the universal constant G, which is already present in (1) and (2). That is, we can set $m^P = m^A$. There is no need to distinguish between active and passive gravitational masses; it is sufficient to work with just gravitational mass m^G and to write (1) as

$$\mathbf{F}_{12} = -G \frac{m_1^G m_2^G}{r^2} \hat{\mathbf{r}}. \tag{4}$$

Comment

Evidently, the same reasoning applies to the notions of active and passive electric charge. Thus, if one were to write Coulomb's law for the electrostatic force between two charges in vacuum as

$$\mathbf{F}_{12} = k \frac{q_1^A q_2^P}{r^2} \hat{\mathbf{r}}, \tag{5}$$

where k is a universal constant, a discussion similar to the above would lead to

$$\frac{q_2^A}{q_2^P} = \frac{q_1^A}{q_1^P}. \tag{6}$$

Consequently, the ratio of active to passive charge is a universal constant that can be included in k in (5); it is sufficient to consider just electric charge q.

Question 2.4

The inertial mass of a particle is, by definition, the mass that appears in Newton's second law. Consider free fall of a particle with gravitational mass m^G and inertial mass m^I near the surface of a homogeneous planet having gravitational mass M^G and radius R. Express the gravitational acceleration a of the particle in terms of these quantities. (Neglect any frictional forces.)

Solution

The equation of motion is

$$m^I a = F, \tag{1}$$

where F is the gravitational force exerted by the planet

$$F = G \frac{M^G m^G}{R^2} \tag{2}$$

(see Question 11.17). Thus

$$a = \frac{m^G}{m^I} \frac{GM^G}{R^2}. \tag{3}$$

Comments

(i) In many treatments of this topic the factor m^G/m^I in (3) is absent because it is tacitly assumed that the gravitational and inertial masses are equal.

(ii) Equation (3) is approximate insofar as it neglects atmospheric drag (see Question 3.13) and motion of the planet toward the falling object (see Question 11.23). Nevertheless, it is an important idealization. The first significant work in this connection was by Galileo, who enunciated an empirically based result that, in the absence of drag, all bodies fall with the same gravitational acceleration. This is sometimes referred to as Galileo's law of free fall.

(iii) Galileo's law, together with (3), encouraged the hypothesis that gravitational and inertial masses can be taken to be the same, $m^G = m^I$, and one need consider only mass. This is the weak equivalence principle, which plays an important role in the formulation of the general theory of relativity.

(iv) Because of its importance, numerous experiments have been performed to test Galileo's law, and hence the weak equivalence principle. Modern experiments show[3] "that bodies fall with the same acceleration to a few parts in 10^{13}." See also Question 2.5.

Question 2.5

In Question 4.3 an expression is derived for the period T of a simple pendulum, tacitly assuming equality of the inertial and gravitational masses m^I and m^G of the bob. Study this calculation and then adapt it to apply when m^I and m^G are allowed to be different, thereby obtaining the dependence of T on these masses.

Solution

In terms of m^I and m^G, the equation of motion (2) of Question 4.3 is

$$m^I \frac{d^2 s}{dt^2} \mathbf{n} = -m^G g \sin\theta \, \mathbf{n}, \tag{1}$$

where $g = GM^G/R^2$ (see (2) of Question 2.4) and other symbols have the same meaning as in Question 4.3. Then, for small oscillations ($|\theta| \ll 1$) we see from (1), that (4) of Question 4.3 is replaced by

$$\frac{d^2\theta}{dt^2} + \frac{m^G}{m^I} \frac{g}{\ell} \theta = 0, \tag{2}$$

where ℓ is the length of the pendulum. Thus, we obtain the desired expression for the period

$$T = 2\pi \sqrt{\frac{m^I}{m^G} \frac{\ell}{g}}. \tag{3}$$

When $m^I = m^G$ this reduces to the result in Question 4.3.

---

[3] C. M. Will, "Relativity at the centenary," Physics World, vol. 18, pp. 27–32, January 2005.

Comments

(i) Newton used the result (3) in conjunction with experiments on pendulums to test the equality, in modern terminology, of inertial and gravitational mass.[4] He was aware that this test could be performed more accurately with pendulums than by using 'Galileo's free-fall experiment' and (3) of Question 2.4. Newton evidently attached importance to these pendulum experiments and often referred to them. He used two identical pendulums with bobs consisting of hollow wooden spheres suspended by threads 11 feet in length. By placing equal weights of various substances in the bobs, Newton observed that the pendulums always swung together over long periods of time. He concluded that "... by these experiments, in bodies of the same weight, I could manifestly have discovered a difference of matter less than the thousandth part of the whole, had any such been."[4] The accuracy of pendulum experiments was later improved to one part in 10^5 by Bessel.

(ii) Newton also showed how astronomical data could be used to test the equality of inertial and gravitational mass.[4] Modern lunar laser-ranging measurements provide an accuracy of a few parts in 10^{13}, while planned satellite-based experiments (where an object is in perpetual free fall) may improve this to one part in 10^{15}, and perhaps even a thousand-fold beyond that.[3]

(iii) The equality $m^{\text{P}} = m^{\text{A}}$ of passive and active gravitational masses in Question 2.3 is based on a theoretical condition (Newton's third law) that is presumably exact. By contrast, the accuracy of the equality $m^{\text{I}} = m^{\text{G}}$ of inertial and gravitational masses is limited by the accuracy of the experiments that test it.

Question 2.6

By applying the second and third laws of motion to the interaction between two particles in the absence of any third object, show how one can obtain an operational definition of relative mass.

Solution

Let F_{21} be the magnitude of the force exerted by particle 2 on particle 1, and similarly for F_{12}. The equations of motion of the two particles are

$$F_{21} = m_1 a_1, \qquad F_{12} = m_2 a_2, \tag{1}$$

where the m_i are the masses and the a_i are the magnitudes of the accelerations. According to the third law

$$F_{21} = F_{12}. \tag{2}$$

From (1) and (2) we have

$$\frac{m_2}{m_1} = \frac{a_1}{a_2}. \tag{3}$$

---

[4] S. Chandrasekhar, *Newton's Principia for the common reader*. Oxford: Clarendon Press, 1995. Sections 10 and 103.

Comments

(i) Equation (3) provides an operational definition of relative mass: one can, in principle, determine the mass m_2 of a particle relative to an arbitrarily selected mass m_1 by measuring the magnitudes of their accelerations at some instant, in the absence of any external disturbance.

(ii) It is clear that the Lagrangian L of a system can always be multiplied by an arbitrary constant without affecting the Lagrange equations – see (20) in Chapter 1. For a system of non-interacting particles, where $\mathsf{L} = \sum \frac{1}{2} m v^2$, this reflects the fact that the unit of mass is arbitrary and only relative masses have significance.

Question 2.7

Use a three-particle interaction to show that mass is an additive quantity.

Solution

The equations of motion for three particles interacting in the absence of any other objects are

$$m_1 \mathbf{a}_1 = \mathbf{F}_{21} + \mathbf{F}_{31}, \qquad m_2 \mathbf{a}_2 = \mathbf{F}_{12} + \mathbf{F}_{32}, \qquad m_3 \mathbf{a}_3 = \mathbf{F}_{13} + \mathbf{F}_{23}. \tag{1}$$

According to the third law, $\mathbf{F}_{21} = -\mathbf{F}_{12}$, etc., and so by adding equations (1) we have

$$m_1 \mathbf{a}_1 + m_2 \mathbf{a}_2 + m_3 \mathbf{a}_3 = 0. \tag{2}$$

Suppose particles 1 and 2 are stuck together rigidly to form a single particle. Then $\mathbf{a}_1 = \mathbf{a}_2 = \mathbf{a}_c$, the acceleration of the composite particle due to its interaction with particle 3, and (2) yields

$$(m_1 + m_2) \mathbf{a}_c = -m_3 \mathbf{a}_3. \tag{3}$$

Let m_c denote the mass of the composite particle. According to the previous question, for the two-particle interaction of masses m_c and m_3,

$$m_c \mathbf{a}_c = -m_3 \mathbf{a}_3. \tag{4}$$

It follows from (3) and (4) that

$$m_c = m_1 + m_2. \tag{5}$$

Comment

In thermodynamics a distinction is made between two types of variable. First, there are quantities that are additive when two systems are combined. For example, their volumes, the number of particles, etc. Such variables are referred to as extensive. Secondly, there are quantities such as temperature and pressure that are unchanged when two identical systems are combined – these are intensive variables. According to (5), the mass of a system is an extensive variable.

Question 2.8

Consider a 'mass dipole' consisting of two particles having opposite masses[‡] m (> 0) and $-m$. Describe its motion in the following cases:

(a) The dipole is initially at rest in empty inertial space.
(b) The constituents of the dipole in (a) have electric charge q_1 and q_2.
(c) The charged mass dipole of (b) is placed vertically (with the negative mass above the positive mass) in the Earth's gravitational field. Assume that the distance d between the particles is negligible in comparison with the distance r to the centre of the Earth.

Solution

Opposite-mass particles repel each other (this follows from the law of gravitation, see (4) of Question 2.3). Also, for a negative-mass particle the force $\mathbf{F}$ and the acceleration $\mathbf{a}$ in $\mathbf{F} = m\mathbf{a}$ point in opposite directions.

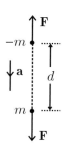

(a) In empty inertial space the only force acting on neutral particles a distance d apart is the gravitational repulsion $F = Gm^2/d^2$. In response, each particle accelerates at the same rate $a = Gm/d^2$ in the direction shown: the negative mass pursues the positive mass and d remains constant. The motion eventually becomes relativistic – see Question 15.13.

(b) The net force is the sum of the gravitational and electrostatic forces:

$$F = \frac{Gm^2 + kq_1q_2}{d^2} \qquad (k = 1/4\pi\epsilon_0). \qquad (1)$$

For like charges ($q_1q_2 > 0$), or for unlike charges ($q_1q_2 < 0$) with $q_1q_2 > -Gm^2/k$, the force $\mathbf{F}$ is repulsive and the motion is the same as in (a) with acceleration

$$a = \frac{Gm^2 + kq_1q_2}{md^2}. \qquad (2)$$

But, for unlike charges with $q_1q_2 < -Gm^2/k$ the force is attractive. The directions of $\mathbf{F}$ and $\mathbf{a}$ are reversed: the positive mass pursues the negative mass.

(c) Since $d \ll r$ the total force on each mass has the same magnitude, and the resulting acceleration of a vertical dipole is

$$a = \frac{F}{m} = \frac{GM}{r^2} + \frac{Gm^2 + kq_1q_2}{md^2}, \qquad (3)$$

[‡]Negative-mass particles have never been observed. It is, nevertheless, interesting and instructive to consider the dynamics of such objects.

where M is the Earth's mass. Again, for like charges or for unlike charges with $q_1 q_2 > -q_c^2$ where

$$q_c^2 = \frac{Gm^2}{k}\left(1 + \frac{Md^2}{mr^2}\right), \qquad (4)$$

the forces **F** are directed as shown in the first diagram, and the dipole accelerates toward the Earth at a rate a. But for unlike charges with $q_1 q_2 < -q_c^2$, the forces are reversed, as shown in the second diagram, and the dipole accelerates away from the Earth. Each particle accelerates at the same rate (3), and so d remains constant. The acceleration increases to the asymptotic value (2) as r increases. For unlike charges with $q_1 q_2 = -q_c^2$, the acceleration $a = 0$ and the dipole remains at rest relative to the Earth.

Comments

(i) Despite its strange dynamical properties, a mass dipole would not violate any of the laws of physics.[5] For example, despite the acceleration in empty space, energy is conserved because the total kinetic energy $\frac{1}{2}mv^2 + \frac{1}{2}(-m)v^2$ is always zero.

(ii) The acceleration a in (2) and (3) can be controlled (in both magnitude and direction) by altering the charges q_1 and q_2. The dipole is an 'anti-gravity glider'[5] that can fall, hover, or rise in a gravitational field.

(iii) In a frame that is accelerating at a rate a, the total force on each particle is zero because the respective translational forces, $-ma$ and $-(-m)a$, cancel the forces $F = ma$ and $-ma$ on each particle. Thus, the mass dipole is at rest in this frame. It follows that the dipole is unstable with respect to any relative motion of the particles toward or away from each other. It would be necessary to have some feedback mechanism to counter any such drift.

(iv) One can consider variations of the above, such as a mass dipole in which both inertial masses m^I are positive, and the gravitational masses m^G and $-m^G$ have opposite signs. Or one can consider interactions that point in the same direction, as in a predator-prey problem.

(v) The preceding questions just touch on the rather mysterious concept of mass. Access to the extensive literature on this subject is provided in an article by Roche.[6] In the theory of special relativity, mass has the property that it can vary in space and time if so-called 'impure' forces are present (see Question 15.11). Perhaps future, richer theories will reveal further properties of mass.

Question 2.9

Discuss the following statement in relation to Lagrange's equations: 'In the equation of motion $\mathbf{F} = m\mathbf{a}$ the units must be the same on both sides'.

---

[5] R. H. Price, "Negative mass can be positively amusing," American Journal of Physics, vol. 61, pp. 216–217, 1993.
[6] J. Roche, "What is mass?," European Journal of Physics, vol. 26, pp. 225–242, 2005.

Solution

In the Lagrange equations for a system of particles (see Chapter 1)

$$\frac{d}{dt}\frac{\partial \mathsf{L}}{\partial \dot{q}_i} = \frac{\partial \mathsf{L}}{\partial q_i}, \qquad (1)$$

the units on each side are clearly the same. The right-hand side of (1) gives the (generalized) forces F_i and the left-hand side the rates of change $\dot{p}_i$ of the (generalized) momenta (see Chapter 1). Therefore, the above statement follows.

Comments

(i) The generalization of this statement is:

$$\left.\begin{array}{l}\text{'All equations in physics (including all physical}\\ \text{laws) have the same units on both sides'.}\end{array}\right\} \qquad (2)$$

That is, one has an example of a 'relativity principle': the laws of physics are equally valid in all systems of units.

(ii) The statement (2) is the basis for dimensional analysis, which has far-reaching consequences in physics.[7] Some simple examples follow.

Question 2.10

Solutions to physical problems often involve functions like

$$\cos u, \qquad \sin u, \qquad e^u, \qquad \ln u, \qquad \cdots \quad , \qquad (1)$$

where the argument u is a scalar that depends on physical quantities such as time, frequency, mass, etc. Explain why u must be dimensionless (that is, independent of the system of units used for mass, length and time).

Solution

The result follows by inspection of the Taylor expansions of the functions in (1). (We can, if we wish, take these expansions to be defining relations of the functions.[8]) For example,

$$e^u = 1 + \frac{u}{1!} + \frac{u^2}{2!} + \cdots \qquad \text{(for all } u\text{)}. \qquad (2)$$

It follows that $1, u, u^2, \cdots$ must have the same physical dimensions, and therefore u is dimensionless.

---

[7]   G. I. Barenblatt, *Scaling, self-similarity, and intermediate asymptotics.* Cambridge: Cambridge University Press, 1996.
[8]   J. M. Hyslop, *Real variable.* London: Oliver and Boyd, 1960.

Comments

(i) It is a good idea to check whether the results of a calculation satisfy the above condition. Thus, an expression like $e^{t/m}$ (where t is time and m is mass) is clearly unacceptable.

(ii) The earliest standards for space, time and mass were related to the human body and human activities. With the introduction of the SI system of units in the nineteenth century, the metre was defined by the length of a platinum-iridium bar, the kilogram by the mass of a platinum-iridium cylinder (both preserved under carefully controlled conditions), and the second was related to the rotation of the Earth. In the twentieth century the metre and second were redefined in terms of physical and atomic constants. The kilogram is therefore an anachronism in that it is still based on a physical object, and it seems likely that the kilogram will be redefined in a more convenient and accurate way, possibly by relating it to Planck's constant. An absorbing account of this topic has been given in Ref. [9]. (Planck's constant is already used in a system of units – see Question 2.17.)

Question 2.11

Use dimensional analysis to determine the dependence of the period T of a simple pendulum on its mass m, weight w and length ℓ.

Solution

Here, we neglect any dependence of T on the amplitude of oscillation; this is discussed in Question 2.12. We also assume that the desired function of three variables $T = T(m, w, \ell)$ is a power-law relation

$$T = k m^\alpha w^\beta \ell^\gamma, \qquad (1)$$

where k, α, β, γ are dimensionless constants. We require that the physical dimensions of each side of (1) be the same, that is

$$[T] = [m]^\alpha [w]^\beta [\ell]^\gamma. \qquad (2)$$

Here, $[Q]$ denotes the dimensions of the quantity Q (Maxwell's notation). In terms of the fundamental units of mass (M), length (L) and time (T) we have[‡] $[T] = T$, $[m] = M$, $[w] = MLT^{-2}$ (w being a force = mass × acceleration), and $[\ell] = L$. Thus, (2) can be written

$$M^0 L^0 T = M^\alpha (MLT^{-2})^\beta L^\gamma, \qquad (3)$$

which provides three equations in the unknowns α, β and γ:

[‡]We use T in two senses (a period and also a fundamental unit); which meaning is intended is clear from the context.

---

[9]   I. Robinson, "Redefining the kilogram," Physics World, vol. 17, pp. 31–35, May 2004.

$$\alpha+\beta=0, \qquad \beta+\gamma=0, \qquad -2\beta=1. \qquad (4)$$

Hence $\alpha=-\beta=\gamma=\tfrac{1}{2}$, and (1) becomes

$$T=k\sqrt{\frac{m\ell}{w}}. \qquad (5)$$

Comments

(i) The requirement (2), of equality of dimensions in a physical equation, is the essence of the method of dimensional analysis. It is a consequence of the necessity for physical laws and results to be independent of our arbitrary choice of units for mass, length, time, etc. The numerical values of physical quantities such as velocity, momentum and force do depend on the choice of units but physical laws expressing the relations between these quantities do not. Thus, for example, the law $F=ma$ is valid in any system of units.

(ii) The assumption made in (1) that the desired form is a power-law monomial in the independent variables, is typical of dimensional analysis (see also the following examples). This use of power-law relations should not be regarded as a weakness of the method. In fact, power-law (or scaling) relationships "give evidence of a very deep property of the phenomena under consideration – their *self-similarity*: such phenomena reproduce themselves, so to speak, in time and space."[7] Further, it can be proved that the dimension of any physical quantity Q is given by a power-law monomial: for example, in the M, L, T class of units

$$[Q]=M^a L^b T^c, \qquad (6)$$

where a, b, and c are dimensionless constants.[7]

(iii) In the above example, dimensional analysis provides enough independent equations to solve for the three unknown quantities α, β and γ. Often, this is not the case (see the following questions).

(iv) With $w=mg$, (5) becomes

$$T=k\sqrt{\frac{\ell}{g}}. \qquad (7)$$

(Strictly, $w=m^G g$ and $m=m^I$, where m^G and m^I are the gravitational and inertial masses, so that (5) is

$$T=k\sqrt{\frac{m^I\,\ell}{m^G\,g}}. \qquad (8)$$

According to the weak equivalence principle, $m^I=m^G$ and (8) reduces to (7); see Question 2.5.)

(v) The constant k in (7) has to be determined from a detailed dynamical analysis. This shows that k is, in fact, a function of the amplitude of oscillation (the maximum arc-length s) with the simple limit $k \to 2\pi$ as $s \to 0$. In the next question we examine what happens if we try to use dimensional analysis to obtain also the dependence of T on the amplitude of oscillation.

Question 2.12

Use dimensional analysis to determine the dependence of the period T of a simple pendulum on its mass m, weight w, length ℓ and arc-length of swing s.

Solution

Instead of (1) of Question 2.11 we now have a power-law relation in four variables:
$$T = k m^\alpha w^\beta \ell^\gamma s^\delta. \tag{1}$$

Hence
$$M^0 L^0 T^1 = M^\alpha (MLT^{-2})^\beta L^\gamma L^\delta, \tag{2}$$

and so
$$\alpha + \beta = 0, \qquad \beta + \gamma + \delta = 0, \qquad -2\beta = 1. \tag{3}$$

These yield $\alpha = -\beta = \frac{1}{2}$ and $\gamma = \frac{1}{2} - \delta$. Consequently, (1) becomes
$$T = k \sqrt{\frac{m\ell}{w}} \left(\frac{s}{\ell}\right)^\delta, \tag{4}$$

where δ is an undetermined number.

Comments

(i) Because s/ℓ is a dimensionless quantity, we cannot determine the dependence of T on it by using dimensional analysis. In fact, it is clear that we can replace the factor $(s/\ell)^\delta$ by a power series in (s/ℓ) without disturbing the dimensional balance of (4). Thus, the most general form allowed on dimensional grounds is

$$T = \sqrt{\frac{m\ell}{w}} \phi\left(\frac{s}{\ell}\right) = \sqrt{\frac{\ell}{g}} \phi\left(\frac{s}{\ell}\right), \tag{5}$$

where ϕ is an undetermined function of the amplitude s/ℓ of the oscillations. A numerical calculation of ϕ is given in Question 5.18. In the limit $s/\ell \to 0$, $\phi \to 2\pi$, and it is only in this limit that it is reasonable to assume that T is independent of s (as was done in Question 2.11).

(ii) Thus, in the present question dimensional analysis has reduced an unknown function of four variables $T = T(m, w, \ell, s)$ to an unknown function of one variable $\phi(s/\ell)$. Despite this inability of the method to reduce a result beyond a function of one (or more) dimensionless quantities in most cases, dimensional analysis is a powerful and useful technique, particularly in its application to more complex phenomena (such as the next question). Often, the forms provided by dimensional analysis provide clues on how to perform a more detailed theoretical analysis or how to analyze experimental results. In fact, "using dimensional analysis, researchers have been able to obtain remarkably deep results that have sometimes changed entire branches of science.... The list of great names involved runs from Newton and Fourier, to Maxwell, Rayleigh and Kolmogorov."[7]

Question 2.13

A liquid having density ρ and surface tension σ drips slowly from a vertical tube of external radius r. Use dimensional arguments to analyze the dependence of the mass m of a drop on ρ, σ, r and g (the gravitational acceleration).

Solution

Assume that the mass of a drop is given by the power-law relation

$$m = k\rho^\alpha \sigma^\beta r^\gamma g^\delta, \tag{1}$$

where k, α, β, γ and δ are dimensionless constants. Recall that surface tension is a force per unit length: $[\sigma] = MLT^{-2}/L = MT^{-2}$. Thus, dimensional balance in (1) requires

$$ML^0T^0 = (ML^{-3})^\alpha (MT^{-2})^\beta L^\gamma (LT^{-2})^\delta. \tag{2}$$

Therefore

$$\alpha + \beta = 1, \qquad -3\alpha + \gamma + \delta = 0, \qquad -2\beta - 2\delta = 0, \tag{3}$$

which yield for α, β and γ in terms of δ

$$\alpha = 1 + \delta, \qquad \beta = -\delta, \qquad \gamma = 3 + 2\delta. \tag{4}$$

From (1) and (4) we have

$$m = k\rho r^3 (\rho r^2 g/\sigma)^\delta. \tag{5}$$

Comments

(i) We see again that the existence of a dimensionless combination – in this case $\rho r^2 g/\sigma$ – means that the power-law dependence on this quantity (the number δ in (5)) cannot be determined by dimensional arguments. In fact, we can generalize (5) to

$$m = \rho r^3 \phi(\rho r^2 g/\sigma), \tag{6}$$

where ϕ is an unknown function, without disturbing the dimensional balance. Equation (6) is the most general form allowed by dimensional requirements.

(ii) Because $\rho = m/V$, where V is the volume of a drop, (6) can be inverted to read

$$\sigma r/mg = F(V/r^3), \tag{7}$$

where F is an unknown function. Measurements[10] show that $F(u)$ decreases slowly from 0.2647 at $u = 2$ to 0.2303 at $u = 18$. The above results are the basis for Harkins and Brown's drop-weight method for measuring the surface tension

---

[10] See, for example, A. W. Porter, *The method of dimensions*. London: Methuen, 3rd edn, 1946. Chap. 3.

of a liquid.[11] Note that $F \neq (2\pi)^{-1}$ and therefore it is not correct to make the approximation $mg = 2\pi r\sigma$, as would follow if the surface tension acted vertically around the outer radius of the tube at the instant that a drop breaks away: the phenomenon is more complicated than that.

Question 2.14

A sphere of radius R moves with constant velocity v through a fluid of density ρ and viscosity η. The fluid exerts a frictional force F on the sphere. Use dimensional arguments to study the dependence of F on ρ, R, v and η.

Solution

Assume that
$$F = k\rho^\alpha R^\beta v^\gamma \eta^\delta, \tag{1}$$
where k, α, β, γ and δ are dimensionless constants. Recall that viscosity is the proportionality between a tangential force per unit area and a velocity gradient. So
$$[\eta] = (MLT^{-2} \div L^2)/(LT^{-1} \div L) = ML^{-1}T^{-1}. \tag{2}$$
Then dimensional balance in (1) requires
$$MLT^{-2} = (ML^{-3})^\alpha L^\beta (LT^{-1})^\gamma (ML^{-1}T^{-1})^\delta. \tag{3}$$
Hence
$$\alpha + \delta = 1, \qquad -3\alpha + \beta + \gamma - \delta = 1, \qquad \gamma + \delta = 2, \tag{4}$$
and we can express α, β and γ in terms of one unknown δ:
$$\alpha = 1 - \delta, \qquad \beta = \gamma = 2 - \delta. \tag{5}$$
Thus, (1) becomes
$$F = k\rho R^2 v^2 (\eta/\rho R v)^\delta. \tag{6}$$

Comments

(i) The existence of the dimensionless quantity $\eta/\rho R v$ means that the power-law dependence on this number in (6) cannot be determined by dimensional reasoning. Clearly, we can generalize (6) to
$$F = \rho R^2 v^2 \phi \left(\rho R v / \eta\right), \tag{7}$$
where ϕ is an unknown function. Equation (7) is the most general form allowed by dimensional requirements. Thus, dimensional analysis has enabled us to reduce

---

[11] See, for example, F. C. Champion and N. Davy, *Properties of matter*. London: Blackie, 3rd edn, 1959. Chap. 7.

an unknown function of four variables to a function of one variable. The dimensionless quantity $2\rho R v/\eta$ is known as the Reynolds number, and it is an essential parameter that governs this phenomenon. The function ϕ is rather complicated in general, although a reasonable approximation can be given that applies over a fairly wide range of Reynolds numbers (see Question 3.8). For 'low' Reynolds numbers a dynamical analysis shows that $\phi(u) \to 6\pi/u$, and hence (7) becomes

$$F = 6\pi \eta R v, \tag{8}$$

which is Stokes's law. For 'higher' Reynolds numbers $\phi \approx 0.2\pi$ and (7) gives

$$F = 0.2\pi \rho R^2 v^2. \tag{9}$$

The meanings of 'low' and 'high' are explained in Question 3.8

(ii) The Reynolds number also enters naturally in dimensional analysis of other phenomena. Consider, for example, the steady flow of fluid through a long, cylindrical pipe. The constant decrease of pressure per unit length of pipe, dp/dx, depends on the fluid density ρ, viscosity η, pipe diameter D, and the fluid velocity v (averaged over the cross-section of the pipe). By a calculation similar to that leading to (7) one finds

$$\frac{dp}{dx} = \frac{\rho v^2}{D} \phi(\rho D v/\eta), \tag{10}$$

where ϕ is an undetermined function. Except for the transition region between laminar and turbulent flow, a single function ϕ represents all experimental data.[7]

Question 2.15

A planet moves in a circular orbit of radius R around a star of mass M. Use dimensional analysis to determine the dependence of the period T of the motion on M, R and G (the universal constant of gravitation).

Solution

We make an analogy with the dimensional analysis of the simple pendulum. The gravitational acceleration experienced by the planet is $g = GM/R^2$ and therefore from (7) of Question 2.11, a dimensionally acceptable expression for the period is

$$T = k\sqrt{\frac{R}{g}} = k\sqrt{\frac{R^3}{GM}}, \tag{1}$$

where k is a dimensionless constant.

Comment

A detailed calculation shows that, in general, planetary orbits are elliptical and

$$T = 2\pi\sqrt{\frac{a^3}{G(M+m)}}, \qquad (2)$$

where m is the mass of the planet and a is the length of the semi-major axis of the ellipse (see Question 10.11). The result $T^2 \propto a^3$ is known as Kepler's third law. According to (2), T depends also on m. In the limit $m/M \to 0$ and for circular orbits ($a = R$), (2) reduces to (1) with $k = 2\pi$.

Question 2.16

Let R be the radius of a shock wave front a time t after a nuclear explosion has released an amount of energy E in an atmosphere of initial density ρ. Use dimensional analysis to determine the dependence of R on E, ρ and t.

Solution

If we assume that

$$R = kE^\alpha \rho^\beta t^\gamma, \qquad (1)$$

where k, α, β and γ are dimensionless constants, then dimensional balance requires

$$M^0 L T^0 = (ML^2T^{-2})^\alpha (ML^{-3})^\beta T^\gamma. \qquad (2)$$

Thus $\alpha = -\beta = \frac{1}{5}$ and $\gamma = \frac{2}{5}$, and (1) becomes

$$R = k\left(\frac{Et^2}{\rho}\right)^{\frac{1}{5}}. \qquad (3)$$

Comment

The above is a well-known result due to G. I. Taylor, who also showed that $k \approx 1$ and who used (3) to determine the energy of a nuclear explosion from a series of high-speed photographs of the fireball.[7,12]

Question 2.17

By taking power-law combinations of the three fundamental constants $\hbar$, c and G (the reduced Planck constant, the speed of light in vacuum and the universal constant of gravitation, respectively), construct quantities with the units of (a) mass, (b) length, and (c) time.

---

[12] M. Longair, *Theoretical concepts in physics (An alternative view of theoretical reasoning in physics)*, pp. 169–170. Cambridge: Cambridge University Press, 2nd edn, 2003.

Solution

From the Planck relation $E = \hbar\omega$, for example, the units of $\hbar$ are those of energy × time; that is, Js. So $[\hbar] = ML^2T^{-1}$. Also, $[c] = LT^{-1}$ and $[G] = M^{-1}L^3T^{-2}$. Let M_p, L_p and T_p denote the desired mass, length and time.

(a) Write
$$M_\mathrm{p} = \hbar^\alpha c^\beta G^\gamma. \tag{1}$$
Then
$$ML^0T^0 = (ML^2T^{-1})^\alpha (LT^{-1})^\beta (M^{-1}L^3T^{-2})^\gamma. \tag{2}$$
That is,
$$\alpha - \gamma = 1, \qquad 2\alpha + \beta + 3\gamma = 0, \qquad -\alpha - \beta - 2\gamma = 0. \tag{3}$$
Hence, $\alpha = \beta = -\gamma = \tfrac{1}{2}$ and (1) is
$$M_\mathrm{p} = \sqrt{\hbar c/G}. \tag{4}$$

(b) Similarly, $L_\mathrm{p} = \hbar^\alpha c^\beta G^\gamma$ yields $\alpha = \gamma$, $2\alpha + \beta + 3\gamma = 1$ and $\alpha + \beta + 2\gamma = 0$. Hence, $\alpha = \beta = -\tfrac{1}{3}\gamma = \tfrac{1}{2}$ and
$$L_\mathrm{p} = \sqrt{\hbar G/c^3}. \tag{5}$$

(c) With $T_\mathrm{p} = \hbar^\alpha c^\beta G^\gamma$ we have $\alpha = \gamma$, $2\alpha + \beta + 3\gamma = 0$, $\alpha + \beta + 2\gamma = -1$. That is, $\alpha = \gamma = \tfrac{1}{5}\beta = \tfrac{1}{2}$ and
$$T_\mathrm{p} = \sqrt{\hbar G/c^5}. \tag{6}$$

Comments

(i) The system of units defined by (4)–(6) was introduced by Planck in 1899 when he discovered his quantum of action, and they are named after him: the Planck mass M_p, length L_p and time T_p. If the laws of physics containing $\hbar$, c and G (that is, quantum mechanics and special and general relativity) are universal, then M_p, L_p and T_p comprise an absolute (or natural) system of units. For this reason Planck remarked that the new units would be " ... independent of particular bodies or substances, would necessarily retain their significance for all times and for all cultures, including extraterrestrial and non-human ones, and can therefore be designated as 'natural units' ... ."[13]

(ii) For more than half a century after their introduction the Planck units were largely ignored, or even regarded in a negative light. However, beginning in the 1950s, a number of works appeared that considered the possible physical significance of the Planck values. For example, it was suggested that the Planck mass M_p is an upper limit for the mass spectrum of elementary particles and a lower limit for the mass

---

[13] M. Planck, "Über irreversible strahlungsvorgänge," Sitzungsberichte der Preussischen Akademie der Wissenschaften, vol. 5 . Mittheilung, pp. 440–480, 1899.

of a black hole. The Planck density ($M_\mathrm{p}/L_\mathrm{p}^3 = 5.16 \times 10^{96}\,\mathrm{kg\,m^{-3}}$) was proposed as an upper limit for the density of matter. An outline of these developments, together with references to the literature, has been given in Ref. [14].

(iii) In terms of MKS units the Planck values are

$$M_\mathrm{p} = 2.18 \times 10^{-8}\,\mathrm{kg}, \quad L_\mathrm{p} = 1.62 \times 10^{-35}\,\mathrm{m} \quad \text{and} \quad T_\mathrm{p} = 5.39 \times 10^{-44}\,\mathrm{s}. \quad (7)$$

(iv) Planck units for charge and temperature can be defined by including also the permittivity of free space, ϵ_0, and Boltzmann's constant, k, in our set of fundamental constants. Then

$$Q_\mathrm{p} = \sqrt{4\pi\epsilon_0 \hbar c} = 1.88 \times 10^{-18}\,\mathrm{C}, \quad \theta_\mathrm{p} = M_\mathrm{p} c^2 / k = 1.42 \times 10^{32}\,\mathrm{K}. \quad (8)$$

(v) Physical quantities can readily be expressed in terms of Planck units. For example, the Planck acceleration $L_\mathrm{p}/T_\mathrm{p}^2 = 5.58 \times 10^{51}\,\mathrm{m\,s^{-2}}$, and therefore an acceleration of $1\,\mathrm{m\,s^{-2}}$ is equal to $1.79 \times 10^{-52} L_\mathrm{p}/T_\mathrm{p}^2$. The Planck velocity is $L_\mathrm{p}/T_\mathrm{p} = c$; the Planck energy is $E_\mathrm{p} = M_\mathrm{p} c^2 = 1.96 \times 10^9\,\mathrm{J}$; the Planck density is $\rho_\mathrm{p} = M_\mathrm{p}/L_\mathrm{p}^3 = 5.16 \times 10^{96}\,\mathrm{kg\,m^{-3}}$; and so on.

(vi) There are other systems of absolute units for mass, length and time that can be constructed from fundamental constants, such as the classical system based on e (the electronic charge), c and G. Some difficult questions remain concerning numerical relations (such as mass ratios) between the absolute systems.[15]

(vii) In general, it is an interesting activity to use fundamental (and other) constants to construct certain physical quantities because these invariably play a central role in various phenomena. For example:

☞ The quantity $\lambda = \hbar/mc$ has the unit of length. It is known as the Compton wavelength (of a particle with mass m) because, with m equal to an electron mass m_e, it first appeared in the theory of the Compton effect for scattering of a photon by an electron. With m equal to a meson mass, λ gives the range of the nuclear force in Yukawa's theory of this force.

☞ The ratio $h/e = 2\phi_0$ has the units of magnetic flux; ϕ_0 is the so-called flux quantum and it plays an essential role in understanding superconductors.

☞ The quantity $R = h/e^2$ has the units of electrical resistance. It occurs in the theory of the quantum Hall effect and provides a standard for resistance.

☞ The ratio $e\hbar/m_e = 2\mu_\mathrm{B}$ has the units of magnetic dipole moment; μ_B is known as the Bohr magneton and it provides a scale for the magnetic moments of atoms and the intrinsic moment of an electron.

---

[14] K. A. Tomilin, "Natural systems of units." At http://web.ihep.su/library/pubs/tconf99/ps/tomil.pdf.

[15] F. Wilczek, "On absolute units, II: challenges and responses," Physics Today, vol. 59, pp. 10–11, January 2006.

3
One-dimensional motion

The examples in this chapter deal with a particle of mass m that moves in one dimension (along the x-axis) and is acted on by a force F that may in general be a function of x, $v = \dot{x}$, and t. Note that the actual force is the vector $F\hat{\mathbf{x}}$, the velocity is $v\hat{\mathbf{x}}$, and so on; in one dimension it is convenient to omit reference to the unit vector $\hat{\mathbf{x}}$. The motion is non-relativistic and so m is constant. At time $t = 0$ the particle is at x_0 and has velocity v_0.

Question 3.1

If F is constant (independent of x, t and v) determine the velocity $v(t)$ and the trajectory $x(t)$ in terms of F, m, v_0, and x_0.

Solution

The equation of motion is

$$m\frac{dv}{dt} = F. \tag{1}$$

We integrate both sides of (1) with respect to t, between the limits $t = 0$ and t. Then, because $\left(\dfrac{dv}{dt}\right) dt = dv$, and m and F are constants, we have

$$m \int_{v_0}^{v(t)} dv = F \int_0^t dt. \tag{2}$$

Thus, the velocity at time t is given by

$$v(t) = v_0 + Ft/m. \tag{3}$$

By definition, $v = dx/dt$ and therefore integration of both sides of (3) with respect to t yields

$$\int_{x_0}^{x(t)} dx = \int_0^t \left(v_0 + Ft/m\right) dt. \tag{4}$$

Thus, the position at time t is

$$x(t) = x_o + v_o t + Ft^2/2m\,. \tag{5}$$

Equations (3) and (5) are the desired solutions. They are linear and quadratic functions of t, respectively, which are depicted below (for v_o, x_o, $F > 0$).

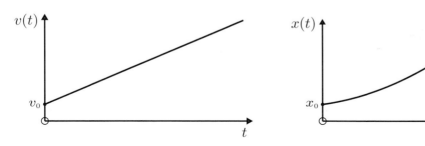

Comments

(i) The two unknown quantities $v(t)$ and $x(t)$ appear as the upper limits of the integrals in (2) and (4). This is typical of problems that can be solved by simple integration (see examples below).

(ii) The future behaviour of the particle is completely determined if the quantities F, m, x_o and v_o on the right-hand side of (5) are known. In this sense classical mechanics is deterministic. (See, however, Questions 2.1 and 2.2.)

(iii) In the sixteenth century, Galileo Galilei performed experiments on the distance moved by an object when it undergoes a constant acceleration. For this purpose he used brass balls rolling down inclined planes. The acceleration was quite low (about $0.1\,\mathrm{m\,s^{-2}}$) and he was able to time the motion to sufficient accuracy by using a simple water clock (a large vessel that drained through a narrow tube into a beaker). He found that the distance travelled is proportional to the square of the weight of the water that flowed into the beaker. This result, which is in accord with (5), was counter to the prevailing conventional wisdom, which held that the distance should be proportional to the time of travel. Galileo's experiment is thought by some to be among the most beautiful that have been performed in physics.[1] The theory of this experiment is given in Question 12.20.

(iv) Equation (5) has many simple applications, such as the following. For a stone that is dropped into a well of unknown depth D, one can express D in terms of the time t elapsed until the splash is heard, the acceleration g due to gravity, and the speed V of sound in air: according to (5), $D = \frac{1}{2}gt_1^2$ for the falling stone and $D = Vt_2$ for the sound wave, so that $t = t_1 + t_2 = (2D/g)^{1/2} + D/V$ and therefore

$$D = \frac{V^2}{2g}\left(\sqrt{1 + 2gt/V} - 1\right)^2.$$

Measurements of V, g and t give D.

---

[1] R. P. Crease, "The most beautiful experiment," Physics World, vol. 15, pp. 19–20, September 2002.

(v) Equations (3) and (5) are approximate (non-relativistic) solutions which apply for $v \ll c$, where c is the speed of light in vacuum: thus, for a particle that starts from rest ($v_0 = 0$) they are valid for $t \ll mc/F$ (see Question 15.13). Often, the time scale mc/F in a particular problem is large compared to the time interval(s) of interest, and so (3) and (5) are acceptable approximations.

Question 3.2

If F is the time-dependent force $F = A - Bt$, where A and B are positive constants, determine the velocity $v(t)$ and the trajectory $x(t)$ in terms of A, B, m, v_0, and x_0. Sketch the graphs of $F(t)$, $v(t)$ and $x(t)$ versus t for $v_0 = 0$ and $x_0 > 0$.

Solution

In place of (2) in Question 3.1, we now have

$$m \int_{v_0}^{v(t)} dv = \int_0^t (A - Bt)\, dt, \tag{6}$$

and therefore

$$v(t) = v_0 + At/m - Bt^2/2m. \tag{7}$$

Then

$$\int_{x_0}^{x(t)} dx = \int_0^t \left(v_0 + At/m - Bt^2/2m\right) dt, \tag{8}$$

and so

$$x(t) = x_0 + v_0 t + At^2/2m - Bt^3/6m. \tag{9}$$

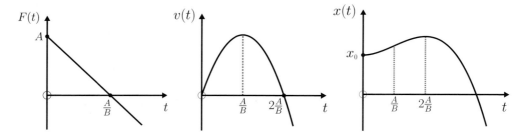

Comments

(i) For $t < A/B$ the force is positive and the particle is accelerated in the positive x-direction; for $t > A/B$ the force is negative and the particle is decelerated, coming to rest at $t = 2A/B$. For $t > 2A/B$ it moves in the negative x-direction.

(ii) When the force is zero (at $t = A/B$) the velocity is a maximum and the displacement has a point of inflection; when the velocity is zero (at $t = 2A/B$) the displacement along the positive x-axis is a maximum.

Question 3.3

(a) A particle is subject to an oscillatory force $F = F_0 \cos(\omega t + \phi)$, where F_0, ω and ϕ are positive constants. Calculate the velocity $v(t)$ and the trajectory $x(t)$ in terms of m, F_0, ω, ϕ, v_0 and x_0.

(b) Discuss and plot the possible graphs of $v(t)$ and $x(t)$ versus t.

Solution

(a) The equation of motion

$$m\frac{dv}{dt} = F_0 \cos(\omega t + \phi), \tag{1}$$

can be integrated with respect to t to yield

$$v(t) = v_0 + v_c\{\sin(\omega t + \phi) - \sin\phi\}, \tag{2}$$

where $v_c = F_0/m\omega$, has the units of velocity. Integration of (2) with respect to t gives

$$x(t) = x_0 + (v_c/\omega)[\{(v_0/v_c) - \sin\phi\}\omega t + \cos\phi - \cos(\omega t + \phi)]. \tag{3}$$

(b) The trajectory (3) has the interesting property that it may be either bounded ($x(t)$ finite) or unbounded ($|x(t)|$ infinite at $t = \infty$), depending on whether the term in ωt is present. If

$$v_0 = v_c \sin\phi, \tag{4}$$

then (3) becomes

$$x(t) = x_0 + (v_c/\omega)\{\cos\phi - \cos(\omega t + \phi)\}, \tag{5}$$

which is a bounded motion: $x(t) - x_0$ oscillates between $(v_c/\omega)(\cos\phi - 1)$ and $(v_c/\omega)(\cos\phi + 1)$. On the other hand, if $v_0 \neq v_c \sin\phi$ then the motion given by (3) is unbounded: if $v_0 > v_c \sin\phi$ then $x(t) \to \infty$ as $t \to \infty$ (as in the fourth figure below), while if $v_0 < v_c \sin\phi$ then $x(t) \to -\infty$ as $t \to \infty$. Plots of $v(t)$ given by (2) and $x(t)$ given by (5) for bounded motion are shown below. Here, we have set $x_0 = 0$ and taken $\phi = \pi/4$.

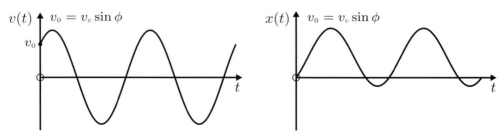

Plots of $v(t)$ given by (2) and $x(t)$ given by (3) for unbounded motion are shown below for $x_0 = 0$ and $\phi = \pi/4$. The position $x(t)$ oscillates about the term proportional to ωt in (3), and this is indicated by the dotted straight line in the plot of $x(t)$. Note that the velocity (2) is always bounded: it oscillates between $v_0 - v_c \sin\phi - v_c$ and $v_0 - v_c \sin\phi + v_c$.

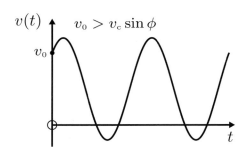
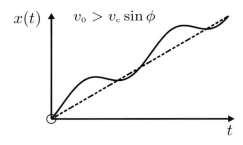

Comments

(i) This question illustrates again the sensitivity to initial conditions that can be present in classical mechanics and has a bearing on determinism (see Questions 2.1 and 2.2). In general, sensitivity to small perturbations, particularly in non-linear phenomena (see Chapter 13), makes tasks such as long-range weather forecasting problematic.

(ii) All one-dimensional problems involving time-dependent forces $F(t)$ (such as Questions 3.2 and 3.3) can be solved by integration:

$$m \int_{v_0}^{v(t)} dv = \int_0^t F(t)\, dt. \tag{6}$$

Question 3.4

A particle starts from rest at x_0 (> 0) in an attractive inverse-cube force field $F = -k/x^3$ (k is a positive constant). Show that the time taken to reach the origin is

$$T_0 = \sqrt{mx_0^4/k}. \tag{1}$$

Solution

The equation of motion

$$m\frac{dv}{dt} = -\frac{k}{x^3} \tag{2}$$

cannot be integrated as it stands. Instead, make the replacement

$$\frac{dv}{dt} = \frac{dx}{dt}\frac{dv}{dx} = v\frac{dv}{dx}, \tag{3}$$

to obtain

$$mv\, dv = -\frac{k}{x^3}\, dx, \tag{4}$$

and then integrate between corresponding limits:

$$m \int_0^{v(x)} v\, dv = -k \int_{x_0}^{x} \frac{dx}{x^3}. \tag{5}$$

Hence
$$\frac{1}{2}mv^2(x) = \frac{1}{2}k\left(\frac{1}{x^2} - \frac{1}{x_0^2}\right), \tag{6}$$

and so
$$v(x) = \frac{dx}{dt} = -\sqrt{\frac{k}{mx_0^2}}\frac{\sqrt{x_0^2 - x^2}}{x}. \tag{7}$$

In (7) a negative sign has been used in extracting a square root because we are considering motion in the negative x-direction, that is $v(x) \leq 0$. Integration of (7) with respect to t gives
$$\int_{x_0}^{x(t)} \frac{x\,dx}{\sqrt{x_0^2 - x^2}} = -\sqrt{\frac{k}{mx_0^2}} \int_0^t dt. \tag{8}$$

Now
$$\int \frac{x\,dx}{\sqrt{x_0^2 - x^2}} = -\sqrt{x_0^2 - x^2}, \tag{9}$$

and therefore (8) yields
$$\sqrt{x_0^2 - x^2(t)} = \sqrt{\frac{k}{mx_0^2}}\, t. \tag{10}$$

Thus, we obtain the trajectory
$$x(t) = x_0\sqrt{1 - \frac{k}{mx_0^4}t^2} \qquad (t \leq \sqrt{mx_0^4/k}), \tag{11}$$

where we have taken a positive root because the particle is to the right of the origin ($x > 0$). According to (11), $x = 0$ when t is given by (1).

Comments

(i) The velocity $v(t)$, obtained by differentiating (11), is
$$v(t) = -\frac{k}{mx_0^3}t\left(1 - \frac{k}{mx_0^4}t^2\right)^{-\frac{1}{2}}. \tag{12}$$

Graphs of (11) and (12) are shown below.

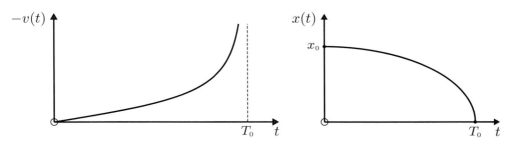

(ii) After passing the origin (with infinite speed) the particle comes to rest at $-x_0$ (at time $t = 2T_0$), and it then retraces its motion to again reach x_0. Thus, the complete motion is oscillatory with amplitude x_0 (the particle oscillates between x_0 and $-x_0$) and period $T = 4T_0$. The two figures above show the first quarter-cycle of the motion, and it is left as an exercise for the reader to sketch $v(t)$ and $x(t)$ for the full oscillatory motion.

(iii) Three-dimensional motion in an inverse-cube force field is solved in Question 8.11.

Question 3.5

A particle in an attractive inverse-square force field $F = -k/x^2$ (k is a positive constant) is projected along the x-axis with velocity v_0 (> 0) from the point x_0 (> 0). Determine its trajectory.

Solution

In the equation of motion
$$m\frac{dv}{dt} = -\frac{k}{x^2}, \tag{1}$$
make the replacement (3) of Question 3.4 to obtain
$$mv\,dv = -\frac{k}{x^2}\,dx, \tag{2}$$
and then integrate between corresponding limits:
$$m\int_{v_0}^{v(x)} v\,dv = -k\int_{x_0}^{x} \frac{dx}{x^2}. \tag{3}$$

Hence
$$\tfrac{1}{2}m\{v^2(x) - v_0^2\} = k\left(\frac{1}{x} - \frac{1}{x_0}\right), \tag{4}$$

and so
$$v(x) = \pm v_0\left\{1 + \frac{2k}{mv_0^2}\left(\frac{1}{x} - \frac{1}{x_0}\right)\right\}^{\frac{1}{2}}, \tag{5}$$
where the upper (lower) sign applies to outward (inward) motion.

Before proceeding it is helpful to note that, according to (5), the asymptotic value of the velocity v_∞ (the velocity at $x = \infty$) is
$$v_\infty = v_0\left\{1 - \frac{2k}{mx_0v_0^2}\right\}^{\frac{1}{2}}, \tag{6}$$

which exists (is real) if
$$v_0 \geq v_e = \sqrt{2k/mx_0}, \tag{7}$$

in which case
$$v_\infty = \sqrt{v_0^2 - v_e^2}. \tag{8}$$
Thus, there are three distinct cases to consider: (a) $v_0 = v_e$ (the particle 'comes to rest at infinity'), (b) $v_0 < v_e$ (the particle comes to rest at a finite value of x and then returns to x_0), and (c) $v_0 > v_e$ (the particle 'reaches infinity with a finite speed' given by (8)). We consider these in turn.

(a) $v_0 = v_e$

This is the simplest case to study because when $v_0 = v_e$, (5) simplifies to
$$\frac{dx}{dt} = v_0 \sqrt{\frac{x_0}{x}}. \tag{9}$$

So
$$\int_{x_0}^{x(t)} \sqrt{x}\, dx = v_0 \sqrt{x_0} \int_0^t dt, \tag{10}$$

and therefore
$$x(t) = x_0 \left(1 + \frac{3v_0}{2x_0} t\right)^{\frac{2}{3}}. \tag{11}$$

According to (9) and (11) the velocity is
$$v(t) = v_0 \left(1 + \frac{3v_0}{2x_0} t\right)^{-\frac{1}{3}}. \tag{12}$$

Clearly, $x \to \infty$ and $v \to 0$ as $t \to \infty$.

The calculations for the next two cases, while in essence the same as the preceding one, involve more algebra because an integral is more complicated. From (5) we have
$$\frac{dx}{dt} = \pm v_0 \sqrt{\frac{\alpha x + \beta}{x}}, \tag{13}$$
where $\alpha = 1 - \gamma^{-2}$, $\beta = x_0/\gamma^2$, and
$$\gamma = \frac{v_0}{v_e}. \tag{14}$$

(It is the fact that now $\gamma \neq 1$ that complicates the ensuing algebra.) To integrate (13) we require the following two integrals:

$$\int \sqrt{\frac{x}{\alpha x + \beta}}\, dx = \begin{cases} \dfrac{\sqrt{\alpha x^2 + \beta x}}{\alpha} + \dfrac{\beta}{\alpha\sqrt{-\alpha}} \sin^{-1}\sqrt{\dfrac{\alpha x + \beta}{\beta}} & \text{if } \alpha < 0 \\[2ex] \dfrac{\sqrt{\alpha x^2 + \beta x}}{\alpha} - \dfrac{\beta}{\alpha\sqrt{\alpha}} \sinh^{-1}\sqrt{\dfrac{\alpha x}{\beta}} & \text{if } \alpha > 0. \end{cases} \tag{15}$$

(b) $v_0 < v_e$

Here, $\gamma < 1$ and so $\alpha < 0$. We use $(15)_1$ to integrate (13). After some algebra we find the equation of the trajectory in its inverse form:

$$t = \frac{\gamma}{1-\gamma^2}\frac{x_0}{v_0}\left[\gamma + \frac{1}{\sqrt{1-\gamma^2}}\sin^{-1}\gamma \mp \left\{\sqrt{(\gamma^2-1)X^2 + X}\right.\right.$$
$$\left.\left. + \frac{1}{\sqrt{1-\gamma^2}}\sin^{-1}\sqrt{(\gamma^2-1)X+1}\right\}\right], \tag{16}$$

where the upper (lower) sign applies for the outward (inward) motion, and

$$X = x/x_0. \tag{17}$$

(c) $v_0 > v_e$

Now, $\gamma > 1$ and so $\alpha > 0$. We use $(15)_2$ to integrate (13), where we choose the upper sign because the particle is always moving in the positive x-direction. This gives the equation of the trajectory in its inverse form:

$$t = \frac{\gamma}{\gamma^2-1}\frac{x_0}{v_0}\left[-\gamma + \frac{1}{\sqrt{\gamma^2-1}}\sinh^{-1}\sqrt{\gamma^2-1} + \sqrt{(\gamma^2-1)X^2+X}\right.$$
$$\left. - \frac{1}{\sqrt{\gamma^2-1}}\sinh^{-1}\sqrt{(\gamma^2-1)X}\,\right]. \tag{18}$$

Comments

(i) The solutions (11) and (12) for $v_0 = v_e$ are the monotonic functions shown below. Here, $v_\infty = 0$ according to (8).

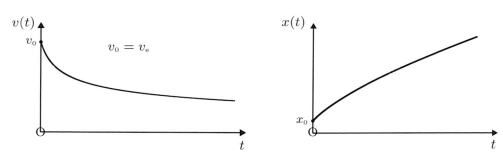

(ii) When $v_0 < v_e$ the particle reaches a maximum position x_m at $t = t_m$, and then returns to its starting point x_0 at $t = 2t_m$. Now $v = 0$ at x_m and therefore (13) yields

$$x_m = -\beta/\alpha = (1-\gamma^2)^{-1}x_0. \tag{19}$$

From (16) and (19) we have

$$t_m = \left(\frac{x_0}{v_0}\right)\frac{\gamma}{1-\gamma^2}\left\{\gamma + \frac{1}{\sqrt{1-\gamma^2}}\sin^{-1}\gamma\right\}. \tag{20}$$

Note that $x_m \to \infty$ as $v_0 \to v_e$, as one expects. Curves of $v(t)$ and $x(t)$ obtained from (13) and (16) are shown below for $v_0 = 0.9 v_e$.

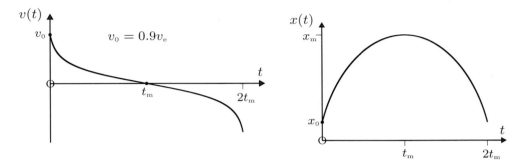

(iii) If v_0 is small compared to v_e, one can make a 'constant-force approximation' by replacing the position-dependent force $F = -k/x^2$ with its value at x_0, that is with $F_0 = -k/x_0^2$. Then, (5) of Question 3.1 gives

$$x(t) = x_0 + v_0 t + F_0 t^2/2m. \qquad (21)$$

Because the force is over-estimated in this approximation, (21) is a lower bound to the trajectory obtained from (16). This is illustrated below where the dotted curves are obtained from (21) and the solid curves from (16). As one expects, (21) is a poor approximation when v_0 is comparable to v_e. (Note that the vertical scales differ in these plots; the values of x_m are obtained from (19).) The approximation (21) is commonly used in the theory of projectiles moving near the Earth's surface (see Question 7.1). The numerical values in the diagrams are for the Earth ($x_0 = 6370$ km and $v_e = 11.2$ km s^{-1}).

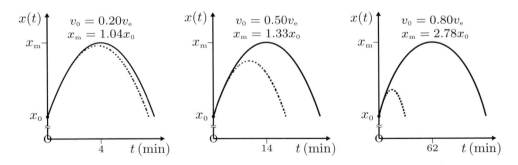

(iv) When $v_0 > v_e$ the motion is unbounded. Plots of $v(t)$ and $x(t)$ obtained from (13) and (18) are shown below for $v_0 = 2v_e$. The graph of $x(t)$ appears to be close to a linear relationship and it is interesting to investigate this further. There are simple upper and lower bounds, $x_U(t)$ and $x_L(t)$, for $x(t)$. The former is given by a

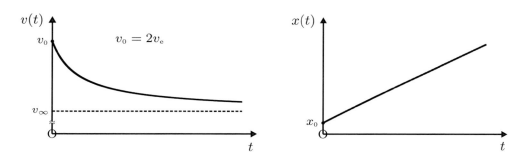

'force-free' approximation

$$x_{\rm U}(t) = x_0 + v_0 t, \tag{22}$$

(see (5) of Question 3.1 with $F = 0$) while

$$\begin{aligned} x_{\rm L}(t) &= x_0 + v_\infty t \\ &= x_0 + (1 - \gamma^{-2})^{\frac{1}{2}} v_0 t, \end{aligned} \tag{23}$$

where we have used (8). Now as $t \to 0$, $x(t) \to x_{\rm U}(t)$, and as $t \to \infty$, $x(t) \to x_{\rm L}(t)$. This is illustrated for the Earth and for $v_0 = 3v_{\rm e}$ in the figure below. The two dotted straight lines are $x_{\rm U}(t)$ and $x_{\rm L}(t)$. With increasing γ, the lower bound (23) approaches the upper bound (22) and the solution $x(t)$ is increasingly squeezed between the two bounds. The result is that $x(t)$ is very nearly linear.

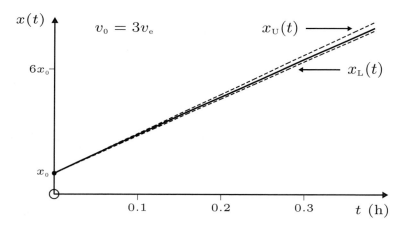

Question 3.6

A force $F = -F_0 e^{-x/\lambda}$ (where F_0 and λ are positive constants) acts on a particle that is initially at $x_0 = 0$ and moving with velocity v_0 (> 0). Determine its velocity $v(x)$ and sketch the three possible graphs of $v(x)$ versus x.

Solution

The equation of motion

$$m\frac{dv}{dt} = -F_0 e^{-x/\lambda}, \qquad (1)$$

can be written (using (3) of Question 3.4) as

$$mv\,dv = -F_0 e^{-x/\lambda} dx, \qquad (2)$$

and then integrated between corresponding limits:

$$m\int_{v_0}^{v(x)} v\,dv = -F_0 \int_{x_0}^{x} e^{-x/\lambda} dx. \qquad (3)$$

This gives

$$\frac{1}{2}m\{v^2(x) - v_0^2\} = F_0 \lambda\left(e^{-x/\lambda} - 1\right) \qquad (4)$$

and thus

$$v(x) = \pm\left\{v_0^2 + (2F_0\lambda/m)\left(e^{-x/\lambda} - 1\right)\right\}^{\frac{1}{2}}, \qquad (5)$$

where the upper (lower) sign applies for motion in the positive (negative) x-direction. Consider first the upper sign in (5). The asymptotic velocity (the limiting velocity as $x \to \infty$), namely

$$v_\infty = \sqrt{v_0^2 - 2F_0\lambda/m}, \qquad (6)$$

exists (is real) if

$$v_0 \geq v_e = \sqrt{2F_0\lambda/m}, \qquad (7)$$

in which case

$$v_\infty = \sqrt{v_0^2 - v_e^2}. \qquad (8)$$

If, however, $v_0 < v_e$ then the particle comes to rest at a finite value of x given by (see (5) with $v(x) = 0$)

$$x_m = -\lambda \ln(1 - v_0^2/v_e^2). \qquad (9)$$

After this, the particle accelerates to the left and $v(x)$ is obtained by taking the lower sign in (5). The particle reaches the origin with velocity $-v_0$. There are three possible graphs of $v(x)$ versus x corresponding to ☞ $v_0 = v_e$, ☞ $v_0 > v_e$ and ☞ $v_0 < v_e$:

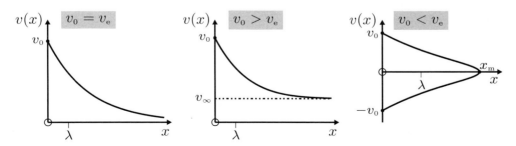

Comments

(i) All one-dimensional problems involving position-dependent forces $F(x)$ (such as the preceding three questions) can be solved by integration. First, use (3) of Question 3.4 and then integrate the equation of motion:

$$m \int_{v_0}^{v(x)} v \, dv = \int_{x_0}^{x} F(x) \, dx. \tag{10}$$

This yields $v(x)$, which can be integrated with respect to t to obtain the trajectory in the inverse form $t = t(x)$.

(ii) Equation (10) is an example of the work–energy theorem: the change in kinetic energy of a particle is equal to the work done by the force acting on the particle during the motion from x_0 to x. Use of (10) is also equivalent to using the law of conservation of energy – the change in kinetic energy is the negative of the change in the potential energy of the particle. See also (6) of Question 3.4 and (4) of Questions 3.5 and 3.6. The general case is treated in Chapter 5.

Question 3.7

Find the velocity $v(t)$ and position $x(t)$ of a particle of mass m that is subject to a frictional force (retarding force or drag) proportional to the velocity.

Solution

The drag can be written $F_f = -\alpha v$, where α is a positive constant, and hence the equation of motion is

$$m \frac{dv}{dt} = -\alpha v. \tag{1}$$

This can be rearranged and integrated to yield

$$\int_{v_0}^{v(t)} \frac{dv}{v} = -\frac{1}{\tau} \int_0^t dt, \tag{2}$$

where v_0 (> 0) is the initial velocity and

$$\tau = m/\alpha \tag{3}$$

is a characteristic time. From (2) we obtain

$$v(t) = v_0 e^{-t/\tau}. \tag{4}$$

Setting $v = dx/dt$ in (4) and integrating with respect to t, we have

$$\int_{x_0}^{x(t)} dx = v_0 \int_0^t e^{-t/\tau} dt, \tag{5}$$

and hence

$$x(t) = x_0 + v_0 \tau \left(1 - e^{-t/\tau}\right). \tag{6}$$

Comments

(i) It is evident from (6) that for small t ($t \ll \tau$), $x \approx x_o + v_o t$ and for large t ($t \gg \tau$), $x \to x_\infty = x_o + v_o \tau$. Thus, the motion is bounded: the particle moves a finite distance $v_o \tau$.

(ii) Graphs of $v(t)$ and $x(t)$ are shown below.

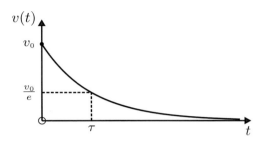
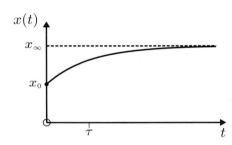

Question 3.8

Find the velocity $v(t)$ and position $x(t)$ of a particle of mass m that is subject to a frictional force (retarding force or drag) proportional to the square of the velocity.

Solution

The equation of motion is

$$m\frac{dv}{dt} = -\beta v^2, \tag{1}$$

where β is a positive constant. Thus

$$\int_{v_0}^{v(t)} \frac{dv}{v^2} = -\frac{\beta}{m} \int_0^t dt, \tag{2}$$

and hence

$$v(t) = \frac{v_0}{1 + t/\tau}, \tag{3}$$

where

$$\tau = m/\beta v_0 \tag{4}$$

is a characteristic time. Integration of (3) with respect to t yields

$$x(t) = x_0 + v_0 \tau \ln(1 + t/\tau). \tag{5}$$

Comments

(i) Here, $x(t)$ is unbounded – the drag is too weak at low velocity to restrict the particle to a finite interval.

(ii) Graphs of the solutions (3) and (5) are shown below.

44 Solved Problems in Classical Mechanics

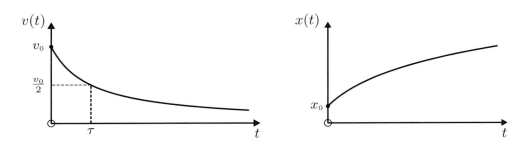

(iii) In general, the frictional force F_f due to motion in a resistive medium is more complicated than the linear and quadratic expressions considered above. For motion of a sphere of radius R in a medium with density ρ_m, it can be shown that[2]

$$F_f = \tfrac{1}{2}\pi C_d \rho_m R^2 v^2. \tag{6}$$

Here, the 'drag coefficient' C_d depends in a complicated way on the Reynolds number $\mathsf{Re} = 2\rho_m R v/\eta$, where η is the dynamic viscosity of the medium. An approximate formula for C_d, which applies if $\mathsf{Re} \lesssim 2 \times 10^5$ is[3]

$$C_d(\mathsf{Re}) \approx \frac{24}{\mathsf{Re}} + \frac{6}{1 + \sqrt{\mathsf{Re}}} + 0.4. \tag{7}$$

(For $\mathsf{Re} \approx 2 \times 10^5$ there is a change from laminar to turbulent flow near the sphere's surface, resulting in a sudden decrease in C_d.) Thus, for small Reynolds number (say $\mathsf{Re} < 1$), $C_d \approx 24/\mathsf{Re}$ and (6) reduces to Stokes's law

$$F_f = 6\pi \eta R v. \tag{8}$$

That is, the linear drag considered in Question 3.7. For larger Reynolds number (typically $10^3 < \mathsf{Re} < 2 \times 10^5$) equation (7) gives $C_d \approx 0.4$ and hence

$$F_f = 0.2\pi \rho_m R^2 v^2, \tag{9}$$

which is the quadratic dependence used above. Linear drag applies, for example, to a sediment settling in water (see Question 3.10), whereas quadratic drag applies to a stone or a sky diver falling through the atmosphere. (We mention that (6) is, strictly speaking, restricted to motion with constant velocity v; if v varies then there are additional effects that can be significant.[4])

Question 3.9

Determine the velocity $v(t)$ and position $x(t)$ of a particle of mass m that is projected vertically upwards with initial velocity v_0 in a uniform gravitational field in a medium with a frictional force proportional to the velocity.

---

[2]   See, for example, L. D. Landau and E. M. Lifshitz, *Fluid dynamics*. Oxford: Pergamon, 1959.
[3]   See, for example, F. M. White, *Viscous fluid flow*. New York: McGraw-Hill, 1974.
[4]   C. Pozridikis, *Introduction to theoretical and computational fluid dynamics*. Oxford: Oxford University Press, 1997.

Solution

Choose the x-axis to be vertically upwards and suppose that the particle is projected from the origin. The equation of motion is

$$m\frac{dv}{dt} = -mg - \alpha v, \qquad (1)$$

where g (> 0) is the gravitational acceleration, assumed to be constant. Note that (1) applies to both the initial (upward) and the subsequent downward motion (that is, for $v > 0$ and $v < 0$, respectively). From (1) we have

$$\int_{v_0}^{v(t)} \frac{dv}{1 + v/v_t} = -g\int_0^t dt, \qquad (2)$$

where

$$v_t = mg/\alpha \qquad (3)$$

is a characteristic velocity. Performing the integration in (2) and solving for $v(t)$ we find

$$v(t) = v_0 e^{-t/\tau} - v_t\left(1 - e^{-t/\tau}\right), \qquad (4)$$

where

$$\tau = v_t/g = m/\alpha \qquad (5)$$

is a characteristic time. Integration of (4) with respect to t yields

$$x(t) = (v_0 + v_t)\,\tau\left(1 - e^{-t/\tau}\right) - v_t t. \qquad (6)$$

Comments

(i) If $g = 0$ then $v_t = 0$ and (4) reduces to (4) of Question 3.7.
(ii) The significance of v_t is apparent from (4): $v(t) \to -v_t$ as $t \to \infty$. Thus, v_t is the asymptotic velocity for the downward motion. This property is evident already in (1): the particle stops accelerating when $v = -v_t$. For this reason, v_t is referred to as the terminal velocity.
(iii) Graphs of $v(t)$ and $x(t)$ are shown below.

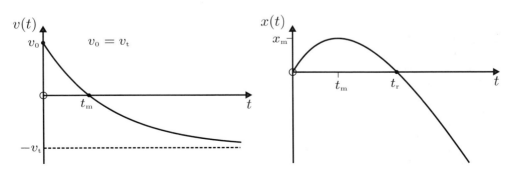

The time to reach the maximum height x_m is obtained by setting $v = 0$ in (4):

$$t_m = \tau \ln(1 + v_0/v_t). \tag{7}$$

From (5)–(7) we have

$$x_m = (v_0 v_t/g) - (v_t^2/g) \ln(1 + v_0/v_t). \tag{8}$$

The time t_r to return to the origin is obtained by setting $x = 0$ in (6), that is by the non-zero root of the transcendental equation

$$t = (1 + v_0/v_t)\,\tau\left(1 - e^{-t/\tau}\right). \tag{9}$$

(iv) The time $t_d = t_r - t_m$ for the downward part of the motion, from $x = x_m$ back to the origin, is longer than the time t_m for the upward motion. This is a general result, independent of the expression for the frictional force: the net force (and hence the acceleration) for the upward motion is greater than that for the downward motion. Therefore, $t_m < t_d$. In the above case, and with $u = v_0/v_t$, we have from (6) and (7) that[‡]

$$x(2t_m) = v_t \tau \left[1 + u - (1+u)^{-1} - 2\ln(1+u)\right] > 0 \tag{10}$$

for $u > 0$. Therefore, $2t_m < t_r$ and consequently $t_m < t_d$.

(v) It is interesting to compare x_m with $x_0 = v_0^2/2g$, the maximum height reached in the absence of friction. With $u = v_0/v_t$ we have from (8)

$$g(x_0 - x_m)/v_t^2 = \tfrac{1}{2}u^2 - u + \ln(1+u) > 0 \tag{11}$$

for $u > 0$. Thus, $x_m < x_0$, as one expects. This is also a general result – independent of the expression for the frictional force – which follows from energy considerations. To see this, integrate the equation of motion with respect to x. Then, if W and F_f denote the weight of the particle and the magnitude of the frictional force, we have

$$\int_0^{x_0} W\,dx = \tfrac{1}{2}mv_0^2 \quad \text{and} \quad \int_0^{x_m}(W + F_f)\,dx = \tfrac{1}{2}mv_0^2$$

for motion without and with friction, respectively. So,

$$W x_0 = W x_m + \int_0^{x_m} F_f\,dx, \tag{12}$$

and hence $x_0 > x_m$.

(vi) Similarly, we compare t_m with $t_0 = v_0/g$, the time of ascent (or descent) in the absence of friction. From (7) and (5), and with $u = v_0/v_t > 0$,

$$g(t_0 - t_m)/v_t = u - \ln(1+u) > 0. \tag{13}$$

Thus, $t_m < t_0$; the time of ascent is decreased by friction. This is also a general result, obtained by integrating the equation of motion with respect to t:

[‡]The inequalities in (10), (11) and (13) follow from the fact that $f(u) > g(u)$ for $u > 0$ if $f(0) = g(0)$ and $f'(u) > g'(u)$.

$$\int_0^{t_o} W\,dt = mv_0 \quad \text{and} \quad \int_0^{t_m}(W+F_f)\,dt = mv_0$$

for motion without and with friction, respectively. Thus

$$Wt_o = Wt_m + \int_0^{t_m} F_f\,dt, \qquad (14)$$

and so $t_o > t_m$. For the time of descent t_d no such general result exists.[5]

(vii) In the equation of motion (1) we have neglected the buoyancy $m'g$, where m' is the mass of the medium displaced by the 'particle'. Taking this into account means that g in (1)–(3) and (5) should be replaced by an effective gravitational acceleration

$$g_e = g\left(1 - m'/m\right) = g\left(1 - \rho_m/\rho\right), \qquad (15)$$

where ρ and ρ_m are the densities of the 'particle' and the medium, respectively. Thus, g_e can be either positive or negative. Interesting numerical results have been presented that illustrate the roles of the frictional and buoyancy forces for a grain of sand moving in water.[6]

Question 3.10

Consider a uniform suspension of spherical particles of radius $a = 1.40 \times 10^{-3}$ mm and density $\rho = 4.10 \times 10^3$ kg m^{-3} in a lake of depth $D = 2.00$ m. The density and viscosity of the water are $\rho_m = 1.00 \times 10^3$ kg m^{-3} and $\eta = 1.00 \times 10^{-3}$ Pa s, respectively. Take the acceleration g due to gravity to be 9.80 m s^{-2}.

(a) Calculate the percentage of particles still in suspension after 12 h.
(b) How long does it take for the lake to be clear of suspended particles?

Solution

(a) Assume that the particle velocities are sufficiently low so that linear friction applies (see Question 3.8). At the terminal velocity v_∞ the frictional force on a particle is equal to the net downward force (the weight minus the buoyancy):

$$6\pi\eta a v_\infty = \tfrac{4}{3}\pi a^3 (\rho - \rho_m) g. \qquad (1)$$

Thus

$$v_\infty = \frac{2a^2(\rho - \rho_m)g}{9\eta} = \frac{2 \times (1.40 \times 10^{-6})^2 \times 3.10 \times 10^3 \times 9.80}{9 \times 1.00 \times 10^{-3}}$$
$$= 1.32 \times 10^{-5}\,\text{m s}^{-1},$$

---

[5] J. M. Lévy-Leblond, "Solution to the problem on pg 15," American Journal of Physics, vol. 51, p. 88, 1983.
[6] P. Timmerman and J. P. van der Weele, "On the rise and fall of a ball with linear or quadratic drag," American Journal of Physics, vol. 67, pp. 538–546, 1999.

which is low enough to justify the initial assumption (see Question 3.8). The time taken for the particles to accelerate from rest to their terminal velocity is of order

$$\tau = m/6\pi\eta a = 2a^2\rho/9\eta \approx 2 \times 10^{-6}\,\text{s}$$

(see (3) of Question 3.7). Therefore, we assume that the particles fall with constant speed v_∞ until they reach the bottom. It follows that all particles that were initially within a distance

$$D_\text{t} = v_\infty t = 1.32 \times 10^{-5} \times 12 \times 3600 = 0.57\,\text{m}$$

from the bottom will settle out during the 12 h period. Then, the percentage of particles still in suspension after this time is

$$\frac{D - D_\text{t}}{D} \times 100 = \frac{2.00 - 0.57}{2.00} \times 100 \approx 72\%\,.$$

(b) The time taken for all the particles to reach the bottom is

$$\frac{D}{v_\infty} = \frac{2.00}{1.32 \times 10^{-5}} = 1.52 \times 10^5\,\text{s} \approx 42\,\text{h}\,.$$

Question 3.11

A small spherical drop of density ρ carries a charge q'. Its terminal velocity when falling through air of density ρ_a is v_∞. When a uniform vertical electrostatic field E is applied, its terminal velocity for upward motion is v'_∞. These velocities are small enough for the drag to be linear (see Question 3.8). Prove that for a drop of fixed mass

$$q' = 9\pi\,\frac{\eta}{E}\sqrt{\frac{2\eta v_\infty}{g(\rho - \rho_\text{a})}}\,(v_\infty + v'_\infty)\,, \tag{1}$$

where η is the viscosity of air.

Solution

The net downward force on a drop of volume V falling in the absence of an electric field is

$$F_\text{d} = (\rho - \rho_\text{a})Vg - \alpha v\,. \tag{2}$$

At the terminal velocity for downward motion, $F_\text{d} = 0$ and therefore

$$\alpha v_\infty = (\rho - \rho_\text{a})Vg\,. \tag{3}$$

When an electric field is applied to make the drop rise, the net upward force is

$$F_\text{u} = q'E - (\rho - \rho_\text{a})Vg - \alpha v\,. \tag{4}$$

At the terminal velocity for upward motion, $F_\text{u} = 0$ and therefore

$$\alpha v'_\infty = q'E - (\rho - \rho_\text{a})Vg\,. \tag{5}$$

By taking the ratio of (5) and (3), and adding one to both sides, we obtain

$$q' = \frac{(\rho - \rho_\mathrm{a})Vg}{E}\left(\frac{v_\infty + v'_\infty}{v_\infty}\right). \tag{6}$$

The volume V is determined from (3) and the expression for α given by Stokes's law, namely $\alpha = 6\pi\eta R$ where R is the radius of the drop (see Question 3.8). Thus

$$V = 9\pi\sqrt{\frac{2\eta^3 v_\infty^3}{g^3(\rho - \rho_\mathrm{a})^3}}. \tag{7}$$

Equations (6) and (7) yield (1). Note that in (1) both v_∞ and v'_∞ are positive and therefore $q' > 0$ (< 0) if the electric field points up (down).

Comments

(i) Equation (1) forms the basis of Millikan's famous oil-drop experiment to determine the magnitude of the electronic charge e. In about 1910 Millikan measured the terminal speeds v_∞ and v'_∞ for charged oil droplets (about 1μm in diameter[‡]) in weakly ionized air and found[8]

$$e = 4.774 \times 10^{-10}\,\mathrm{esu} \quad (\text{or } 1.591 \times 10^{-19}\,\mathrm{C}).$$

This is about 1% lower than the modern value. It was later shown that Millikan's data, when corrected for a small error in the measured viscosity of air, gives[9]

$$e = 4.807 \times 10^{-10}\,\mathrm{esu} \quad (\text{or } 1.602 \times 10^{-19}\,\mathrm{C}).$$

Drops carrying charges of both signs were observed in Millikan's experiments. Individual drops were studied sometimes for hours at a time, during which changes in the drop's charge would occasionally occur.

(ii) Initially, Millikan obtained a rough value of e from measurements on water droplets. Errors due to evaporation led him and a doctoral student[10] to consider non-volatile alternatives such as oil. Accurately spherical drops were produced using an atomizer, and some of these acquired a charge as they travelled through the nozzle. Further ionization was achieved by introducing α-particles (from a radium source) into the observation chamber. Sometimes the upward speed of the drop was observed to change discontinuously from v'_∞ to v''_∞, when it either captured or released an ion (whose mass was negligible compared to m). Then, (1) becomes

$$q'' = 9\pi\frac{\eta}{E}\sqrt{\frac{2\eta v_\infty}{g(\rho - \rho_\mathrm{a})}}\,(v_\infty + v''_\infty). \tag{8}$$

[‡]For diameters comparable to the mean free path of the air molecules, a modified form of Stokes's law is used.[7]

---

[7] R. A. Millikan, "The isolation of an ion, a precision measurement of its charge, and the correction of Stokes's law," The Physical Review, vol. XXXII, pp. 349–397, 1911.

[8] R. A. Millikan, "On the elementary electrical charge and the Avogadro constant," The Physical Review, vol. 2, pp. 109–143, 1913.

[9] See, for example, H. Semat, *Introduction to atomic and nuclear physics*. London: Chapman and Hall, 1954.

[10] H. Fletcher, "My work with Millikan on the oil-drop experiment," Physics Today, vol. 35, pp. 43–46, June 1982.

From (1) and (8) it follows that
$$\frac{q''}{q'} = \frac{v_\infty + v''_\infty}{v_\infty + v'_\infty}, \tag{9}$$
and similarly for further changes in the drop's charge and speed. Measurements taken on thousands of oil drops revealed that the values of q''/q', q'''/q'', etc. could always be expressed as the ratio of two small integers, thereby supporting the hypothesis of the discrete nature of electric charge.

(iii) In 1923, Millikan was awarded the Nobel prize for his work on the elementary electric charge and the photoelectric effect. In his acceptance speech[11] he stressed that " ... the particular dimensions of the apparatus and the voltage of the battery were the element which turned possible failure into success. Indeed, nature here was very kind. She left only a narrow range of field strengths within which such experiments as these are at all possible. They demand that the droplets be large enough so that the minute dancing movements, the 'Brownian movements', are nearly negligible, that the droplets be round and homogeneous, light and non-evaporable, that the distance of fall be long enough to make the timing accurate, and that the field be strong enough to more than balance gravity by its upward pull on a drop carrying but one or two electrons. Scarcely any other combination of dimensions, field strengths and materials could have yielded the results obtained. Had the electronic charge been one-tenth its actual size, or the sparking potential in air a tenth of what it actually is, no such experimental facts as here presented would ever have been seen." Millikan's oil-drop experiment was placed third in a survey to choose 'the most beautiful experiment in physics'.[1]

Question 3.12

Determine the velocity and position of a particle of mass m that is projected vertically upwards with initial velocity v_0 in a medium with a frictional force that is quadratic in the velocity. Assume the gravitational field is uniform.

Solution

It is necessary to consider separately the upward and downward parts of the motion.

Upward motion

The equation of motion for the upward journey is
$$m\frac{dv}{dt} = -mg - \beta v^2, \tag{1}$$
where β is a positive constant. This equation can be rearranged and integrated

---

[11] G. Holton, "Electrons or subelectrons? Millikan, Ehrenhaft and the role of preconceptions," in *History of twentieth century physics. Proceedings of the international school of physics - Enrico Fermi.* (C. Weiner, ed.), New York, Academic Press, pp. 277–278, 1977.

$$\int_{v_0}^{v(t)} \frac{dv}{1+(v/v_t)^2} = -g\int_0^t dt, \qquad (2)$$

where

$$v_t = \sqrt{mg/\beta} \qquad (3)$$

is a characteristic velocity. From (2) we obtain

$$v(t) = v_t \tan\left[\tan^{-1}(v_0/v_t) - t/\tau\right], \qquad (4)$$

where

$$\tau = v_t/g = \sqrt{m/\beta g} \qquad (5)$$

is a characteristic time. It is convenient to express (4) in terms of t_m, the time to reach the maximum height. Now, $v(t_m) = 0$ and (4) gives

$$t_m = \tau \tan^{-1}(v_0/v_t). \qquad (6)$$

From (4) and (6) we have the simple result

$$v(t) = v_t \tan(t_m - t)/\tau \qquad (t \le t_m). \qquad (7)$$

By integrating (7) with respect to t, and taking $x_0 = 0$, we obtain

$$x(t) = v_t \tau \ln\left[\frac{\cos(t_m - t)/\tau}{\cos t_m/\tau}\right] \qquad (t \le t_m). \qquad (8)$$

The maximum height $x_m = x(t_m)$ is

$$x_m = v_t \tau \ln\left(\cos t_m/\tau\right)^{-1} = \tfrac{1}{2} v_t \tau \ln\left(1 + v_0^2/v_t^2\right), \qquad (9)$$

where in the last step we have used (6) and the trigonometric relation

$$\cos(\tan^{-1}\theta) = (1+\theta^2)^{-1/2}.$$

Downward motion

Instead of (1) we now have

$$m\frac{dv}{dt} = -mg + \beta v^2. \qquad (10)$$

The initial condition is $v = 0$ at $t = t_m$, and so (10) yields

$$\int_0^{v(t)} \frac{dv}{1-(v/v_t)^2} = -g\int_{t_m}^t dt \qquad (11)$$

instead of (2). Thus

$$v(t) = -v_t \tanh(t - t_m)/\tau \qquad (t \geq t_m). \tag{12}$$

We see that $v \to -v_t$ as $t \to \infty$, meaning that v_t is the terminal velocity. Integration of (12) with respect to t and use of the initial condition $x = x_m$ at $t = t_m$ gives

$$x(t) = v_t \tau \ln\left[\frac{\sqrt{1 + v_0^2/v_t^2}}{\cosh(t - t_m)/\tau}\right] \qquad (t \geq t_m). \tag{13}$$

Equations (7), (8), (12) and (13) are the desired solutions.

Comments

(i) Graphs of these solutions are shown below.

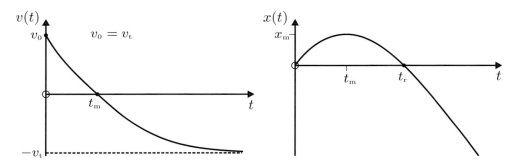

The time $t_d = t_r - t_m$ for the downward part of the motion (from $x = x_m$ back to the origin) is obtained by setting $x(t_r) = 0$ in (13). Thus,

$$t_d = \tau \cosh^{-1}\sqrt{1 + v_0^2/v_t^2} = \frac{v_t}{g}\ln\left\{\frac{v_0}{v_t} + \sqrt{1 + v_0^2/v_t^2}\right\}. \tag{14}$$

One can show from (6) and (14) that $t_m < t_d$. It is also evident from (9) that $x_m < x_0 = v_0^2/2g$, the maximum height reached in the absence of friction. Both of these conclusions agree with the general results deduced in Question 3.9.

(ii) The speed with which the particle returns to the starting point is, from (12) with $t = t_r$,

$$v_r = -v_t \tanh t_d/\tau. \tag{15}$$

Now, $\tanh(\ln \theta) = (\theta^2 - 1)/(\theta^2 + 1)$ and so (14) and (15) yield the simple result

$$v_r = -v_0(1 + v_0^2/v_t^2)^{-1/2}. \tag{16}$$

Thus, the fractional loss in kinetic energy due to dissipation,

$$\frac{\Delta E}{E} = \frac{\frac{1}{2}mv_0^2 - \frac{1}{2}mv_r^2}{\frac{1}{2}mv_0^2} = \frac{1}{(1 + v_t^2/v_0^2)}, \tag{17}$$

is small when $v_0 \ll v_t$, and approaches 1 when $v_0 \gg v_t$.

Question 3.13

A sphere of radius R and density ρ falls through a medium of constant density ρ_m that exerts a quadratic drag $F_f = 0.2\pi\rho_m R^2 v^2$ on it (see (9) of Question 3.8). Prove that the time taken for the sphere to fall from rest through a height H is given by

$$t = \left(\frac{1}{\sqrt{2u}}\cosh^{-1} e^u\right) t_0, \qquad (1)$$

where

$$u = 3\rho_m H / 20\rho R \qquad (2)$$

and $t_0 = \sqrt{2H/g_e}$ is the time taken in the absence of drag. Here, $g_e = g(1 - \rho_m/\rho)$ is the effective gravitational acceleration (see Question 3.9).

Solution

The equation of motion is

$$m\frac{dv}{dt} = mg_e - \beta v^2, \qquad (3)$$

where $\beta = 0.2\pi\rho_m R^2$ and we have chosen the positive x-axis downward. Integration of (3) with the initial condition $v_0 = 0$ yields

$$v(t) = v_t \tanh t/\tau, \qquad (4)$$

where $v_t = \sqrt{mg_e/\beta}$ and $\tau = \sqrt{m/\beta g_e}$. Integration of (4) with the initial condition $x_0 = 0$ yields

$$x(t) = v_t \tau \ln(\cosh t/\tau). \qquad (5)$$

Setting $x = H$ and solving (5) for t gives

$$t = \sqrt{\frac{m}{\beta g_e}} \cosh^{-1} e^{\beta H/m}. \qquad (6)$$

Using $m = \frac{4}{3}\pi R^3 \rho$ and the above expression for β in (6), we obtain (1).

Comments

(i) For 'small' values of u there is a simple approximation to (1) that can be obtained as follows. We use the logarithmic form

$$\cosh^{-1}\theta = \ln\left\{\theta + \sqrt{(\theta-1)(\theta+1)}\right\} \qquad (7)$$

and the expansions

$$e^u = 1 + u + \tfrac{1}{2}u^2 + \cdots, \qquad \ln(1+\delta) = \delta - \tfrac{1}{2}\delta^2 + \tfrac{1}{3}\delta^3 - \cdots \qquad (8)$$

to write

$$\cosh^{-1} e^u = \sqrt{2u} + \tfrac{1}{6}\sqrt{2u^3} + O(u^{\frac{5}{2}}). \qquad (9)$$

Thus, for small u equation (1) can be approximated by

$$t = (1 + \tfrac{1}{6}u)t_0. \qquad (10)$$

As required, $t \to t_o$ in the limit $\rho_m \to 0$ (i.e. $\beta \to 0$ in (3)). The solution (1) and the approximation (10) are compared in the figure below. We see that the linear form (10) (the dotted line) is reasonable up to surprisingly large values of u (≈ 5), and it is therefore applicable to a variety of macroscopic objects falling through the Earth's atmosphere.

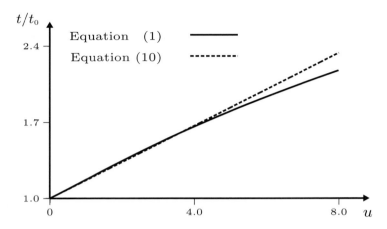

(ii) Consider an iron ball of mass $0.50\,\text{kg}$ falling from a height $H = 100\,\text{m}$ through air. Here, $\rho_m = 1.29\,\text{kg}\,\text{m}^{-3}$ and $\rho = 7870\,\text{kg}\,\text{m}^{-3}$. Also, $R = (3m/4\pi\rho)^{\frac{1}{3}} = 2.48 \times 10^{-2}\,\text{m}$ and hence (2) gives $u = 0.0993$. According to (10), the correction to the time of free fall t_o ($\approx 4.52\,\text{s}$ for $g = 9.80\,\text{m}\,\text{s}^{-2}$) is about 1.66% or $0.075\,\text{s}$. By contrast, for a 50-kg iron ball, $u = 0.0214$ and the correction is about $0.016\,\text{s}$. Thus, the larger ball strikes the ground about $0.059\,\text{s}$ before the smaller one. That there is such an effect was known already to Galileo.[12]

(iii) In making the above estimate, we have neglected the very short initial stage where the Reynolds number is less than about 10^3 (see Question 3.8).

(iv) Note that one cannot make the above effect indefinitely large by using an ever smaller sphere to increase u in (2) because the quadratic approximation for the drag eventually becomes inapplicable when the Reynolds number is too small (see Question 3.8).

(v) All one-dimensional problems involving velocity-dependent forces $F(v)$ (such as in Questions 3.7–3.13) can be solved by integration:

$$m \int_{v_0}^{v(t)} \frac{dv}{F(v)} = \int_0^t dt. \qquad (11)$$

This yields $t = t(v)$; if it can be inverted to provide $v = v(t)$, as in the above questions, then one can integrate with respect to t to obtain $x = x(t)$.

(vi) In the above questions we have neglected changes in the gravitational acceleration and air density with height. We now consider a numerical calculation in which these are taken into account.

---

[12] G. Galilei, *Dialogues concerning two new sciences*, p. 65. New York: Dover, 1914.

Question 3.14

A sphere of radius R and density ρ falls from rest from a high altitude H through the Earth's atmosphere. For an isothermal atmosphere, the density varies with height x above sea level according to $\rho_m = \rho_0 e^{-x/X}$. Assume a quadratic drag $F_f = 0.2\pi\rho_m R^2 v^2$ and obtain numerical solutions for the velocity $v(t)$ and height $x(t)$ for H equal to (a) $5\,\mathrm{km}$, (b) $10\,\mathrm{km}$, (c) $15\,\mathrm{km}$, and (d) $20\,\mathrm{km}$. Take $R = 2.00\,\mathrm{cm}$, $\rho = 5.00 \times 10^3\,\mathrm{kg\,m^{-3}}$, $\rho_0 = 1.29\,\mathrm{kg\,m^{-3}}$, $X = 7.46 \times 10^3\,\mathrm{m}$, Earth's radius $R_e = 6.37 \times 10^6\,\mathrm{m}$ and at sea level $g_0 = 9.80\,\mathrm{m\,s^{-2}}$.

Solution

Choosing the origin at sea level and the positive x-axis vertically upwards, the equation of motion can be written as

$$\frac{dv}{dt} = -\frac{g_0}{(1+x/R_e)^2} + \lambda_0^{-1} v^2 e^{-x/X}, \tag{1}$$

where $\lambda_0 = m/\beta_{\text{sea level}} = 20\rho R/3\rho_0 = 517\,\mathrm{m}$ is a characteristic length. Equation (1) takes account of variations in both the gravitational acceleration and the air density with altitude (although in the present example the former has only a slight effect). The *Mathematica* notebook given at the end of this question is used to solve (1) and plot graphs of $v(t)$ and $x(t)$ versus t. The desired numerical solutions are presented in the figures below, where each curve is plotted against t up to the instant of impact with the ground.

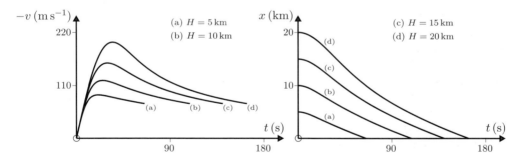

Comments

(i) It is interesting to compare these solutions with those in Questions 3.12 and 3.13 that apply when the fall is from a low altitude. In the latter case we could assume a constant air density, and the calculated speed of the downward motion increases monotonically with t, tending to an asymptotic value v_t (the terminal speed). When the fall is from a greater altitude and the air density varies, the above figure for $v(t)$ shows that the speed is not monotonic: instead, each of the curves has a maximum value. In this particular example, the impact speed and terminal speed of the sphere (see (3) of Question 3.12) are comparable (about $70\,\mathrm{m\,s^{-1}}$).

56 Solved Problems in Classical Mechanics

(ii) It is also interesting to compare the gravity and drag contributions to the total acceleration in (1). This is done for $H = 20$ km in the figure below. The gravity contribution is almost constant but the drag varies strongly. After an initial period of downward acceleration, the sphere decelerates during the remainder of its fall.

(iii) The plot of the Reynolds number (Re $= 2\rho_m Rv/\eta$) for a fall from $H = 20.0$ km shows that Re $< 2 \times 10^5$, and therefore a quadratic drag is a good approximation here (see Question 3.8).

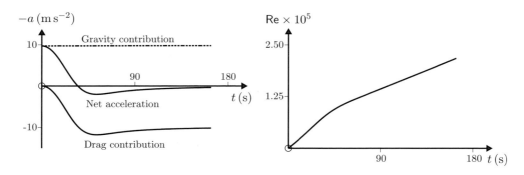

```
In[1]:= g₀ = 9.8; Rₑ = 6.37 × 10⁶; H = 5 × 10³; λ₀ = 517; X = 7.46 × 10³;

       T = √(2H/g₀);  (* First approximation for the time of fall *)

       Sol = NDSolve[{x''[t] + g₀/(1 + x[t]/Rₑ)² - x'[t]² e^(-x[t]/X)/λ₀ == 0,
                      x[0] == H, x'[0] == 0}, {x[t], x'[t]}, {t, 0, 5 T}];

       FallTime = FindRoot[x[t] == 0/.Sol, {t, T}];

       T_fall = t/.FallTime;  (* Actual time of fall *)

       Plot[Evaluate[x[t]/.Sol], {t, 0, T_fall}]

       Plot[Evaluate[-x'[t]/.Sol], {t, 0, T_fall}, PlotRange → All]
```

Question 3.15

Jerk is defined as the rate of change of acceleration: $\mathbf{j} = d\mathbf{a}/dt = d^2\mathbf{v}/dt^2 = d^3\mathbf{r}/dt^3$. Consider a particle moving in one dimension with constant jerk j. Determine the position $x(t)$ of the particle in terms j and the initial values x_0, v_0, and a_0.

Solution

We start with $j = da/dt$ (a constant), and integrate with respect to t. Thus

$$a(t) = a_o + jt. \tag{1}$$

Because $a = dv/dt$ and $v = dx/dt$, further integration with respect to t yields

$$v(t) = v_o + a_o t + \tfrac{1}{2} j t^2 \tag{2}$$

$$x(t) = x_o + v_o t + \tfrac{1}{2} a_o t^2 + \tfrac{1}{6} j t^3. \tag{3}$$

Comments

(i) According to (3), to evaluate $x(t)$ we require, in addition to j, three initial conditions, namely x_o, v_o and a_o. This is because $j = \dot{a} = \dddot{x}$ is a third-order equation.
(ii) The force $ma(t) = m(a_o + jt)$ is a linear function of t.
(iii) If the jerk is zero, then (1)–(3) reduce to the results for constant acceleration.
(iv) The above example is an introduction to the kinematics of jerk. This topic has been discussed in detail in Refs. [13] and [14].
(v) In general, jerk is of interest and importance in three respects. First, the physiological effects of jerk are important: "jerk is the most easily sensed derivative of displacement" and in amusement park rides, for example, it "is the jerk rather than the acceleration or velocity that makes the rides both exciting and uncomfortable."[15] The human body can withstand a jerk of about $2 \times 10^4 \, \text{ms}^{-3}$ (about $2000\,g$ per second).[15] Second, jerk is of theoretical interest[13,16,17] and third, it is important in engineering design.[13,18]

Question 3.16

Consider the third-order equation of motion

$$ma = F(t) + m\tau \frac{da}{dt}, \tag{1}$$

where τ is a positive constant having the dimension of time. Show that physically acceptable solutions to (1) can be expressed as

$$ma(t) = \frac{1}{\tau} \int_t^\infty e^{-(t'-t)/\tau} F(t') \, dt' \qquad (-\infty < t < \infty). \tag{2}$$

(Hint: Consider the behaviour as $t \to \infty$ of the solution to (1).)

---

[13] S. H. Schot, "Jerk: the time rate of change of acceleration," American Journal of Physics, vol. 46, pp. 1090–1094, 1978.
[14] T. R. Sandin, "The jerk," Physics Teacher, vol. 28, pp. 36–40, 1990.
[15] J. M. Wilson, "More jerks," Physics Teacher, vol. 27, p. 7, 1989.
[16] H. P. W. Gottlieb, "Simple nonlinear jerk functions with periodic solutions," American Journal of Physics, vol. 66, pp. 903–906, 1998.
[17] S. J. Linz, "Newtonian jerky dynamics: some general properties," American Journal of Physics, vol. 66, pp. 1109–1114, 1998.
[18] W. F. D. Theron, "Bouncing due to the 'infinite jerk' at the end of a circular track," American Journal of Physics, vol. 63, pp. 950–955, 1995.

Solution

Multiply (1) by $e^{-t/\tau}$. Then

$$\frac{d}{dt}(ae^{-t/\tau}) = -\frac{e^{-t/\tau}}{m\tau}F(t),$$

and integration between $t = 0$ and t yields

$$a(t) = e^{t/\tau}\left[a_0 - \frac{1}{m\tau}\int_0^t e^{-t'/\tau}F(t')\,dt'\right]. \qquad (3)$$

For arbitrary a_0 the solution (3) diverges as $t \to \infty$. This violates the following condition:

$$\left.\begin{array}{l}\text{Acceleration cannot increase indefinitely unless physical}\\ \text{forces act that supply the required energy.}\end{array}\right\} \qquad (4)$$

To avoid this difficulty we require that the quantity in square brackets in (3) must vanish in the limit $t \to \infty$. That is, we impose the condition

$$a_0 = \frac{1}{m\tau}\int_0^\infty e^{-t'/\tau}F(t')\,dt' \qquad (5)$$

on the initial acceleration. Equations (3) and (5) yield (2). Note that in the limit $\tau \to 0$, (2) reduces to the usual equation of motion $ma(t) = F(t)$ because $F(t')$ in the integrand of (2) can be replaced by $F(t)$ in this limit.

Comments

(i) The third-order equation (1) applies to non-relativistic motion of a charged particle in an external force-field $F(t)$. For such motion

$$\tau = e^2/6\pi\epsilon_0 c^3, \qquad (6)$$

and (1) is the one-dimensional form of the Abraham–Lorentz equation.[19] (Here, e is the charge of the particle, ϵ_0 is the permittivity of free space and c is the speed of light in vacuum.) The term involving τ in (1) is associated with radiative reaction: it is a self-force due to radiation of energy by an accelerated charged particle. Equation (1) is a reasonable approximation only when the effects of radiative reaction are small.[19]

(ii) Solutions to (1) that violate (4) are known as 'self-accelerated' or 'run-away' solutions. The simplest example is given by (3) with $F = 0$, that is

$$a(t) = a_0 e^{t/\tau}. \qquad (7)$$

According to (7), a particle set into free motion with a non-zero initial acceleration a_0 would continue to accelerate indefinitely. Such behaviour is not observed in nature.

---

[19] See, for example, J. D. Jackson, *Classical electrodynamics*. New York: Wiley, 3rd edn, 1998. Chap. 16.

(iii) The solutions (2) are a subset of (3) that exclude the run-away solutions. In particular, for $F = 0$, (2) yields $a(t) = 0$ rather than (7). In general, replacing the third-order differential equation (1) by the second-order integro-differential equation (2) has the effect of excluding the unphysical run-away solutions.

(iv) According to (2) the acceleration at time t is determined by the force experienced by the particle at each future instant of time. Thus, the present motion of the particle is determined by the force that will act on it at all future times. This property of the Abraham–Lorentz equation clearly violates causality and is known as preacceleration.

(v) Preacceleration is appreciable only for times of order τ. Consider, for example, a pulse
$$F(t) = A\delta(t), \tag{8}$$
where A is a constant and $\delta(t)$ is the Dirac delta function. This function has the properties
$$\delta(t) = 0 \quad \text{if } t \neq 0 \tag{9}$$
and
$$\int_{-\infty}^{\infty} g(t)\delta(t)\, dt = g(0). \tag{10}$$

The pulse (8) exerts a force on the particle only at the instant $t = 0$. From (2) and (8)–(10) we obtain
$$a(t) = \begin{cases} (A/m\tau)e^{t/\tau} & t < 0 \\ 0 & t > 0. \end{cases} \tag{11}$$

If the particle is initially at rest ($v = 0$ at $t = -\infty$) then integration of (11) gives
$$v(t) = \begin{cases} (A/m)e^{t/\tau} & t < 0 \\ A/m & t \geq 0. \end{cases} \tag{12}$$

These solutions are sketched below.

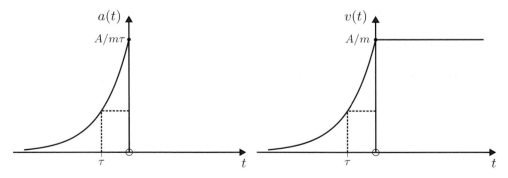

(vi) For familiar charged particles τ is very small. For example, for an electron (6) yields $\tau = 6.26 \times 10^{-24}$ s. Thus, the Abraham–Lorentz equation violates microscopic, and not macroscopic, causality. It should be emphasized that, in reality, classical theory is not applicable on such time scales.[19]

4
Linear oscillations

Linear oscillations occur when a particle is subject to a restoring force that is proportional to the position vector relative to a fixed point in inertial space. Known as the harmonic oscillator, this system has considerable theoretical importance and widespread applications: "... the physics of the harmonic oscillator – that is, Galileo's pendulum – which made it possible to regulate the flow of time, leads far beyond a mere device for making accurate clocks. These oscillators have been found to be the basis not only of what we hear as the sound of music and see as the colours of light but, via the quantum theory, of what we understand as the fabric of the universe." They are "... the simplest, yet most fundamental physics system in nature ... ."[1]

So it is not surprising that in the presentation of elementary dynamics it is traditional to devote a separate chapter to even the one-dimensional harmonic oscillator. Here, one encounters important topics such as harmonic oscillations, frequency and periodicity; linearity, superposition, completeness, and Fourier methods; the harmonic approximation and the theory of small oscillations; damping, relaxation, and quality factors; forced oscillations and resonance. The following questions deal with some of these topics, and we return to the oscillator again in Chapter 9.

Question 4.1

A one-dimensional restoring force $F = -kx$ (k is a positive constant) acts on a particle of mass m. Determine the velocity $v(t)$ and the position $x(t)$ in terms of m, k, and the initial conditions v_0 and x_0. Give two methods of solution: **1.** Using the technique of Questions 3.4–3.6. **2.** By direct solution of the equation of motion.

Solution

1. We use the technique adopted in Questions 3.4–3.6 for one-dimensional problems with position-dependent forces $F(x)$. Thus, we express the equation of motion

$$m\frac{dv}{dt} = -kx \tag{1}$$

---

[1] R. G. Newton, *Galileo's pendulum: From the rhythm of time to the making of matter*, p. 2. Cambridge, Massachusetts: Harvard University Press, 2004.

as
$$mv\frac{dv}{dx} = -kx, \qquad (2)$$

and then integrate both sides with respect to x. Consequently,
$$m\int_{v_0}^{v(x)} v\,dv = -k\int_{x_0}^{x} x\,dx, \qquad (3)$$

and hence
$$\tfrac{1}{2}mv^2(x) - \tfrac{1}{2}mv_0^2 = -\tfrac{1}{2}kx^2 + \tfrac{1}{2}kx_0^2. \qquad (4)$$

We solve this equation for $v(x)$:
$$v(x) = \frac{dx}{dt} = \pm\omega\sqrt{A^2 - x^2}, \qquad (5)$$

where the upper (lower) sign refers to motion to the right (left), and
$$\omega = \sqrt{k/m}, \qquad (6)$$
$$A = \sqrt{x_0^2 + v_0^2/\omega^2} \qquad (7)$$

are constants having the units of (time)$^{-1}$ and length, respectively. Integration of (5) with respect to t gives
$$\int_{x_0}^{x(t)} \frac{dx}{\sqrt{A^2 - x^2}} = \pm\omega\int_0^t dt, \qquad (8)$$

and hence
$$\cos^{-1}(x/A) - \cos^{-1}(x_0/A) = \mp\omega t. \qquad (9)$$

Equations (9) can be inverted to obtain $x(t)$; both results can be subsumed in the single equation
$$x(t) = A\cos(\omega t + \phi), \qquad (10)$$

where $\phi = \mp\cos^{-1}(x_0/A)$ is a constant. Then,
$$v(t) = -\omega A\sin(\omega t + \phi), \qquad (11)$$

and ϕ is given in terms of the initial conditions by
$$\phi = -\tan^{-1}(v_0/\omega x_0). \qquad (12)$$

Equations (10) and (11), with A and ϕ given by (7) and (12), are the desired solutions.

2. We write (1) as
$$\frac{d^2x}{dt^2} + \omega^2 x = 0, \tag{13}$$

and recognize that (13) is a linear, homogeneous, ordinary differential equation with constant coefficients. We then apply the standard technique for solving such an equation: we attempt a solution of the form
$$x(t) = e^{qt}, \tag{14}$$

where the constant q is to be determined. Substitution of (14) in (13) gives
$$(q^2 + \omega^2)x = 0. \tag{15}$$

Thus, the condition for (14) to be a solution to (13) is that (15) should be satisfied for all t. Now $x \neq 0$ for all t, and so we require
$$q^2 + \omega^2 = 0. \tag{16}$$

This is the so-called characteristic (or indicial) equation, and its roots are
$$q = \pm i\omega, \tag{17}$$

where i is the imaginary number $\sqrt{-1}$. Thus, both $e^{i\omega t}$ and $e^{-i\omega t}$ are solutions to (13), and the general solution is the linear combination
$$x(t) = a_1 e^{i\omega t} + a_2 e^{-i\omega t}, \tag{18}$$

where a_1 and a_2 are arbitrary constants. In terms of the initial conditions,
$$a_1 = \tfrac{1}{2}(x_0 - iv_0/\omega), \qquad a_2 = \tfrac{1}{2}(x_0 + iv_0/\omega). \tag{19}$$

Although the solution (18) looks different from the previous solution (10), their equivalence is readily established. Because $e^{i\theta} = \cos\theta + i\sin\theta$ and $\cos(\alpha + \beta) = \cos\alpha\cos\beta - \sin\alpha\sin\beta$, (18) can be written
$$x(t) = a\cos\omega t + b\sin\omega t \tag{20}$$
$$= A\cos(\omega t + \phi), \tag{21}$$

where $A = \sqrt{a^2 + b^2}$, $\phi = -\tan^{-1} b/a$, and
$$a = a_1 + a_2 = x_0, \qquad b = i(a_1 - a_2) = v_0/\omega. \tag{22}$$

Comments

(i) Equations (18), (20) and (21) are three equivalent ways of writing the general solution to the simple harmonic equation of motion (13), and which of these one uses in practice is essentially a matter of convenience. They are non-relativistic approximations valid in the limit $\omega A \ll c$, the speed of light in vacuum (see Question 15.14).

(ii) The solutions (10) and (11) are depicted below for $t \geq 0$. The particle oscillates between $x = -A$ and $x = A$ (the motion is bounded with amplitude A), and the velocity varies between $-\omega A$ and ωA. The intercepts of $x(t)$ on the positive t-axis are given by $t_n = \{(n - \frac{1}{2})\pi - \phi\}/\omega$ with $n = 1, 2, 3, \cdots$. The difference $t_{n+2} - t_n = T$ is the period (the time for one cycle of the motion), and so

$$T = 2\pi/\omega. \tag{23}$$

The velocity has the same period. The number of cycles per second, T^{-1}, is the frequency $f = \omega/2\pi$, and $\omega = 2\pi f$ is the angular frequency. The angle $\omega t + \phi$ in (10) is the phase at time t, and ϕ is the initial phase, given in terms of the initial conditions by (12).

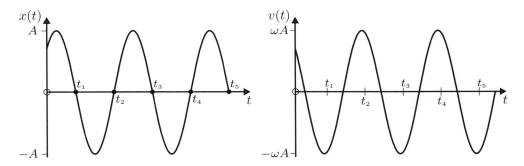

(iii) According to (4), the energy $E = \frac{1}{2}mv^2 + \frac{1}{2}kx^2$ of the oscillator is a constant that can be expressed, using (10) and (11), as

$$E = \tfrac{1}{2}kA^2 = \tfrac{1}{2}m\omega^2 A^2. \tag{24}$$

Question 4.2

As for Question 4.1, but with a force $F = kx$, where k is a positive constant. (Thus, we are examining the effect of a change in the sign of the force.)

Solution

The equation of motion is

$$m\frac{dv}{dt} = kx. \tag{1}$$

Consequently, instead of the characteristic equation (16) of Question 4.1, we now have $q^2 - k/m = 0$ and the real roots $q = \pm\tau^{-1}$, where $\tau = \sqrt{m/k}$ is a characteristic time. Thus, the general solution to (1) is

$$x(t) = a_1 e^{t/\tau} + a_2 e^{-t/\tau}, \tag{2}$$

where the constants a_1 and a_2 are given in terms of the initial conditions x_0 and v_0 by

$$a_1 = \tfrac{1}{2}(x_0 + v_0\tau), \qquad a_2 = \tfrac{1}{2}(x_0 - v_0\tau). \qquad (3)$$

In terms of hyperbolic functions, (2) is

$$x(t) = x_0 \cosh t/\tau + v_0\tau \sinh t/\tau. \qquad (4)$$

Comments

The motion is not oscillatory and it is sensitive to the initial conditions:

(i) If the particle is initially at the origin ($x_0 = 0$) then

$$x(t) = v_0\tau \sinh t/\tau, \qquad v(t) = v_0 \cosh t/\tau. \qquad (5)$$

Thus, $x \to \infty$ ($-\infty$) if $v_0 > 0$ (< 0).

(ii) On the other hand, if the particle is initially at rest ($v_0 = 0$) then

$$x(t) = x_0 \cosh t/\tau, \qquad v(t) = \frac{x_0}{\tau} \sinh t/\tau, \qquad (6)$$

and $x \to \infty$ ($-\infty$) if $x_0 > 0$ (< 0).

(iii) In addition to these unbounded ('run-away') motions there is a bounded motion in the special case $v_0 = -x_0/\tau$. Then, according to (2) and (3),

$$x(t) = x_0 e^{-t/\tau}, \qquad v(t) = -\frac{x_0}{\tau} e^{-t/\tau}. \qquad (7)$$

Thus, $x \to 0$ and $v \to 0$ as $t \to \infty$ (the particle 'comes to rest at the origin after an infinite time').

(iv) The solutions (5)–(7) are depicted below for $v_0 \geq 0$:

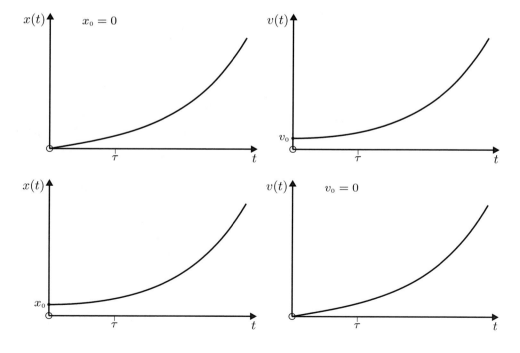

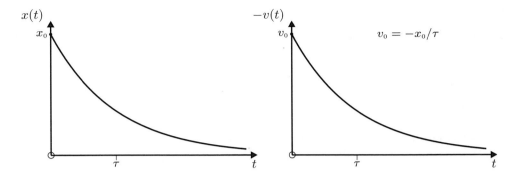

(v) The differences we have found for motion in the force fields $F = -kx$ and $F = kx$ are readily understood. For the former, $x = 0$ is a point of stable equilibrium, whereas for the latter it is a point of unstable equilibrium.

Question 4.3

Show that for small oscillations the period of a pendulum of length ℓ subject to a constant gravitational acceleration g is

$$T = 2\pi \sqrt{\frac{\ell}{g}}. \qquad (1)$$

Solution

The bob of the pendulum has mass m, and its position is given by the arc-length s measured from the equilibrium position O as shown in the figure. At time t the cord of the pendulum makes an angle θ with the vertical. The forces acting on m are the tension $\mathbf{T}$ in the string and the weight $m\mathbf{g}$: the component of the weight along the unit tangent vector $\mathbf{n}$ (which points in the direction of increasing θ) is $-mg\sin\theta$. Thus, the equation of motion is

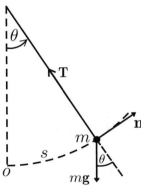

$$m\frac{d^2 s}{dt^2}\mathbf{n} = -mg\sin\theta\,\mathbf{n}. \qquad (2)$$

Now $s = \ell\theta$ (where θ is in radians), and so

$$\frac{d^2\theta}{dt^2} + \frac{g}{\ell}\sin\theta = 0. \qquad (3)$$

In the limit $|\theta| \ll 1$, $\sin\theta \approx \theta$ and (3) simplifies to

$$\frac{d^2\theta}{dt^2} + \frac{g}{\ell}\theta = 0. \qquad (4)$$

This is just the equation of a simple harmonic oscillator (see (13) of Question 4.1) with angular frequency $\omega = \sqrt{g/\ell}$ and period $T = 2\pi/\omega$.

Comments

(i) The solution to (4) that satisfies initial conditions $\theta = \theta_0$ and $d\theta/dt = 0$ for $t = 0$ (a pendulum released from rest at θ_0) is

$$\theta(t) = \theta_0 \cos\sqrt{\frac{g}{\ell}}\, t. \tag{5}$$

(ii) The solution to the non-linear differential equation (3) involves elliptical integrals that require numerical evaluation in general (see Question 5.18).

(iii) The harmonic approximation used above applies to a large variety of mechanical systems. For small displacements from the point $x = 0$ (say) one can make a series expansion for the force:

$$F(x) = F(0) + \left(\frac{dF}{dx}\right)_0 x + \frac{1}{2}\left(\frac{d^2 F}{dx^2}\right)_0 x^2 + \cdots . \tag{6}$$

If $F(0) = 0$ then $x = 0$ is a point of equilibrium. If also $(dF/dx)_0 < 0$ then the linear term is a restoring force: $x = 0$ is a point of stable equilibrium, and for small oscillations (6) allows a harmonic approximation $F \approx -kx$, where $k > 0$.

(iv) Simple harmonic oscillations occur also in non-mechanical systems such as electric circuits. In a circuit comprising an inductance L and capacitance C, the voltages $V_L = -L\,dI/dt$ and $V_C = q/C$ must be equal, where $q(t)$ is the charge on the capacitor and $I = dq/dt$ is the current in the circuit. Thus, we have the simple harmonic equation

$$\frac{d^2 q}{dt^2} + \frac{q}{LC} = 0, \tag{7}$$

and q and I oscillate (in quadrature) with angular frequency

$$\omega = \frac{1}{\sqrt{LC}}. \tag{8}$$

(v) The pendulum is a simple mechanical system that has played an important role in physics,[1] and we will encounter it in several other questions.

Question 4.4

A particle of mass m, which is constrained to move along a curve in the vertical plane, performs simple harmonic oscillations with an amplitude-independent period

$$T = 2\pi\sqrt{\frac{\ell}{g}}, \tag{1}$$

where ℓ is a constant length. Determine the curve $s = s(\theta)$ on which the particle moves.

Solution

Let $s(\theta)$ be the arc-length of the curve, measured from a point of equilibrium O, as shown in the figure. Here, θ, which is the angle between a tangent to the curve and the horizontal, is also the angle between a perpendicular to the curve and the vertical. Thus, the tangential component of the equation of motion is

$$m\frac{d^2s}{dt^2} = -mg\sin\theta. \tag{2}$$

It follows that if

$$s = \ell \sin\theta, \tag{3}$$

then

$$\frac{d^2s}{dt^2} + \frac{g}{\ell}s = 0, \tag{4}$$

and the particle performs simple harmonic oscillations with period given by (1). Unlike the case of the simple pendulum (motion on the arc of a circle $s = \ell\theta$), there is no restriction on s for this result to hold.

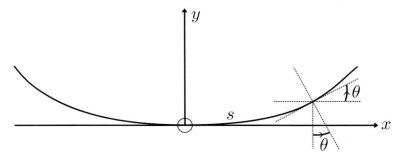

Comments

(i) It is useful to express the solution (3) in terms of parametric equations for $x(\theta)$ and $y(\theta)$. Now

$$dx = \cos\theta\, ds = \ell\cos^2\theta\, d\theta \quad \text{and} \quad dy = \sin\theta\, ds = \ell\cos\theta\sin\theta\, d\theta,$$

and so

$$\left.\begin{array}{l} x(\theta) = \ell\displaystyle\int_0^\theta \cos^2\theta\, d\theta = \tfrac{1}{4}\ell(2\theta + \sin 2\theta) \\[2ex] y(\theta) = \ell\displaystyle\int_0^\theta \cos\theta\sin\theta\, d\theta = \tfrac{1}{4}\ell(1 - \cos 2\theta). \end{array}\right\} \tag{5}$$

These are the parametric equations of a cycloid, which is the path traced out by a point on the circumference of a circle that rolls without slipping along a straight line. The cycloid defined by (5) is shown in the figure below, where the dotted curve is an arc of a circle of radius ℓ. The shape of the cycloid depends only on ℓ; it is unaffected by the gravitational acceleration g.

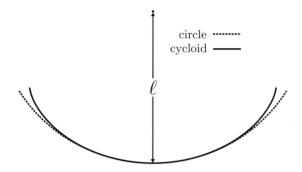

(ii) Oscillations having an amplitude-independent period are referred to as isochronous. They were studied in the seventeenth century by Huyghens, who showed how a cycloidal pendulum could be constructed by having the string wind up on a constraint curve as the pendulum oscillates.

(iii) The reasoning leading to equations (5) also provides the solution to the tautochrone problem for a bead sliding on a frictionless, curved, vertical wire. If the wire has the shape of a cycloid then the time taken to reach the bottom of the wire will be independent of the position at which the bead is released from rest.

Question 4.5

A particle of mass m is subject to a one-dimensional restoring force $F_r = -kx$ (k is a positive constant) and a frictional force proportional to the velocity[‡]: $F_f = -\alpha v$ (α is a positive constant). Determine the position $x(t)$ in terms of m, k, α and the initial conditions v_0 and x_0.

Solution

The equation of motion is
$$m\frac{dv}{dt} = -kx - \alpha v, \tag{1}$$
which we rewrite as
$$\frac{d^2x}{dt^2} + \frac{2}{\tau}\frac{dx}{dt} + \omega_0^2 x = 0. \tag{2}$$
Here
$$\omega_0 = \sqrt{k/m} \quad \text{and} \quad \tau = 2m/\alpha \tag{3}$$
have the dimensions of $(\text{time})^{-1}$ and time, respectively. Equation (2) is the modification of (13) in Question 4.1 to include friction. (We have now used a subscript on ω to

---

[‡]Quadratic drag is treated in Question 13.13.

distinguish it from another angular frequency that is encountered below.) We solve (2) by the same technique that was used in Question 4.1. Thus, the trial function $x = e^{qt}$ is a solution to (2) if q satisfies the characteristic equation

$$q^2 + \frac{2}{\tau} q + \omega_0^2 = 0. \qquad (4)$$

Consequently, the general solution to (2) is

$$x(t) = a_1 e^{q_1 t} + a_2 e^{q_2 t}, \qquad (5)$$

where

$$q_1 = -\frac{1}{\tau}\left(1 - \sqrt{1 - \omega_0^2 \tau^2}\right) \quad \text{and} \quad q_2 = -\frac{1}{\tau}\left(1 + \sqrt{1 - \omega_0^2 \tau^2}\right). \qquad (6)$$

There are three cases according to whether $1 - \omega_0^2 \tau^2$ is positive, negative or zero.

(a) Overdamped ($\omega_0 \tau < 1$)

Here, q_1 and q_2 are real, and (5) and (6) yield

$$x(t) = a_1 e^{-(1-\sqrt{1-\omega_0^2\tau^2})t/\tau} + a_2 e^{-(1+\sqrt{1-\omega_0^2\tau^2})t/\tau}, \qquad (7)$$

where a_1 and a_2 are given in terms of the initial conditions by

$$a_1 = \tfrac{1}{2} x_0 + \frac{1}{2\sqrt{1 - \omega_0^2 \tau^2}} (x_0 + v_0 \tau) \qquad (8)$$

$$a_2 = \tfrac{1}{2} x_0 - \frac{1}{2\sqrt{1 - \omega_0^2 \tau^2}} (x_0 + v_0 \tau). \qquad (9)$$

(b) Underdamped ($\omega_0 \tau > 1$)

Now, q_1 and q_2 are complex: $q_1 = -\frac{1}{\tau} + i\omega_\mathrm{d}$, $q_2 = -\frac{1}{\tau} - i\omega_\mathrm{d}$, where

$$\omega_\mathrm{d} = \sqrt{\omega_0^2 - 1/\tau^2}. \qquad (10)$$

is real. Thus, (5) gives

$$x(t) = e^{-t/\tau}\left(a_1 e^{i\omega_\mathrm{d} t} + a_2 e^{-i\omega_\mathrm{d} t}\right). \qquad (11)$$

We can express the factor in brackets in (11) as $A\cos(\omega_\mathrm{d} t + \phi)$, where A and ϕ are constants (see Question 4.1), and so

$$x(t) = A e^{-t/\tau} \cos(\omega_\mathrm{d} t + \phi). \qquad (12)$$

In terms of the initial conditions, A and ϕ are given by

$$A = \sqrt{x_0^2 + (x_0 + v_0\tau)^2/\omega_\mathrm{d}^2 \tau^2} \quad \text{and} \quad \tan\phi = -(x_0 + v_0\tau)/(x_0 \omega_\mathrm{d} \tau). \qquad (13)$$

(c) Critically damped ($\omega_0\tau = 1$)

The roots q_1 and q_2 in (6) are now equal, and (5) is no longer the general solution. It is easily seen that $te^{-t/\tau}$ is also a solution to (2) when $\omega_0\tau = 1$, and the general solution is the linear combination

$$x(t) = (a_1 + a_2 t)e^{-t/\tau}. \tag{14}$$

The constants in (14) are determined by

$$a_1 = x_0 \quad \text{and} \quad a_2 = (x_0 + v_0\tau)/\tau. \tag{15}$$

Comments

(i) In the plots of $x(t)$ versus t shown below we have taken $x_0 > 0$ and $v_0 < 0$; the particle is initially moving towards the origin.

(a) Overdamped ($\omega_0\tau < 1$)

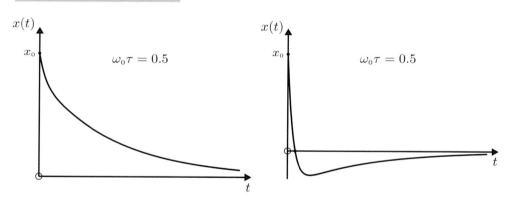

(b) Underdamped ($\omega_0\tau > 1$)

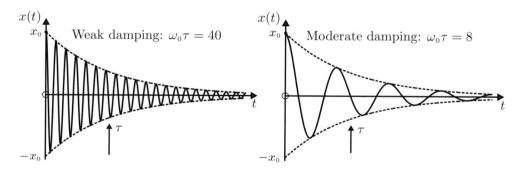

(c) Critically damped ($\omega_0\tau = 1$)

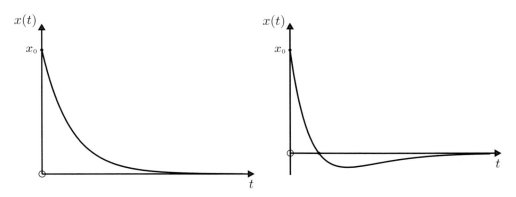

(ii) For overdamped oscillations $x(t)$ decreases monotonically to zero, or it overshoots the origin once, if the initial motion toward the origin is sufficiently rapid, see (a) above. (If we had taken $v_0 > 0$ there would be a maximum $x(t)$, followed by a monotonic decrease.)

(iii) For underdamped oscillations the particle performs harmonic oscillations with exponentially decreasing amplitude $Ae^{-t/\tau}$ and period $T = 2\pi/\omega_\mathrm{d}$. Thus, $x(t)$ oscillates within the envelope $\pm Ae^{-t/\tau}$ indicated by the dotted curves in the figures in (b) above. In the limit $\omega_0\tau \to \infty$ (the weak-damping limit), $\omega_\mathrm{d} \to \omega_0$ the angular frequency of the undamped oscillator, see (10). The relaxation time τ gives the time scale on which $x(t)$ decays by the factor e^{-1} ($\approx 37\%$).

(iv) For critically damped oscillations there are two possibilities, as depicted in (c) above, corresponding to whether or not $x(t)$ has a root. According to (14) and (15) a root occurs at $t = -x_0\tau/(x_0 + v_0\tau)$, which is positive if $v_0 < -x_0/\tau$. Thus, a particle projected towards the origin will overshoot the origin if the initial speed is high enough. After that, its position decays exponentially to zero.

(v) Inclusion of a driving force $F_\mathrm{d} = \gamma v$ in (1), where γ is a positive constant, will change τ from $2m/\alpha$ to $2m/(\alpha - \gamma)$. If $\gamma > \alpha$ then $\tau < 0$, and the solutions (7), (12) and (14) increase exponentially on a time scale τ; the system is dynamically unstable. This is illustrated below for (7) and (12).

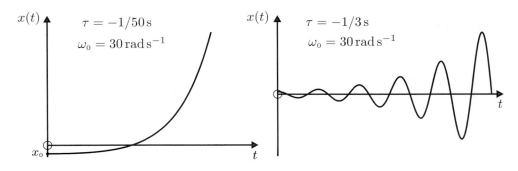

Question 4.6

Show that in the limit of weak damping ($\omega_0\tau \to \infty$) the energy of an underdamped oscillator is given by

$$E(t) = E_0 e^{-2t/\tau}, \tag{1}$$

where $E_0 = \tfrac{1}{2}kA^2 = \tfrac{1}{2}m\omega_0^2 A^2$.

Solution

The energy

$$E = \tfrac{1}{2}m\left(\frac{dx}{dt}\right)^2 + \tfrac{1}{2}kx^2 \qquad (k = m\omega_0^2) \tag{2}$$

is, of course, not constant because there is friction. For an underdamped oscillator $x(t)$ is given by (12) of Question 4.5, and so

$$v(t) = -A\left\{\omega_d \sin(\omega_d t + \phi) + \frac{1}{\tau}\cos(\omega_d t + \phi)\right\}e^{-t/\tau}. \tag{3}$$

In the limit of weak damping the contribution of the second term in (3) to the energy (2) becomes negligible. Also, $\omega_d \to \omega_0$. Thus, as $\omega_0\tau \to \infty$

$$E \to \tfrac{1}{2}A^2\{m\omega_0^2 \sin^2(\omega_0 t + \phi) + k\cos^2(\omega_0 t + \phi)\}e^{-2t/\tau} = \tfrac{1}{2}kA^2 e^{-2t/\tau}. \tag{4}$$

Comment

Plots of $E(t)$ versus t obtained from (2) and (3), and (12) of Question 4.5, without making the above approximation, are shown below. There are points of inflection with horizontal tangents at instants when $v(t) = 0$ because $dE/dt \propto v$. The dotted curves are the approximation (1). As expected, the accuracy of this approximation improves as $\omega_0\tau$ increases.

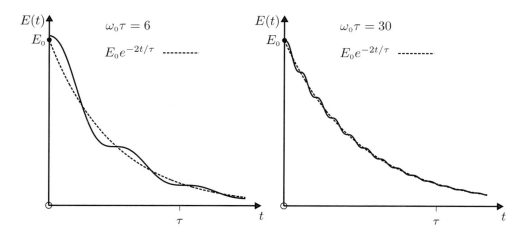

Question 4.7

A particle of mass m is subject to a one-dimensional restoring force $F_r = -kx$ (k is a positive constant), a frictional force proportional to the velocity $F_f = -\alpha v$ (α is a positive constant), and a harmonic driving force $F_d = F_0 \cos\omega t$ (F_0 and ω are constants). Determine the position $x(t)$.

Solution

The equation of motion is

$$m\frac{dv}{dt} = -kx - \alpha v + F_0 \cos\omega t. \tag{1}$$

That is

$$\frac{d^2x}{dt^2} + \frac{2}{\tau}\frac{dx}{dt} + \omega_0^2 x = \frac{F_0}{m}\cos\omega t, \tag{2}$$

where ω_0 and τ are given by (3) of Question 4.5. From the theory of differential equations, we know that the general solution to (2) consists of a complementary function x_c (which is the general solution to (2) with $F_0 = 0$), and a particular integral x_p (which is a particular solution to (2)):

$$x(t) = x_c(t) + x_p(t). \tag{3}$$

The complementary function has already been obtained in Question 4.5, where it is given by (7), (12) or (14), according to whether the oscillator is overdamped, underdamped or critically damped. The two arbitrary constants, which are an essential feature of a general solution to (2), are contained in the $x_c(t)$, and they are to be fixed in terms of the initial conditions x_0 and v_0. We now turn to the task of finding a particular integral $x_p(t)$. (This part of the solution will not involve any arbitrary constants.) The calculation is simplified if we use complex notation. Thus, we write (2) as

$$\frac{d^2x}{dt^2} + \frac{2}{\tau}\frac{dx}{dt} + \omega_0^2 x = \frac{F_0}{m}e^{i\omega t} \tag{4}$$

and look for a solution

$$x_p(t) = ae^{i(\omega t - \vartheta)}, \tag{5}$$

where a and ϑ are real constants. In (4) and (5) the real parts are understood (so $e^{i\omega t}$ implies $\cos\omega t$, etc). Substitution of (5) in (4) yields

$$a\left[\omega_0^2 - \omega^2 + i(2\omega/\tau)\right] = (F_0/m)e^{i\vartheta}. \tag{6}$$

Now equality of two complex numbers $u + iv$ and $re^{i\vartheta}$ requires that $r = \sqrt{u^2 + v^2}$ and $\tan\vartheta = v/u$. It therefore follows immediately from (6) that

$$a = \frac{F_0/m}{\sqrt{(\omega_0^2 - \omega^2)^2 + (2\omega/\tau)^2}} \tag{7}$$

$$\tan\vartheta = \frac{2\omega/\tau}{\omega_0^2 - \omega^2}. \tag{8}$$

This completes the task of finding the general solution to (2). Thus, for example, for an underdamped driven oscillator we have from (3), (5) and (7) above, and (12) of Question 4.5,

$$x(t) = Ae^{-t/\tau}\cos(\omega_\mathrm{d} t + \phi) + \frac{F_0/m}{\sqrt{(\omega_0^2 - \omega^2)^2 + (2\omega/\tau)^2}}\cos(\omega t - \vartheta), \qquad (9)$$

where ϑ is given by (8). The two arbitrary constants A and ϕ in (9) are to be determined by applying the initial conditions x_0 and v_0 to (9) and its derivative dx/dt. Analogous expressions can be obtained for the overdamped and critically damped driven oscillators: one simply replaces the first term in (9) by either (7) or (14) of Question 4.5.

Comments

(i) The complementary function $x_\mathrm{c}(t)$ in (3) is referred to as a transient because eventually it decays exponentially to zero, as in (9) where the decay is on a time scale τ. The particular integral $x_\mathrm{p}(t)$ in (3) is referred to as the steady-state solution; it is the part that remains after transients have died out, and it is given by the second term in (9). Graphs of $x(t)$ given by (9) for an underdamped driven oscillator are presented below for $\boxed{\omega < \omega_0}$ and $\boxed{\omega > \omega_0}$. In these we have taken $F_0/m = 1.0\,\mathrm{m\,s^{-2}}$, $\omega_0 = 1.0\,\mathrm{rad\,s^{-1}}$, $\tau = 20.0\,\mathrm{s}$, $x_0 = 0$ and $v_0 = 0$. These plots show the initial distortion produced by the transient, and how the motion tends to the steady-state solution $x_\mathrm{p}(t)$ (indicated by a dotted curve in each case). In the example shown, the solution $x(t)$ is more distorted and takes more cycles to reach the steady state for the case $\omega > \omega_0$.

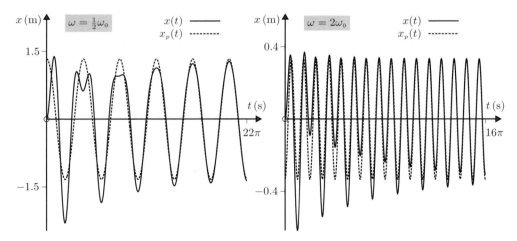

(ii) Graphs of the amplitude a in (7), in units of $F_0/m\omega_0^2$, and the phase ϑ in (8) versus ω/ω_0 (the ratio of the driving frequency to the natural frequency) are shown below for various values of $\omega_0\tau$.

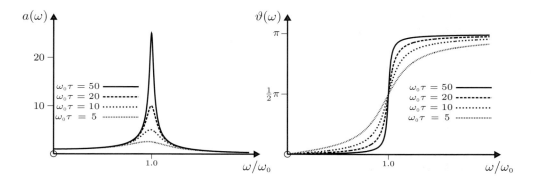

(iii) The graph of $a(\omega)$ illustrates the phenomenon of amplitude resonance. The amplitude is a maximum at a frequency ω_r, known as the amplitude resonance frequency. From (7) and $da/d\omega = 0$ we obtain $\omega_r = \sqrt{\omega_0^2 - 2/\tau^2}$. This differs from the frequency $\omega_d = \sqrt{\omega_0^2 - 1/\tau^2}$ of the transient of an underdamped oscillator and the natural frequency ω_0: in fact, $\omega_r < \omega_d < \omega_0$. In the weak-damping limit, ω_d and ω_r tend to ω_0.

(iv) The phase ϑ is always positive, meaning that the response $x(t)$ always lags the driving force $F(t)$, see (4) and (5).

(v) Damped, driven oscillations occur also in non-mechanical systems such as electric circuits. In a circuit comprising an inductance L, resistance R, and capacitance C connected in series across an oscillator producing an emf $V(t) = V_0 \cos \omega t$, the circuit equation is

$$\frac{d^2 q}{dt^2} + \frac{R}{L}\frac{dq}{dt} + \frac{q}{LC} = \frac{V_0}{L}\cos \omega t, \qquad (10)$$

where q is the charge on the capacitor and dq/dt is the current. Comparing this with (4), we see that results for the series LRC circuit can be obtained directly from those for the mechanical system by making the following substitutions

$$x \to q, \qquad \tau \to 2L/R, \qquad \omega_0 \to 1/\sqrt{LC}, \qquad F_0/m \to V_0/L. \qquad (11)$$

(vi) As a special case of (9), consider a driven, undamped oscillator having the initial conditions: $x_0 = 0$, $v_0 = 0$. With $\tau = \infty$ in (9), these initial conditions require

$$x(t) = \frac{F_0/m}{\omega_0^2 - \omega^2}(\cos \omega t - \cos \omega_0 t). \qquad (12)$$

If the driving frequency ω is close to the natural frequency ω_0, then (12) can be approximated as

$$x(t) = \frac{F_0/m}{\omega_0(\omega_0 - \omega)} \sin\{\tfrac{1}{2}(\omega_0 + \omega)t\} \sin\{\tfrac{1}{2}(\omega_0 - \omega)t\}. \qquad (13)$$

Here we have used the identity $\cos A - \cos B = 2\sin \tfrac{1}{2}(A+B)\sin \tfrac{1}{2}(B-A)$. According to (13), $x(t)$ oscillates at a frequency equal to the average of ω_0 and ω, and with an amplitude that is modulated at the lower frequency $\frac{1}{2}(\omega_o - \omega)$. The so-called beat frequency is double this difference since the amplitude peaks twice every cycle. This is illustrated in the graph below, which is for $\omega_o = 1.1\,\mathrm{rad\,s^{-1}}$ and $\omega = 1.0\,\mathrm{rad\,s^{-1}}$. The phenomenon of beats occurs whenever the driving frequency is close to the natural frequency ω_o.

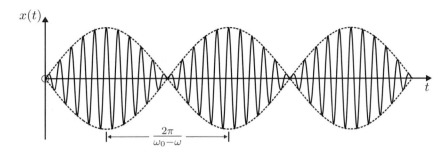

Question 4.8

For a damped, driven oscillator in the steady state, calculate the average values of the kinetic energy $K = \frac{1}{2}mv^2$, the potential energy $V = \frac{1}{2}kx^2$, and the total energy $E = K + V$.

Solution

In the steady state

$$x(t) = a\cos(\omega t - \vartheta) \qquad \text{and} \qquad v(t) = -a\omega\sin(\omega t - \vartheta), \qquad (1)$$

where a and ϑ are given by (7) and (8) of Question 4.7. Then,

$$K = \tfrac{1}{2}m\omega^2 a^2 \sin^2(\omega t - \vartheta) \qquad \text{and} \qquad V = \tfrac{1}{2}m\omega_o^2 a^2 \cos^2(\omega t - \vartheta). \qquad (2)$$

Now the average values of $\cos^2\theta$ and $\sin^2\theta$ over a complete cycle are equal. Also $\cos^2\theta + \sin^2\theta = 1$, and therefore

$$\langle \cos^2(\omega t - \vartheta) \rangle = \langle \sin^2(\omega t - \vartheta) \rangle = \tfrac{1}{2}, \qquad (3)$$

where the angular brackets denote an average over one cycle. From (2) and (3) and (7) of Question 4.7 we have

$$\langle K \rangle = \frac{F_0^2}{4m} \frac{\omega^2}{(\omega_o^2 - \omega^2)^2 + (2\omega/\tau)^2} \qquad (4)$$

$$\langle V \rangle = \frac{F_0^2}{4m} \frac{\omega_o^2}{(\omega_o^2 - \omega^2)^2 + (2\omega/\tau)^2} \qquad (5)$$

$$\langle E \rangle = \frac{F_0^2}{4m} \frac{\omega^2 + \omega_o^2}{(\omega_o^2 - \omega^2)^2 + (2\omega/\tau)^2}. \qquad (6)$$

Comments

(i) The average kinetic energy in (4) is a maximum when

$$\frac{d\langle K \rangle}{d\omega} = 0,$$

that is when $\omega = \omega_0$. Thus, resonance of $\langle K \rangle$ occurs at the natural frequency of the oscillator. By contrast, resonance of the average potential energy $\langle V \rangle$ occurs at

$$\omega_r = \sqrt{\omega_0^2 - 2/\tau^2},$$

which is the same as the frequency for amplitude resonance (see Question 4.7).

(ii) In the weak-damping limit ($\omega_0 \tau \gg 1$) the average energy (6) is appreciable only for ω close to ω_0. Thus, we can approximate $\omega_0^2 - \omega^2$ by $2\omega_0(\omega_0 - \omega)$, and (6) becomes

$$\langle E \rangle = \frac{F_0^2 \tau^2}{8m} \frac{1}{\tau^2(\omega_0 - \omega)^2 + 1}. \tag{7}$$

The frequency dependence in (7) is specified by a Lorentz function

$$(X^2 + 1)^{-1}, \quad \text{where} \quad X = \tau(\omega_0 - \omega). \tag{8}$$

The Lorentz function is plotted below: it drops to half its maximum value at $X = \pm 1$, and therefore the full width of the curve at half-maximum is $\Delta X = 2$. Then, (8) gives

$$\tau \Delta \omega = 2, \tag{9}$$

where $\Delta \omega$ is the full width of the resonance curve (7) for $\langle E \rangle$: thus, the resonance curve becomes narrower (broader) as τ increases (decreases). An oscillator that responds only to a narrow band of frequencies ($\Delta \omega$ small) will have transients that persist for a long time (τ large) – it will take a long time to reach the steady state after a driving force is applied – and vice versa if $\Delta \omega$ is large.

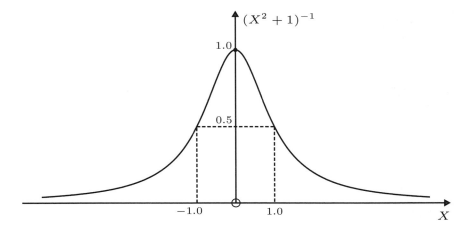

Question 4.9

Consider the following experiment. A glass aspirator is closed with a stopper through which passes a vertical, precision-made glass tube of uniform cross-section. A closely fitting, smooth, steel ball is placed in the tube. If the ball is displaced from its equilibrium position and released, it performs underdamped oscillations. Assuming a frictional force F_f that is linear in the velocity, show that in the limit of weak damping the period of oscillation is given by

$$T = 2\pi\sqrt{\frac{mV}{KA^2}}, \tag{1}$$

where m is the mass of the ball, A is the cross-sectional area of the tube, and V and K are the volume and bulk modulus of the gas in the aspirator.

Solution

Consider a displacement x of the ball from its equilibrium position, and let ΔP be the corresponding change in the gas pressure. The restoring force on the ball is $F_r = A\Delta P$. For small oscillations $\Delta P = -K\Delta V/V$, where V is the equilibrium volume of the gas and $\Delta V = Ax$. With $F_f = -\alpha dx/dt$, where α is a positive constant, the equation of motion is

$$m\frac{d^2x}{dt^2} = F_r + F_f = -\frac{KA^2}{V}x - \alpha\frac{dx}{dt}. \tag{2}$$

That is,

$$\frac{d^2x}{dt^2} + \frac{\alpha}{m}\frac{dx}{dt} + \frac{KA^2}{mV}x = 0, \tag{3}$$

which is the equation of a damped, simple harmonic oscillator. The solution for underdamped oscillations is (see (12) of Question 4.5 with $\phi = 0$)

$$x(t) = x_0 e^{-t/\tau}\cos 2\pi t/T, \tag{4}$$

where $\tau = m/2\alpha$ is the relaxation time, and the period T is given by

$$T = 2\pi\sqrt{\frac{mV}{KA^2}}\left(1 - \frac{mV}{KA^2\tau^2}\right)^{-1/2}. \tag{5}$$

It is apparent from (5) that underdamped oscillations require $\tau > \sqrt{mV/KA^2}$. For weak damping, $\tau \gg \sqrt{mV/KA^2}$, and (5) reduces to (1).

Comments

(i) This experiment was devised by Rüchardt[2] in 1929, and it has become one of the standard methods for measuring the ratio of the specific heat of a gas at

---

[2] E. Rüchardt, "Eine einfache methode zur bistimmung von c_p/c_v," Physikalische Zeitschrift, vol. 30, pp. 58–59, 1929.

constant pressure to the specific heat at constant volume, $\gamma = C_p/C_v$. For this one assumes that the oscillations are adiabatic and the gas is ideal; then in (1), K is the adiabatic bulk modulus $K_a = \gamma P$, where P is the equilibrium pressure of the gas. One can take account of corrections due to non-adiabatic conditions, molecular interactions, and departures from weak damping.[3]

(ii) Various traces of $x(t)$ versus t (as captured on a digital oscilloscope)[3] are shown for three different volumes V of a polyatomic gas ($C_2C\ell F_5$). An interesting feature is the emergence of an isothermal 'tail' (for which T is given by (1) with $K = K_i = P$). The start of this tail is indicated by an arrow labelled **1** in diagrams (b) and (c); the amplitude at which it first appears increases as the volume V increases. The arrow labelled **L** indicates the end of the (almost) adiabatic oscillations.

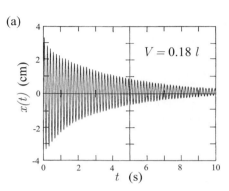

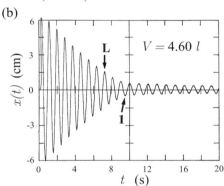

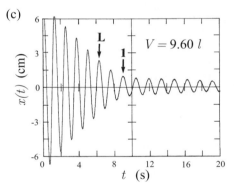

Question 4.10

One definition of the quality factor Q of an underdamped oscillator is

$$Q = \tfrac{1}{2}\omega_d \tau. \qquad (1)$$

Express Q in terms of the amplitudes x_n and x_{n+1} of successive oscillations.

Solution

For the underdamped oscillator the amplitudes of successive oscillations are given by

---

[3] O. L. de Lange and J. Pierrus, "Measurement of bulk moduli and ratio of specific heats of gases using Rüchardt's experiment," American Journal of Physics, vol. 68, pp. 265–270, 2000.

$$x_n = Ae^{-t_n/\tau} \qquad \text{and} \qquad x_{n+1} = Ae^{-(t_n+T)/\tau}, \qquad (2)$$

where $T = 2\pi/\omega_d$ is the period (see (12) of Question 4.5). Thus,

$$x_n/x_{n+1} = e^{T/\tau} = e^{2\pi/\omega_d \tau}. \qquad (3)$$

From (1) and (3) we have the desired result

$$Q = \frac{\pi}{\ln(x_n/x_{n+1})}. \qquad (4)$$

Comments

(i) The quantity $\ln(x_n/x_{n+1})$ is known as the logarithmic decrement δ. Thus, $Q = \pi/\delta$.

(ii) It is clear that for heavily (lightly) damped oscillators Q is small (large). In the weak-damping limit, $Q = \frac{1}{2}\omega_0 \tau$.

(iii) Values of Q range from about 10 to 100 for mechanical systems (such as springs and loudspeakers), to about 10^3 for musical instruments, and about 10^4 for a microwave cavity. Excited atoms and nuclei are very lightly damped ($Q \approx 10^7$ and 10^{12}), and gas lasers even less so ($Q \approx 10^{14}$).

(iv) Values of Q can be extracted from measurements of $x(t)$ by using (4). An interesting example of a mechanical system (a confined gas) is the Rüchardt experiment discussed in the previous question, where the oscillation is due to motion of part of the boundary of the container. From traces of $x(t)$, such as those shown in Question 4.9, and using (4), one can determine Q. Results for several gases, plotted 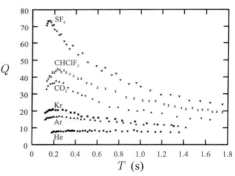 as functions of the period of oscillation T (or $\sqrt{V}$ – see (1) of Question 4.9), are shown above. We see that within this mechanical system we can classify heavily damped gases (such as He), medium-damped gases (such as CO_2), and lightly damped gases (polyatomic gases). The range of Q ($\sim$ 7 to 75) is typical of the range mentioned above for mechanical systems. The results in the figure can be understood in terms of a competition between two loss mechanisms, namely heat flows in the gas and friction of the moving part of the boundary.[4]

Question 4.11

Establish the relationship between the quality factor $Q = \frac{1}{2}\omega_d \tau$, of an underdamped oscillator, and two other definitions that are used, namely

---

[4] O. L. de Lange and J. Pierrus, "The quality factor of low-frequency oscillations in gases," Transactions of the Royal Society of South Africa, vol. 58, pp. 115–117, 2003.

$$Q' = 2\pi(\text{energy at the start of a cycle})/(\text{energy lost during that cycle}), \quad (1)$$

$$Q'' = 2\pi \langle \text{energy stored} \rangle / \langle \text{energy loss per cycle} \rangle. \quad (2)$$

In (2) the angular brackets denote an average over one cycle.

Solution

In terms of the energy $E(t)$ of the oscillator, (1) and (2) are

$$Q' = 2\pi \frac{E(t)}{E(t) - E(t+T)} \quad (3)$$

$$Q'' = 2\pi \frac{\int_t^{t+T} E(t)\,dt}{\int_t^{t+T} \{E(t) - E(t+T)\}\,dt}, \quad (4)$$

where $T = 2\pi/\omega_\mathrm{d}$ is the period of underdamped oscillations. Now, $E(t)$ is given by (2) and (3) of Question 4.6 and (12) of Question 4.5. It can be written

$$E(t) = f(t) e^{-2t/\tau}, \quad (5)$$

where the function $f(t)$ consists of three terms involving $\cos^2(\omega_\mathrm{d} t + \phi)$, $\sin^2(\omega_\mathrm{d} t + \phi)$, and $\sin 2(\omega_\mathrm{d} t + \phi)$. For our purposes the essential feature is that $f(t)$ has period $\tfrac{1}{2}T$, and so

$$f(t+T) = f(t). \quad (6)$$

It follows from (3)–(6) that

$$Q' = Q'' = \frac{2\pi}{1 - e^{-2T/\tau}} = \frac{2\pi}{1 - e^{-2\pi/Q}}. \quad (7)$$

Note that in obtaining (7) we have not assumed the weak-damping limit.

Comments

(i) In the limit of weak damping ($Q \gg 1$), (7) shows that Q' and $Q'' \to Q + \pi$. For small values of Q (when $\omega_0 \tau$ is close to unity – see (10) of Question 4.5), Q' and Q'' are approximately 2π. The graph of (7) is plotted alongside, where the dotted line is the asymptote $Q' = Q + \pi$.

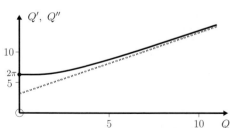

(ii) For weak damping, Q is also related to the width of the resonance curve. From (9) of Question 4.8 we have

$$\frac{\Delta \omega}{\omega_0} = \frac{2}{\omega_0 \tau} \approx \frac{2}{\omega_\mathrm{d} \tau} = \frac{1}{Q}. \quad (8)$$

With increasing Q the resonance curve becomes narrower.

Question 4.12

An undamped oscillator is driven at its resonance frequency ω_0 by a harmonic force $F = F_0 \sin\omega_0 t$. The initial conditions are $x_0 = 0$ and $v_0 = 0$.

(a) Determine $x(t)$.
(b) If the breaking strength of the 'spring' of the oscillator is $5F_0$, deduce an equation from which the time t_b taken to reach the breaking point can be calculated in terms of ω_0.

Solution

(a) The equation of motion
$$\frac{d^2x}{dt^2} + \omega_0^2 x = \frac{F_0}{m}\sin\omega_0 t \tag{1}$$
has the complementary function
$$x_c(t) = A\sin(\omega_0 t + \phi), \tag{2}$$
where A and ϕ are constants that are to be determined from the initial conditions. A particular integral for (1) is
$$x_p(t) = -\frac{F_0}{2m\omega_0^2}\omega_0 t \cos\omega_0 t. \tag{3}$$
The general solution $x = x_c + x_p$ satisfies the initial condition $x_0 = 0$ if $\phi = 0$. That is,
$$x(t) = A\sin\omega_0 t - \frac{F_0}{2m\omega_0^2}\omega_0 t \cos\omega_0 t. \tag{4}$$
The time derivative of (4) is the velocity
$$v(t) = A\omega_0\cos\omega_0 t - \frac{F_0}{2m\omega_0}\cos\omega_0 t + \frac{F_0}{2m\omega_0}\omega_0 t\sin\omega_0 t. \tag{5}$$
Thus, the initial condition $v_0 = 0$ requires $A = F_0/2m\omega_0^2$, and (4) becomes
$$x(t) = \frac{F_0}{2m\omega_0^2}(\sin\omega_0 t - \omega_0 t\cos\omega_0 t). \tag{6}$$

(b) At the breaking point $F = kx_b = \pm 5F_0$. Now, $k = m\omega_0^2$, and so the displacement at the breaking point is given by
$$x_b = \pm 5F_0/m\omega_0^2. \tag{7}$$
From (6) and (7), t_b is the smallest positive root of the transcendental equations
$$\tfrac{1}{2}(\sin\omega_0 t - \omega_0 t\cos\omega_0 t) = \pm 5. \tag{8}$$

Comment

Using *Mathematica*'s `FindRoot` function to solve (8), we find that $t_b = 11.90/\omega_0$. This is illustrated in the plot below. Note that t_b is a discontinuous function of x_b and hence of the breaking strength.

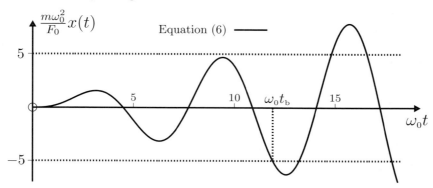

Question 4.13

Suppose a jerk‡ force $F = -\gamma d^3 x/dt^3$ (where γ is a constant) is applied to the damped, driven oscillator of Question 4.7.

(a) Show that the amplitude $a(\omega)$ and phase $\vartheta(\omega)$ of the steady-state oscillations are given by

$$a(\omega) = \frac{F_0/m}{\sqrt{(\omega_0^2 - \omega^2)^2 + \omega_0^2 Q^{-2}(\omega - 2\omega^3/\omega_c^2)^2}}, \tag{1}$$

and

$$\tan\vartheta = \frac{\omega_0(\omega - 2\omega^3/\omega_c^2)}{Q(\omega_0^2 - \omega^2)}, \tag{2}$$

where $\omega_0^2 = k/m$, $Q = \frac{1}{2}\omega_0\tau$ and $\omega_c^2 = 4m/\gamma\tau$.

(b) Suppose $\gamma > 0$. Show that the amplitude of the steady-state oscillations is increased by the jerk force provided $\omega < \omega_c$.

(c) Plot graphs of $a(\omega)$, in units of $F_0/m\omega_0^2$, versus ω/ω_0 and $\vartheta(\omega)$ versus ω/ω_0 for $Q = 10$ and $0.9 \leq \omega/\omega_0 \leq 1.1$, when 1. $\omega_c/\omega_0 = 0.8$, 2. $\omega_c/\omega_0 = 2.0$, 3. $\omega_c/\omega_0 = 2.5$, and 4. $\omega_c/\omega_0 = 25$.

Solution

(a) The equation of motion (1) of Question 4.7 is now modified to read

$$m\frac{dv}{dt} = -kx - \alpha v - \gamma\frac{d^3 v}{dt^3} + F_0\cos\omega t. \tag{3}$$

‡See Question 3.15.

84 Solved Problems in Classical Mechanics

That is,
$$\frac{\gamma}{m}\frac{d^3x}{dt^3} + \frac{d^2x}{dt^2} + \frac{2}{\tau}\frac{dx}{dt} + \omega_0^2 x = \frac{F_0}{m}\cos\omega t, \quad (4)$$
where $\tau = 2m/\alpha$. As before, the steady-state solution has the form
$$x_{\mathrm{p}}(t) = ae^{i(\omega t - \vartheta)}, \quad (5)$$
where the amplitude a and phase ϑ are real constants that are to be determined (see Question 4.7). Substituting (5) in (4) gives
$$a\left[\omega_0^2 - \omega^2 + i(2\omega/\tau - \omega^3\gamma/m)\right] = (F_0/m)e^{i\vartheta}, \quad (6)$$
and so
$$a = \frac{F_0/m}{\sqrt{(\omega_0^2 - \omega^2)^2 + (2\omega/\tau - \omega^3\gamma/m)^2}}, \quad \text{and} \quad \tan\vartheta = \frac{2\omega/\tau - \omega^3\gamma/m}{\omega_0^2 - \omega^2}. \quad (7)$$

Equations (1) and (2) follow.
(b) It is clear from $(7)_1$ that the effect of the jerk is to increase the amplitude a provided $(2\omega/\tau - \omega^3\gamma/m)^2 < (2\omega/\tau)^2$. That is, $(\gamma\omega^2/m)^2(\omega^2 - 4m/\gamma\tau) < 0$, which is possible if $\gamma > 0$ and $\omega < \omega_{\mathrm{c}}$.
(c) Note that the resonance curve for $\omega_{\mathrm{c}}/\omega_0 = 25$ approximates that of an 'ordinary' damped, driven oscillator (the effect of the jerk is negligible).

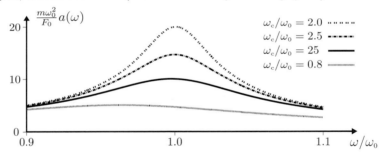

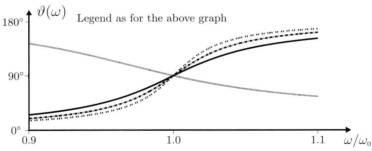

Comment

The graphs of $a(\omega)$ illustrate that the displacement amplitude is increased by the jerk when $\omega < \omega_{\mathrm{c}}$ (compare the curves for $\omega_{\mathrm{c}}/\omega_0 = 2.0$ and $\omega_{\mathrm{c}}/\omega_0 = 2.5$ with that for $\omega_{\mathrm{c}}/\omega_0 = 25$), and decreased when $\omega > \omega_{\mathrm{c}}$ (see the curve for $\omega_{\mathrm{c}}/\omega_0 = 0.8$).

Question 4.14

An object of mass m is subject to a one-dimensional restoring force $F_r = -kx$ (k is a positive constant) and a frictional force of constant magnitude $F_f = \mu N$, where μ is the coefficient of kinetic friction between the mass and the horizontal surface on which it slides. Here, N, the normal reaction force, is equal to the weight of the object. Assuming the initial conditions $x_0 = A\ (>0)$ and $v_0 = 0$, determine the displacement $x(t)$ and velocity $v(t)$ of the object during the first cycle of its motion.

Solution

If $kA \leq \mu_s N$ (μ_s is the coefficient of static friction) the object will not move. For $kA > \mu_s N$, the equation of motion is

$$m\frac{d^2 x}{dt^2} = -kx \mp F_f, \qquad (1)$$

where the upper (lower) sign is for $v > 0$ ($v < 0$). With $F_f = \mu mg$, (1) gives

$$\frac{d^2 x}{dt^2} + \omega^2 x = \mp \mu g, \qquad (2)$$

where $\omega = \sqrt{k/m}$. A particular solution of (2) is

$$x = \mp \mu g/\omega^2. \qquad (3)$$

The complementary function (the solution of the homogeneous equation (2) with $\mu = 0$) is given by (20) of Question 4.1. The general solution of the inhomogeneous equation (2) is the sum of the particular solution (3) and the complementary function. Thus

$$x(t) = a_1 \cos\omega t + b_1 \sin\omega t + \mu g/\omega^2 \qquad (v < 0), \qquad (4)$$
$$x(t) = a_2 \cos\omega t + b_2 \sin\omega t - \mu g/\omega^2 \qquad (v > 0), \qquad (5)$$

where a_1 and b_1 (a_2 and b_2) are constants that depend on the initial conditions for motion to the left (right). Equation (4) and the initial conditions $x_0 = A$, $v_0 = 0$ yield

$$a_1 = A - \mu g/\omega^2, \qquad b_1 = 0. \qquad (6)$$

Thus

$$x(t) = (A - \mu g/\omega^2) \cos\omega t + \mu g/\omega^2, \qquad (7)$$
$$v(t) = -\omega (A - \mu g/\omega^2) \sin\omega t. \qquad (8)$$

Equations (7) and (8) give the displacement and velocity of the object during the first half-cycle. The direction of motion of the object reverses after $v = 0$. From (8) this occurs at time $t_1 = \pi/\omega$. According to (7):

$$x(t_1) = -A + 2\mu g/\omega^2, \tag{9}$$

which shows that the amplitude decreases by $2\mu g/\omega^2$ in the first half-cycle. In the second half-cycle, the motion is determined by (5) with the initial conditions

$$x(\pi/\omega) = -A + 2\mu g/\omega^2, \qquad v(\pi/\omega) = 0. \tag{10}$$

It follows that
$$a_2 = A - 3\mu g/\omega^2, \qquad b_2 = 0. \tag{11}$$

Therefore, in the second half-cycle:

$$x(t) = (A - 3\mu g/\omega^2)\cos\omega t - \mu g/\omega^2, \tag{12}$$

$$v(t) = -\omega(A - 3\mu g/\omega^2)\sin\omega t. \tag{13}$$

Thus, $v = 0$ at time $t_2 = 2\pi/\omega = T$, the period of the oscillations, and

$$x(t_2) = A - 4\mu g/\omega^2. \tag{14}$$

Comments

(i) Equation (14) shows that the amplitude decreases in the second half-cycle by the same amount as in the first, namely $2\mu g/\omega^2$. Continuing in this way, we see that there is a constant decrease of the amplitude by $2\mu g/\omega^2$ every half-cycle of the motion. The above results are easily generalized to obtain the displacement and velocity for the nth half-cycle ($n = 1, 2, 3, \cdots$)

$$x_n(t) = \left(A - (2n-1)\frac{\mu g}{\omega^2}\right)\cos\omega t + (-1)^{n+1}\frac{\mu g}{\omega^2}, \tag{15}$$

$$v_n(t) = -\omega\left(A - (2n-1)\frac{\mu g}{\omega^2}\right)\sin\omega t. \tag{16}$$

Plots of (15) and (16) are given below, and they show the linear decrease of the amplitude of successive oscillations. This contrasts with the exponential decay when the damping force depends linearly on v (see Question 4.5).

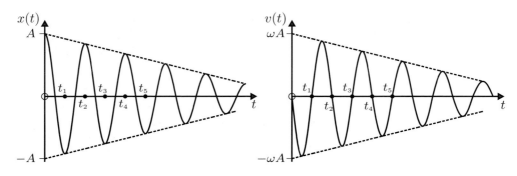

(ii) Notice that the frequency of the oscillation is unaffected by the frictional force. This is not the case for viscous damping, see (10) of Question 4.5.
(iii) The motion will cease at the peak x_n of the nth half-cycle for which the restoring force $k|x_n|$ is less than the static friction force $\mu_s N$.
(iv) The equation of motion (1) can be rewritten in an interesting form

$$\frac{d^2x}{dt^2} + \omega^2 x = F(t)/m, \qquad (17)$$

where

$$F(t) = (-1)^n \mu m g, \qquad (18)$$

and n (the number of a half-cycle) is the greatest integer less than $\omega t/\pi$. Here, the force $F(t)$ is interpreted as a square-wave driving force at the natural frequency ω. For example, for a spring–mass system, $F(t)$ could be due to the back-and-forth motion of the wall to which the spring is attached. The motions of the wall and the mass are out of phase by π: when the mass moves to the left the wall moves to the right with acceleration μg. This is known as 'negative forcing' and relates to the interesting concept of negative damping. Motion of the support is common in many applications. Another example is the 'flip-flop' pendulum.[5]

Question 4.15

For one-dimensional motion of a single particle, a plot of $\dot{x}$ (or $p = m\dot{x}$) versus x is known as a phase trajectory.

(a) Determine the phase trajectory of a simple harmonic oscillator.
(b) For an underdamped oscillator the dimensionless coordinate $\bar{x} = x/A$ is related to the dimensionless time $\bar{t} = \omega_0 t$ by (see Question 4.5)

$$\bar{x} = e^{-\bar{t}/\omega_0 \tau} \cos(\omega_d \bar{t}/\omega_0 + \phi), \qquad (1)$$

where $\omega_d/\omega_0 = \sqrt{1 - \omega_0^{-2}\tau^{-2}}$. Use *Mathematica* to plot the phase trajectory for $\phi = 0$, $\omega_0 \tau = 10$, and $\bar{t} \geq 0$.
(c) Express the solution $x(t)$ for a driven, underdamped oscillator (see (9) of Question 4.7) in dimensionless form. Then modify the notebook for (b) and plot the phase trajectory for $\phi = 0$, $\omega_0 \tau = 2$, $\omega/\omega_0 = 2$, $F_0/mA\omega_0^2 = 1.0$ and $\bar{t} \geq 0$.
(d) The representation of phase trajectories may be simplified by sampling them stroboscopically at a suitable frequency ω_s. Thus, the coordinates to be plotted are determined at regular time intervals of $2\pi/\omega_s$. The resulting diagram is known as a Poincaré section.

   1. What is the Poincaré section of a simple harmonic oscillator if $\omega_s = \omega$?
   2. Extend the above notebook to obtain the Poincaré section of the driven oscillator in (c), when ω_s equals the driving frequency ω.

---

[5] R. D. Peters and T. Pritchett, "The not-so-simple harmonic oscillator," *American Journal of Physics*, vol. 65, pp. 1067–1073, 1977.

88 Solved Problems in Classical Mechanics

Solution

(a) For a simple harmonic oscillator: $x = A\cos(\omega t + \phi)$ and $\dot{x} = -\omega A \sin(\omega t + \phi)$ (see Question 4.1). Therefore, the phase trajectory is the ellipse

$$(x/A)^2 + (\dot{x}/\omega A)^2 = 1. \tag{2}$$

(b) The phase trajectory for the solution (1) at $\bar{t} \geq 0$ and the notebook are:

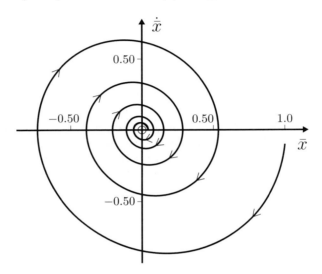

```
In[1]:= (* IN THIS NOTEBOOK t & τ ARE DIMENSIONLESS & RELAXATION TIMES RESPECTIVELY *)
    ω0 = 1; τ = 10; γ = 2; A = 0; tmax = 10π;           (* USE FOR (b) *)
    (* ω0 = 1; τ = 2; γ = 2; A = 1; tmax = 10π; t0 = 0; *)    (* USE FOR (c) *)

    θ = ArcTan[ 2γ / (ω0 τ (1 - γ²)) ];

    X[t_] := Exp[-t / (ω0 τ)] Cos[√(1 - 1/(ω0² τ²)) t] + A / √((1 - γ²)² + 4γ²/(ω0² τ²)) Cos[γ t - θ];

    V[t_] = D[X[t], t]; t0 = 0;
    DataTable = {{N[X[t0], 12], N[V[t0], 12]}};
    Do[ t = t0 + 2n π/γ; X1 = N[X[t], 12]; V1 = N[V[t], 12];
        DataTable = Append[DataTable, {X1, V1}], {n, 1, 9} ];   Clear[t]
    ParametricPlot[{X[t], V[t]}, {t, 0, tmax}, PlotRange → {{-0.5, 1.5},
        {-1.5, 1.}}, AspectRatio → 0.5, PlotPoints → 100,
        Epilog → {PointSize[Medium], Point[DataTable]}]
    ListPlot[DataTable, Joined → False, PlotRange → {{-1.5, 1.5},
        {-1.5, 1.5}}, Frame → True, FrameLabel → {"X", "Xdot", "", ""},
        PlotStyle → {RGBColor[1, 0, 0], PointSize[0.01]}]
```

(c) Equation (9) of Question 4.7 can be expressed as

$$\bar{x}(\bar{t}) = e^{-\bar{t}/\omega_0 \tau} \cos\left(\frac{\omega_d}{\omega_0}\bar{t} + \phi\right) + \frac{F_0/mA\omega_0^2}{\sqrt{(1-\gamma^2)^2 + (2\gamma/\omega_0\tau)^2}} \cos(\gamma\bar{t} - \vartheta), \quad (3)$$

where $\gamma = \omega/\omega_0$ and $\vartheta = \tan^{-1} 2\gamma/(\omega_0\tau(1-\gamma^2))$. For $\bar{t} \geq 0$ and the given parameters, the phase trajectory below is obtained from the above notebook. The four points plotted on the phase trajectory were calculated at the times indicated.

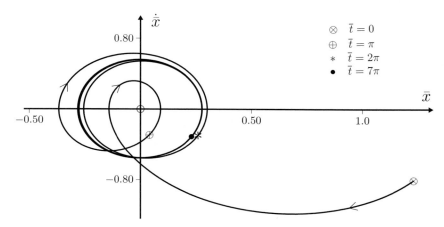

When the transient has died out ($\bar{t} \gg \omega_0\tau$), the phase trajectory approximates an ellipse.

(d) **1.** The Poincaré section is a point at $(\bar{x}_0, \dot{\bar{x}}_0)$.

2. For $0 \leq \bar{t} \leq 7\pi$ and $\omega_s = \omega$, the Poincaré section obtained from (3) is shown below. We remark that no additional points appear in the Poincaré section for $\bar{t} > 7\pi$. The point plotted as a • indicates the steady state; the other three points correspond to the transient. (The times at which these samples were taken are given in the previous diagram.)

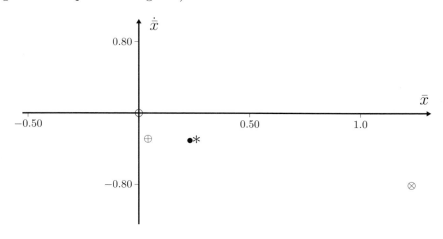

Comments

(i) The $(x, \dot{x})$ plane is known as the phase space for one-dimensional motion of a particle. For three-dimensional motion the phase space $(x, y, z, \dot{x}, \dot{y}, \dot{z})$ is six-dimensional, and for an N-particle system the dimensionality is $6N$.

(ii) The phase trajectories of conservative systems (such as the oscillator in (a)) form closed paths. This is not true for dissipative systems, such as the damped oscillator in (b), where the phase trajectory is a logarithmic spiral, and all initial conditions eventually result in a particle at rest at $x = 0$. The origin is an example of an 'attractor'; that is, a point (or set of points) in phase space to which the system is attracted in the presence of damping.

(iii) In more complicated systems the phase space may be divisible into regions each having its own attractor. Such regions are called 'basins of attraction'.

(iv) The phase trajectory for the damped driven oscillator in (c) shows the initial transient and subsequent steady-state behaviour. The ellipse associated with the latter is referred to as a 'limit cycle' – it is an attractor in the above sense.

(v) The simplification of phase space provided by Poincaré sections[6] can be an important tool in analyzing and describing the physics of chaotic systems (see also Chapter 13).

Question 4.16

Consider the damped oscillator of Question 4.5. Apply the transformation

$$x(t) = e^{-t/\tau} y(t) \tag{1}$$

to the equation of motion, and hence deduce a Lagrangian for the damped oscillator:

$$\mathsf{L} = (\tfrac{1}{2} m \dot{x}^2 - \tfrac{1}{2} m \omega_0^2 x^2) e^{2t/\tau}. \tag{2}$$

Solution

We start with the equation of motion for the damped oscillator

$$\ddot{x} + (2/\tau)\dot{x} + \omega_0^2 x = 0, \tag{3}$$

where $\omega_0^2 = k/m$ and $\tau = 2m/\alpha$ (see Question 4.5). The transformation (1) changes (3) to

$$\ddot{y} + \omega_\mathrm{d}^2 y = 0, \tag{4}$$

where $\omega_\mathrm{d}^2 = \omega_0^2 - \tau^{-2}$. Equation (4) describes a simple harmonic oscillator, which is a conservative system with Lagrangian

$$\widetilde{\mathsf{L}} = K - V = \tfrac{1}{2} m \dot{y}^2 - \tfrac{1}{2} m \omega_\mathrm{d}^2 y^2. \tag{5}$$

---

[6] G. L. Baker and J. P. Gollub, *Chaotic dynamics (an introduction)*, pp. 23–27. Cambridge: Cambridge University Press, 1990.

If we now use (1) to express (5) in terms of x and $\dot{x}$ we find a Lagrangian for the damped oscillator:

$$\mathsf{L} = (\tfrac{1}{2}m\dot{x}^2 - \tfrac{1}{2}m\omega_0^2 x^2)e^{2t/\tau} + \frac{d}{dt}\left(\frac{m}{2\tau}x^2 e^{2t/\tau}\right). \tag{6}$$

In general, a term of the form $dG(x,t)/dt$ in L makes no contribution in the Lagrange equation

$$\frac{d}{dt}\frac{\partial \mathsf{L}}{\partial \dot{x}} - \frac{\partial \mathsf{L}}{\partial x} = 0. \tag{7}$$

So, we can ignore the last term in (6) and take (2) as the Lagrangian.

Comments

(i) The reader can easily verify that (2) and (7) yield the equation of motion (3).
(ii) In the limit $\tau \to \infty$, (2) reduces to the familiar Lagrangian $\mathsf{L} = \tfrac{1}{2}m\dot{x}^2 - \tfrac{1}{2}m\omega_0^2 x^2$ for an undamped oscillator.
(iii) Lagrangians such as (2), that do not have the conventional form $\mathsf{L} = K - V$, are referred to as generalized Lagrangians. They are useful in the study of dissipative systems.[7,8]
(iv) Note that the Hamiltonian $\mathsf{H} = \dot{x}\,\partial \mathsf{L}/\partial \dot{x} - \mathsf{L}$ (see Question 11.33) for the damped oscillator is given by

$$\mathsf{H} = (\tfrac{1}{2}m\dot{x}^2 + \tfrac{1}{2}m\omega_0^2 x^2)e^{2t/\tau} = E e^{2t/\tau}, \tag{8}$$

where E is the energy. So, here the Hamiltonian is not equal to the energy (except in the limit $\tau \to \infty$).[8]
(v) The Lagrangian (2) also describes an undamped oscillator with time-dependent mass $m(t) = m e^{2t/\tau}$. In this case the energy $E = \tfrac{1}{2}m(t)\dot{x}^2 + \tfrac{1}{2}m(t)\omega_0^2 x^2$ is equal to H.
(vi) A general method to determine a Lagrangian $\mathsf{L}(x,\dot{x},t)$ associated with an equation of motion $m\ddot{x} = F(x,\dot{x},t)$ was provided by Darboux in 1894.[9]

---

[7] L. Y. Bakar and H. G. Kwatny, "Generalized Lagrangian and conservation law for the damped harmonic oscillator," American Journal of Physics, vol. 49, pp. 1062–1065, 1981.

[8] D. H. Kobe and G. Reali, "Lagrangians for dissipative systems," American Journal of Physics, vol. 54, pp. 997–999, 1986.

[9] See, C. Leubner and P. Krumm, "Lagrangians for simple systems with variable mass," European Journal of Physics, vol. 11, pp. 31–34, 1990.

5
Energy and potentials

The questions in this chapter deal with the important topic of the mechanical energy of a particle. Two simple examples (Questions 5.2 and 5.3) are used to motivate the general formulation of mechanical energy associated with position-dependent force-fields $\mathbf{F}(\mathbf{r})$. We remind the reader of some standard notation that is employed here. A Cartesian vector $\mathbf{A} = A_x\hat{\mathbf{x}} + A_y\hat{\mathbf{y}} + A_z\hat{\mathbf{z}}$ is abbreviated as $\mathbf{A} = (A_x, A_y, A_z)$; thus, we write $\mathbf{r} = (x, y, z)$ for a position vector, $d\mathbf{r} = (dx, dy, dz)$ for an infinitesimal displacement vector, $\mathbf{F} = (F_x, F_y, F_z)$ for a force, and so on.

Question 5.1

Consider a particle of mass m acted on by a force $\mathbf{F}$ in an inertial frame. Prove that

$$dK = \mathbf{F} \cdot d\mathbf{r}, \qquad (1)$$

where $d\mathbf{r}$ is the change in the position vector of the particle in a time dt, and dK is the corresponding change in the kinetic energy $K = \tfrac{1}{2}mv^2$.

Solution

The rate of change of kinetic energy is

$$\frac{dK}{dt} = \frac{d}{dt}\left(\tfrac{1}{2}m\mathbf{v}\cdot\mathbf{v}\right) = m\frac{d\mathbf{v}}{dt}\cdot\mathbf{v} = \mathbf{F}\cdot\frac{d\mathbf{r}}{dt}, \qquad (2)$$

and (1) follows.

Comments

(i) According to (1), the change in kinetic energy is equal to the work done on the particle by the force $\mathbf{F}$. This result is known as the work–energy theorem. It applies to forces that can be dependent on position, velocity and time, $\mathbf{F} = \mathbf{F}(\mathbf{r}, \mathbf{v}, t)$. For finite changes, (1) yields

$$K_\mathrm{f} - K_\mathrm{i} = \int_{\mathbf{r}_\mathrm{i}}^{\mathbf{r}_\mathrm{f}} \mathbf{F}\bigl(\mathbf{r}, \mathbf{v}(\mathbf{r}), t(\mathbf{r})\bigr) \cdot d\mathbf{r}, \qquad (3)$$

where the kinetic energies K_i and K_f are evaluated at the initial and final positions $\mathbf{r}_i$ and $\mathbf{r}_f$, respectively.

(ii) According to (2), the rate of change of kinetic energy is equal to the power $\mathbf{F} \cdot \mathbf{v}$ expended on the particle by the force $\mathbf{F}$.

(iii) In a non-inertial frame, (1) is $dK = \mathbf{F}_e \cdot d\mathbf{r}$ where the effective force $\mathbf{F}_e$ is defined in (25) of Chapter 1.

Question 5.2

For a one-dimensional force $\mathbf{F} = F(x)\hat{\mathbf{x}}$, prove that

$$d(K+V) = 0, \tag{1}$$

where

$$V(x) = -\int F(x)\,dx. \tag{2}$$

Solution

Here, $\mathbf{F} \cdot d\mathbf{r} = \bigl(F(x),\,0,\,0\bigr) \cdot \bigl(dx,\,dy,\,dz\bigr) = F(x)\,dx$, and hence (1) of Question 5.1 gives

$$dK = F(x)\,dx = d\int F(x)\,dx, \tag{3}$$

which is (1) with $V(x)$ given by (2).

Question 5.3

For a central, isotropic force $\mathbf{F} = F(r)\hat{\mathbf{r}}$, where $r = \sqrt{x^2+y^2+z^2}$, prove that (1) of Question 5.2 holds with

$$V(r) = -\int F(r)\,dr. \tag{1}$$

Solution

Write $d\mathbf{r}$ in terms of radial and transverse components: $d\mathbf{r} = dr\,\hat{\mathbf{r}} + d\mathbf{r}_\perp$, where $d\mathbf{r}_\perp$ is perpendicular to the radial unit vector $\hat{\mathbf{r}}$. Then

$$\mathbf{F} \cdot d\mathbf{r} = F(r)\hat{\mathbf{r}} \cdot (dr\,\hat{\mathbf{r}} + d\mathbf{r}_\perp) = F(r)\,dr, \tag{2}$$

because $\hat{\mathbf{r}} \cdot \hat{\mathbf{r}} = 1$ and $\hat{\mathbf{r}} \cdot d\mathbf{r}_\perp = 0$. Hence (1) of Question 5.1 yields

$$dK = F(r)\,dr = d\int F(r)\,dr, \tag{3}$$

which is (1) of Question 5.2 with $V(r)$ given by (1) above.

Comments

(i) Note the algebraic similarity between the one-dimensional and the isotropic three-dimensional cases by comparing (3) with (3) of Question 5.2. (See also Chapter 8.)

(ii) The function V is referred to as the potential energy of the particle, and $E = T+V$ is its mechanical energy (hereafter referred to simply as energy).

(iii) The potential energy in (1) is spherically symmetric (in terms of spherical polar coordinates r, θ and ϕ, it is independent of the angles θ and ϕ). The corresponding force $F(r)\hat{\mathbf{r}}$ is always directed towards or away from the origin, and its magnitude $|F(r)|$ is constant on any sphere of radius r centred on the origin. The force associated with a spherically symmetric potential $V(r)$ is always central and isotropic.

(iv) The following questions deal with the energy of a particle in an arbitrary position-dependent force field $\mathbf{F}(\mathbf{r})$.

Question 5.4

Show that the relations between $\mathbf{F}(\mathbf{r})$ and $V(\mathbf{r})$ in the preceding two questions are particular cases of

$$\mathbf{F}(\mathbf{r}) = -\boldsymbol{\nabla} V(\mathbf{r}). \tag{1}$$

Here, $\boldsymbol{\nabla}$ is the gradient operator, given in Cartesian coordinates by

$$\boldsymbol{\nabla} = \left(\frac{\partial}{\partial x}, \frac{\partial}{\partial y}, \frac{\partial}{\partial z}\right).$$

Solution

From (2) of Question 5.2 and $\mathbf{F} = F(x)\hat{\mathbf{x}}$ we have

$$\mathbf{F} = -\frac{dV(x)}{dx}\hat{\mathbf{x}}. \tag{2}$$

Similarly, (1) of Question 5.3 and $\mathbf{F} = F(r)\hat{\mathbf{r}}$ yield

$$\mathbf{F} = -\frac{dV(r)}{dr}\hat{\mathbf{r}}. \tag{3}$$

These are (1) for the special cases $V = V(x)$ and $V = V(r)$, respectively.

Question 5.5

Show that if a force $\mathbf{F}(\mathbf{r})$ is 'derivable from a scalar potential $V(\mathbf{r})$' in the sense of (1) of Question 5.4, then

$$\frac{d}{dt}(K+V) = 0. \tag{1}$$

Solution

Here, $V = V(x, y, z)$ and therefore

$$dV = \frac{\partial V}{\partial x} dx + \frac{\partial V}{\partial y} dy + \frac{\partial V}{\partial z} dz \equiv (\nabla V) \cdot d\mathbf{r} = -\mathbf{F} \cdot d\mathbf{r} \qquad (2)$$

because of (1) of Question 5.4. But $\mathbf{F} \cdot d\mathbf{r} = dK$ (see Question 5.1) and so (2) yields

$$d(K + V) = 0. \qquad (3)$$

Comments

(i) A force $\mathbf{F}(\mathbf{r})$ that satisfies $\mathbf{F}(\mathbf{r}) = -\nabla V(\mathbf{r})$ is called conservative because, according to (1), it conserves the energy of a particle on which it acts. Thus, we have the law of conservation of mechanical energy for a particle in a conservative force field:

$$E_f = E_i. \qquad (4)$$

(ii) According to (2), the potential energy is given, to within an arbitrary constant, by the line integral of the force:

$$V(\mathbf{r}) = -\int \mathbf{F} \cdot d\mathbf{r}. \qquad (5)$$

(iii) Not all forces $\mathbf{F}(\mathbf{r})$ are conservative, as the following example shows.

Question 5.6

Prove that $\mathbf{F} = (y, -x, 0)$ is not conservative.

Solution

Use *reductio ad absurdum*: Assume that $\mathbf{F}$ is conservative. Then, $\mathbf{F} = -\nabla V$ requires

$$\frac{\partial V(x, y)}{\partial x} = -y \quad \text{and} \quad \frac{\partial V(x, y)}{\partial y} = x.$$

The solutions to these two equations are

$$V(x, y) = -xy + f(y) \quad \text{and} \quad V(x, y) = xy + g(x),$$

respectively, which is clearly impossible. Therefore, $\mathbf{F}$ is not conservative.

Comment

In general, we require a (necessary and sufficient) condition to test whether a given force $\mathbf{F}(\mathbf{r})$ is conservative. The following two questions provide this.

Question 5.7

Prove that a necessary condition for $\mathbf{F}(\mathbf{r})$ to be conservative is

$$\boldsymbol{\nabla} \times \mathbf{F}(\mathbf{r}) = 0. \qquad (1)$$

That is, prove

$$\mathbf{F} = -\boldsymbol{\nabla} V(\mathbf{r}) \Longrightarrow \boldsymbol{\nabla} \times \mathbf{F} = 0, \qquad (2)$$

where an arrow $\Longrightarrow$ means 'implies'. Do this in two ways:

(a) By using the Cartesian form for $\boldsymbol{\nabla}$.
(b) By applying Stokes's theorem:‡

$$\int_S (\boldsymbol{\nabla} \times \mathbf{F}) \cdot d\mathbf{S} = \oint_C \mathbf{F} \cdot d\mathbf{r}, \qquad (3)$$

where S is a surface of arbitrary shape bounded by a closed curve C.

Solution

(a) In Cartesian coordinates (see Question 5.4) we have

$$\boldsymbol{\nabla} \times (\boldsymbol{\nabla} V) = \begin{vmatrix} \hat{\mathbf{x}} & \hat{\mathbf{y}} & \hat{\mathbf{z}} \\ \dfrac{\partial}{\partial x} & \dfrac{\partial}{\partial y} & \dfrac{\partial}{\partial z} \\ \dfrac{\partial V}{\partial x} & \dfrac{\partial V}{\partial y} & \dfrac{\partial V}{\partial z} \end{vmatrix}$$

$$= \left(\dfrac{\partial^2 V}{\partial y \partial z} - \dfrac{\partial^2 V}{\partial z \partial y},\ \dfrac{\partial^2 V}{\partial z \partial x} - \dfrac{\partial^2 V}{\partial x \partial z},\ \dfrac{\partial^2 V}{\partial x \partial y} - \dfrac{\partial^2 V}{\partial y \partial x} \right) = 0 \qquad (4)$$

if the order of the partial derivatives is unimportant.

(b) If $\mathbf{F} = -\boldsymbol{\nabla} V$, then Stokes's theorem, (3), yields

$$\int_S (\boldsymbol{\nabla} \times \mathbf{F}) \cdot d\mathbf{S} = -\oint_C \boldsymbol{\nabla} V \cdot d\mathbf{r} = -\oint dV(\mathbf{r}) = 0, \qquad (5)$$

because $V(\mathbf{r})$ is a single-valued function. In (5), S is an arbitrary surface and therefore it follows that $\boldsymbol{\nabla} \times \mathbf{F} = 0$ everywhere.

Comment

To prove Stokes's theorem we start by evaluating $\oint \mathbf{F} \cdot d\mathbf{r}$ around an infinitesimal, closed rectangular path δC in the xy-plane:

$$(x, y, z) \to (x + dx, y, z) \to (x + dx, y + dy, z) \to (x, y + dy, z) \to (x, y, z).$$

If we number the corners of this rectangle 1, 2, 3, and 4, then

‡For readers who not familiar with this theorem, a proof is provided in the Comment.

$$\oint_{\delta C} \mathbf{F} \cdot d\mathbf{r} = \left(\int_{1\to 2} + \int_{2\to 3} - \left\{ \int_{1\to 4} + \int_{4\to 3} \right\} \right) \mathbf{F} \cdot d\mathbf{r}$$
$$= F_x(x,y,z)\,dx + F_y(x+dx,y,z)\,dy - F_y(x,y,z)\,dy - F_x(x,y+dy,z)\,dx$$
$$= \left(\frac{\partial F_y}{\partial x} - \frac{\partial F_x}{\partial y} \right) dx dy$$
$$= (\boldsymbol{\nabla} \times \mathbf{F})_z\, dx dy$$
$$= (\boldsymbol{\nabla} \times \mathbf{F}) \cdot \mathbf{n}\, dS\,, \tag{6}$$

where $\mathbf{n}$ is a unit vector vector perpendicular to a rectangular element of area dS. (There is a sign convention, a right-hand rule, implicit in (6), relating the direction in which δC is traversed and the direction of $\mathbf{n}$, see below.) Equation (6) is independent of the choice of coordinates, and applies to an element of any orientation. An arbitrary finite surface S with boundary C can be subdivided into infinitesimal rectangular elements δC_i ($i = 1, 2, \cdots$). Then,

$$\oint_C \mathbf{F} \cdot d\mathbf{r} = \sum_i \oint_{\delta C_i} \mathbf{F} \cdot d\mathbf{r}\,, \tag{7}$$

because on common segments of adjacent elements the $d\mathbf{r}$ point in opposite directions and the contributions of $\mathbf{F} \cdot d\mathbf{r}$ to the sum in (7) cancel, whereas no such cancellation occurs on the boundary C. Equations (6) and (7) yield (3). The figure below illustrates the right-hand convention that is assumed here.

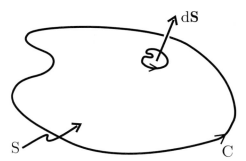

Question 5.8

Use Stokes's theorem to prove that $\boldsymbol{\nabla} \times \mathbf{F}(\mathbf{r}) = 0$ is a sufficient condition for $\mathbf{F}(\mathbf{r})$ to be conservative. That is, prove

$$\boldsymbol{\nabla} \times \mathbf{F} = 0 \implies \mathbf{F} = -\boldsymbol{\nabla} V(\mathbf{r})\,. \tag{1}$$

Solution

The proof is less obvious than the preceding 'necessary' part because one has to prove the existence of the function $V(\mathbf{r})$. If $\boldsymbol{\nabla} \times \mathbf{F} = 0$ *everywhere*, it follows from Stokes's theorem that

$$\oint_C \mathbf{F} \cdot d\mathbf{r} = 0 \qquad (2)$$

for *all* closed curves C. According to (2):

$$\int_1 \mathbf{F} \cdot d\mathbf{r} = \int_2 \mathbf{F} \cdot d\mathbf{r}, \qquad (3)$$

where 1 and 2 are any two paths joining two points A and B. Therefore, the line integral between any two such points is independent of the path followed from A to B, and depends only on the endpoints A and B. Thus, $\mathbf{F} \cdot d\mathbf{r}$ must be the differential of some single-valued scalar function $V(\mathbf{r})$ (we say it is a perfect differential):

$$\mathbf{F} \cdot d\mathbf{r} = -dV(\mathbf{r}), \qquad (4)$$

where a minus sign has been inserted to conform with (2) of Question 5.5. But

$$dV(\mathbf{r}) = (\boldsymbol{\nabla} V) \cdot d\mathbf{r} \qquad (5)$$

(see Question 5.5). In (4) and (5), $d\mathbf{r}$ is arbitrary and therefore $\mathbf{F} = -\boldsymbol{\nabla} V$.

Comments

(i) It follows from the above that the conditions (1) of Question 5.4, (1) of Question 5.7, and (2) above are equivalent in the following sense:

$$\begin{array}{c} \mathbf{F} = -\boldsymbol{\nabla} V(\mathbf{r}) \\ \swarrow \qquad \searrow \\ \boldsymbol{\nabla} \times \mathbf{F}(\mathbf{r}) = 0 \iff \oint_C \mathbf{F} \cdot d\mathbf{r} = 0. \end{array} \qquad (6)$$

In (6), the gradient and curl equations must hold at *all* points $\mathbf{r}$ in the force field $\mathbf{F}(\mathbf{r})$, and C is *any* closed curve.

(ii) The class of all position-dependent forces $\mathbf{F}(\mathbf{r})$ consists of two sub-classes: Those that are conservative, $\mathbf{F}^c$, and those that are non-conservative, $\mathbf{F}^n$. For these, the following statements hold:

	$\mathbf{F}^c$	$\mathbf{F}^n$
$\mathbf{F} = -\boldsymbol{\nabla} V$ everywhere?	Yes	No
$\boldsymbol{\nabla} \times \mathbf{F} = 0$ everywhere?	Yes	No
$\oint_C \mathbf{F} \cdot d\mathbf{r} = 0$ for all closed curves C?	Yes	No

(iii) If a force consists of both conservative and non-conservative parts, $\mathbf{F} = \mathbf{F}^c + \mathbf{F}^n$, then

$$E_f = E_i + \int_{\mathbf{r}_i}^{\mathbf{r}_f} \mathbf{F}^n \cdot d\mathbf{r}. \tag{7}$$

That is, the change in mechanical energy is equal to the work done by non-conservative forces during the motion from $\mathbf{r}_i$ to $\mathbf{r}_f$.

(iv) In $\mathbf{F} = -\boldsymbol{\nabla} V$, an arbitrary constant can be added to the potential V without changing the force $\mathbf{F}$. (Usually, the arbitrary constant implicit in V is fixed by making a convenient choice for the zero of V – frequently V is taken to be zero at infinity. This choice is, of course, immaterial in the conservation law (4) of Question 5.5.) Non-uniqueness of a physical quantity such as V is the 'tip of an iceberg', and it is discussed further in Question 5.23.

(v) These results have applications in other areas as well, notably in electromagnetism. For example, in electrostatics the curl of the electric field is always zero, and therefore this field is derivable from a scalar potential: $\mathbf{E}(\mathbf{r}) = -\boldsymbol{\nabla}\phi(\mathbf{r})$. Thus the electrostatic force $\mathbf{F} = q\mathbf{E}$ is conservative.

(vi) A vector field $\mathbf{F}$ whose curl is zero everywhere is referred to as irrotational; if $\boldsymbol{\nabla} \times \mathbf{F} \neq 0$ the field is rotational. Thus, we have seen that all irrotational fields can be derived from a scalar potential.

(vii) In general, the electric field $\mathbf{E}$ is rotational (according to Faraday's law $\boldsymbol{\nabla} \times \mathbf{E} = -\partial \mathbf{B}/\partial t$), and the question arises: in what way can $\mathbf{E}$, and hence the electric force $q\mathbf{E}$, be expressed in terms of 'potentials'? This topic involves an interesting discussion of so-called solenoidal fields (fields whose divergence is zero everywhere) and it is considered in Question 5.23.

(viii) The notion of the path or 'history' independence of a physical quantity in some parameter space (which is usually first encountered in the theory of conservative forces, as in (2) and (3) above) has wider ramifications in physics. For example, heat energy Q is a path-dependent quantity (dQ is an imperfect differential), whereas internal energy U and entropy S are path independent (dU and dS are perfect differentials). Consequently, U and S are unique functions of the relevant physical parameters, whereas Q is not; one can speak of the energy and entropy content of a system, but not of its heat content.

(ix) Calculations involving path-independent quantities (such as the work done by a conservative force, and hence the potential energy – see (5) of Question 5.5) can often be simplified by selecting a convenient path.

Question 5.9

Prove that a frictional force is non-conservative.

Solution

A frictional force $\mathbf{F}$ always points in the opposite direction to the infinitesimal displacement $d\mathbf{r}$. Consequently, $\mathbf{F} \cdot d\mathbf{r} < 0$ everywhere and

$$\oint_C \mathbf{F} \cdot d\mathbf{r} < 0. \tag{1}$$

Comment

Often we encounter non-conservative forces in a macroscopic description of phenomena (such as the conversion of mechanical energy into heat energy in dissipative processes), whereas the underlying microscopic processes are, in fact, conservative – for example, energy is conserved in the interactions of the molecules of the media experiencing friction.

Question 5.10

Show that the irrotational condition $\nabla \times \mathbf{F} = 0$ for a central force $\mathbf{F} = F(\mathbf{r})\hat{\mathbf{r}}$ to be conservative can be expressed as

$$\frac{1}{x}\frac{\partial F}{\partial x} = \frac{1}{y}\frac{\partial F}{\partial y} = \frac{1}{z}\frac{\partial F}{\partial z}. \tag{1}$$

Solution

Recall that $\hat{\mathbf{r}} = \mathbf{r}\,r^{-1} = (x, y, z)/r$, where $r = \sqrt{x^2 + y^2 + z^2}$. Then

$$\nabla \times (F\hat{\mathbf{r}}) = \begin{vmatrix} \hat{\mathbf{x}} & \hat{\mathbf{y}} & \hat{\mathbf{z}} \\ \dfrac{\partial}{\partial x} & \dfrac{\partial}{\partial y} & \dfrac{\partial}{\partial z} \\ \dfrac{x}{r}F & \dfrac{y}{r}F & \dfrac{z}{r}F \end{vmatrix}$$

$$= \frac{1}{r}\left(z\frac{\partial F}{\partial y} - y\frac{\partial F}{\partial z},\ x\frac{\partial F}{\partial z} - z\frac{\partial F}{\partial x},\ y\frac{\partial F}{\partial x} - x\frac{\partial F}{\partial y}\right), \tag{2}$$

which is zero everywhere if (1) is satisfied.

Comments

(i) In general, (1) is not satisfied. For example, the central force $kxyz\hat{\mathbf{r}}$ (k is a constant) is rotational, and therefore not conservative.

(ii) Equations (1) possess the important solution $F = F(r)$. For this,

$$\frac{1}{x}\frac{\partial F(r)}{\partial x} = \frac{1}{x}\frac{\partial r}{\partial x}\frac{dF(r)}{dr} = \frac{1}{r}\frac{dF(r)}{dr}, \tag{3}$$

and similarly

$$\frac{1}{y}\frac{\partial F(r)}{\partial y} = \frac{1}{r}\frac{dF}{dr} \quad \text{and} \quad \frac{1}{z}\frac{\partial F(r)}{\partial z} = \frac{1}{r}\frac{dF}{dr}. \tag{4}$$

Thus, (1) is satisfied and consequently all central, isotropic forces $F(r)\hat{\mathbf{r}}$ are conservative – as we already know from Question 5.3. This class of forces is important because it applies to several types of interaction, including gravitational, electrostatic, and certain molecular and nuclear interactions; see Question 5.19.

(iii) In general, it is clear that the line integral of a central, isotropic force $F(r)\hat{\mathbf{r}}$ around any closed curve is zero (see Question 5.3), and therefore it follows directly from Stokes's theorem (see Question 5.7) that such forces are irrotational. Equations (2)–(4) demonstrate this by an explicit calculation in Cartesian coordinates. In other coordinate systems the calculations are longer.

Question 5.11

Consider a time-dependent force that can be expressed as the gradient of a scalar:

$$\mathbf{F}(\mathbf{r},t) = -\boldsymbol{\nabla} V(\mathbf{r},t). \tag{1}$$

Show that

$$\frac{d}{dt}(K+V) = \frac{\partial V}{\partial t}. \tag{2}$$

Solution

Here, $V = V(x,y,z,t)$ and so

$$dV = \frac{\partial V}{\partial x}dx + \frac{\partial V}{\partial y}dy + \frac{\partial V}{\partial z}dz + \frac{\partial V}{\partial t}dt$$

$$\equiv (\boldsymbol{\nabla} V) \cdot d\mathbf{r} + \frac{\partial V}{\partial t}dt$$

$$= -\mathbf{F} \cdot d\mathbf{r} + \frac{\partial V}{\partial t}dt$$

$$= -dK + \frac{\partial V}{\partial t}dt,$$

and (2) follows. (In the last step we have used (1) of Question 5.1.)

Comment

Even if a force $\mathbf{F}(\mathbf{r},t)$ that depends explicitly on time can be obtained from the gradient of a scalar, as in (1), it will not conserve the mechanical energy $K+V$. Thus, for example, the motion of a charged particle in a time-dependent electric field $\mathbf{E} = -\boldsymbol{\nabla}\phi(\mathbf{r},t)$ is not conservative. (It can be shown that the total energy of the particle and the field is conserved.)

Question 5.12

A one-dimensional force $\mathbf{F} = -kx\hat{\mathbf{x}}$, where k is a constant, acts on a particle of mass m.

(a) Calculate the potential energy $V(x)$ of the particle.
(b) Sketch the possible graphs of $V(x)$. Use these graphs and conservation of energy to discuss the possible motions of the particle.

Solution

(a) Use (2) of Question 5.2. Then

$$V(x) = k\int x\,dx = \tfrac{1}{2}kx^2. \qquad (1)$$

In (1) we have omitted an arbitrary constant, and this means we have chosen the zero of potential at $x = 0$.

(b) On the graphs below we have also drawn horizontal lines to denote the constant energy

$$E = \tfrac{1}{2}mv^2 + V(x) \qquad (2)$$

of the particle. We consider separately the cases $k > 0$ and $k < 0$.

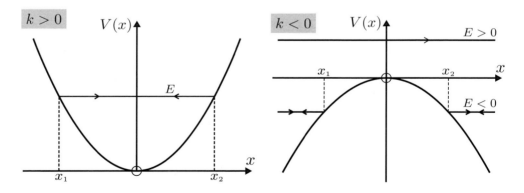

$k > 0$

If $k > 0$, the force $\mathbf{F} = -kx\hat{\mathbf{x}}$ is a Hooke's-law-type force: it is linear in x and it is a restoring force ($\mathbf{F}$ is always directed towards the origin O). In general, the energy of the particle can never be less than the potential energy because, according to (2), that would mean $v^2 < 0$ and hence imaginary speed, which is impossible. Thus, $E \geq 0$. If $E = 0$, the particle is at rest at O; it is in stable equilibrium there (stable because if given a slight displacement away from O, the force tends to return it to O). If $E > 0$, the particle is confined to regions where

$$V(x) \leq E. \tag{3}$$

When $V(x) = E$, that is $\frac{1}{2}kx^2 = E$, the particle is at rest, see (2). The roots of this equation are

$$x_1 = -\sqrt{2E/k} \quad \text{and} \quad x_2 = \sqrt{2E/k}. \tag{4}$$

Consider a particle initially at $x = x_1$. The force is to the right and the particle accelerates towards O, where it reaches its maximum speed $\sqrt{2E/m}$. After passing O, the force is to the left and the particle is decelerated, coming to rest at $x = x_2$. The particle then retraces its motion, eventually coming to rest at $x = x_1$. And so on. The motion is periodic, with amplitude equal to $\sqrt{2E/k}$. The period $T = 2\pi\sqrt{m/k}$ (see Question 4.1). The regions $x < x_1$ and $x > x_2$ are referred to as classically forbidden because in classical mechanics a particle with energy E cannot enter them; x_1 and x_2 are known as the classical turning points.

$k < 0$

Consider a particle at $x < 0$ and moving to the right. If $E > 0$, the particle is decelerated and attains its minimum speed $v = \sqrt{2E/m}$ at the origin. After passing O it is accelerated to the right. The motion is unbounded. If $E = 0$, the particle comes to rest at O, where it is in unstable equilibrium (unstable because any displacement will result in the particle being accelerated away from O). If $E < 0$, the particle comes to rest at $x = x_1$ and is then accelerated to the left. The particle is reflected by the potential and again the motion is unbounded.

Comments

(i) The trajectories for this example are calculated in Questions 4.1 and 4.2.
(ii) The above question is a simple illustration of the use of energy diagrams to obtain a qualitative picture of one-dimensional motion. By plotting a graph of $V(x)$ and drawing on it horizontal lines to represent the energy E, one can distinguish between bounded and unbounded motion, identify classical turning points, classically forbidden regions, and points of stable and unstable equilibrium. The next three questions provide further illustration of this method.

Question 5.13

A particle of mass m is subject to a one-dimensional force $\mathbf{F} = (-kx + bx^3)\hat{\mathbf{x}}$, where k and b are positive constants.

(a) Sketch the energy diagram and use it to discuss the motion.
(b) Determine the frequency of small oscillations about a point of stable equilibrium.

Solution

(a) The potential energy is (see Question 5.2)

$$V(x) = -\int (-kx + bx^3)\, dx = \tfrac{1}{2}kx^2 - \tfrac{1}{4}bx^4, \tag{1}$$

where we have chosen the zero of potential at $x = 0$. To plot $V(x)$, note the following: $V = 0$ at $x = 0$ and $x = \pm\sqrt{2k/b}$. Also, $V \to -\infty$ as $x \to \pm\infty$, and $dV/dx = kx - bx^3 = 0$ at $x = 0$ and $x = \pm\sqrt{k/b}$. At the latter two points there are maxima with

$$V_{\max} = \tfrac{1}{2}k(k/b) - \tfrac{1}{4}b(k^2/b^2) = k^2/4b. \tag{2}$$

Thus, we have the energy diagram:

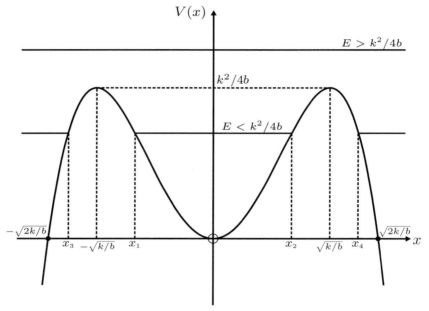

The origin O is a point of stable equilibrium: a particle with $E = 0$ placed at O will remain there, and will oscillate about O if given a small displacement. The points $x = \pm\sqrt{k/b}$ are also points of equilibrium: a particle with $E = k^2/4b$ can be at rest at these points, but the equilibrium is unstable with respect to any disturbance. There is a critical value of the energy, namely

$$E_c = k^2/4b, \tag{3}$$

below which the potential can bind the particle. If $E > E_c$, the motion is always unbounded. If $E < E_c$, there are four classical turning points given by the roots of the quartic equation

$$-\tfrac{1}{4}bx^4 + \tfrac{1}{2}kx^2 = E. \tag{4}$$

If a particle with energy $E < E_c$ is located between the points x_1 and x_2 shown in the figure, it will perform a bounded motion, oscillating between x_1 and x_2. The points x_3 and x_4 are positions at which a particle outside the 'well' is reflected by the potential (unbounded motion). As an example, if $E = \frac{3}{4}E_c = 3k^2/16b$, then the solutions to (4) are

$$x_1 = -x_2 = -\sqrt{k/2b}\,, \qquad x_3 = -x_4 = -\sqrt{3k/2b}\,. \tag{5}$$

(b) When $E \to 0$ the turning points x_1 and $x_2 \to 0$, and the oscillations become 'small'. Then the cubic term in the force can be neglected in comparison with the linear term, and $\mathbf{F} = -kx\hat{\mathbf{x}}$. The resulting equation of motion

$$\frac{d^2x}{dt^2} + \frac{k}{m}x = 0\,, \tag{6}$$

is just the equation of the simple harmonic oscillator studied in Question 4.1. The angular frequency of the oscillations is $\omega = \sqrt{k/m}$.

Comment

This system is known as an anharmonic oscillator because the cubic term bx^3 in the force causes a departure from harmonic behaviour – see Chapter 13. Anharmonic effects contribute to many phenomena, such as the thermal expansion of a solid and molecular vibrations.

Question 5.14

Consider a simple pendulum consisting of a mass m supported by a massless, rigid rod of length ℓ.

(a) Determine the potential energy $V(\theta)$ in terms of the angular position θ of the pendulum.
(b) Sketch the energy diagram and discuss the possible types of motion.

Solution

(a) Consider the trajectory from O to A. The tension in the rod is everywhere perpendicular to OA, and so it does no work on the mass m. The work done by the gravitational force mg is independent of the path followed from O to A, and it is convenient to evaluate this work along OBA. On OB no work is done, and along BA the gravitational force and displacement vector are anti-parallel. Thus, (2) of Question 5.2 gives for the potential energy relative to O

$$V(h) = -\int_0^h (-mg)dx = mgh\,.$$

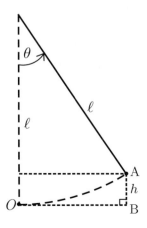

In terms of θ, $h = \ell - \ell\cos\theta$ and so

$$V(\theta) = mg\ell(1 - \cos\theta). \tag{1}$$

(b) It is sufficient to sketch the energy diagram for $-\pi \leq \theta \leq \pi$; outside this range the potential repeats itself. There is a point of stable equilibrium at $\theta = 0$ (a pendulum with $E = 0$ hanging vertically downward) and two points of unstable equilibrium at $\theta = \pm\pi$ (at both points $E = 2mg\ell$ and the pendulum is balanced vertically upward). There is a critical value of the energy $E_c = 2mg\ell$. For $E < E_c$ the motion is bounded: the pendulum is an oscillator with classical turning points at θ_1 and θ_2 $(= -\theta_1)$ given by the roots of

$$mg\ell(1 - \cos\theta) = E. \tag{2}$$

For $E > E_c$ there are no classical turning points, and the pendulum behaves like a rigid rotor; the value of $\theta(t)$ is unbounded. This behaviour is apparent in the energy diagram:

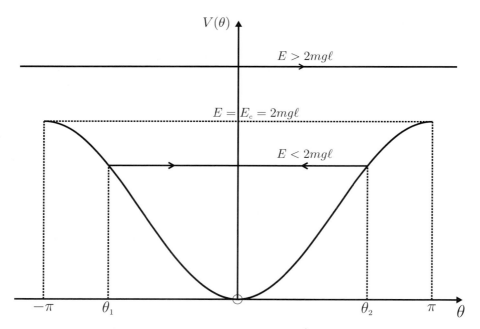

Comment

If $E \ll E_c$ then θ_2 is small. So $\cos\theta \approx 1 - \tfrac{1}{2}\theta^2$ and (1) can be approximated by

$$V \approx \tfrac{1}{2}mg\ell\theta^2 = \tfrac{1}{2}(mg/\ell)x^2. \tag{3}$$

That is, a harmonic approximation with force constant $k = mg/\ell$, angular frequency $\omega = \sqrt{k/m} = \sqrt{g/\ell}$ and period $T = 2\pi\sqrt{\ell/g}$.

Question 5.15

A particle of mass m is acted on by a one-dimensional force

$$\mathbf{F} = \left(b \sin \frac{2\pi x}{\lambda} \right) \hat{\mathbf{x}}, \tag{1}$$

where b and λ are positive constants. Sketch the energy diagram and discuss the possible types of motion.

Solution

From (2) of Question 5.2 and (1) we have

$$V(x) = -b \int \sin \frac{2\pi x}{\lambda} \, dx = \frac{b\lambda}{2\pi} \cos \frac{2\pi x}{\lambda}. \tag{2}$$

(Here, an arbitrary constant in the potential has been set equal to zero.) The energy diagram is shown below. If $E > b\lambda/2\pi$, the motion is unbounded. A particle initially moving to the right (say) will continue its motion indefinitely. The velocity of such a particle has maxima at $x = (n + \frac{1}{2})\lambda$ and minima at $x = n\lambda$ where $n = 0, \pm 1, \cdots$. If $E < b\lambda/2\pi$, the particle is trapped in one of the 'wells' and performs a periodic motion, oscillating between classical turning points such as x_1 and x_2, given by the roots of

$$\frac{b\lambda}{2\pi} \cos \frac{2\pi x}{\lambda} = E. \tag{3}$$

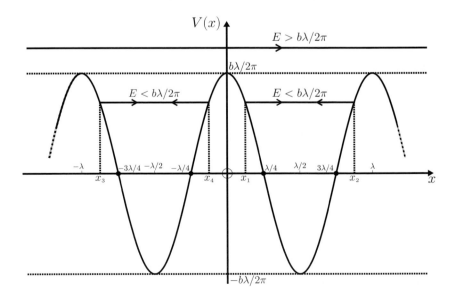

There are points of stable equilibrium at $x = (n + \frac{1}{2})\lambda$ and unstable equilibrium at $x = n\lambda$. A particle with energy $E = -b\lambda/2\pi$ is at rest at one of the points of stable equilibrium; a particle with energy $E = b\lambda/2\pi$ will come to rest at one of the points of unstable equilibrium. For small oscillations about a point of stable equilibrium (i.e. for $|E| \ll b\lambda/2\pi$) we can write $x = (n + \frac{1}{2})\lambda + X$, where $|X| \ll \frac{1}{4}\lambda$, and approximate (2) as[‡]

$$V(x) = -\frac{b\lambda}{2\pi}\left(1 - \frac{2\pi^2}{\lambda^2}X^2\right). \qquad (4)$$

Then, $F = -(2\pi b/\lambda)X$ and one has a simple harmonic equation of motion

$$\frac{d^2 X}{dt^2} + \frac{2\pi b}{m\lambda}X = 0, \qquad (5)$$

with angular frequency $\omega = \sqrt{2\pi b/m\lambda}$ and period $T_o = \sqrt{2\pi m\lambda/b}$. For larger values of $|X|$ there are anharmonic corrections to (5).

Comment

We can create a one-dimensional 'crystal' by having impenetrable barriers ($V = \infty$) at $x = \pm N\lambda$ (N an integer) in the above model. Then, particles with energy $E > b\lambda/2\pi$ are free to wander throughout the crystal, while those with $E < b\lambda/2\pi$ are trapped in the wells around $x = (n + \frac{1}{2})\lambda$.

Question 5.16

(a) For one-dimensional motion of a particle of mass m acted upon by a force $F(x)$, obtain the formal solution to the trajectory $x(t)$ in the inverse form

$$t(x) = \int_{x_0}^{x} \sqrt{\frac{m}{2\{E - V(x)\}}}\, dx, \qquad (1)$$

where $V(x)$ is the potential energy and x_0 is the position at $t = 0$.

(b) Use (1) to obtain the trajectory if F is a constant.

Solution

(a) According to Question 5.2, a one-dimensional force $F(x) = -dV/dx$ conserves the energy

$$E = \frac{1}{2}m\left(\frac{dx}{dt}\right)^2 + V(x). \qquad (2)$$

It follows that

$$dt = \sqrt{\frac{m}{2\{E - V(x)\}}}\, dx. \qquad (3)$$

Integration of (3) between $t = 0$ and t, and $x = x_0$ and x yields (1).

[‡]Use $\cos\theta \approx 1 - \frac{1}{2}\theta^2$.

(b) If $F(x)$ is constant, then $V = -Fx$ and (1) gives

$$t(x) = \sqrt{\frac{m}{2}} \int_{x_0}^{x} \frac{dx}{\sqrt{E+Fx}} = \frac{\sqrt{2m}}{F}\left(\sqrt{E+Fx} - \sqrt{E+Fx_0}\right). \tag{4}$$

In terms of the initial conditions the energy is

$$E = \frac{1}{2}mv_0^2 - Fx_0. \tag{5}$$

From (4) and (5) we have

$$x(t) = x_0 + v_0 t + Ft^2/2m, \tag{6}$$

which is the familiar solution to this simple problem (see Question 3.1).

Comment

In the step leading to (3), we have assumed a positive root; that is, we have ignored the possibility of a negative sign in (1). This negative sign is related to the time reversal $t \to -t$: for each trajectory $x(t)$, the time-reversed solution $x(-t)$ is also a possible trajectory.

Question 5.17

A particle of mass m acted upon by a one-dimensional force $F(x)$ performs periodic motion, oscillating between classical turning points x_1 and x_2. Show that the period of oscillation is

$$T = \int_{x_1}^{x_2} \sqrt{\frac{2m}{V(x_2) - V(x)}}\, dx, \tag{1}$$

where $V(x)$ is the potential energy.

Solution

At a classical turning point, $E = V(x)$. Therefore, $V(x_2) = E$, and (3) of the previous question yields

$$dt = \sqrt{\frac{m}{2\{V(x_2) - V(x)\}}}\, dx. \tag{2}$$

If the particle is at x_1 at time t_1 and x_2 at time t_2, integration of (2) gives

$$t_2 - t_1 = \int_{x_1}^{x_2} \sqrt{\frac{m}{2\{V(x_2) - V(x)\}}}\, dx. \tag{3}$$

The period is $T = 2(t_2 - t_1)$, and hence we obtain (1).

Question 5.18

For each of the potentials in Questions 5.13–5.15, use (1) of the previous question and a numerical integration to determine the period T of the oscillations as a function of the amplitude. Display the results graphically. (Hint: For Questions 5.13 and 5.15, use a suitable dimensionless amplitude.)

Solution

Anharmonic potential

We substitute the anharmonic potential $V(x) = \tfrac{1}{2}kx^2 - \tfrac{1}{4}bx^4$ of Question 5.13 into (1) of the previous question. Making the change of variable $u = x/x_2$ and putting $A = \sqrt{bx_2^2/k}$ results in the convenient form

$$T = \frac{2T_0}{\pi} \int_0^1 \frac{du}{\sqrt{1 - \tfrac{1}{2}A^2 - u^2 + \tfrac{1}{2}A^2 u^4}}, \tag{1}$$

where $T_0 = 2\pi\sqrt{m/k}$ is the period in the harmonic limit $b = 0$. Here A is a dimensionless amplitude for the anharmonic oscillations: because $x_2 < \sqrt{k/b}$ (see Question 5.13), it follows that $A < 1$. A numerical integration of (1) using the following *Mathematica* notebook yields the plot of T/T_0 versus A shown below.

```
In[1]:= Amin = 0; Amax = 0.995;

       f[A_] := Evaluate[Chop[Integrate[ (2/π) / Sqrt[1 - (1/2)A^2 - u^2 + (1/2)A^2 u^4], {u, 0, 1}]]];

       Plot[f[A], {A, Amin, Amax}, PlotRange → {{0, 1}, {0, 3}}]
```

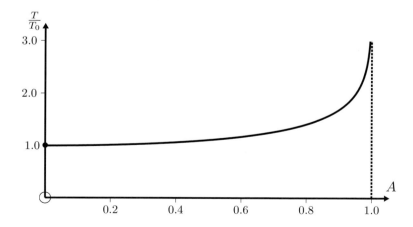

Pendulum potential

For the pendulum in Question 5.14, $V(\theta) = mg\ell(1 - \cos\theta)$ and $\theta = x/\ell$. Then,

$$T = \sqrt{\frac{8\ell}{g}} \int_0^{\theta_0} \frac{d\theta}{\sqrt{\cos\theta - \cos\theta_0}}, \tag{2}$$

where θ_0 is the angular amplitude of the oscillations. The integral in (2) is a complete elliptic integral of the first kind. The limiting value for small θ_0 is

$$T_0 = 4\sqrt{\frac{\ell}{g}} \int_0^{\theta_0} \frac{d\theta}{\sqrt{\theta_0^2 - \theta^2}} = 2\pi \sqrt{\frac{\ell}{g}}, \tag{3}$$

the familiar expression for the period of a pendulum performing small oscillations. In terms of T_0, (2) is

$$T = T_0 \frac{\sqrt{2}}{\pi} \int_0^{\theta_0} \frac{d\theta}{\sqrt{\cos\theta - \cos\theta_0}}, \tag{4}$$

$$= \frac{2T_0}{\pi \sin\frac{1}{2}\theta_0} \int_0^{\frac{1}{2}\theta_0} \frac{d\theta}{\sqrt{1 - \sin^2\theta / \sin^2\frac{1}{2}\theta_0}}. \tag{5}$$

Equation (5) can be expressed in terms of *Mathematica*'s EllipticF function. A numerical evaluation of T as a function of θ_0 can be done using the first cell of the following notebook, and it yields the graph shown below.

```
In[1]:= (* USE THIS CELL FOR THE PENDULUM POTENTIAL *)
        θmin = 0; θmax = 0.995 π;
        f[θ_] := Abs[ 2/(π Sin[θ/2]) EllipticF[θ/2, 1/(Sin[θ/2])^2 ]];
        Plot[Evaluate[f[θ] ], {θ, θmin, θmax}, PlotRange → {{0, π}, {0, 4}}]
        Clear[θ]

In[2]:= (* USE THIS CELL FOR THE OSCILLATORY POTENTIAL *)
        θmin = 0; θmax = 0.49995 π;
        g[θ_] := 1/Sin[θ] Abs[ 2/π EllipticF[θ, 1/Sin[θ]^2 ]];
        Plot[Evaluate[g[θ] ], {θ, θmin, θmax}, PlotRange → {{0, π/2}, {0, 3}}]
```

There has been interest in obtaining simple approximations to (4), such as[1]

$$T = -T_0 \frac{\ln(\cos\frac{1}{2}\theta_0)}{1 - \cos\frac{1}{2}\theta_0}. \tag{6}$$

Values of (6) are indicated by the dashed curve in the figure.

---

[1] F. M. S. Lima and P. Arun, "An accurate formula for the period of a simple pendulum oscillating beyond the small angle regime," American Journal of Physics, vol. 74, pp. 892–895, 2006.

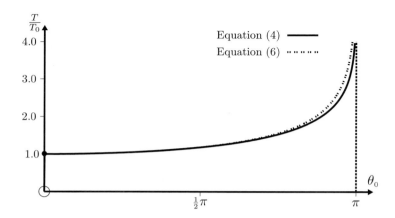

Oscillatory potential

For the potential in Question 5.15 we consider oscillations in the well centred on $x = \frac{1}{2}\lambda$. Shifting the origin of coordinates to $\frac{1}{2}\lambda$ by setting $x = X + \frac{1}{2}\lambda$ gives

$$V(X) = \frac{-b\lambda}{2\pi} \cos \frac{2\pi X}{\lambda} \qquad \left(-\tfrac{1}{2}\lambda < X < \tfrac{1}{2}\lambda\right). \qquad (7)$$

Substituting (7) into (1) of Question 5.17 yields

$$T = \frac{2T_0}{\pi \sin \theta_2} \int_0^{\theta_2} \frac{d\theta}{\sqrt{1 - \sin^2\theta / \sin^2\theta_2}}. \qquad (8)$$

Here, $T_0 = \sqrt{2\pi m \lambda / b}$ is the period of small (harmonic) oscillations, $\theta = \pi X/\lambda$ and so $\theta_2 = \pi X_2/\lambda$ with X_2 equal to the amplitude of the oscillations ($0 < X_2 < \tfrac{1}{2}\lambda$). Using cell 2 in the above notebook to evaluate (8) gives the following graph:

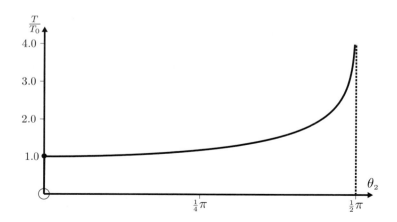

Question 5.19

Calculate the potential $V(r)$ for each of the following central, isotropic forces.

(a) $\mathbf{F} = -\dfrac{k}{r^2}\hat{\mathbf{r}}$, (b) $\mathbf{F} = \dfrac{V_0}{r^2}\left(1 + \dfrac{r}{\lambda}\right)e^{-r/\lambda}\,\hat{\mathbf{r}}$, (c) $\mathbf{F} = \left(\dfrac{A}{r^{\alpha+1}} - \dfrac{B}{r^{\beta+1}}\right)\hat{\mathbf{r}}$.

Here k, V_0, λ, A, B, α, and β are constants.

Solution

According to Question 5.3, a central, isotropic force $\mathbf{F} = F(r)\hat{\mathbf{r}}$ is conservative and the potential is given by

$$V(r) = -\int F(r)\,dr\,. \tag{1}$$

We apply this result to each of the given forces.

(a)
$$V(r) = k\int \frac{dr}{r^2} = -\frac{k}{r}\,. \tag{2}$$

(b)
$$V(r) = -V_0 \int \frac{1}{r^2}\left(1 + \frac{r}{\lambda}\right)e^{-r/\lambda}\,dr$$
$$= V_0 \int \left(\frac{d}{dr}\frac{e^{-r/\lambda}}{r}\right)dr = \frac{V_0}{r}e^{-r/\lambda}\,. \tag{3}$$

(c)
$$V(r) = -\int \left(\frac{A}{r^{\alpha+1}} - \frac{B}{r^{\beta+1}}\right)dr = \frac{C}{r^{\alpha}} - \frac{D}{r^{\beta}}\,, \tag{4}$$

where $C = A/\alpha$ and $D = B/\beta$. In (2)–(4) we have chosen the zero of potential at $r = \infty$. In (4) we have supposed that $\alpha, \beta > 0$.

Comments

Equations (2)–(4) represent famous potentials:

(i) The potential (2) is often referred to as the Coulomb potential. With a suitable choice of the constant k it represents either the gravitational or the electrostatic potential for the interaction of two particles:

$$V(r) = -G\frac{m_1 m_2}{r} \quad\text{or}\quad V(r) = \frac{1}{4\pi\epsilon_0}\frac{q_1 q_2}{r}\,. \tag{5}$$

(ii) The potential in (3) is known as the Yukawa potential because it occurs in Yukawa's theory of the nuclear force between two nucleons. In this theory the force is due to the exchange of a particle (a pion) between interacting nucleons, and the constant λ is positive and inversely proportional to the mass m of the exchanged particle. The value of λ is finite and it defines the range of the force. In the limit $m \to 0$, that is $\lambda \to \infty$, the Yukawa potential (3) becomes the Coulomb potential (2). The latter is therefore a potential with infinite range. It is believed that in gravity and electromagnetism the relevant exchanged particles (the graviton and photon, respectively) are massless, and consequently it is the Coulomb potential, and not the Yukawa potential, that applies in these cases.

(iii) With suitable positive values of α and β, (4) is known as a Lennard-Jones potential, and it is used to describe molecular interactions. For example, with $\alpha = 12$ and $\beta = 6$ one has the Lennard-Jones 6-12 potential. If C and D are positive, there is a point of stable equilibrium at $r_e = (2C/D)^{1/6}$. This could represent a molecular bond length with dissociation energy equal to $-V_{\min} = D^2/4C$.

(iv) Energy diagrams can also be used to analyze motion in a spherically symmetric potential $V(r)$, provided care is taken with the contribution of the kinetic energy, see Chapter 8.

(v) The potentials (2)–(4) are illustrated in the figures below. In the Lennard-Jones potential, the force is repulsive if $r < r_e$ and attractive if $r > r_e$.

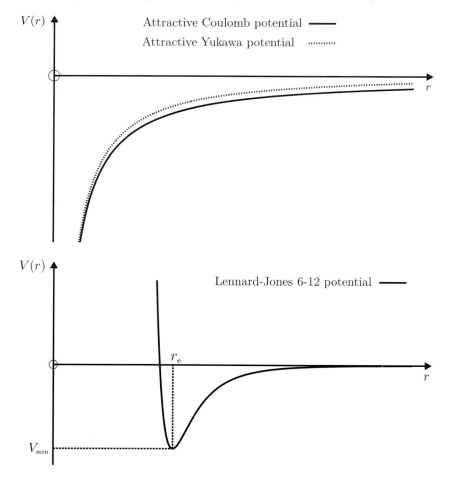

Question 5.20

Prove that the force
$$\mathbf{F} = -(k_1 x,\ k_2 y,\ k_3 z), \tag{1}$$
where the k_i are constants, is conservative and determine the potential-energy function.

Solution

$$\nabla \times \mathbf{F} = -\begin{vmatrix} \hat{\mathbf{x}} & \hat{\mathbf{y}} & \hat{\mathbf{z}} \\ \dfrac{\partial}{\partial x} & \dfrac{\partial}{\partial y} & \dfrac{\partial}{\partial z} \\ k_1 x & k_2 y & k_3 z \end{vmatrix}$$

$$= -\left(k_3 \frac{\partial z}{\partial y} - k_2 \frac{\partial y}{\partial z},\ k_1 \frac{\partial x}{\partial z} - k_3 \frac{\partial z}{\partial x},\ k_2 \frac{\partial y}{\partial x} - k_1 \frac{\partial x}{\partial y} \right)$$

$$= 0.$$

Thus, $\mathbf{F}$ is conservative and there exists a scalar function $V(\mathbf{r})$ such that

$$\begin{aligned} V(\mathbf{r}) &= -\int \mathbf{F} \cdot d\mathbf{r} = k_1 \int x\,dx + k_2 \int y\,dy + k_3 \int z\,dz \\ &= \tfrac{1}{2} k_1 x^2 + \tfrac{1}{2} k_2 y^2 + \tfrac{1}{2} k_3 z^2. \end{aligned} \tag{2}$$

Here, we have chosen the zero of potential energy at the origin.

Comments

(i) If the k_i are positive then the linear force (1) is a restoring force (a Hooke's-law-type force) and the system is known as a three-dimensional anisotropic harmonic oscillator. If the k_i are all equal it is a three-dimensional isotropic harmonic oscillator with a spherically symmetric potential

$$V(r) = \tfrac{1}{2} k r^2. \tag{3}$$

These important systems are discussed in Chapters 7 and 8.

(ii) The above analysis can clearly be extended to any force of the separable type:

$$\mathbf{F} = \big(F_1(x),\ F_2(y),\ F_3(z) \big), \tag{4}$$

and it yields

$$V(\mathbf{r}) = V_1(x) + V_2(y) + V_3(z), \tag{5}$$

where

$$V_1 = -\int F_1(x)\,dx, \qquad V_2 = -\int F_2(y)\,dy, \qquad V_3 = -\int F_3(z)\,dz. \tag{6}$$

Question 5.21

Show that the force $\mathbf{F} = -k(yz, xz, xy)$, where k is a constant, is conservative and determine the potential-energy function.

Solution

$$\nabla \times \mathbf{F} = -k \begin{vmatrix} \hat{\mathbf{x}} & \hat{\mathbf{y}} & \hat{\mathbf{z}} \\ \dfrac{\partial}{\partial x} & \dfrac{\partial}{\partial y} & \dfrac{\partial}{\partial z} \\ yz & xz & xy \end{vmatrix}$$

$$= -k\left(\frac{\partial}{\partial y}xy - \frac{\partial}{\partial z}xz,\ \frac{\partial}{\partial z}yz - \frac{\partial}{\partial x}xy,\ \frac{\partial}{\partial x}xz - \frac{\partial}{\partial y}yz\right)$$

$$= 0.$$

Consequently, $\mathbf{F}$ is conservative and there exists a scalar $V(\mathbf{r})$ such that $\mathbf{F} = -\nabla V$. Thus

$$\frac{\partial V}{\partial x} = kyz, \qquad \frac{\partial V}{\partial y} = kxz, \qquad \frac{\partial V}{\partial z} = kxy. \qquad (1)$$

To solve these equations for $V(x, y, z)$ we integrate each in turn. From $(1)_1$ we obtain

$$V(x, y, z) = kxyz + f(y, z), \qquad (2)$$

where $f(y, z)$ is to be determined. From (2) and $(1)_2$ we have

$$\frac{\partial f(y, z)}{\partial y} = 0, \qquad (3)$$

and hence

$$f(y, z) = g(z), \qquad (4)$$

where $g(z)$ is to be determined. From $(1)_3$, (2) and (4) we have

$$\frac{dg(z)}{dz} = 0. \qquad (5)$$

Thus, $g(z)$ is a constant, which can be set equal to zero. Equations (2) and (4) yield the potential

$$V(x, y, z) = kxyz. \qquad (6)$$

Comment

It is useful to check that the given force is obtained from the negative gradient of the potential-energy function that one calculates. In the above example we see, by inspection, that the negative gradient of (6) is $-k(yz, xz, xy)$.

Question 5.22

Prove that a necessary condition for a vector field $\mathbf{F}(\mathbf{r})$ to be derivable from a vector potential $\mathbf{A}(\mathbf{r})$ is
$$\nabla \cdot \mathbf{F} = 0. \tag{1}$$

That is, prove
$$\mathbf{F} = \nabla \times \mathbf{A}(\mathbf{r}) \Longrightarrow \nabla \cdot \mathbf{F} = 0, \tag{2}$$
where the arrow $\Longrightarrow$ means 'implies'. Do this in two ways:

(a) By using the Cartesian form of ∇.
(b) By applying Stokes's theorem (see Question 5.7) and Gauss's theorem:
$$\oint_S \mathbf{F} \cdot d\mathbf{S} = \int_V \nabla \cdot \mathbf{F}\, dV. \tag{3}$$

Here, S is a surface enclosing a volume V, and $d\mathbf{S}$ and dV are infinitesimal elements of S and V: the direction of $d\mathbf{S}$ at each point of the surface is along the outward normal to S.

Solution

(a) If $\mathbf{F} = \nabla \times \mathbf{A}$, then in Cartesian coordinates

$$\nabla \cdot \mathbf{F} = \nabla \cdot \left(\frac{\partial A_z}{\partial y} - \frac{\partial A_y}{\partial z},\ \frac{\partial A_x}{\partial z} - \frac{\partial A_z}{\partial x},\ \frac{\partial A_y}{\partial x} - \frac{\partial A_x}{\partial y} \right)$$

$$= \frac{\partial^2 A_z}{\partial x \partial y} - \frac{\partial^2 A_y}{\partial x \partial z} + \frac{\partial^2 A_x}{\partial y \partial z} - \frac{\partial^2 A_z}{\partial y \partial x} + \frac{\partial^2 A_y}{\partial z \partial x} - \frac{\partial^2 A_x}{\partial z \partial y}$$

$$= 0$$

if the order of the partial derivatives can be interchanged (that is, if the second partial derivatives of the components A_i are continuous functions).

(b) Divide a closed surface S into two 'caps' S_1 and S_2 bounded by a common closed curve C, as shown in the next figure. According to Stokes's theorem:

$$\int_{S_1} \mathbf{F} \cdot d\mathbf{S}_1 = \int_{S_1} (\nabla \times \mathbf{A}) \cdot d\mathbf{S}_1 = \oint_C \mathbf{A} \cdot d\mathbf{r} = \int_{S_2} (\nabla \times \mathbf{A}) \cdot d\mathbf{S}_2 = \int_{S_2} \mathbf{F} \cdot d\mathbf{S}_2,$$

where the sense in which C is traversed and the directions of $d\mathbf{S}_1$ and $d\mathbf{S}_2$ are fixed by the right-hand rule. Therefore[‡]

$$\oint_S \mathbf{F} \cdot d\mathbf{S} = \int_{S_1} \mathbf{F} \cdot d\mathbf{S}_1 + \int_{S_2} \mathbf{F} \cdot (-d\mathbf{S}_2) = 0,$$

[‡]Note that $d\mathbf{S}_2$ is along an inward normal, as shown in the figure, and therefore the element to be used in Gauss's theorem is $-d\mathbf{S}_2$.

which, by Gauss's theorem (3) means that

$$\int_V \nabla \cdot \mathbf{F}\, dV = 0.$$

Because S_1 and S_2 are arbitrary, so is the volume V that they enclose. It follows that $\nabla \cdot \mathbf{F} = 0$.

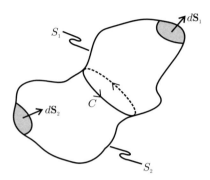

Question 5.23

(a) Use Gauss's theorem and Stokes's theorem to prove that $\nabla \cdot \mathbf{F}(\mathbf{r}) = 0$ is a sufficient condition for $\mathbf{F}(\mathbf{r})$ to be derivable from a vector potential $\mathbf{A}(\mathbf{r})$. That is, show that

$$\nabla \cdot \mathbf{F} = 0 \Longrightarrow \mathbf{F} = \nabla \times \mathbf{A}(\mathbf{r}). \tag{1}$$

(b) Also, obtain a formula for $\mathbf{A}$ in terms of integrals of the components of $\mathbf{F}$. (Hint: Show that for this purpose it is sufficient to consider a two-dimensional form such as $\mathbf{A} = (A_x, A_y, 0)$.)

Solution

(a) The initial part of the proof involves the inverse of the reasoning used in part (b) of Question 5.22. If $\nabla \cdot \mathbf{F} = 0$ everywhere then it follows from Gauss's theorem that

$$\oint_S \mathbf{F} \cdot d\mathbf{S} = 0, \tag{2}$$

for all closed surfaces S. That is, for 'caps' S_1 and S_2 that share a common bounding curve C, as depicted in the above figure, we have

$$\int_{S_2} \mathbf{F} \cdot d\mathbf{S}_2 = \int_{S_1} \mathbf{F} \cdot d\mathbf{S}_1, \tag{3}$$

meaning that the flux of $\mathbf{F}$ through a cap is unchanged by any deformation of the cap that leaves the bounding curve C unaltered. Therefore, the fluxes in (3) can depend only on the curve C and not on other details of S_1 and S_2: they can be expressed as the line integral around C of some vector field $\mathbf{A}(\mathbf{r})$:

$$\int_{S_i} \mathbf{F} \cdot d\mathbf{S}_i = \oint_C \mathbf{A} \cdot d\mathbf{r} \qquad (i = 1, 2) \tag{4}$$

$$= \int_{S_i} (\nabla \times \mathbf{A}) \cdot d\mathbf{S}_i \qquad (i = 1, 2), \tag{5}$$

where the last step relies on Stokes's theorem. The surfaces S_i in (5) are arbitrary and therefore

$$\mathbf{F} = \nabla \times \mathbf{A}. \tag{6}$$

(b) It is clear that the vector field $\mathbf{A}$ introduced in (4) is not unique: the change

$$\mathbf{A} \to \mathbf{A} + \nabla \chi, \tag{7}$$

where $\chi(\mathbf{r})$ is an arbitrary single-valued function, leaves the line integral in (4) unchanged because

$$\oint_C \nabla \chi \cdot d\mathbf{r} = \oint_C d\chi = 0. \tag{8}$$

Equation (7) can be used to transform any three-dimensional vector potential to a two-dimensional form: for example, the choice

$$\chi = -\int A_z \, dz \tag{9}$$

removes the z-component of $\mathbf{A}$. This enables one to obtain a simple formula for $\mathbf{A}$ as follows. With $\mathbf{A} = (A_x, A_y, 0)$ the components of (6) are

$$F_x = -\frac{\partial A_y}{\partial z}, \qquad F_y = \frac{\partial A_x}{\partial z}, \qquad F_z = \frac{\partial A_y}{\partial x} - \frac{\partial A_x}{\partial y}. \tag{10}$$

Integration of $(10)_1$ and $(10)_2$ gives

$$A_x = \int_{z_0}^{z} F_y \, dz + f(x, y) \quad \text{and} \quad A_y = -\int_{z_0}^{z} F_x \, dz + g(x, y), \tag{11}$$

where f and g are arbitrary functions of x and y, and z_0 is a constant. Then, $(10)_3$ becomes

$$F_z(x, y, z) = -\int_{z_0}^{z} \left(\frac{\partial F_x}{\partial x} + \frac{\partial F_y}{\partial y} \right) dz - \frac{\partial f}{\partial y} + \frac{\partial g}{\partial x}. \tag{12}$$

The solenoidal property $\nabla \cdot \mathbf{F} = 0$ means that the integrand in (12) is equal to $-\partial F_z / \partial z$, and therefore (12) yields

$$\frac{\partial f}{\partial y} - \frac{\partial g}{\partial x} = -F_z(x, y, z_0). \tag{13}$$

If we set the arbitrary function f equal to zero, then

$$g = \int_{x_0}^{x} F_z(x, y, z_0)\, dx .\tag{14}$$

From (11) and (14) we obtain the desired formula

$$\mathbf{A} = \left(\int_{z_0}^{z} F_y(x, y, z)\, dz ,\ -\int_{z_0}^{z} F_x(x, y, z)\, dz + \int_{x_0}^{x} F_z(x, y, z_0)\, dx ,\ 0 \right).\tag{15}$$

The reader can readily check that, provided $\boldsymbol{\nabla} \cdot \mathbf{F} = 0$, the curl of (15) yields $\mathbf{F}(x, y, z)$. Having found one vector potential, such as (15), we can generate an infinite number of them via the transformation (7).

Comments

(i) In Questions 5.7, 5.8, 5.22 and 5.23 we have proved the following pair of equivalences for a vector field[‡] $\mathbf{F}(\mathbf{r})$:

$$\boldsymbol{\nabla} \times \mathbf{F} = 0 \iff \mathbf{F} = -\boldsymbol{\nabla} V(\mathbf{r}) \tag{16}$$

$$\boldsymbol{\nabla} \cdot \mathbf{F} = 0 \iff \mathbf{F} = \boldsymbol{\nabla} \times \mathbf{A}(\mathbf{r}) .\tag{17}$$

If $\boldsymbol{\nabla} \times \mathbf{F} = 0$ everywhere then $\mathbf{F}$ is called irrotational; if $\boldsymbol{\nabla} \cdot \mathbf{F} = 0$ everywhere then $\mathbf{F}$ is called solenoidal. Thus, an irrotational field is always derivable from a scalar potential, and a solenoidal field is always derivable from a vector potential.

(ii) The reader may find it instructive to compare the proofs of the existence of $V(\mathbf{r})$, see Question 5.8, and $\mathbf{A}(\mathbf{r})$, see above. The former follows from the path independence of a line integral of $\mathbf{F}$ between two points (the integral depends only on the endpoints); the latter follows from the independence of a surface integral of $\mathbf{F}$ on the details of the surface (the integral depends only on the curve bounding the surface).

(iii) The results (16) and (17) have important applications in the theories of mechanics, electrodynamics, hydrodynamics, and elasticity.

(iv) In Question 5.8 we have already mentioned the role of (16) in the theory of conservative forces, such as the electrostatic force $\mathbf{F} = q\mathbf{E}(\mathbf{r})$. For time-dependent forces associated with the electric and magnetic fields $\mathbf{E}(\mathbf{r}, t)$ and $\mathbf{B}(\mathbf{r}, t)$, (16) and (17) are used as follows. A fundamental property of the magnetic field is its solenoidal nature $\boldsymbol{\nabla} \cdot \mathbf{B} = 0$ (Gauss's law for magnetic fields). Consequently, $\mathbf{B}$ is derivable from a vector potential: $\mathbf{B} = \boldsymbol{\nabla} \times \mathbf{A}(\mathbf{r}, t)$. This, together with Faraday's law $\boldsymbol{\nabla} \times \mathbf{E} = -\partial \mathbf{B}/\partial t$, implies that the vector $\mathbf{E} + \partial \mathbf{A}/\partial t$ is irrotational and may therefore be derived from a scalar potential $\phi(\mathbf{r}, t)$:

$$\mathbf{E} = -\boldsymbol{\nabla} \phi(\mathbf{r}, t) - \frac{\partial \mathbf{A}(\mathbf{r}, t)}{\partial t} .\tag{18}$$

[‡] $\mathbf{F}$ could represent a force field, a velocity field in a fluid, an electric field, a magnetic field, etc.

Thus, the electric and magnetic forces ($\mathbf{F}_E = q\mathbf{E}$ and $\mathbf{F}_B = q\mathbf{v} \times \mathbf{B}$) exerted on a charge q are given in terms of a scalar potential $\phi(\mathbf{r}, t)$ and a vector potential $\mathbf{A}(\mathbf{r}, t)$ by

$$\mathbf{F}_E = -q\left(\boldsymbol{\nabla}\phi + \frac{\partial \mathbf{A}}{\partial t}\right), \qquad \mathbf{F}_B = q\mathbf{v} \times (\boldsymbol{\nabla} \times \mathbf{A}). \qquad (19)$$

(v) Forces such as the electrostatic force $-q\boldsymbol{\nabla}\phi(\mathbf{r})$ possess an obvious invariance: they are unaffected by the addition of an arbitrary constant to ϕ. This invariance is known as a 'global' gauge invariance (global because ϕ is changed by the same amount at all spatial points and instants of time).

(vi) The time-dependent forces (19) possess a more elaborate invariance: they are unaffected by the simultaneous replacements

$$\mathbf{A} \to \mathbf{A} + \boldsymbol{\nabla}\chi(\mathbf{r},t) \qquad \text{and} \qquad \phi \to \phi - \frac{\partial \chi}{\partial t}, \qquad (20)$$

where $\chi(\mathbf{r}, t)$ is an arbitrary[†] scalar function. This invariance – which is known as a 'local gauge invariance' ('local' because the potentials can be changed by different amounts at different spatial and temporal points) – is a key ingredient in a theory of fundamental interactions.[2,3]

Question 5.24

(a) Use the formula (15) of the previous question to find a vector potential for a uniform, static field $\mathbf{F} = (0, 0, F)$, where F is a constant.
(b) Determine the effect of the transformation $\mathbf{A} \to \mathbf{A} + \boldsymbol{\nabla}\chi$ for $\chi = F_1 xy$, where F_1 is a constant.

Solution

(a) Equation (15) yields

$$\mathbf{A} = \left(0, \int_{x_0}^{x} F\, dx, 0\right) = (0, Fx, 0), \qquad (1)$$

where we have omitted a constant term $-Fx_0 \hat{\mathbf{y}}$.

(b) The gradient of $\chi = F_1 xy$ is $\boldsymbol{\nabla}\chi = (F_1 y, F_1 x, 0)$. If this is added to (1) we obtain a second, transformed vector potential

$$\mathbf{A} = (F_1 y, Fx + F_1 x, 0). \qquad (2)$$

Note that the choice $F_1 = -\tfrac{1}{2}F$ in (2) gives

$$\mathbf{A} = \left(-\tfrac{1}{2}Fy, \tfrac{1}{2}Fx, 0\right) = \tfrac{1}{2}\mathbf{F} \times \mathbf{r}. \qquad (3)$$

[†]It should be single-valued if the flux of $\mathbf{B}$ is to be unaffected by $(20)_1$.

---

[2] See, for example, G. t' Hooft, "Gauge theories of the forces between elementary particles," Scientific American, vol. 242, pp. 90–116, June 1980.
[3] L. O'Raifeartaigh, *The dawning of gauge theory*. Princeton: Princeton University Press, 1997.

Comments

(i) Equations (1) and (3) are often used for the vector potential of a uniform magnetostatic field, and they find application in the quantum theory of a charged particle in such a field.[4]

(ii) The vector potential is an intriguing quantity. Consider the magnetic field **B** of an ideal solenoid: this field is zero outside the solenoid and uniform inside. Therefore, charged particles moving outside the solenoid should be unaffected when the field **B** is turned on. However, when such a solenoid[‡] is placed between the slits of a double-slit experiment using electrons (with the axis of the solenoid parallel to the slits), it is observed that the interference pattern shifts when **B** is turned on. This is an example of the well-known Aharonov–Bohm effect[5] and its explanation relies on the fact that the vector potential **A** is not zero outside the solenoid, as is evident from the fact that the circulation of **A** around a closed curve encircling the solenoid is equal to the flux of **B** through the solenoid. According to quantum mechanics, the phase of the electronic wavefunction depends on the line integral of **A**, and it is this that accounts for the observed effect on the interference pattern.[5] A similar quantum-mechanical effect exists for the electric field **E** and the associated scalar potential ϕ. Thus, there are effects on charged particles moving in field-free regions.

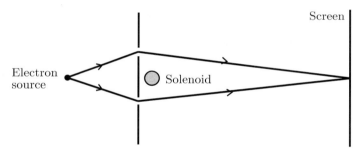

Question 5.25

Use (15) of Question 5.23 to obtain a vector potential

$$\mathbf{A} = (Fxz,\ Fyz,\ 0), \qquad (1)$$

for the field $\mathbf{F} = (-y,\ x,\ 0)F$, where F is a constant. What scalar χ in $\mathbf{A} \to \mathbf{A} + \nabla\chi$ will change (1) into

$$\mathbf{A} = (Fxz + Fyz,\ Fxz + Fyz,\ Fxy)? \qquad (2)$$

[‡] In practice, a magnetized iron 'whisker' (about $1\,\mu\mathrm{m}$ in diameter) is used.

---

[4] See, for example, O. L. de Lange and R. E. Raab, *Operator methods in quantum mechanics*. Oxford: Clarendon Press, 1991.

[5] See, for example, M. Peshkin and A. Tonomura, *The Aharonov–Bohm effect*. Berlin: Springer, 1989.

Solution

Equation (15) gives $\mathbf{A} = \left(\int_{z_0}^{z} Fx\, dz,\ \int_{z_0}^{z} Fy\, dz,\ 0\right)$ which, apart from a constant, is (1). By inspection, $\chi = Fxyz$ changes (1) into (2).

Question 5.26

Consider the non-uniform field

$$\mathbf{B} = (0,\ -\alpha y,\ 1 + \alpha z)B, \tag{1}$$

where α and B are positive constants.

(a) State why $\mathbf{B}$ is derivable from a scalar potential $\psi(y, z)$ and show that

$$\psi(y, z) = \{z + \tfrac{1}{2}\alpha(z^2 - y^2)\}B. \tag{2}$$

(b) State why $\mathbf{B}$ is derivable from a vector potential $\mathbf{A}$ and construct an expression for $\mathbf{A}$.

(c) Show that the field lines of B in the yz-plane are given by

$$y = \frac{C}{1 + \alpha z}, \tag{3}$$

where C is a constant. For this purpose, use the following argument: the field lines of $\mathbf{B}$ (given by $\phi(y, z) =$ constant) are orthogonal to the equipotential curves ($\psi(y, z) =$ constant), and therefore the functions ϕ and ψ must satisfy the Cauchy–Riemann equations[6]

$$\frac{\partial \phi}{\partial y} = \frac{\partial \psi}{\partial z},\qquad \frac{\partial \phi}{\partial z} = -\frac{\partial \psi}{\partial y}. \tag{4}$$

(d) Sketch the field lines and equipotential curves in the yz-plane for $\alpha = 0.1\,\mathrm{m}^{-1}$ and $B = 1.0\,\mathrm{T}$.

Solution

(a) The field (1) is irrotational ($\boldsymbol{\nabla} \times \mathbf{B} = 0$) and therefore, according to Question 5.8, there exists a scalar function $\psi(y, z)$ such that

$$\mathbf{B} = \boldsymbol{\nabla}\psi. \tag{5}$$

The components of (5) are

$$\frac{\partial \psi}{\partial y} = -\alpha y B \quad \text{and} \quad \frac{\partial \psi}{\partial z} = (1 + \alpha z)B. \tag{6}$$

Integration of $(6)_1$ shows that $\psi = -\tfrac{1}{2}\alpha y^2 B + f(z)$, and then $(6)_2$ requires $df(z)/dz = (1 + \alpha z)B$. Thus, we obtain the result (2).

---

[6] See, for example, J. Irving and N. Mullineux, *Mathematics in physics and engineering*. New York: Academic Press, 1959. Chap. VIII.

124 *Solved Problems in Classical Mechanics*

(b) The field (1) is solenoidal ($\nabla \cdot \mathbf{B} = 0$) and therefore, according to Question 5.23, there exists a vector function $\mathbf{A}(\mathbf{r})$ such that $\mathbf{B} = \nabla \times \mathbf{A}$. An expression for $\mathbf{A}$ can be found by substituting (1) in (15) of Question 5.23. Then, with $z_o = 0$,

$$\mathbf{A} = (-\alpha yz, \ x, \ 0)B. \tag{7}$$

$\mathbf{A}$ can be transformed, using the gauge function $\chi = -xyB$, to align it along the x-axis:

$$\mathbf{A} \to \mathbf{A} + \nabla\chi = -(y + \alpha yz, \ 0, \ 0)B. \tag{8}$$

(c) According to (4) and (6) we have

$$\frac{\partial \phi}{\partial y} = (1 + \alpha z)B \quad \text{and} \quad \frac{\partial \phi}{\partial z} = \alpha y B. \tag{9}$$

Therefore, $\phi(y,z) = y(1+\alpha z)B$, and the field lines $\phi =$ constant are given by (3).

(d) In the diagram, the solid curves are the field lines (3) for the values of C shown. The dotted curves are the equipotentials $z + \frac{1}{2}\alpha(z^2 - y^2) = D$ (see (2)): the values of D were selected to produce a set of evenly spaced equipotentials. The scale for both axes is $[-4.2, 4.2]$ with y and z in units of α^{-1}.

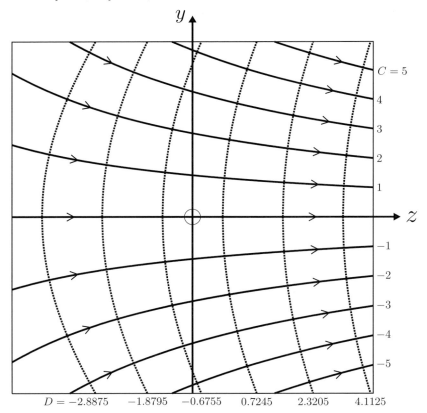

Comments

(i) Equation (3) can also be deduced as follows. Along a field line $\phi(y,z)$ is constant and therefore
$$\frac{\partial \phi}{\partial y} dy + \frac{\partial \phi}{\partial z} dz = 0. \tag{10}$$
Also, along such a line the slope of the tangent dy/dz is equal to $B_y/B_z = -\alpha y/(1+\alpha z)$. So,
$$(1+\alpha z)dy + \alpha y\, dz = 0. \tag{11}$$
From (10) and (11), $\phi \sim (1+\alpha z)y$ and (3) follows.

(ii) It is interesting to consider the solenoidal, irrotational field (1) in relation to the well-known Helmholtz theorem.[7] According to this theorem, a vector field which vanishes suitably at infinity is specified uniquely in terms of its divergence and curl, and all solenoidal, irrotational fields are zero. There is no contradiction here because (1) does not satisfy the boundary condition at infinity. (A field which is confined to a finite region of space is determined also by its normal component on the boundary.)

(iii) The motion of a charged particle in the magnetic field (1) is of interest. It is analyzed in Question 7.23.

Question 5.27

Consider the vector field
$$\mathbf{F} = \begin{cases} \dfrac{k}{a^2}(-y,\ x,\ 0) & \text{if } \sqrt{x^2+y^2} \leq a, \\ \dfrac{k}{x^2+y^2}(-y,\ x,\ 0) & \text{if } \sqrt{x^2+y^2} > a, \end{cases} \tag{1}$$
where k and a are constants. Discuss the representation of $\mathbf{F}$ in terms of (a) a vector potential $\mathbf{A}(\mathbf{r})$, and (b) a scalar potential $V(\mathbf{r})$.

Solution

(a) It is easily shown that $\nabla \cdot \mathbf{F} = 0$ everywhere and therefore $\mathbf{F}$ is solenoidal. It follows from Question 5.23 that $\mathbf{F}$ can then be derived from a vector potential $\mathbf{A}(\mathbf{r})$ according to $\mathbf{F} = \nabla \times \mathbf{A}$. An explicit expression for $\mathbf{A}$ can be obtained by substituting (1) in (15) of Question 5.23:
$$\mathbf{A}(\mathbf{r}) = \begin{cases} \dfrac{k}{a^2}(xz,\ yz,\ 0) & \text{if } \sqrt{x^2+y^2} \leq a, \\ \dfrac{k}{x^2+y^2}(xz,\ yz,\ 0) & \text{if } \sqrt{x^2+y^2} > a. \end{cases} \tag{2}$$

---

[7] See, for example, P. Morse and H. Feshbach, *Methods of theoretical physics*, Part I. New York: McGraw-Hill, 1953. Ch. I.

(b) A short calculation shows that

$$\nabla \times \mathbf{F} = \begin{cases} \dfrac{2k}{a^2}(0,\,0,\,1) & \text{if } \sqrt{x^2+y^2} \leq a\,, \\ 0 & \text{if } \sqrt{x^2+y^2} > a\,. \end{cases} \qquad (3)$$

Thus, $\mathbf{F}$ is not irrotational and it cannot be expressed everywhere in terms of a scalar potential $V(\mathbf{r})$. However, $\mathbf{F}$ is irrotational in all space external to an infinite cylinder of radius a centred on the z-axis. So for $\sqrt{x^2+y^2} > a$, $\mathbf{F} = \nabla V(\mathbf{r})$ and therefore

$$\frac{\partial V}{\partial x} = -\frac{ky}{x^2+y^2}\,, \qquad \frac{\partial V}{\partial y} = \frac{kx}{x^2+y^2}\,.$$

That is,

$$V(\mathbf{r}) = -k\tan^{-1}(x/y)\,. \qquad (4)$$

Comments

(i) Since $\nabla \times \mathbf{F} \neq 0$ everywhere, $\mathbf{F}$ is not conservative: for $\sqrt{x^2+y^2} \leq a$, a scalar potential $V(\mathbf{r})$ does not exist. By Stokes's theorem and (3),

$$\oint_C \mathbf{F} \cdot d\mathbf{r} = 2\pi k \neq 0\,, \qquad (5)$$

for any closed curve C lying outside the above cylinder and encircling the z-axis; this is associated with the fact that the potential (4) is not single-valued. These remarks apply even in the limit $a \to 0$, when $\mathbf{F}$ is irrotational everywhere except on the z-axis, where $\mathbf{F}$ is singular.

(ii) The above results have application in the theory of the magnetostatic field of an infinitely long, straight, current-carrying wire having radius a. With $k = \mu_0 I/2\pi$, and in cylindrical coordinates, we recognize (1) as the magnetostatic field

$$\mathbf{B}(\mathbf{r}) = \begin{cases} \dfrac{\mu_0}{2\pi}\dfrac{I}{a}\dfrac{\rho}{a}\hat{\phi} & \text{if } \rho \leq a\,, \\ \dfrac{\mu_0}{2\pi}\dfrac{I}{\rho}\hat{\phi} & \text{if } \rho > a\,. \end{cases} \qquad (6)$$

(Equations (3) and (5) are now just the differential and integral forms of Ampère's law.) Therefore, such a field can be represented by the vector potential (2) everywhere, or by a magnetic scalar potential (4) in the region $\rho > a$ external to the wire.

6

Momentum and angular momentum

For a particle there are two fundamental dynamical quantities that can be constructed from its mass m, position vector $\mathbf{r}$ and velocity $\mathbf{v}$. They are the momentum $\mathbf{p} = m\mathbf{v}$ and angular momentum $\mathbf{L} = \mathbf{r} \times \mathbf{p}$. This chapter contains various questions dealing with these quantities for a particle and also for simple objects such as a uniform sphere. Further questions can be found in later chapters, including the extension to systems of interacting particles and rigid bodies.

Question 6.1

What can be stated regarding the momentum of a free (isolated) particle relative to an inertial frame of reference?

Solution

In an inertial frame $d\mathbf{p}/dt = \mathbf{F}$, and for a free particle $\mathbf{F} = 0$. Therefore

$$d\mathbf{p}/dt = 0 \text{ and so } \mathbf{p} = \text{constant}. \tag{1}$$

Comments

(i) Equation (1) is the simplest form of a fundamental law of nature, namely the law of conservation of momentum.
(ii) In general, if $F_i = 0$ then p_i is constant. That is, for each component of the force that is zero, the corresponding component of the momentum is conserved. For example, when a charged particle moves in the electric field of a parallel-plate capacitor, the two components of the momentum parallel to the plates are conserved.
(iii) Within the framework of Newtonian dynamics, (1) is an obvious consequence of the frame being inertial. It is less evident that the law applies also to the total momentum of any collection of interacting particles for which the total external force is zero (see Question 10.2), a result that is assumed in some of the following questions.
(iv) Conservation of momentum of an isolated system is associated with invariance (or symmetry) of such a system under spatial translation in inertial space (see

Question 14.7). It is an example of a more general result known as Noether's theorem, according to which each invariance of a system implies the existence of a corresponding conserved quantity.
(v) Conservation of momentum of an isolated system holds even in processes for which Newtonian mechanics fails. There are no known violations of the law of conservation of momentum.

Question 6.2

A rocket of mass $m_1 + m_2$ is launched with a velocity whose horizontal and vertical components are u_x and u_y. At the highest point in its path the rocket explodes into two parts of mass m_1 and m_2 that separate in a horizontal direction in the original plane of motion. Show that the fragments strike the ground at a distance apart given by

$$D = \frac{u_y}{g}\sqrt{\frac{2(m_1 + m_2)K}{m_1 m_2}}, \tag{1}$$

where K is the kinetic energy produced by the explosion. (Neglect air resistance, the mass of the explosive, any spinning motion of the fragments, and assume g is constant.)

Solution

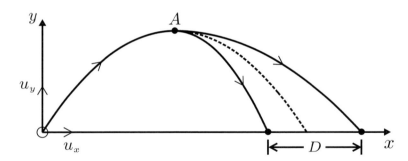

Let v_1 and v_2 be the horizontal components of the velocities of the fragments after the explosion (v_1 and v_2 are constants). Then

$$D = |v_2 - v_1|t, \tag{2}$$

where $t =$ the time for the fragments to reach the ground from the apex A $=$ the time for the rocket to reach A from O. Now $u_y - gt = 0$ and so

$$t = u_y/g. \tag{3}$$

To calculate $v_2 - v_1$, use conservation of momentum and energy at A:

$$m_1 v_1 + m_2 v_2 = (m_1 + m_2)u_x \tag{4}$$

$$\tfrac{1}{2}m_1 v_1^2 + \tfrac{1}{2}m_2 v_2^2 = \tfrac{1}{2}(m_1 + m_2)u_x^2 + K. \tag{5}$$

Then, use (4) to eliminate u_x from (5):

$$\tfrac{1}{2}m_1v_1^2 + \tfrac{1}{2}m_2v_2^2 = \tfrac{1}{2}(m_1+m_2)\left(\frac{m_1v_1+m_2v_2}{m_1+m_2}\right)^2 + K,$$

which simplifies to

$$v_2^2 - 2v_1v_2 + v_1^2 = 2\frac{m_1+m_2}{m_1m_2}K. \tag{6}$$

So

$$|v_2 - v_1| = \sqrt{\frac{2(m_1+m_2)K}{m_1m_2}}. \tag{7}$$

Equations (2), (3) and (7) yield (1).

Comments

(i) This calculation can be simplified by using an inertial frame moving horizontally with speed u_x. Relative to this frame, (4) and (5) are simply

$$m_1v_1 + m_2v_2 = 0, \qquad \tfrac{1}{2}m_1v_1^2 + \tfrac{1}{2}m_2v_2^2 = K. \tag{8}$$

Thus, $v_1 = -m_2v_2/m_1$ and

$$v_2 = \pm\sqrt{2\frac{m_1}{m_2}\frac{K}{m_1+m_2}}, \tag{9}$$

and hence (7).

(ii) If we take the positive root in (9) then the horizontal speed of fragment 2 in the original frame is greater than u_x, while that of fragment 1 is

$$u_x - \sqrt{2\frac{m_2}{m_1}\frac{K}{m_1+m_2}}, \tag{10}$$

which is less than u_x. So fragment 1 strikes the ground to the right of A, or to the left of A, or directly below A, depending on whether K is less than, greater than, or equal to $\dfrac{m_1}{2m_2}(m_1+m_2)u_x^2$.

Question 6.3

Consider oblique impact of a smooth sphere on a fixed plane. According to Newton's experimental law of impact, the velocity of rebound is proportional to the velocity of approach. Express this statement in the form of an equation. Write down also the equation required by conservation of momentum.

Solution

The figure shows the sphere at the instant of impact: P is the point of impact, C is the centre of the sphere, and PC is perpendicular to the plane. The velocities of the sphere before and after impact (**u** and **v**) make angles α and β with PC. The velocities of approach and rebound are, respectively, $u\cos\alpha$ and $-v\cos\beta$, directed along CP. So Newton's experimental law of impact states that

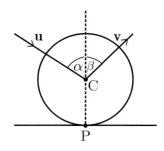

$$v\cos\beta = eu\cos\alpha, \qquad (1)$$

where e is a positive constant. Because the sphere is smooth, the interaction between the sphere and plane is along the perpendicular PC. Therefore, the component of the momentum of the sphere parallel to the plane is conserved, and this provides an additional condition

$$v\sin\beta = u\sin\alpha. \qquad (2)$$

Comments

(i) In this question (and in the rest of this chapter) we have assumed that the sphere is not spinning. Furthermore, because the surface of the sphere is smooth (frictionless), no spin is generated by the impact. Spin can produce some intriguing effects that are well known and important in sports such as golf, tennis and billiards (see Chapter 12).

(ii) The constant e is known as the coefficient of restitution:[‡] it depends on the materials of the colliding bodies, and often has the useful property that for a given pair of bodies it is constant over a range of velocities.

(iii) If $e = 1$ the collision is termed elastic: according to (1) and (2), $v = u$ and so the kinetic energy of the ball is conserved. If $e = 0$ the collision is totally inelastic: the ball sticks to the plane, and all its kinetic energy is lost. For most materials e lies between these extremes and collisions are inelastic: the ball loses part of its kinetic energy. Some approximate values of e are 0.5 for wooden balls, 0.6 for steel balls, 0.9 for ivory billiard balls, and 0.95 for hard rubber balls. For a squash ball and racquet $e \approx 0.3$. The introduction of the coefficient of restitution e allows one to account for the effects of inelasticity in an empirical way.

(iv) The total energy of the sphere, plane and surroundings is conserved. The energy Q released by the impact equals the loss in kinetic energy:

$$Q = K_i - K_f = \tfrac{1}{2}mu^2 - \tfrac{1}{2}mv^2 = \tfrac{1}{2}mu^2(1-e^2)\cos^2\alpha. \qquad (3)$$

That is, a fraction $(1-e^2)\cos^2\alpha$ of the initial kinetic energy is converted into other forms of energy (such as heat and sound). Collisions of macroscopic objects

---

[‡] Use of the symbol e should not be confused with the base of the natural logarithm. Which is intended should be clear from the context.

are invariably inelastic – there is a loss in kinetic energy associated with internal changes of the objects. By contrast, elastic collisions are of importance in microscopic (atomic, nuclear and particle) physics. In fact, it is even possible to have superelastic collisions ($e > 1$) where the kinetic energy of one of the particles (e.g. an electron) is increased at the expense of the internal energy of a target particle (e.g. an excited atom).

Question 6.4

A ball falls from a height H onto a fixed horizontal plane. The coefficient of restitution is e and the height reached by the ball in the nth rebound is H_n.

(a) Show that
$$H_n = e^{2n} H \qquad (n = 1, 2, \cdots). \tag{1}$$

(b) If the ball travels a total distance D before coming to rest after a time T, deduce that
$$D = \frac{1+e^2}{1-e^2} H \quad \text{and} \quad T = \frac{1+e}{1-e}\sqrt{\frac{2H}{g}}. \tag{2}$$

(Neglect air resistance and any variation in g.)

Solution

(a) Let u be the speed with which the ball first impacts the plane. Conservation of energy requires $\frac{1}{2}mu^2 = mgH$, and so $u^2 = 2gH$. According to Newton's law of impact the ball rebounds with speed eu and therefore it rises to a height $H_1 = e^2 u^2/2g = e^2 H$. Similarly, $H_2 = e^2 H_1 = e^4 H$, and so on.

(b) The total distance travelled is
$$D = H + 2(H_1 + H_2 + \cdots) = H + 2(e^2 + e^4 + \cdots)H, \tag{3}$$

which is $(2)_1$. The time taken to fall through a height H_n is $\sqrt{2H_n/g} = e^n\sqrt{2H/g}$, and therefore the total time for which the ball bounces is
$$T = \sqrt{2H/g} + 2(e + e^2 + e^3 + \cdots)\sqrt{2H/g}, \tag{4}$$

which is $(2)_2$.

Comments

(i) It is apparent from (1) that $e = \sqrt{H_1/H}$, where H_1 is the height of the first rebound. This relation provides a simple means of measuring e.

(ii) In the elastic limit $e \to 1$ and therefore $D, T \to \infty$ as one expects. Graphs of (1) and (2) are plotted below.

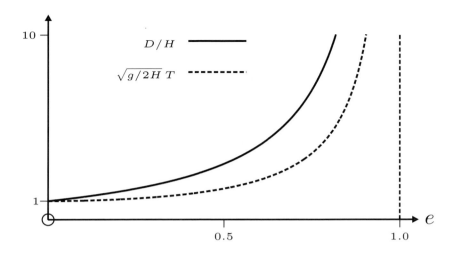

Question 6.5

A smooth ball is projected with speed u at an angle θ ($< \frac{1}{2}\pi$) to a horizontal surface. It bounces across the surface in a series of hops.‡ The coefficient of restitution is e.

(a) Show that the range R_n and maximum height H_n of the nth hop are

$$R_n = e^{n-1} \frac{u^2}{g} \sin 2\theta \quad \text{and} \quad H_n = e^{2(n-1)} \frac{u^2}{2g} \sin^2 \theta. \tag{1}$$

(b) Show that the trajectory of the nth hop is given in terms of H_n and R_n by

$$y_n(x) = \frac{4H_n}{R_n^2}(x - X_n)(R_n - x + X_n) \qquad n = 1, 2, \cdots. \tag{2}$$

Here, $X_1 = 0$ and $X_n = \sum_{i=1}^{n-1} R_i$ for $n = 2, 3, \cdots$.

(c) Write a *Mathematica* notebook to calculate the trajectory† for the first 8 hops of a ball for which $\theta = 45°$ and $e = 0.75$. Use *Mathematica*'s `Manipulate` command to simulate the ball bouncing across the surface.

Solution

(a) For the first hop ($n = 1$) the initial velocity of the ball is $(u\cos\theta,\ u\sin\theta)$. For the second hop, conservation of momentum (parallel to the surface) and Newton's law of impact require that the initial velocity* is $(u\cos\theta,\ eu\sin\theta)$. In general, for

---

‡Before attempting this question, the reader should be familiar with Question 7.1.
†It is convenient here to express x and y in units of u^2/g.
*That is, the velocity immediately after the first impact with the surface.

the nth hop the initial velocity is $(u\cos\theta,\ e^{n-1}u\sin\theta)$. With this initial velocity, (7) of Question 7.1 yields (1).
(b) The result follows directly from (10) of Question 7.1.
(c) The following notebook yields the trajectory plotted below. (In the animation, the speed at which the ball moves along its trajectory can be adjusted using the up arrows ⩙ and down arrows ⩗ on the `Manipulate` slider control.)

```
In[1]:= θ = 45 π/180; e = 3/4; nmax = 8; xrange = 0; yrange = Sin[θ]^2/2;

list1 = {}; list2 = {};

R[n_] := e^(n-1) Sin[2θ];

X[1] = 0;

X[n_] := Sum[R[i], {i, 1, n - 1, 1}];

Y[x_, n_] := 1/2 Sec[θ]^2 (x - X[n]) (R[n] - x + X[n]);

n = 1;

While[n ≤ nmax,
  root = x /. Solve[Y[x, n] == 0, x];
  xlow[n_] := First[root];
  xhigh[n_] := Last[root];
  gr[n] = Plot[Y[x, n], {x, xlow[n], xhigh[n]}];
  list1 = Append[list1, gr[n]];
  list2 = Append[list2, Table[{x, Y[x, n]}, {x, xlow[n],
      xhigh[n], (xhigh[n] - xlow[n])/100}]];
  xrange = xrange + R[n];
  n = n + 1];

list3 = DeleteDuplicates[Flatten[list2, 1]];

num = Length[list3];

traj[i_] := Graphics[{PointSize[Medium], Point[Flatten[
    Take[list3, {i, i}]]]}, PlotRange → {{0, xrange},
    {-yrange/100, yrange}}, AspectRatio → 0.7];

Manipulate[Show[{list1, traj[i]}, PlotRange → {{0, xrange},
    {-yrange/100, yrange}}], {i, 1, num, 1}]
```

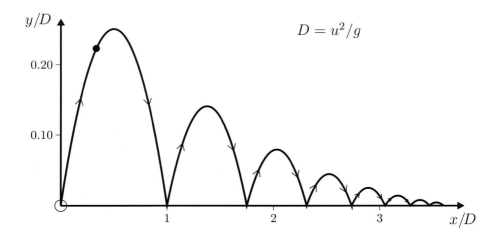

Comments

(i) The horizontal distance $R = R_1 + R_2 + \cdots$ that a smooth ball bounces before it slides on the surface is

$$R = (1 + e + e^2 + \cdots)\frac{u^2}{g}\sin 2\theta = \frac{1}{1-e}\frac{u^2}{g}\sin 2\theta, \qquad (3)$$

and the time for which it bounces is

$$T = \frac{R}{u\cos\theta} = \frac{2}{1-e}\frac{u}{g}\sin\theta. \qquad (4)$$

(ii) The ratio of the kinetic energy K_n of the ball at the start of the nth hop to the initial kinetic energy K_1 is

$$K_n/K_1 = \cos^2\theta + e^{2(n-1)}\sin^2\theta. \qquad (5)$$

Thus, K_n decreases to its final value $K_1\cos^2\theta$ in a time T, at which point the ball starts sliding and its kinetic energy is constant.

Question 6.6

A smooth billiard ball is projected from a point A on the edge of a circular billiard table in a direction making an angle ϕ with the radius to A. The coefficient of restitution is e. The ball makes q impacts with the wall before returning to A. Show that

(a) $\quad \phi = 0 \quad$ if $\quad q = 1$, $\hfill (1)$

(b) $\quad \tan^2\phi = e^3/(1 + e + e^2) \quad$ if $\quad q = 2$, $\hfill (2)$

(c) $\quad \tan^2\phi = e^3 \quad$ if $\quad q = 3$. $\hfill (3)$

Solution

(a) If $q = 1$ there is just one impact before the ball returns to A: the ball moves along a diameter and back, and $\phi = 0$.

(b) If $q = 2$ the path back to A is a triangle. In terms of the angles shown, the condition for the ball to reach A after impacts at B and C is

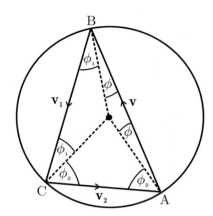

$$2(\phi + \phi_1 + \phi_2) = \pi. \qquad (4)$$

For the impact at B, conservation of momentum (along the tangent at B) and Newton's law of impact require $v_1 \sin\phi_1 = v \sin\phi$ and $v_1 \cos\phi_1 = ev \cos\phi$. That is,

$$\tan\phi_1 = e^{-1} \tan\phi. \qquad (5)$$

Similarly, for the impact at C,

$$\tan\phi_2 = e^{-1} \tan\phi_1 = e^{-2} \tan\phi. \qquad (6)$$

From (4) we have $\tan(\phi_1 + \phi_2) = \tan(\tfrac{1}{2}\pi - \phi)$. That is,

$$\frac{\tan\phi_1 + \tan\phi_2}{1 - \tan\phi_1 \tan\phi_2} = \frac{1}{\tan\phi}. \qquad (7)$$

From (5), (6) and (7) we obtain

$$\frac{e^{-1}\tan\phi + e^{-2}\tan\phi}{1 - e^{-3}\tan^2\phi} = \frac{1}{\tan\phi}, \qquad (8)$$

and hence (2). Note that for $e < 1$, (6) yields $\phi_2 > \phi_1 > \phi$.

(c) If $q = 3$ then the path back to A is a quadrilateral. Instead of (4) we have

$$2(\phi + \phi_1 + \phi_2 + \phi_3) = 2\pi, \qquad (9)$$

where ϕ_1 and ϕ_2 are given by (5) and (6), and

$$\tan\phi_3 = e^{-3} \tan\phi. \qquad (10)$$

From (9), $\phi + \phi_1 = \pi - \phi_2 - \phi_3$ and so $\tan(\phi + \phi_1) = -\tan(\phi_2 + \phi_3)$. Thus,

$$\frac{\tan\phi + \tan\phi_1}{1 - \tan\phi \tan\phi_1} = -\frac{\tan\phi_2 + \tan\phi_3}{1 - \tan\phi_2 \tan\phi_3}. \qquad (11)$$

By substituting (5), (6) and (10) in (11), the solution is found to be (3).

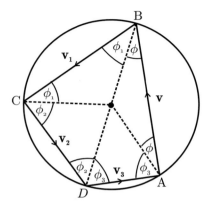

Comments

(i) Graphs of the solutions (2) and (3) are shown below.

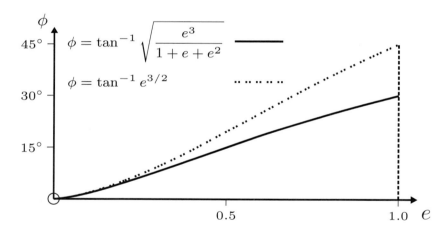

In the limit $e \to 1$ the triangle in (b) becomes equilateral and the quadrilateral in (c) becomes square, and the motions are periodic and reversible.

(ii) In general, the condition that the ball returns to A after q impacts with the wall is

$$2(\phi + \sum_{i=1}^{q} \phi_i) = (q-1)\pi, \quad \text{where} \quad \tan\phi_q = e^{-q}\tan\phi. \tag{12}$$

(iii) The reader may wish to consider further questions, such as, if $e \neq 1$, is the motion periodic, and is it reversible (if we reverse the velocity just before the ball returns to A, does it retrace its path back to A)?

Question 6.7

Consider a head-on collision of two spheres. Use Newton's law of impact (see Question 6.3) and conservation of momentum to express the velocities after impact in terms of the masses, the velocities before impact, and the coefficient of restitution.

Solution

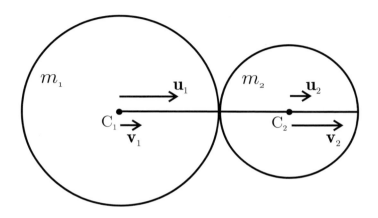

Let u_1 and v_1 be the velocity of sphere 1 along the line of centres C_1C_2 just before and just after the collision, and similarly for u_2 and v_2. According to Newton's law of impact, the relative speed of separation $(v_2 - v_1)$ is proportional to the relative speed of approach $(u_1 - u_2)$:

$$v_2 - v_1 = e(u_1 - u_2), \tag{1}$$

where e is a positive constant. Also, conservation of momentum requires

$$m_1 v_1 + m_2 v_2 = m_1 u_1 + m_2 u_2. \tag{2}$$

Solution of the linear equations (1) and (2) yields the desired expressions for the velocities after impact:

$$v_1 = \frac{m_1 - em_2}{m_1 + m_2} u_1 + \frac{(1+e)m_2}{m_1 + m_2} u_2, \qquad v_2 = \frac{(1+e)m_1}{m_1 + m_2} u_1 + \frac{m_2 - em_1}{m_1 + m_2} u_2. \tag{3}$$

Note that under interchange of the subscripts 1 and 2, $(3)_1 \leftrightarrow (3)_2$.

Comments

(i) For a totally inelastic collision $e = 0$ and the spheres move together after the collision with velocity

$$v_1 = v_2 = \frac{m_1 u_1 + m_2 u_2}{m_1 + m_2}. \tag{4}$$

(ii) For an elastic collision $e = 1$ and the velocities

$$v_1 = \frac{m_1 - m_2}{m_1 + m_2} u_1 + \frac{2m_2}{m_1 + m_2} u_2, \qquad v_2 = \frac{2m_1}{m_1 + m_2} u_1 + \frac{m_2 - m_1}{m_1 + m_2} u_2, \qquad (5)$$

are different. If $m_1 = m_2$ then $v_1 = u_2$ and $v_2 = u_1$: the particles exchange their velocities.

(iii) If m_2 is a stationary target ($u_2 = 0$) then (3) shows that

$$v_1 = \frac{m_1 - em_2}{m_1 + m_2} u_1, \qquad v_2 = \frac{(1+e)m_1}{m_1 + m_2} u_1. \qquad (6)$$

It follows that if $m_1 \gg m_2$ then $v_1 \approx u_1$ (the incident sphere is hardly affected by the collision) and $v_2 \approx (1+e)u_1$. If $m_2 \gg m_1$ then $v_1 \approx -eu_1$ and $v_2 \approx 0$; in an elastic collision the incident sphere rebounds with almost unchanged speed.

(iv) The energy Q released by the collision is given by (1) of Question 6.11.

Question 6.8

Consider an oblique collision between two smooth spheres. On a sketch of the spheres at the instant of impact indicate the velocities just before and just after the collision. Then, write down an equation for Newton's law of impact (see Question 6.3) and also the equations required by conservation of momentum.

Solution

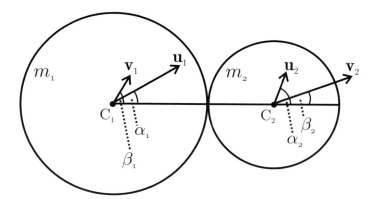

The figure shows the velocities of the two spheres just before and just after impact, and the angles that these velocities make with the line of centres $C_1 C_2$. Here, Newton's law of impact applies to the relative speed of separation and the relative speed of approach along the line of centres:

$$v_2 \cos \beta_2 - v_1 \cos \beta_1 = e(u_1 \cos \alpha_1 - u_2 \cos \alpha_2). \qquad (1)$$

Also, there is conservation of momentum along the line of centres

$$m_1 v_1 \cos \beta_1 + m_2 v_2 \cos \beta_2 = m_1 u_1 \cos \alpha_1 + m_2 u_2 \cos \alpha_2. \tag{2}$$

Furthermore, because the spheres are smooth, their interaction is along $C_1 C_2$ and therefore, for each sphere separately, the component of momentum perpendicular to $C_1 C_2$ is conserved

$$v_1 \sin \beta_1 = u_1 \sin \alpha_1, \qquad v_2 \sin \beta_2 = u_2 \sin \alpha_2. \tag{3}$$

Comment

If the masses m_1 and m_2, and the initial values u_1, u_2, α_1, α_2, and the coefficient of restitution e are known, then (1)–(3) can be used to solve for the four unknowns v_1, v_2, β_1 and β_2 – that is, for the velocities $\mathbf{v}_1$ and $\mathbf{v}_2$ after an oblique impact. See Question 6.9.

Question 6.9

For the oblique collision of two smooth spheres, use Newton's law of impact and conservation of momentum ((1)–(3) of the previous question) to express the velocities $\mathbf{v}_1$ and $\mathbf{v}_2$ after impact in terms of the components of the velocities before impact.

Solution

Choose a coordinate system with x- and y-axes in the plane of $\mathbf{u}_1$ and $\mathbf{u}_2$, and with the x-axis along the line of centres $C_1 C_2$ (see the above figure). Then (1) and (2) are two simultaneous linear equations for the x-components, $v_1 \cos \beta_1$ and $v_2 \cos \beta_2$, of $\mathbf{v}_1$ and $\mathbf{v}_2$. Also, their y-components are given by (3). Thus, we have the desired expressions

$$\mathbf{v}_1 = \hat{\mathbf{x}} \frac{(m_1 - em_2) u_1 \cos \alpha_1 + (1+e) m_2 u_2 \cos \alpha_2}{(m_1 + m_2)} + \hat{\mathbf{y}} u_1 \sin \alpha_1 \tag{1}$$

$$\mathbf{v}_2 = \hat{\mathbf{x}} \frac{(1+e) m_1 u_1 \cos \alpha_1 + (m_2 - em_1) u_2 \cos \alpha_2}{(m_1 + m_2)} + \hat{\mathbf{y}} u_2 \sin \alpha_2. \tag{2}$$

Note that under interchange of subscripts 1 and 2, (1) $\leftrightarrow$ (2).

Comments

(i) The magnitudes and directions of $\mathbf{v}_1$ and $\mathbf{v}_2$ can be obtained from (1) and (2) in the usual way:

$$|\mathbf{v}_i| = \sqrt{v_{ix}^2 + v_{iy}^2} \quad \text{and} \quad \tan \beta_i = v_{iy}/v_{ix} \quad (i = 1, 2). \tag{3}$$

(ii) For a head-on collision $\alpha_1 = \alpha_2 = 0$, and (1) and (2) reduce to $(3)_1$ and $(3)_2$ of Question 6.7.

(iii) If m_2 is a stationary target (that is, if $u_2 = 0$) then (1) and (2) reduce to

$$\mathbf{v}_1 = \hat{\mathbf{x}} \frac{m_1 - em_2}{m_1 + m_2} u_1 \cos\alpha_1 + \hat{\mathbf{y}} u_1 \sin\alpha_1 \qquad (4)$$

$$\mathbf{v}_2 = \hat{\mathbf{x}} \frac{(1+e)m_1}{m_1 + m_2} u_1 \cos\alpha_1 . \qquad (5)$$

We see that after the collision m_2 moves along the x-axis (the line of centres $C_1 C_2$): this is expected since the force on impact is along $C_1 C_2$. Furthermore, if $m_1/m_2 = e$ (as, for example, in an elastic collision between equal masses) and $\alpha_1 \neq 0$, then $\mathbf{v}_1$ is perpendicular to $\mathbf{v}_2$.

Question 6.10

In an oblique collision between two smooth spheres a mass m_1 strikes a stationary target of mass m_2. The initial and final velocities $\mathbf{u}$ and $\mathbf{v}$ of m_1 make angles α and β with the line of centres at the instant of impact. The coefficient of restitution is e.

(a) Suppose $m_1/m_2 \geq e$. The deflection $\delta = \beta - \alpha$ of m_1 has a maximum value $\delta_{\max}$ for some $\alpha = \alpha_m$. Show that

$$\alpha_m = \tan^{-1} \sqrt{\frac{m_1 - em_2}{m_1 + m_2}} \qquad (1)$$

$$\delta_{\max} = \tan^{-1} \sqrt{\frac{m_1 + m_2}{m_1 - em_2}} - \tan^{-1} \sqrt{\frac{m_1 - em_2}{m_1 + m_2}} . \qquad (2)$$

(Hint: Use $(3)_2$ and (4) of Question 6.9.)

(b) What happens if $m_1/m_2 < e$?

Solution

(a) From $(3)_2$ and (4) of Question 6.9 (where we omit the subscript 1 on α and β) we see that the directions of motion of m_1 before and after impact are related by

$$\tan\beta = \frac{m_1 + m_2}{m_1 - em_2} \tan\alpha, \qquad (3)$$

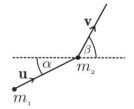

where $0 < \alpha < \frac{1}{2}\pi$ for an oblique impact. For the angle of deflection $\delta = \beta - \alpha$ we have

$$\tan\delta = \frac{\tan\beta - \tan\alpha}{1 + \tan\alpha \tan\beta} = \frac{(1+e)m_2 \tan\alpha}{m_1 - em_2 + (m_1 + m_2)\tan^2\alpha} . \qquad (4)$$

If $m_1/m_2 > e$ then $0 < \delta < \frac{1}{2}\pi$. Consequently, for δ to be a maximum, $\tan\delta$ must be a maximum and therefore $d\tan\delta/d\alpha = 0$. Differentiating (4) with respect to α gives

$$\tan \alpha_m = \sqrt{\frac{m_1 - em_2}{m_1 + m_2}}, \tag{5}$$

and hence (1) for the angle at which m_1 must be projected in order to obtain maximum deflection. From (3) and (5)

$$\tan \beta_m = \sqrt{\frac{m_1 + m_2}{m_1 - em_2}}. \tag{6}$$

Equations (5) and (6) show that the maximum deflection $\delta_{\max} = \beta_m - \alpha_m$ is given by (2). Note that as $m_1/em_2 \to 1_+$, $\alpha_m \to 0_+$ in (5) and $\beta_m \to \tfrac{1}{2}\pi$ in (6): thus $\delta_{\max} \to \tfrac{1}{2}\pi$.

(b) If $m_1/m_2 < e$ then $\tan \beta < 0$ in (3), and so $\tfrac{1}{2}\pi < \beta < \pi$. The maximum deflection is obtained in the limit $\alpha \to 0_+$, where $\beta \to \pi$. So

$$\delta_{\max} = \pi \tag{7}$$

if $m_1/m_2 < e$. That is, backward scattering yields the largest deflection. We see that $\delta_{\max}$ is discontinuous at $m_1/m_2 = e$.

Comment

Graphs of $\delta_{\max}$ versus m_1/m_2, calculated from (2) and for various values of e, are plotted below. The limiting value of 90° at $m_1/m_2 = e$ is evident in each case. If $m_1/m_2 < e$ then $\delta_{\max} = 180°$.

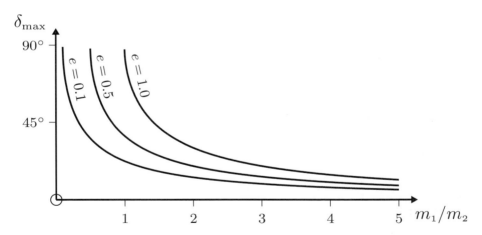

Question 6.11

Consider a collision between two particles in which their masses m_1 and m_2 are unchanged. Use conservation of momentum, conservation of energy and Newton's law of impact to show that the energy Q released by the collision is

$$Q = (1-e^2)\tfrac{1}{2}\mu(\mathbf{u}_2 - \mathbf{u}_1)^2, \qquad (1)$$

where $\mathbf{u}_1$ and $\mathbf{u}_2$ are the initial velocities and

$$\mu = \frac{m_1 m_2}{(m_1 + m_1)}. \qquad (2)$$

Solution

Conservation of momentum and energy require that

$$\mathbf{P} = m_1\mathbf{v}_1 + m_2\mathbf{v}_2 = m_1\mathbf{u}_1 + m_2\mathbf{u}_2 \quad \text{and} \quad K_\text{f} + Q = K_\text{i}, \qquad (3)$$

where $K_\text{i} = \tfrac{1}{2}m_1 u_1^2 + \tfrac{1}{2}m_2 u_2^2$ and $K_\text{f} = \tfrac{1}{2}m_1 v_1^2 + \tfrac{1}{2}m_2 v_2^2$ are the initial and final kinetic energies. Also, the law of impact requires that the relative speeds before and after the collision are related by the coefficient of restitution:

$$|\mathbf{v}_1 - \mathbf{v}_2| = e|\mathbf{u}_1 - \mathbf{u}_2|. \qquad (4)$$

By squaring $(3)_1$ we obtain the identity

$$P^2 = 2(m_1 + m_2)K_\text{f} - m_1 m_2(\mathbf{v}_1 - \mathbf{v}_2)^2 = 2(m_1 + m_2)K_\text{i} - m_1 m_2(\mathbf{u}_1 - \mathbf{u}_2)^2. \qquad (5)$$

The desired result (1) follows directly from $(3)_2$, (4) and (5).

Comments

(i) It is apparent from (1) that $Q = 0$ if $e = 1$: no energy is released in an elastic collision. Also, $Q > 0$ if $e < 1$: kinetic energy is lost in an inelastic collision. If $e > 1$ then $Q < 0$: some internal energy of the colliding particles is released and this produces an increase in their kinetic energy.

(ii) The quantity μ defined in (2) has the dimension of mass and it is less than either m_1 or m_2. It is known as the reduced mass and is an important quantity in the theory of the two-body problem (see Chapter 10).

Question 6.12

Three particles A, B and C with masses $m_\text{A} = 2m_\text{B} = m_\text{C}$ are arranged (in that order) in a straight line. Initially, B and C are at rest a distance L apart, and A is projected towards B with speed u. The particles then undergo head-on elastic collisions. Show that A and B collide twice and that the time interval between these two collisions is

$$\Delta t = 12L/7u. \qquad (1)$$

Solution

For the first collision between A and B, the latter is a stationary target and (6) of Question 6.7 (with $e = 1$) yields for the velocities after the collision:

$$v'_A = \frac{m_A - m_B}{m_A + m_B} u = \frac{u}{3}, \qquad v'_B = \frac{2m_A}{m_A + m_B} u = \frac{4u}{3}. \tag{2}$$

B then collides with the stationary target C, after which the velocity of B is

$$v''_B = \frac{m_B - m_C}{m_B + m_C} v'_B = -\frac{4u}{9}. \tag{3}$$

Equations $(2)_1$ and (3) show that A and B are now moving in opposite directions and so they will collide again. The time elapsed is

$$\Delta t = L/v'_B + (L - v'_A \Delta t)/|v''_B|. \tag{4}$$

That is,

$$\Delta t = \frac{L}{v'_B} \frac{v'_B + |v''_B|}{v'_A + |v''_B|}. \tag{5}$$

Equations (2), (3) and (5) yield (1).

Question 6.13

N identical, stationary, rigid spheres S_n ($n = 1, 2, \cdots, N$), each of radius a, are placed along the x-axis with their centres at positions x_n, such that $x_1 = 2a$ and the distances between the surfaces of adjacent spheres is $b/2^n$. At time $t = 0$ an identical sphere S_0, moving in the positive x-direction with speed v, collides elastically with S_1. Determine the eventual outcome of this event if the ensuing collisions are also elastic.

Solution

The centres of the spheres are at $x_1 = 2a$, $x_2 = 4a + b/2$, $x_3 = 6a + 3b/4$, $\cdots$. That is, at

$$x_n = 2na + (1 - 2^{-n+1})b, \quad \text{where } n = 1, 2, \cdots, N. \tag{1}$$

The outcome of the sequence of elastic collisions that occurs for $t \geq 0$ is the following: S_0 stops at $\bar{x}_0 = x_1 - 2a = 0$ at time $t = 0$; S_1 stops at $\bar{x}_1 = x_2 - 2a = 2a + b/2$ at time $t = b/2v$; S_2 stops at $\bar{x}_2 = x_3 - 2a = 4a + 3b/4$ after a further time $b/4v$; $\cdots$; and S_N is eventually ejected from the array. Therefore, after a time

$$\frac{b}{2v} + \frac{b}{4v} + \cdots + \frac{b}{2^N v} = \left(1 - \frac{1}{2^N}\right)\frac{b}{v} \tag{2}$$

there will be an array of N stationary spheres S_n ($n = 0, 1, \cdots, N - 1$) with their centres at

$$\bar{x}_n = x_{n+1} - 2a = 2na + (1 - 2^{-n})b, \tag{3}$$

and the sphere S_N moving with speed v in the positive x-direction.

Comments

(i) In the limit $N \to \infty$ the outcome after a finite time b/v is an infinite array of stationary spheres S_0, S_1, $\cdots$ with centres located at $\bar{x}_n$ given by (3).

(ii) The time-reversed version of this infinite model has been posed as a simple example of a 'spontaneously self-excited' system that demonstrates indeterminism in classical dynamics.[1]

Question 6.14

The angular momentum **L** of a particle is defined in terms of its position vector **r** and momentum $\mathbf{p} = m\dot{\mathbf{r}}$ by

$$\mathbf{L} = \mathbf{r} \times \mathbf{p}. \tag{1}$$

Show that in an inertial frame

$$\frac{d\mathbf{L}}{dt} = \mathbf{\Gamma}, \tag{2}$$

where

$$\mathbf{\Gamma} = \mathbf{r} \times \mathbf{F} \tag{3}$$

and **F** is the force acting on the particle.

Solution

The rule for differentiating a product of two functions applies also to a vector product such as (1), because the latter is a linear combination of products $r_i p_j$. Therefore,

$$\frac{d}{dt}(\mathbf{r} \times \mathbf{p}) = \dot{\mathbf{r}} \times \mathbf{p} + \mathbf{r} \times \dot{\mathbf{p}}. \tag{4}$$

Now, $\dot{\mathbf{r}} \times \mathbf{p} = 0$ because $\dot{\mathbf{r}}$ and $\mathbf{p} = m\dot{\mathbf{r}}$ are parallel vectors. Also, in an inertial frame $\dot{\mathbf{p}} = \mathbf{F}$. Thus, (4) yields (2).

Comments

(i) The quantity $\mathbf{\Gamma}$ defined in (3) is known as the torque or moment of a force. It is a familiar quantity. For example, in the use of a spanner, **r** and **F** are perpendicular and the magnitude of the torque is rF; in a torque wrench the value of the maximum applied torque can be pre-set. In general, torque plays an important role in the statics and dynamics of rigid bodies (see Chapter 12).

(ii) Equation (2) is the rotational counterpart of the equation of motion $d\mathbf{p}/dt = \mathbf{F}$. There is, however, an important difference between vectors appearing in these two equations. The vectors **L** and $\mathbf{\Gamma}$ in (1) and (3) depend on **r**, and therefore they depend on our choice of coordinate origin; by contrast, the momentum **p** and force

---

[1] J. P. Laraudogoitia, "On indeterminism in classical dynamics," European Journal of Physics, vol. 18, pp. 180–181, 1997.

F are independent of this choice. The values of origin-dependent quantities such as **L** and **Γ** have meaning only with respect to a specified choice of coordinate origin O: to emphasize this they are often referred to as the angular momentum about O and the torque about O. Origin dependence/independence of physical quantities and origin independence of physical laws is an interesting topic, and some examples are given in Chapter 14.

(iii) Sometimes a 'mixed' definition of angular momentum is used, where the momentum and position vectors are with respect to different frames: for example, momentum $\mathbf{p} = m\dot{\mathbf{r}}$ relative to the inertial frame with origin at O, and position vector $\mathbf{r} - \mathbf{D}$ with respect to a frame with origin Q at $\mathbf{D}$ relative to O (see also Comment (iv) in Question 11.3). Then,

$$\mathbf{L}_Q = (\mathbf{r} - \mathbf{D}) \times \mathbf{p}. \tag{5}$$

Clearly, since $\mathbf{r} - \mathbf{D}$ does not depend on the position of O, $\mathbf{L}_Q$ is an origin-independent vector. If Q is fixed relative to the inertial frame then $\dot{\mathbf{D}} = 0$ and one has the following extension of (2):

$$\frac{d\mathbf{L}_Q}{dt} = (\mathbf{r} - \mathbf{D}) \times \mathbf{F}. \tag{6}$$

(iv) The origin dependence of torque can be used to simplify certain calculations by making a convenient choice of coordinate origin (see Chapter 12).

(v) According to (2), if the torque acting on a particle in an inertial frame is zero then the angular momentum of the particle is conserved. This is the simplest statement of the law of conservation of angular momentum, which is a fundamental law of nature. Its extension to a system of interacting particles is considered in Question 11.2.

(vi) In general, if $\Gamma_i = 0$ then L_i is constant. That is, for each component of the torque that is zero, the corresponding component of the angular momentum is conserved.

(vii) Conservation of angular momentum is associated with invariance under rotations of an isolated system in inertial space (see Questions 14.18 and 14.19).

(viii) Conservation of angular momentum holds even in processes for which Newtonian dynamics fails. There are no known violations of this law.

Question 6.15

Prove that the angular momentum $\mathbf{L} = \mathbf{r} \times \mathbf{p}$ of a particle is conserved in a central force field

$$\mathbf{F} = F(\mathbf{r})\hat{\mathbf{r}}. \tag{1}$$

Solution

The torque exerted on the particle is

$$\mathbf{\Gamma} = \mathbf{r} \times \mathbf{F} = F(\mathbf{r})\mathbf{r} \times \hat{\mathbf{r}} = 0, \tag{2}$$

because $\mathbf{r} = r\hat{\mathbf{r}}$ and $\hat{\mathbf{r}}$ are parallel vectors. It follows from this and Question 6.14 that $d\mathbf{L}/dt = 0$ and therefore $\mathbf{L}$ is constant.

Comments

(i) Central forces do not necessarily conserve energy. There is, however, an important sub-class of these forces that conserves both energy and angular momentum, namely central, isotropic forces[‡] (see Question 5.10)

$$\mathbf{F} = F(r)\hat{\mathbf{r}}. \qquad (3)$$

Force	Conserves **L**?	Conserves E?
Central: $F(\mathbf{r})\hat{\mathbf{r}}$	Yes	Not necessarily
Central, isotropic: $F(r)\hat{\mathbf{r}}$	Yes	Yes

Important examples of (3) are the gravitational and Coulomb (electrostatic) forces, and for these the trajectories are usually expressed in terms of the conserved energy E and angular momentum L (see Question 7.27).

(ii) Two important results follow from the conservation of **L**. First, because **r** is always perpendicular to **L**, it follows that when the direction of **L** is fixed the motion is confined to a fixed plane perpendicular to **L** and through the origin O (the centre of force). This plane is defined by the initial values $\mathbf{r}_0$ and $\mathbf{v}_0$. Secondly, constancy of the magnitude $L = m|\mathbf{r} \times d\mathbf{r}|/dt$ means that

$$|\mathbf{r} \times d\mathbf{r}| = L\,dt/m \qquad (4)$$

is a constant. Now, $|\mathbf{r} \times d\mathbf{r}|$ is equal to the area of the parallelogram formed by the vectors **r** and $d\mathbf{r}$, and it is equal to twice the area dA of the triangle OPQ swept out by the position vector **r** in time dt. That is,

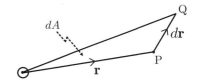

$$dA = L\,dt/2m, \qquad (5)$$

meaning that the position vector sweeps out equal areas in equal times. This result is known as Kepler's second law and it applies to all central forces.[†]

(iii) The conservation of energy and angular momentum by central, isotropic fields is valid also in quantum mechanics, albeit with a twist: because of the uncertainty principle only one component (L_z say) and the magnitude L^2 can be measured simultaneously – the components L_x and L_y are indeterminate. The fact that E, L^2, and L_z are conserved is an essential ingredient for understanding the periodic table of the chemical elements.

[‡]In spherical polar coordinates $F(\mathbf{r}) = F(r,\theta,\phi)$; if F does not depend explicitly on the angles θ and ϕ then $\mathbf{F} = F\hat{\mathbf{r}}$ is isotropic as well as central.

[†]The law was discovered by Kepler in relation to planetary motion about the Sun (see also Question 10.11).

Question 6.16

A particle of mass m attached to one end of a thin, light, inextensible string moves with speed v_0 in a circle of radius r_0 in free space. Calculate the work required to reduce the radius from r_0 to r by pulling the other end of the string through a smooth tube that is perpendicular to the plane of the circle. Express the result in terms of m, v_0, r_0 and r.

Solution

According to the work–energy theorem (see Question 5.1) the work done on the particle is equal to the change in kinetic energy:

$$W = K_f - K_i = \tfrac{1}{2}mv^2 - \tfrac{1}{2}mv_0^2, \tag{1}$$

where v is the speed when the radius is r. The tension in the string is a central force (it is always radial) and so the angular momentum of the particle is conserved as the radius decreases (see Question 6.15):

$$mvr = mv_0 r_0. \tag{2}$$

Use of (2) to eliminate v from (1) gives

$$W = \left(\frac{r_0^2}{r^2} - 1\right)\tfrac{1}{2}mv_0^2. \tag{3}$$

Note that $W < 0$ if $r > r_0$, meaning that work is done by the particle if the radius is allowed to increase.

Question 6.17

Suppose that in the previous question the string is winding up on the outside of the tube (instead of being pulled into it), so that the particle spirals around the tube. The initial length of the string is r_0 and the initial velocity is v_0 perpendicular to the string. The outer radius of the tube is a. Choose x- and y-axes in the plane of motion, with origin O at the initial point of contact between the string and tube, and the string initially along the y-axis.

(a) Show that the trajectory is given in parametric form by

$$x(\theta) = a(1 - \cos\theta) + (r_0 - a\theta)\sin\theta \tag{1}$$
$$y(\theta) = a\sin\theta + (r_0 - a\theta)\cos\theta, \tag{2}$$

where $\theta(t)$ is the angle subtended at the centre of the tube by the arc of the string wound onto the tube.

(b) Prove that the speed $v(\theta)$ of the particle is constant. (Hint: Show that the tension **T** in the string is always perpendicular to the trajectory.)

148   *Solved Problems in Classical Mechanics*

(c) Show that
$$\theta(t) = \left(1 - \sqrt{1 - \frac{2av_0 t}{r_0^2}}\right) \frac{r_0}{a}. \tag{3}$$
How long does it take for the particle to reach the tube?

(d) Show that the velocity of the particle is given by
$$\mathbf{v}(\theta) = (v_0 \cos\theta, \; -v_0 \sin\theta, \; 0). \tag{4}$$

(e) Show that the magnitude of the angular momentum is
$$L(\theta) = mv_0(r_0 - a\theta + a\sin\theta). \tag{5}$$

(f) Show that the tension in the string (that is, the force on the particle) is
$$\mathbf{T} = -\frac{mv_0^2}{r_0 - a\theta}(\sin\theta, \; \cos\theta, \; 0). \tag{6}$$

(g) Show that, relative to O, the torque acting on the particle is
$$\mathbf{\Gamma} = \frac{2mv_0^2 a}{r_0 - a\theta}(0, \; 0, \; \sin^2\tfrac{1}{2}\theta). \tag{7}$$

Solution

(a) The figure (which is a schematic view of the plane of motion) shows the initial position A of the particle and also a portion APF of the trajectory. The axis of the tube passes through C. When the particle is at P the string has wound around an arc OB – of length $a\theta$ – of the tube, so that $BP = r_0 - a\theta$. It is therefore apparent from the figure that the x- and y-coordinates of P are given by (1) and (2).

(b) From (1) and (2) we have

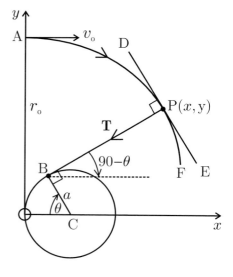

$$\left.\begin{array}{l} \dfrac{dx}{d\theta} = (r_0 - a\theta)\cos\theta \\[6pt] \dfrac{dy}{d\theta} = -(r_0 - a\theta)\sin\theta\,. \end{array}\right\} \tag{8}$$

Therefore, the slope of the tangent DE to the trajectory at P is
$$\frac{dy}{dx} = \frac{dy}{d\theta} \Big/ \frac{dx}{d\theta} = -\tan\theta. \tag{9}$$

The slope of BP is $\cot\theta$, and so the product of these two slopes is -1, meaning that BP is perpendicular to DE. That is, the tension **T** in the string is always perpendicular to the displacement vector $d\mathbf{r}$ of the particle. Thus, $\mathbf{T}$ does no work on the particle, and according to the work–energy theorem the kinetic energy – and therefore the speed $v(\theta)$ – is constant:

$$v(\theta) = v_0. \tag{10}$$

(c) The components of the velocity are

$$\dot{x} = \dot{\theta}\frac{dx}{d\theta} = \dot{\theta}(r_0 - a\theta)\cos\theta, \qquad \dot{y} = \dot{\theta}\frac{dy}{d\theta} = -\dot{\theta}(r_0 - a\theta)\sin\theta, \tag{11}$$

and therefore the speed is

$$v = \sqrt{\dot{x}^2 + \dot{y}^2} = \dot{\theta}(r_0 - a\theta). \tag{12}$$

According to (10) and (12) the angular speed is

$$\frac{d\theta}{dt} = \frac{v_0}{r_0 - a\theta}. \tag{13}$$

By integrating (13) with respect to t, and using the initial condition $\theta = 0$ at $t = 0$, we obtain (3). It follows from (3) that $\theta(t)$ increases monotonically to the value r_0/a in a time

$$\tau = r_0^2/2av_0, \tag{14}$$

at which instant the particle reaches the tube (BP = 0).

(d) From (11) and (13) we have

$$\dot{x} = v_0 \cos\theta, \qquad \dot{y} = -v_0 \sin\theta, \tag{15}$$

and hence (4).

(e) The angular momentum about O is $\mathbf{L} = m\mathbf{r} \times \mathbf{v}$, where $\mathbf{r} = (x, y, 0)$. With x and y given by (1) and (2), and $\mathbf{v}$ by (4), a short calculation shows that

$$\mathbf{L} = -mv_0(r_0 - a\theta + a\sin\theta)\hat{\mathbf{z}}. \tag{16}$$

The magnitude of (16) is (5).

(f) Differentiation of (4) with respect to t and use of (13) yields the acceleration

$$\mathbf{a} = -\frac{v_0^2}{r_0 - a\theta}(\sin\theta,\ \cos\theta,\ 0) \tag{17}$$

and hence the force (6). The magnitude of the tension,

$$T = \frac{mv_0^2}{r_0 - a\theta}, \tag{18}$$

increases without limit as the particle approaches the tube.

(g) The result follows directly from $\mathbf{\Gamma} = \dot{\mathbf{L}}$ and (13) and (16).

Comments

(i) In the above question we have used the trajectory $(x, y, 0)$ to obtain the force on the particle. Other examples of this type appear in Chapter 8.

(ii) The graph of $\theta(t)$ versus t/τ (equation (3)) and two examples of the trajectory (equations (1) and (2)) are shown below.

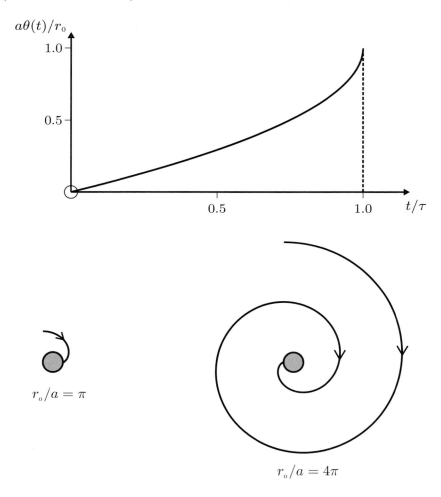

(iii) The graphs below are plots of

$$\frac{L(\theta)}{L_0} = 1 - \frac{a\theta}{r_0} + \frac{a \sin \theta}{r_0} \qquad (19)$$

versus θ for $r_0/a = \pi$ and 8π. There is a point of inflection with horizontal tangent whenever the particle returns to the positive y-axis ($\theta = 2\pi \times$ integer) because the torque about O is zero there – see (7).

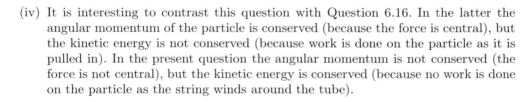

(iv) It is interesting to contrast this question with Question 6.16. In the latter the angular momentum of the particle is conserved (because the force is central), but the kinetic energy is not conserved (because work is done on the particle as it is pulled in). In the present question the angular momentum is not conserved (the force is not central), but the kinetic energy is conserved (because no work is done on the particle as the string winds around the tube).

Question 6.18

Consider a pendulum (see Question 4.3) whose length ℓ varies with time.

(a) Show that the equation of motion in the absence of damping is
$$\ell\ddot{\theta} + 2\dot{\ell}\dot{\theta} + g\sin\theta = 0. \qquad (1)$$

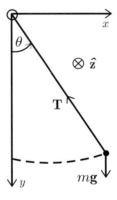

Do this in three different ways:

1. Using $\mathbf{\Gamma} = \dot{\mathbf{L}}$,
2. using $\mathbf{F} = m\mathbf{a}$, and
3. using Lagrange's equation $\dfrac{d}{dt}\left(\dfrac{\partial \mathsf{L}}{\partial \dot{\theta}}\right) - \dfrac{\partial \mathsf{L}}{\partial \theta} = 0$, where $\mathsf{L} = K - V$ is the Lagrangian.

(b) Suppose $\ell(t) = \ell_o + \alpha t$, where ℓ_o and α are constants. Solve (1) for $\theta(t)$ for small θ and subject to the initial conditions $\theta(0) = \theta_o$ and $\dot{\theta}(0) = 0$.
(Hint: The solution can be expressed in terms of Bessel functions.)

Solution

(a) 1. With respect to O, the torque and angular momentum are $\mathbf{\Gamma} = mg\ell\sin\theta\,\hat{\mathbf{z}}$ and $\mathbf{L} = -m\ell^2\dot{\theta}\,\hat{\mathbf{z}}$. The relation $\mathbf{\Gamma} = \dot{\mathbf{L}}$ gives (1).

2. The component equations of $\mathbf{F} = m\mathbf{a}$ are
$$m\ddot{x} + Tx/\ell = 0 \quad \text{and} \quad m\ddot{y} + Ty/\ell - mg = 0. \qquad (2)$$

If we multiply $(2)_1$ by y and $(2)_2$ by x and subtract, the result is

$$y\ddot{x} - x\ddot{y} + gx = 0. \qquad (3)$$

Now, $x = \ell \sin\theta$ and $y = \ell \cos\theta$, and so (3) reduces to (1).

3. The kinetic energy of the pendulum $K = \tfrac{1}{2}mv^2 = \tfrac{1}{2}m(\ell\dot\theta)^2$ and potential energy $V = -mgy = -mg\ell\cos\theta$ yield a Lagrangian $\mathsf{L} = \tfrac{1}{2}m\ell^2\dot\theta^2 + mg\ell\cos\theta$. Equation (1) then follows from Lagrange's equation.

(b) By making use of the chain rule $\dfrac{d\theta}{dt} = \dfrac{d\ell}{dt}\dfrac{d\theta}{d\ell} = \alpha\dfrac{d\theta}{d\ell}$ and approximating $\sin\theta$ by θ, (1) becomes

$$\ell \frac{d^2\theta}{d\ell^2} + 2\frac{d\theta}{d\ell} + \frac{g}{\alpha^2}\theta = 0. \qquad (4)$$

For $\ell(t) = \ell_\circ + \alpha t$ and the given initial conditions, the general solution of (4) is

$$\theta(t) = \pi\sqrt{\frac{g}{\ell}\frac{\ell_\circ}{\alpha}}\left[-J_1\!\left(\frac{2\sqrt{g\ell}}{\alpha}\right)N_2\!\left(\frac{2\sqrt{g\ell_\circ}}{\alpha}\right) + J_2\!\left(\frac{2\sqrt{g\ell_\circ}}{\alpha}\right)N_1\!\left(\frac{2\sqrt{g\ell}}{\alpha}\right)\right]\theta_\circ, \qquad (5)$$

where J_k and N_k are Bessel functions of the first and second kind of order k, respectively.[2]

Comments

(i) The above problem of a simple pendulum with variable length has an interesting history dating back to the beginning of the eighteenth century.[3] Practical questions of the type: 'How long does it take to pull a swinging bucket out of a mine well?' could have motivated these studies.[3] Later, Lorentz posed it as a question in relation to quantum theory at the first Solvay Congress in 1911. Chandrasekhar brought the problem to the attention of Littlewood who subsequently published a paper on it titled 'Lorentz's pendulum problem'.[4]

(ii) If θ is not small then (1) cannot be solved analytically. Numerical solutions are considered below.

Question 6.19

(a) For Question 6.18, use *Mathematica* to find a numerical solution of the equation of motion $\ell\ddot\theta + 2\dot\ell\dot\theta + g\sin\theta = 0$ for $\theta(t)$ when $\ell(t) = \ell_\circ + \alpha t$. Assume $\theta_\circ = 60°$ and $\dot\theta(0) = 0$. Take $g = 9.8\,\text{ms}^{-2}$, $\ell_\circ = 10.0\,\text{m}$, and $\alpha = 0.1\,\text{ms}^{-1}$ (a lengthening pendulum). Repeat this for $\alpha = -0.1\,\text{ms}^{-1}$ (a shortening pendulum).

1. Plot the graphs of $\theta(t)$ for both values of α and for $0 \leq t \leq 85\,\text{s}$.

2. Plot the trajectory of the lengthening pendulum up to $t = 180\,\text{s}$.

---

[2] M. Boas, *Mathematical methods in the physical sciences*, pp. 598–599. Wiley, 3rd edn, 2006.
[3] L. LeCornu, "Mémoire sur le pendule de longueur variable," Acta Mathematica, vol. 19, pp. 201–249, 1895.
[4] J. E. Littlewood, "Lorentz's pendulum problem," Annals of Physics, vol. 21, pp. 233–242, 1963.

(b) In the *Mathematica* notebook change θ_0 to $1°$. Calculate the quarter-periods T_n'' ($n = 1, 2, 3 \cdots$) for the first sixteen quarter-cycles for both values of α and tabulate these results. Plot graphs of the half-periods T_n' versus n on the same axes.
(c) Derive an approximate formula for the half-periods T_n', assuming small oscillations and $|\alpha| \ll \sqrt{\ell_0 g}$.
(d) For the energy $E(t) = K + V$, consider a shortening pendulum with $\ell_0 = 10.0\,\text{m}$ and $\alpha = -1.0\,\text{ms}^{-1}$. Plot graphs of $E(t)/E(0)$ for $0 \le t \le 9\,\text{s}$ and for $\theta_0 = 10°$ and $\theta_0 = 45°$.

Solution

(a) 1. The *Mathematica* notebook below produces the graphs:

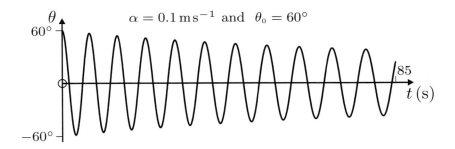

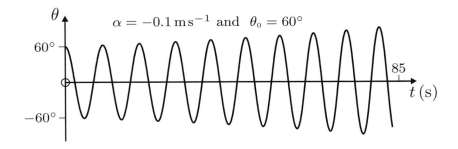

```
In[1]:= g = 9.8; α = 0.1; θ0 = 60 π/180; θ0dot = 0.0;
       L0 = 10.0; Lf = 20.0; T = (Lf - L0)/α;
       Sol = NDSolve[{ (L0 + α t) θ''[t] + 2 α θ'[t] + g Sin[θ[t]] == 0,
           θ[0] == θ0, θ'[0] == θ0dot} , {θ[t], θ'[t]}, {t, 0., T}];
       Plot[Evaluate[180/π θ[t]/.Sol], {t, 0, T},
           PlotRange → {{0, 85}, {-60, 60}}]
```

2. The trajectory of the lengthening pendulum from A ($t = 0$) to B ($t = 180\,\text{s}$) is:

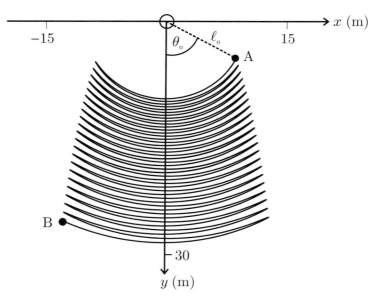

(b) Values of the quarter-periods T_n'' of the nth quarter-cycle (measured from a root of $\theta(t)$ to the contiguous root of $d\theta/dt$) are tabulated below.

$$\alpha = 0.1\,\text{ms}^{-1}$$

n	T_n'' (s)	n	T_n'' (s)	n	T_n'' (s)	n	T_n'' (s)
1	1.600798	5	1.651155	9	1.701506	13	1.751862
2	1.598080	6	1.648435	10	1.698788	14	1.749142
3	1.625977	7	1.676332	11	1.726685	15	1.777039
4	1.623258	8	1.673612	12	1.723965	16	1.774321

$$\alpha = -0.1\,\text{ms}^{-1}$$

n	T_n'' (s)	n	T_n'' (s)	n	T_n'' (s)	n	T_n'' (s)
1	1.572901	5	1.522500	9	1.472197	13	1.421843
2	1.575621	6	1.525269	10	1.474914	14	1.424559
3	1.547727	7	1.497373	11	1.447020	15	1.396667
4	1.550445	8	1.500091	12	1.449738	16	1.399384

Values of the half-period T_n' of the nth half-cycle ($n = 1, 1\tfrac{1}{2}, 2, 2\tfrac{1}{2} \cdots$) were calculated by adding successive quarter-periods.[‡] This yields the following graphs.

[‡] Here, integral values of n correspond to adding quarter-periods from the same half-cycle, whereas half-integral values of n correspond to adding the second quarter-period of the nth half-cycle to the first quarter-period of the $(n+1)$th half-cycle.

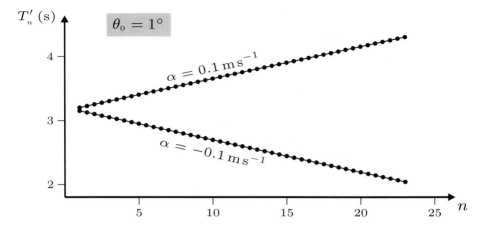

The slopes of these straight lines, obtained by linear regression, are $\pm 0.050355\,\text{s}$.

(c) For small oscillations we can make the harmonic approximation

$$T'_n = \pi\sqrt{\frac{\ell}{g}}. \qquad (1)$$

Here, $\ell = \ell_o + \alpha t$ and $t \approx (n-1)T'_1 = (n-1)\pi\sqrt{\frac{\ell_o}{g}}$. That is,

$$T'_n \approx \pi\sqrt{\frac{\ell_o + \alpha(n-1)\pi\sqrt{\ell_o/g}}{g}} = \pi\sqrt{\frac{\ell_o}{g}}\sqrt{1 + \frac{\alpha(n-1)\pi}{\sqrt{\ell_o g}}}. \qquad (2)$$

If $|\alpha| \ll \sqrt{\ell_o g}$ we can approximate (2) by an expansion to first order in α:

$$T'_n \approx \pi\sqrt{\frac{\ell_o}{g}} + \frac{\pi^2 \alpha}{2g}(n-1). \qquad (3)$$

According to (3), a plot of T'_n versus n should be a straight line with slope $\pi^2\alpha/2g$. For $\alpha = \pm 0.1\,\text{ms}^{-1}$ and $g = 9.8\,\text{ms}^{-2}$ the values of this slope are $\pm 0.050355\,\text{s}$, which are the same (to five figures) as the values obtained from the graphs in (b).

(d) We choose the zero of potential energy at $y = \ell_o$. Then, $E(t) = K + V = \frac{1}{2}m\ell^2\dot\theta^2 + mg(\ell_o - \ell\cos\theta)$ and so

$$\frac{E(t)}{E(0)} = \frac{\frac{1}{2}\ell_o\left(1 + \frac{\alpha t}{\ell_o}\right)^2\dot\theta^2 + g\left(1 - \left(1 + \frac{\alpha t}{\ell_o}\right)\cos\theta\right)}{g(1 - \cos\theta_o)}. \qquad (4)$$

Plots of (4) for a shortening pendulum with $\theta_o = 10°$ and $45°$, and $\alpha = -0.1\,\text{ms}^{-1}$ are shown below.

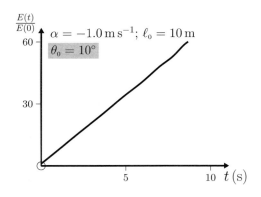
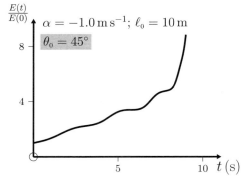

Comments

(i) The small-angle solution expressed in terms of Bessel functions (see (5) of Question 6.18) is reasonable for $|\theta| \lesssim 5°$.

(ii) The trajectory from A to B plotted in part (a) above, was calculated for $\alpha > 0$. A similar calculation shows that for the corresponding $\alpha < 0$ the pendulum moves along the same path but in the reverse direction, from B to A. This reversibility is expected since the system is frictionless.

(iii) The values of the quarter-periods, given for $\theta_\circ = 1°$ in the first of the above tables, for a lengthening pendulum shows that an inward swing (n odd) takes longer than either the preceding or following outward swing. A proof of this result based on properties of the zeros of Bessel functions has been given by Chambers (see Ref. [2]). The reverse is true for a shortening pendulum (see the second table). Interestingly, these feature occur also for large values of $\theta_\circ$. This variation in the quarter-periods is absent in the half-periods, as the above graphs of T'_n show.

(iv) For $|\theta| \lesssim 10°$ the rate at which the external agent does work on the pendulum is approximately constant, as illustrated above.

(v) There are other interesting physical systems for which the equation of motion is the analogue of the small-angle approximation to (1) of Question 6.18. For example, for an undamped linear mechanical oscillator having a time-dependent mass $(m\ddot{x} + \dot{m}\dot{x} + kx = 0)$ and an LC circuit having a time-dependent inductance $(L\ddot{q} + \dot{L}\dot{q} + q/C = 0)$. Here, the counterpart of θ is the displacement x for the mechanical oscillator or the charge q on the capacitor for the electrical oscillator. Note that the term containing $\dot{m}$ (or $\dot{L}$) simulates damping linear in $\dot{x}$ or $\dot{q}$. The solutions of these equations are decreasing (increasing) oscillatory functions, depending on whether $\dot{m}$ or $\dot{L}$ is positive (negative), such as those depicted in part (a) above.

7
Motion in two and three dimensions

This chapter contains a variety of problems that illustrate the use of vector techniques and calculus in the solution and analysis of two- and three-dimensional motion. Most of the questions are solved using a Cartesian system; included among these is the fundamental problem of motion in an inverse-square force, which is more traditionally treated in polar coordinates.

Question 7.1

A particle of mass m is projected in a uniform gravitational field. The initial conditions at time $t = 0$ are[‡]
$$\mathbf{r}_0 = 0 \quad \text{and} \quad \mathbf{v}_0 = \mathbf{u} = (u_1, u_2, u_3). \tag{1}$$

(a) Determine the trajectory $\mathbf{r}(t)$ in terms of $\mathbf{u}$ and $\mathbf{g}$ (the gravitational acceleration), and show that it is a parabola.
(b) Determine the range of the projectile on a horizontal plane, and its maximum height above the plane (in terms of the u_i and g). Hence, determine the maximum range.

Solution

(a) We integrate the equation of motion
$$m\frac{d\mathbf{v}}{dt} = m\mathbf{g}, \tag{2}$$
with respect to t between the limits $t = 0$ and t, and use the initial condition $(1)_2$. Because $\mathbf{g}$ is a constant this yields
$$\int_{\mathbf{u}}^{\mathbf{v}(t)} d\mathbf{v} = \mathbf{g}\int_0^t dt, \quad \text{and so} \quad \mathbf{v}(t) = \mathbf{u}(t) + \mathbf{g}t. \tag{3}$$

To find $\mathbf{r}(t)$ we recall that $d\mathbf{r}/dt = \mathbf{v}$ and integrate (3) with respect to t, taking account of the initial condition $(1)_1$. This gives

---

[‡]We will often use $\mathbf{u}$, rather than $\mathbf{v}_0$, to denote the initial velocity; this simplifies the notation by avoiding the occurrence of double subscripts (as in, for example, v_{01}).

$$\int_0^{r(t)} d\mathbf{r} = \int_0^t (\mathbf{u} + \mathbf{g}t) dt, \qquad \text{and thus} \qquad \mathbf{r}(t) = \mathbf{u}t + \tfrac{1}{2}\mathbf{g}t^2. \qquad (4)$$

To interpret this trajectory it is convenient to orient the coordinate axes so that the y-axis is vertically upward and the initial velocity is in the xy-plane. Thus, $\mathbf{g} = (0, -g, 0)$, where $g > 0$, and $\mathbf{u} = (u_1, u_2, 0)$. Then (4) yields the two parametric equations

$$x = u_1 t, \qquad y = u_2 t - \tfrac{1}{2} g t^2, \qquad (5)$$

and consequently the parabola

$$y = \frac{u_2}{u_1} x - \frac{g}{2 u_1^2} x^2. \qquad (6)$$

(b) This parabola has intercepts at $x = 0$ and $x = 2 u_1 u_2 / g$, and maximum value $y_{\max} = u_2^2 / 2g$ at $x = u_1 u_2 / g$. Thus, the range R on a horizontal plane and the maximum height H of the projectile above this plane are given by

$$R = \frac{2 u_1 u_2}{g}, \qquad H = \frac{u_2^2}{2g}. \qquad (7)$$

Now, $u_1 = u \cos\theta$ and $u_2 = u \sin\theta$, where θ is the angle of projection. Thus,

$$R(\theta) = \frac{u^2}{g} \sin 2\theta. \qquad (8)$$

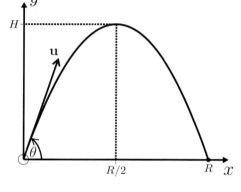

It follows that the range is a maximum when $\sin 2\theta = 1$, that is $\theta = \tfrac{1}{4}\pi$:

$$R_{\max} = \frac{u^2}{g} \quad \text{when} \quad \theta = \tfrac{1}{4}\pi. \qquad (9)$$

Comments

(i) The trajectory (4) is independent of the mass m of the particle. This feature is a result of our tacit use of the weak equivalence principle – that the gravitational and inertial masses m^G and m^I can be taken to be the same. Without this assumption the equation of motion (2) is $m^I \dot{\mathbf{v}} = m^G \mathbf{g}$, and consequently $\mathbf{g}$ is replaced by $(m^G/m^I)\mathbf{g}$ in (3) and (4). The experimental basis for the equivalence principle is discussed in Questions 2.4 and 2.5.

(ii) If the particle is subject to friction then the trajectory is, in general, dependent on m (see Questions 7.7 and 7.8).

(iii) The trajectory (6) can also be expressed in terms of H and R as

$$y = \frac{4H}{R}\left(x - \frac{x^2}{R}\right). \qquad (10)$$

(iv) The parabolic trajectory (6) for motion in a uniform field can be a good approximation to projectile motion in the non-uniform field of an idealized (spherically symmetric) Earth. The latter trajectory is part of an ellipse with one focus at the centre C of the Earth (see Question 8.9):

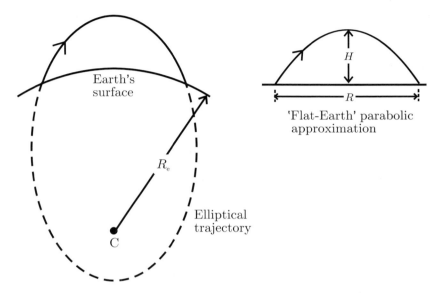

This approximation was known to Newton, and the conditions for its validity are usually given as either $H \ll R_e$ or $R \ll R_e$, or both, where R_e is the Earth's radius. Burko and Price[1] have analyzed in detail the reduction of an exact, elliptical trajectory to an approximate, 'flat-Earth' parabola and shown that it requires two conditions: 1. $H \ll R_e$, 2. The maximum curvature of the trajectory $(g/u_1^2$ where u_1 is the velocity at the apex) should be large compared to the Earth's curvature R_e^{-1}. According to (7) this means $R \ll \sqrt{HR_e}$, which is more restrictive than the condition $R \ll R_e$.

Question 7.2

For the projectile discussed in the previous question, show that the range on a plane inclined at an angle α to the horizontal is

$$R(\alpha) = \frac{2u^2}{g\cos^2\alpha}\cos\theta\sin(\theta-\alpha),\qquad(1)$$

for $-\tfrac{1}{2}\pi \leq \alpha \leq \theta$. Deduce that R is a maximum for

$$\theta = \tfrac{1}{4}\pi + \tfrac{1}{2}\alpha,\qquad(2)$$

and determine $R_{\max}$.

---

[1] L. M. Burko and R. H. Price, "Ballistic trajectory: parabola, ellipse, or what?," American Journal of Physics, vol. 73, pp. 516–520, 2005.

Solution

We resolve the trajectory (4) of Question 7.1 into components x' (parallel to the plane) and y' (perpendicular to the plane) by resolving **u** and **g** along these directions:

$$x' = u\cos(\theta - \alpha)t - \tfrac{1}{2}(g\sin\alpha)t^2 \quad \text{and} \quad y' = u\sin(\theta - \alpha)t - \tfrac{1}{2}(g\cos\alpha)t^2. \qquad (3)$$

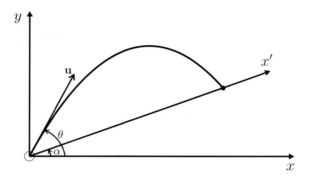

From $(3)_2$ with $y' = 0$, the time taken by the projectile to reach the plane is

$$t = \frac{2u\sin(\theta - \alpha)}{g\cos\alpha}. \qquad (4)$$

Equations $(3)_1$ and (4) yield (1). Use of the trigonometric identity $2\cos A\sin B = \sin(A+B) - \sin(A-B)$ shows that (1) can be expressed as

$$R = \frac{u^2}{g\cos^2\alpha}\bigl[\sin(2\theta - \alpha) - \sin\alpha\bigr]. \qquad (5)$$

According to (5), R is a maximum when $2\theta - \alpha = \tfrac{1}{2}\pi$, which is (2). From (5) and (2) the maximum range is

$$R_{\max}(\alpha) = \frac{u^2}{g}\frac{1 - \sin\alpha}{\cos^2\alpha} = \frac{u^2}{g(1 + \sin\alpha)}. \qquad (6)$$

Comments

(i) Graphs of (2) and (6) for $-\tfrac{1}{2}\pi \le \alpha \le \tfrac{1}{2}\pi$ are shown below. For $\alpha = 0$ the values of θ and $R_{\max}$ are those of (9) in the previous question. For a downward tilt of the plane ($\alpha < 0$) the maximum range is increased and, as one expects, it tends to infinity as $\alpha \to -\tfrac{1}{2}\pi$. An upward tilt decreases the maximum range: increasing α from 0 to $\tfrac{1}{2}\pi$ has the effect of halving $R_{\max}$ (from u^2/g to $u^2/2g$).

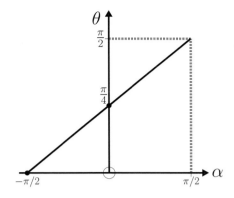
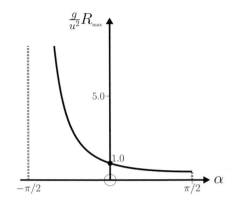

(ii) We can also regard (5) as a relation between u and θ for fixed R and α. The derivative $du/d\theta$ is zero when θ is given by (2): by inspection of (5), $u(\theta)$ is a minimum at this angle. Thus, the same angle θ that maximizes the range R of the projectile for a given launch speed u, also minimizes the required speed u for a given range R:

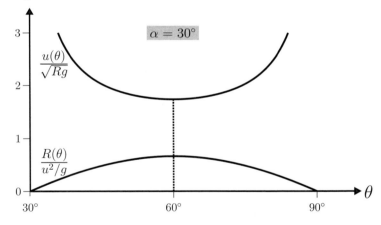

Question 7.3

A projectile is fired from the origin with velocity $\mathbf{u} = (u_1, u_2)$ at time $t = 0$. At the same instant a target is released from rest at the point (X, Y). Assume that $\mathbf{g}$ is constant and neglect air resistance.

(a) Show that the condition for the particles to collide is
$$\tan\theta = Y/X, \tag{1}$$
where θ is the angle of projection.

(b) Determine the value of the initial speed u for which the collision occurs 1. at 'ground level' ($y = 0$), and 2. at the apex of the projectile's trajectory ($y = H$).

Solution

(a) The condition for the two particles to collide is that their y-coordinates be equal at the instant t when their x-coordinates are equal. That is,
$$u_2 t - \tfrac{1}{2}gt^2 = Y - \tfrac{1}{2}gt^2 \quad \text{when} \quad u_1 t = X. \tag{2}$$
Therefore
$$u_2/u_1 = Y/X, \tag{3}$$
which is (1) because $u_1 = u\cos\theta$ and $u_2 = u\sin\theta$.

(b) **1.** For a collision at $y = 0$, (2) requires $Y - \tfrac{1}{2}gX^2/u_1^2 = 0$ and therefore
$$u_1 = \sqrt{gX^2/2Y}. \tag{4}$$
According to (3) and (4) the initial speed is
$$\begin{aligned} u_g &= \sqrt{u_1^2 + u_2^2} \\ &= \sqrt{g(X^2 + Y^2)/2Y}. \end{aligned} \tag{5}$$

2. At the apex $y = u_2^2/2g$ (see $(7)_2$ of Question 7.1), and so for a collision at this point it is necessary that
$$Y - gX^2/2u_1^2 = u_2^2/2g. \tag{6}$$
The solution to the simultaneous equations (3) and (6) is
$$u_1 = \sqrt{gX^2/Y}, \quad u_2 = \sqrt{gY}, \tag{7}$$
and therefore
$$u_a = \sqrt{g(X^2 + Y^2)/Y} = \sqrt{2}\, u_g. \tag{8}$$

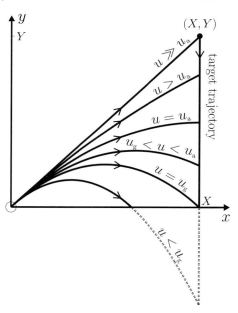

Comments

(i) The result (1) is remarkable in that it is independent of g, u and the masses of the particles. This simplicity is removed if we relax any of the assumptions on which (1) is based (constant g, no drag, equality of inertial and gravitational masses,[‡] simultaneity of starting the motions). The next question deals with the effect of a time delay in releasing the target.

(ii) According to (b) above, the possible points of collision for a given angle of projection depend on the initial speed u, as illustrated in the diagram. If $u \gg u_a$ the projectile travels in very nearly a straight line to the target. The dashed curve is for a collision below the xz-plane.

[‡]See Question 2.5.

Question 7.4

Suppose that the projectile in Question 7.3 is fired at time $t = 0$, while the target is released at $t = t_o$ (which can be positive or negative).

(a) Show that the condition (1) for a collision is modified to read
$$\sin\theta = A\cos\theta + B, \tag{1}$$
where
$$A(t_o) = \frac{Y}{X} - \frac{gt_o^2}{2X} \quad \text{and} \quad B(t_o) = \frac{gt_o}{u}. \tag{2}$$

(b) Show that the relevant solution, $\theta(t_o)$, of (1) is given by
$$\tan\theta = \frac{\frac{Y}{X} - \frac{gt_o^2}{2X} + \frac{gt_o}{u}\sqrt{1 + \left(\frac{Y}{X} - \frac{gt_o^2}{2X}\right)^2 - \frac{g^2 t_o^2}{u^2}}}{1 - \frac{g^2 t_o^2}{u^2}}. \tag{3}$$

(c) Suppose
$$X = 8.0\,\text{m}, \quad Y = 6.0\,\text{m}, \quad u = 11.0\,\text{ms}^{-1}, \quad \text{and} \quad g = 9.8\,\text{ms}^{-2}. \tag{4}$$
Plot the trajectories for **1.** $t_o = 0$, and **2.** $t_o = \pm\frac{1}{5}\sqrt{\frac{2Y}{g}}$.

(d) Plot a graph of $y(t_o) = Y - \frac{1}{2}g(t - t_o)^2$ versus t_o for $-0.5\,\text{s} \le t_o \le 1.05\,\text{s}$. Use the parameters in (4) and $t = X/u_1$ where $u_1(t_o) = u\cos\theta(t_o)$, see Question 7.3. Write a *Mathematica* notebook to determine the value(s) of t_o that correspond to a collision **1.** at $y(t_o) = 0$, and **2.** at the maximum value of $y(t_o)$.

(e) Plot a graph of $\theta(t_o)/\theta_o$ versus t_o for $0° \le \theta(t_o) \le 90°$ and $\theta_o = \tan^{-1}(Y/X)$.

Solution

(a) Instead of (2) in Question 7.3, the condition for a collision at time t is now
$$u_2 t - \tfrac{1}{2}gt^2 = Y - \tfrac{1}{2}g(t - t_o)^2 \quad \text{when} \quad u_1 t = X. \tag{5}$$
That is,
$$u_2 = \left(Y/X - \tfrac{1}{2}gt_o^2/X\right)u_1 + gt_o, \tag{6}$$
which is (1) because $u_1 = u\cos\theta$ and $u_2 = u\sin\theta$.

(b) We convert (1) to a quadratic equation in $\tan\theta$ by writing it as $\tan\theta - A = B\sec\theta$ and squaring both sides of this equation. Then, after using the identity $\sec^2\theta = 1 + \tan^2\theta$, we obtain
$$(1 - B^2)\tan^2\theta - 2A\tan\theta + A^2 - B^2 = 0. \tag{7}$$
The solution to this equation which has the required property that θ increases for small, positive values of t_o is (3): for small values of $|t_o|$, (3) simplifies to the linear relation
$$\tan\theta = Y/X + (gt_o/u)\sqrt{1 + Y^2/X^2}. \tag{8}$$
That is, one should aim higher if the target is released slightly late.

(c) The angle of projection for no time delay is $\theta_0 = \tan^{-1}(Y/X) = 36.9°$. With $t_0 = 0.221\,\text{s}$, we find from (3) that $\theta(t_0) = 45.0°$ and $\theta(-t_0) = 26.6°$. The corresponding trajectories of the projectile, calculated using (6) of Question 7.1, are plotted below. Also shown is the vertical trajectory of the target.

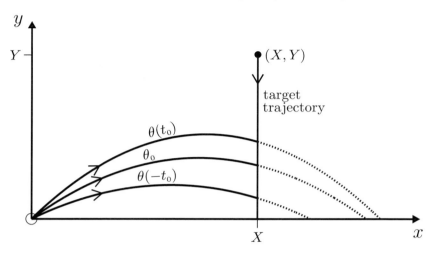

(d) The graph for the y-coordinate of the collision, $y(t_0)$, is:

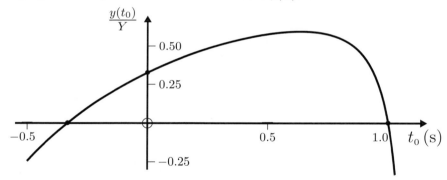

The notebook below gives 1. $t_0 = 1.00\,\text{s}$ and $t_0 = -0.332\,\text{s}$; 2. $t_0 = 0.635\,\text{s}$.

```
In[1]:= X = 8.0; Y = 6.0; g = 9.8; u = 11.0;

θ[t0_] := ArcTan[ (Y/X - g t0²/(2X) + (g t0/u) √(1 + (Y/X - g t0²/(2X))² - g² t0²/u²)) / (1 - g² t0²/u²) ];

t[t0_] := X / (u Cos[θ[t0]]);   f[t0_] := Y - 0.5 g (t[t0] - t0)²;

t0 = t0 /. FindRoot[Y == (1/2) g (t[t0] - t0)², {t0, 0.99}]
t0 = t0 /. FindRoot[f'[t0] == 0, {t0, 0.5}]
```

(e) The following graph is a plot of $\theta(t_0)/\theta_0$ versus t_0, obtained from (3) for the values in (4). The dashed parts of the curve are for collisions below the xz-plane.

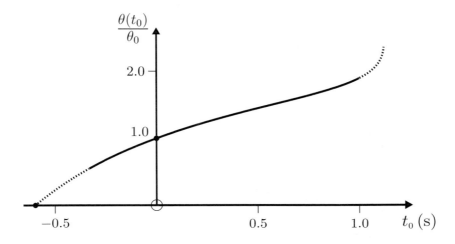

Question 7.5

Consider a point $P(\frac{1}{2}R, Y)$ on the symmetry axis of the parabolic trajectory of the projectile in Question 7.1. Determine the position(s) (x_c, y_c) of closest approach of the projectile to the point P if $Y > 0$.

Solution

According to Pythagoras's theorem the distance D (> 0) between some point $P(X, Y)$ and a point (x, y) on the trajectory satisfies
$$D^2(x) = (x - X)^2 + (y - Y)^2. \tag{1}$$
A minimum D requires $dD/dx = 0$, and therefore $dD^2/dx = 0$: from (1) and with y given by (10) of Question 7.1 we have
$$x^3 - \frac{3R}{2}x^2 + \frac{R^2}{2}\left(1 + \frac{Y}{2H} + \frac{R^2}{16H^2}\right)x - \frac{R^3}{8H}\left(Y + \frac{XR}{4H}\right) = 0. \tag{2}$$
Solution of this cubic equation will give the x-coordinate x_c of the point of closest approach to some point $P(X, Y)$ in the xy-plane. We restrict ourselves to a point P on the symmetry axis, $X = \frac{1}{2}R$. For this value of X we can simplify (2) by making the change of variable $x = w + \frac{1}{2}R$. Then, (2) becomes
$$w^3 - \tfrac{1}{4}R^2(1 - Y/H - R^2/8H^2)w = 0, \tag{3}$$
which has roots
$$w = 0 \quad \text{and} \quad w = \pm\tfrac{1}{2}R\sqrt{1 - Y/H - R^2/8H^2}. \tag{4}$$

166 Solved Problems in Classical Mechanics

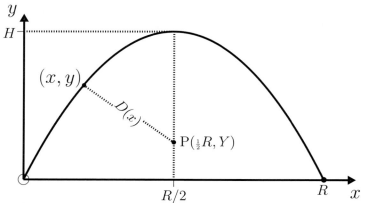

Thus, the possible x-coordinates of the point of closest approach are

$$x_c^o = R/2 \quad \text{and} \quad x_c^\pm = \tfrac{1}{2}R\left[1 \pm \sqrt{1 - Y/H - R^2/8H^2}\right]. \tag{5}$$

The possible y-coordinates of the point of closest approach, obtained by substituting (5) in (10) of Question 7.1, are: $y_c^o = H$ and $y_c^\pm = Y + R^2/8H$.

Comments

(i) Which of the roots in (5) give the x-coordinate(s) of closest approach depends on the value of Y. For

$$Y < H(1 - R^2/8H^2) \tag{6}$$

the relevant roots are $x_c^\pm$ (the values of D for these two roots are equal). For

$$Y \geq H(1 - R^2/8H^2) \tag{7}$$

the solution is x_c^o. It is helpful here to think of circles that can be inscribed within a parabola:

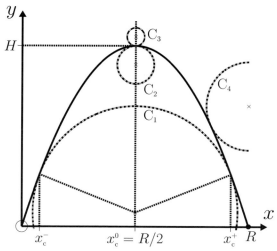

If Y satisfies (6), the inscribed circle C_1 with centre $(\tfrac{1}{2}R, Y)$ shows that the relevant solutions are $x_c^{\pm}$, whereas if Y satisfies (7), we have circles such as C_2 and C_3, and the x-coordinate of closest approach is x_c^o. That is, $D(x)$ has either two minima and a local maximum, or just one minimum:

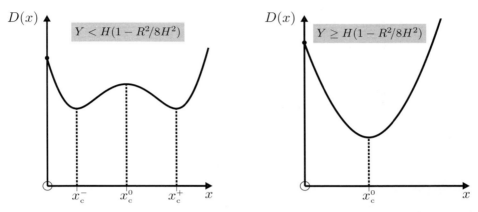

(ii) For points $P(X, Y)$ off the symmetry axis one must solve (2). The interpretation of the results is lengthy and involves an interesting analysis of the roots of a cubic equation. Alternatively, one can obtain a numerical solution with *Mathematica*'s FindRoot function. For $P(R, \tfrac{2}{3}H)$, for example, we find $x_c \approx 0.83R$ and $y_c \approx 0.56H$ (see the semi-circle C_4 above).

Question 7.6

A projectile is fired from a platform that is moving horizontally with velocity $\mathbf{V} = (V, 0, 0)$. The initial velocity of the projectile relative to the platform is $\mathbf{u} = (u_1, u_2, 0)$. Show that the range on a horizontal plane through the platform is

$$R(\theta) = \frac{u^2}{g}\sin 2\theta + \frac{2Vu}{g}\sin\theta, \qquad (1)$$

where $u = \sqrt{u_1^2 + u_2^2}$ and θ is the angle of projection ($0 \le \theta \le \tfrac{1}{2}\pi$). Determine the value of θ that makes R a maximum, and calculate $R_{\max}$ in terms of u, V and g.

Solution

The effect of the motion of the platform is simply to replace the horizontal component u_1 of the initial velocity relative to the Earth with $u_1 + V$. Thus, $(7)_1$ of Question 7.1 becomes

$$R = \frac{2(u_1 + V)u_2}{g}, \qquad (2)$$

which is (1) because $u_1 = u\cos\theta$ and $u_2 = u\sin\theta$. To find the maximum range we first differentiate (1) with respect to θ and set the result equal to zero: then use of the identity $\cos 2\theta = 2\cos^2\theta - 1$ yields the quadratic equation

$$2u\cos^2\theta + V\cos\theta - u = 0. \tag{3}$$

The value of θ that makes R a maximum is given by the positive root of (3), that is

$$\cos\theta = \tfrac{1}{4}\left(-\epsilon + \sqrt{\epsilon^2 + 8}\,\right), \tag{4}$$

where $\epsilon = V/u$. The corresponding value of $\sin\theta = \sqrt{1 - \cos^2\theta}$ is

$$\sin\theta = \sqrt{\tfrac{1}{2} - \tfrac{1}{8}\epsilon^2 + \tfrac{1}{8}\epsilon\sqrt{\epsilon^2 + 8}}. \tag{5}$$

From (1), (4) and (5) we obtain the maximum range

$$R_{\max} = \frac{u^2}{2g}\left(3\epsilon + \sqrt{\epsilon^2 + 8}\,\right)\sqrt{\tfrac{1}{2} - \tfrac{1}{8}\epsilon^2 + \tfrac{1}{8}\epsilon\sqrt{\epsilon^2 + 8}}. \tag{6}$$

Comments

(i) When $\epsilon = 0$ (i.e. $V = 0$), (4) and (6) reduce to (9) of Question 7.1.
(ii) The dependence of θ and $R_{\max}$ on $\epsilon = V/u$ obtained from (4) and (6) is shown in the following figures. Note that for $\epsilon > 0$ the platform is moving towards the target, while for $\epsilon < 0$ it is moving away. The dotted lines represent asymptotic values.

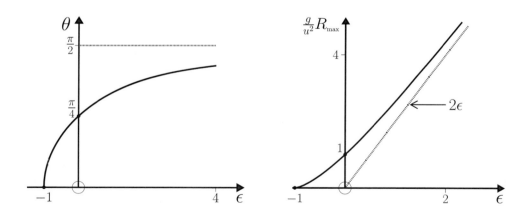

Question 7.7

A projectile of mass m is fired from the origin with initial velocity $\mathbf{u}$ in a uniform gravitational field. Assume a linear drag equal to $-\alpha\mathbf{v}$ (α is a positive constant) due to the surrounding medium, and determine the trajectory $\mathbf{r}(t)$ in terms of m, α, $\mathbf{u}$ and the gravitational acceleration $\mathbf{g}$.

Solution

The equation of motion is

$$m\frac{d\mathbf{v}}{dt} = m\mathbf{g} - \alpha\mathbf{v}. \tag{1}$$

The substitution $\mathbf{v} = \mathbf{g}\tau + \mathbf{w}$, where $\tau = m/\alpha$ is a characteristic time, changes (1) into the homogeneous equation

$$\frac{d\mathbf{w}}{dt} + \frac{\mathbf{w}}{\tau} = 0, \tag{2}$$

whose general solution is $\mathbf{w} = \mathbf{w}_0 e^{-t/\tau}$. Thus, the general solution to (1) is

$$\mathbf{v}(t) = \mathbf{g}\tau + \mathbf{w}_0 e^{-t/\tau}. \tag{3}$$

The initial condition $\mathbf{v}(0) = \mathbf{u}$ requires $\mathbf{w}_0 = \mathbf{u} - \mathbf{g}\tau$, and therefore

$$\mathbf{v}(t) = \mathbf{g}\tau + (\mathbf{u} - \mathbf{g}\tau)e^{-t/\tau}. \tag{4}$$

Integrating (4) with respect to t, subject to the initial condition $\mathbf{r}(0) = 0$, gives the trajectory

$$\mathbf{r}(t) = \mathbf{g}\tau t + \tau(\mathbf{u} - \mathbf{g}\tau)\left(1 - e^{-t/\tau}\right). \tag{5}$$

Comments

(i) The velocity (4) and the trajectory (5) both depend on the mass m. For vanishing drag (i.e. $\tau \to \infty$) they have the mass-independent limits $\mathbf{v} = \mathbf{u} + \mathbf{g}t$ and $\mathbf{r} = \mathbf{u}t + \frac{1}{2}\mathbf{g}t^2$ discussed in Question 7.1.

(ii) The trajectory (5) lies in a vertical plane defined by the vectors $\mathbf{g}$ and $\mathbf{u}$. With axes oriented so that $\mathbf{g} = (0, -g, 0)$ and $\mathbf{u} = (u_1, u_2, 0)$ (see Question 7.1), the motion is in the xy-plane. Then, by eliminating t from the component equations of (5), we have

$$y(x) = g\tau^2 \ln\left(1 - \frac{x}{\tau u_1}\right) + \left(\frac{u_2}{u_1} + \frac{g\tau}{u_1}\right)x. \tag{6}$$

The diagram below illustrates this trajectory for $u = 100\,\text{ms}^{-1}$, $g = 9.8\,\text{ms}^{-2}$, $\theta = \frac{1}{6}\pi$ and various values of τ.

(iii) The above calculation is readily extended to include the effect of a constant crosswind $\mathbf{V}$. The frictional force (which is proportional to the velocity of the projectile relative to the air) is equal to $-\alpha(\mathbf{v} - \mathbf{V})$, and the equation of motion (1) is modified to read

$$m\dot{\mathbf{v}} = m\mathbf{g} - \alpha(\mathbf{v} - \mathbf{V}) = (m\mathbf{g} + \alpha\mathbf{V}) - \alpha\mathbf{v}. \tag{7}$$

Therefore, the solution $\mathbf{r}(t)$ is obtained by making the substitution $\mathbf{g} \to \mathbf{g} + \mathbf{V}/\tau$ in (5). The resulting trajectory lies in an inclined plane defined by $\mathbf{g} + \mathbf{V}/\tau$ and $\mathbf{u}$.

(iv) A linear drag applies only at low Reynolds numbers; more commonplace is the quadratic drag encountered at higher Reynolds numbers (see Question 3.8). This problem cannot be solved analytically, and a numerical solution is given in the next question.

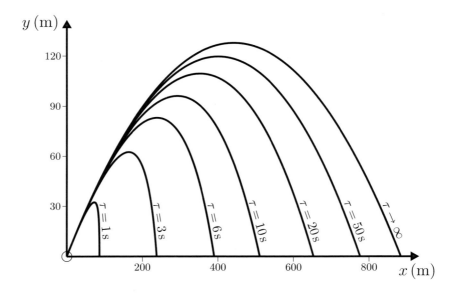

Question 7.8

A projectile of mass m fired from the origin with initial velocity $\mathbf{u}$ in a uniform gravitational field $\mathbf{g}$ experiences a quadratic drag equal to $-\beta \mathbf{v}^2$ (β is a positive constant). Solve the equation of motion $m\ddot{\mathbf{r}} = m\mathbf{g} - \beta|\mathbf{v}|\mathbf{v}$ numerically and obtain the trajectory $y(x)$, taking $u = 100 \text{ ms}^{-1}$, $\theta = \frac{1}{6}\pi$, $g = 9.8 \text{ ms}^{-2}$ and the following values of m/β: 100 m, 250 m, 500 m, 900 m, 1500 m and 4500 m. Plot these trajectories on the same graph and show also the case for zero drag ($m/\beta \to \infty$).

Solution

We solve the equation of motion $\ddot{\mathbf{r}} = \mathbf{g} - \lambda^{-1}|\mathbf{v}|\mathbf{v}$ (where $\lambda = m/\beta$ is a characteristic length) using the following *Mathematica* notebook. This yields the trajectories shown below.

```
In[1]:= u = 100.0; θ = 30 π/180; g = {0, -9.8}; λ0 = 4500; tmax = 2 u Sin[θ]/√Dot[g, g];

       λ[t_] := λ0                                    (* Use λ[t_] for Q 7.8 *)

       (* Y0 = 7460; λ[t_] := λ0 Exp[y[t]/Y0] *)      (* Use λ[t_] for Q 7.9 *)

       x0 = 0; y0 = 0; vx0 = u Cos[θ]; vy0 = u Sin[θ];

       r[t_] := {x[t], y[t]}; v[t_] := √(x'[t]^2 + y'[t]^2);

       EqnMotion = Thread[r''[t] - g + r'[t] v[t]/λ[t] == 0];
```

```
In[2]:= InitCon = Join[Thread[r[0] == {x0, y0}], Thread[r'[0] == {vx0, vy0}]];

        Sol = NDSolve[Join[EqnMotion, InitCon], {x[t], y[t]}, {t, 0, tmax}];

        tmax = t /. FindRoot[y[t] == 0 /. Sol, {t, tmax}];

        ParametricPlot[Evaluate[{x[t], y[t]}] /. Sol, {t, 0, tmax}]
```

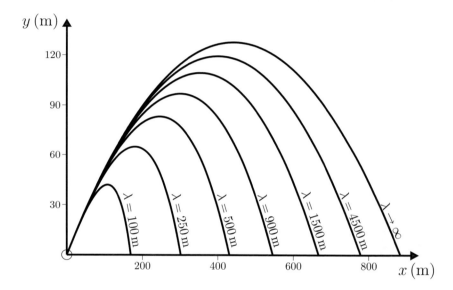

Comment

It is sometimes stated in the literature that the maximum range of a projectile in a resistive medium is attained for an angle of projection θ_m that is always less than the value of 45° for no drag. This seems to be based largely on intuition and the case for linear drag, for which it can be proved[2] that $\theta_m < 45°$. However, a detailed analysis by Price and Romano[3] has shown that in general the behaviour is richer than this. They studied projectile motion subject to a power-law drag force $\mathbf{F}_d = -\gamma v^n \hat{\mathbf{v}}$, and found the following. In general, θ_m depends on the drag parameters n and $k = \gamma v_0^n / mg$ (the ratio of the initial drag force to the weight of the projectile). In the limit of weak drag (k small) they evaluated θ_m to first order in k and found that $\theta_m < 45°$ when n is less than a critical value $n_c \approx 3.41$, whereas $\theta_m > 45°$ for $n > n_c$. The largest value of $\theta_m \approx (45 + 4k)°$ occurred for $n \approx 8$. Computer calculations for larger k showed that n_c increases with k and that θ_m did not exceed about 47°. By contrast, for strong

---

[2] R. H. Price and J. D. Romano, "Comment on 'On the optimal angle of projection in general media', by C. W. Groetsch [Am. J. Phys. 65 (8), 797–799 (1997)]," American Journal of Physics, vol. 66, p. 114, 1998.

[3] R. H. Price and J. D. Romano, "Aim high and go far – optimal projectile launch angles greater than 45°," American Journal of Physics, vol. 66, pp. 109–113, 1998.

drag and $n < n_c$, the values of θ_m are small, meaning that very shallow trajectories are required for maximum range. For 'high' trajectories, these conclusions are altered due to the variation of atmospheric density with altitude (see below).

Question 7.9

Suppose the projectile of the previous question moves through an isothermal atmosphere. The only effect on the equation of motion $\ddot{\mathbf{r}} = \mathbf{g} - \lambda^{-1}|\mathbf{v}|\mathbf{v}$ is to replace λ by $\lambda e^{y/Y}$, where Y is a positive constant (see Question 3.14). Make the necessary change to the *Mathematica* notebook above, and then determine the angle of projection θ (to the nearest degree) that gives maximum range when $g = 9.8\,\mathrm{ms^{-2}}$, $u = 1000\,\mathrm{ms^{-1}}$, $\lambda = 30\,000\,\mathrm{m}$ and $Y = 7460\,\mathrm{m}$.

Solution

From graphs such as the following we find that for the above parameters the maximum range occurs when $\theta = 50°$. To 'go further' one must aim 'higher' (above 45°) to reach a less-dense atmosphere.

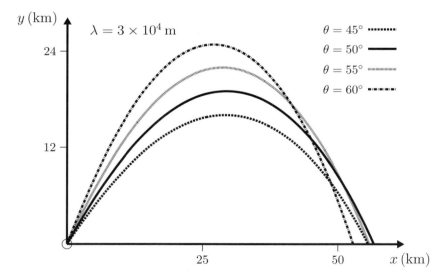

Question 7.10

A straight section of river flows with speed V between parallel banks a distance D apart. A boat crosses the river, travelling with constant speed u relative to the water, in a direction perpendicular to the current. Determine the boat's path relative to the banks (a) if V is equal to a constant V_0, and (b) if V is zero at the banks and increases quadratically to a maximum V_0 in midstream. In each case determine the distance that the boat is carried downstream in crossing the river, and plot the trajectory.

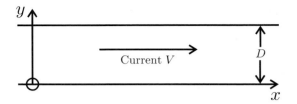

Solution

Choose x- and y-axes fixed relative to the banks as shown, and let the boat start at the origin O. The velocity $(\dot{x}, \dot{y})$ of the boat relative to these axes is the vector sum of its velocity relative to the water and the velocity of the water relative to the axes:

$$\dot{x} = V \quad \text{and} \quad \dot{y} = u. \tag{1}$$

Consequently, $u\,dx = V\,dy$. Now u is a constant, and therefore the path of the boat relative to the banks is given by

$$x = \frac{1}{u}\int_0^{y(x)} V\,dy. \tag{2}$$

(a) If $V = V_0$, then

$$x = V_0 y/u. \tag{3}$$

The boat travels in a straight line relative to the x- and y-axes, and reaches the opposite bank at $x = V_0 D/u$.

(b) The quadratic function

$$V(y) = (4V_0/D^2)(D-y)y \tag{4}$$

is zero at the banks and has a maximum value V_0 at $y = \tfrac{1}{2}D$. From (2) and (4) we have

$$x = (2V_0/uD^2)(D - \tfrac{2}{3}y)y^2. \tag{5}$$

The boat travels along a cubic curve relative to the xy-axes, and reaches the opposite bank at $x = 2V_0 D/3u$.

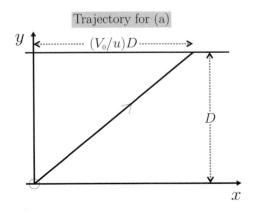

Trajectory for (a)

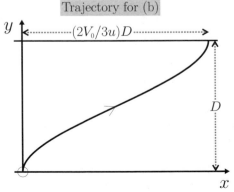

Trajectory for (b)

Question 7.11

A particle of mass m slides on the surface of a fixed, smooth sphere of radius R. The particle starts at the top of the sphere with horizontal initial velocity $\mathbf{v}_0$. Show that it leaves the sphere at an angular position θ_ℓ (measured from the top of the sphere) given by

$$\theta_\ell = \cos^{-1}\left(\frac{2}{3} + \frac{v_0^2}{3Rg}\right). \tag{1}$$

Solution

The forces acting on the particle are its weight $m\mathbf{g}$ and the reaction $\mathbf{N}$ of the sphere, which acts radially since the sphere is smooth. The equation of motion $m\dot{\mathbf{v}} = m\mathbf{g} + \mathbf{N}$ has radial component (see Question 8.1)

$$mv^2/R = mg\cos\theta - N. \tag{2}$$

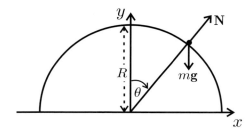

The particle loses contact with the sphere when $\mathbf{N} = 0$. According to (2), this happens at an angle θ_ℓ and velocity v_ℓ related by

$$v_\ell^2 = Rg\cos\theta_\ell. \tag{3}$$

We can calculate $v(\theta)$ by applying the work–energy theorem (see Question 5.1)

$$d(\tfrac{1}{2}mv^2) = \mathbf{F}\cdot d\mathbf{r} = (mg\sin\theta)(Rd\theta). \tag{4}$$

Now m, g and R are constants, and therefore integration of (4) gives

$$\int_{v_0}^{v(\theta)} dv^2 = 2Rg\int_0^\theta \sin\theta\, d\theta. \tag{5}$$

That is,

$$v^2(\theta) = v_0^2 + 2Rg(1 - \cos\theta). \tag{6}$$

From equations (3) and (6) we have

$$\cos\theta_\ell = \left(\frac{2}{3} + \frac{v_0^2}{3Rg}\right). \tag{7}$$

Comments

(i) The angle θ_ℓ depends on g, R and v_0, but not on m.
(ii) The graph of (1) is shown below. Note that when $v_0 \to 0$, $\theta_\ell \to \cos^{-1}(2/3)$ ($\approx 48°$), independent of g and R. Also, when $v_0 = \sqrt{Rg}$ we have $\theta_\ell = 0$. That is, the particle leaves the sphere at the starting point.

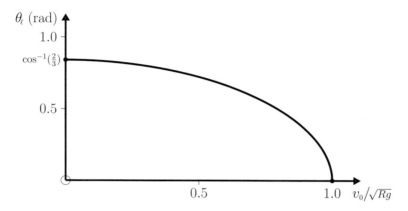

Question 7.12

Suppose that in the previous question a frictional force $\mathbf{F}_f = \mu \mathbf{N}$ acts between the particle and the sphere, where μ is the coefficient of kinetic friction.

(a) Use the work–energy theorem to show that the velocity of the particle, while it is sliding on the sphere, is given by

$$\frac{v(\theta)}{\sqrt{Rg}} = \sqrt{\frac{2\left(2\mu^2 \cos\theta - 3\mu \sin\theta - \cos\theta\right)}{1 + 4\mu^2} + \left(\frac{v_0^2}{Rg} + \frac{2 - 4\mu^2}{1 + 4\mu^2}\right) e^{2\mu\theta}}. \quad (1)$$

(b) From an analysis of numerical plots of $v(\theta)$ show that there exist two possible outcomes to the motion: either the particle comes to rest (sticks) on the sphere, or it continues to slide and eventually leaves the sphere. Deduce that the critical condition that distinguishes these outcomes is

$$v(\theta) = 0 \quad \text{when} \quad dv^2(\theta)/d\theta = 0. \quad (2)$$

(c) Hence, show that the phase diagram, which delineates the two possible motions in the μv_0-plane, is given by

$$\frac{v_{0c}}{\sqrt{Rg}} = \sqrt{\frac{2}{1 + 4\mu^2}\left(2\mu^2 - 1 + \sqrt{1 + \mu^2}\, e^{-2\mu \tan^{-1}\mu}\right)}, \quad (3)$$

and plot this diagram.

Solution

(a) The work done by friction, namely $\mathbf{F}_f \cdot d\mathbf{r} = (-\mu N)(R d\theta)$, must now be included in the work–energy theorem. Here, N is given by (2) of Question 7.11, and so (4) of Question 7.11 changes to

$$\frac{dv^2}{d\theta} - 2\mu v^2 = 2Rg(\sin\theta - \mu\cos\theta). \quad (4)$$

To solve this first-order, inhomogeneous, ordinary differential equation for $v(\theta)$ we multiply both sides by $e^{-2\mu\theta}$, to obtain

$$\frac{d}{d\theta}\{e^{-2\mu\theta}v^2(\theta)\} = 2Rg e^{-2\mu\theta}(\sin\theta - \mu\cos\theta). \quad (5)$$

By integrating (5) between $\theta = 0$ and θ, and using the real and imaginary parts of the integral

$$\int_0^\theta e^{(-2\mu+i)\theta} d\theta = \frac{e^{(-2\mu+i)\theta} - 1}{-2\mu + i}, \quad (6)$$

we find

$$e^{-2\mu\theta}v^2(\theta) - v_0^2 = \frac{2Rg\left(2\mu^2\cos\theta - 3\mu\sin\theta - \cos\theta\right)e^{-2\mu\theta}}{1 + 4\mu^2} + Rg\frac{2 - 4\mu^2}{1 + 4\mu^2}, \quad (7)$$

and hence (1).

(b) Numerical plots of $v(\theta)$ versus v_0, according to (1) and for given μ, show that these plots are of two types:
- ☞ for 'small' v_0 the function $v(\theta)$ decreases from its initial value v_0 to zero at a value $\theta = \theta_s$;
- ☞ for 'larger' v_0 the function $v(\theta)$ has no real roots – instead it decreases to a positive minimum value before increasing again. (The meanings of 'small' and 'larger' are specified below.) These behaviours are illustrated in the following figures, where we have also plotted the velocity $v_\ell(\theta) = \sqrt{Rg\cos\theta}$ required for the particle to leave the sphere (see (3) of Question 7.11):

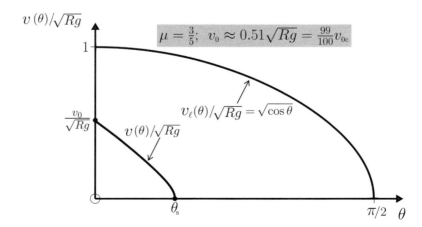

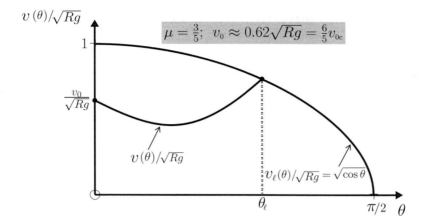

It is clear from these two graphs that the motion of the particle is sensitive to the initial velocity. For 'small' v_0, the particle slows down and comes to rest at $\theta = \theta_s$ where it sticks to the sphere. For 'larger' v_0, the particle initially slows down and then speeds up, and it eventually leaves the sphere at $\theta = \theta_\ell$ where $v(\theta) = v_\ell(\theta)$. It is also clear that the condition that distinguishes these two outcomes occurs at a critical value of the initial velocity $v_0 = v_{0c}$ such that the minimum value of $v^2(\theta)$ occurs on the θ-axis: that is, when (2) is satisfied. The graph of $v(\theta)$ for this critical motion is:

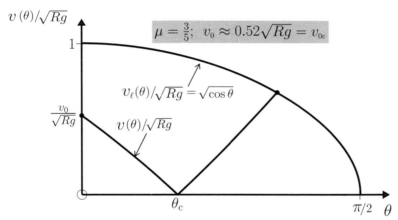

Thus, 'small' initial velocity means $v_0 < v_{0c}$, while 'larger' initial velocity means $v_0 > v_{0c}$. Note that for $v_0 = v_{0c}$ the velocity $v(\theta)$ is not differentiable at the minimum θ_c because $v(\theta) \propto |\theta - \theta_c|$ near θ_c. For $v_0 > v_{0c}$ the velocity is a quadratic function of $\theta - \theta_c$ near the minimum. Also, $\theta_\ell \to 0$ as $v_0 \to \sqrt{Rg}$ (i.e. a particle with initial velocity $\sqrt{Rg}$ leaves the sphere at the starting point $\theta = 0$).

(c) From (4) and (2) we have $\tan\theta_c = \mu$. Then

$$\sin\theta_c = \frac{\mu}{\sqrt{1+\mu^2}} \quad \text{and} \quad \cos\theta_c = \frac{1}{\sqrt{1+\mu^2}}.$$

Substituting these values in (1), and solving (2)$_1$ for v_0, gives the formula (3) for the critical velocity as a function of μ. The resulting phase diagram is:

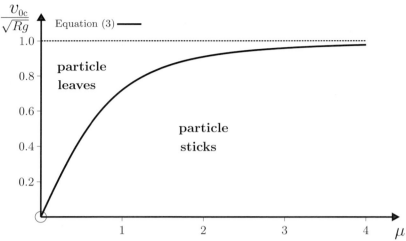

Comments

(i) The angles θ_s (where the particle sticks) and θ_ℓ (where it leaves the sphere) are roots of the equations $v(\theta) = 0$ and $v(\theta) = \sqrt{2Rg\cos\theta}$, respectively, where $v(\theta)$ is given by (1). These roots were computed numerically (see the *Mathematica* notebook below), to obtain the following dependences on v_0 for various values of μ.

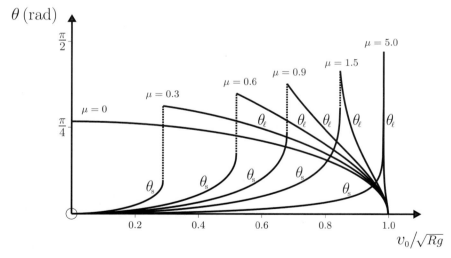

For each μ, the function $\theta_s(v_0)$ is an increasing function of v_0 in its domain ($0 \leq v_0 \leq v_{0c}$) and reaches a maximum value $\tan^{-1}\mu$ at v_{0c}; while the function $\theta_\ell(v_0)$ is a decreasing function of v_0 in its domain ($v_{0c} < v_0 \leq \sqrt{Rg}$) and tends to zero as $v_0 \to \sqrt{Rg}$. We note that θ_s and θ_ℓ depend on v_0, R, g and μ but not on m.

(ii) Close to the phase boundary the motion is highly sensitive to the initial condition v_0: if v_0 is increased infinitesimally from just below v_{0c} to just above v_{0c} then the angle through which the particle slides on the sphere increases by a finite amount $\theta_\ell^{max} - \theta_s^{max}$, as indicated by the vertical dotted lines in the above figure. The dependence of the maximum values of θ_s and θ_ℓ on μ is shown below.

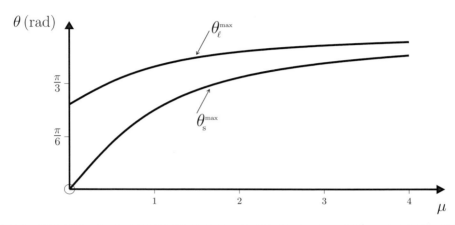

```
In[1]:= μ = 6/10; v0c = √(2/(1 + 4μ²) (2μ² - 1 + √(1 + μ²) e^(-2μ ArcTan[μ]))); v0 = v0c;

       f[θ_] := (2 (2μ² Cos[θ] - 3μ Sin[θ] - Cos[θ]))/(1 + 4μ²)

       g[θ_] := (v0² + 2 (1 - 2μ²)/(1 + 4μ²)) Exp[2μ θ]

       (* DOES THE BRANCH θ_LEAVING *)
       v0min = v0; v0max = 1.0; vstep = (v0max - v0min)/200;
       dat1 =
         Table[{N[v0], N[θ /. Chop[FindRoot[Cos[θ] + 2μ Sin[θ] - (1 + 4μ²)/3 g[θ] == 0,
           {θ, 1.25}]]]}, {v0, v0min, v0max, vstep}];

       (* DOES THE BRANCH θ_STICKING *)
       v0min = 0; v0max = 99999/100000 v0; vstep = (v0max - v0min)/200;
       dat2 =
         Table[{N[v0], N[θ /. Chop[FindRoot[√(f[θ]) + g[θ] == 0, {θ, ArcTan[9μ/10]},
           AccuracyGoal → 16, WorkingPrecision → 40]]]},
         {v0, v0min, v0max, vstep}];

       ListLinePlot[{dat1, dat2}, PlotRange → {{0, 1}, {0, π/2}}]
```

Question 7.13

A particle of mass m slides under gravity on a smooth, vertical circular wire of radius R. At time $t = 0$ the particle is travelling with speed v_0 at the top of the wire ($\theta = 0$).

(a) Show that the magnitude of the force $\mathbf{N}$ that the wire exerts on the particle is given by
$$N(\theta) = mg(3\cos\theta - 2 - v_0^2/Rg), \tag{1}$$
and draw a diagram showing how $\mathbf{N}$ varies with the angle θ in the limit $v_0 \to 0$.

(b) Show that the period T of the motion is given by
$$T = \frac{2R}{v_0} \int_0^\pi \frac{d\theta}{\sqrt{1 + (4Rg/v_0^2)\sin^2\tfrac{1}{2}\theta}}. \tag{2}$$

Evaluate (2) numerically and plot a graph of T (in units of $\sqrt{R/g}$) versus $v_0/\sqrt{Rg}$ for $0 < v_0/\sqrt{Rg} \leq 2$.

Solution

(a) Using the same coordinates as in Question 7.11, the equation of motion $m\dot{\mathbf{v}} = m\mathbf{g} + \mathbf{N}$ has radial component
$$mv^2/R = mg\cos\theta - N. \tag{3}$$

Also, the work–energy theorem (see Question 5.1) requires that the change in kinetic energy of the particle equals the work done by the force $m\mathbf{g} + \mathbf{N}$ acting on it. That is,
$$\tfrac{1}{2}mv^2 - \tfrac{1}{2}mv_0^2 = mgy = mgR(1 - \cos\theta), \tag{4}$$
where v is the speed at angular position θ (see Comment (i) below). Equations (3) and (4) yield (1). Equation (1) shows that in the limit $v_0 \to 0$ the reaction N changes sign at the angle $\theta_c = \cos^{-1}(2/3) \approx 48°$. The direction of $\mathbf{N}$ is always radial because the wire is smooth. The first diagram below illustrates the variation of $\mathbf{N}$ with θ. The arrow at $\theta = 0$ has length mg; at $\theta = \pi$ its length is $5mg$, see (1).

(b) From (4) and with $v = Rd\theta/dt$ we obtain
$$\int_0^{\frac{T}{2}} dt = \frac{R}{v_0} \int_0^\pi \frac{d\theta}{\sqrt{1 + (4Rg/v_0^2)\sin^2\tfrac{1}{2}\theta}}, \tag{5}$$

which is (2). According to (2), T is equal to $\sqrt{R/g}$ times a function of the dimensionless ratio $v_0/\sqrt{Rg}$. The graph of this function is shown below, together with the *Mathematica* notebook used to produce it.

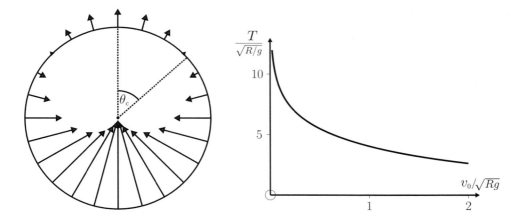

Comments

(i) The work–energy theorem requires that

$$\tfrac{1}{2}mv^2 - \tfrac{1}{2}mv_0^2 = \int_i^f m(\mathbf{g} + \mathbf{N}) \cdot d\mathbf{r}, \qquad (6)$$

where i denotes the initial point ($y = R$) and f is a subsequent position of the particle on the wire, at which the speed is v. For a smooth wire, $\mathbf{N} \cdot d\mathbf{r} = 0$ because $\mathbf{N}$ is perpendicular to the arc $d\mathbf{r}$, and therefore no work is done by the reaction. The gravitational force $m\mathbf{g} = -mg\hat{\mathbf{y}}$ is conservative and so we may choose any convenient path between i and f on which to evaluate the work done by it (see Question 5.8), such as a path along the y-axis and then parallel to the x-axis. Then, $d\mathbf{r} = \hat{\mathbf{x}}dx + \hat{\mathbf{y}}dy$ and $m\mathbf{g} \cdot d\mathbf{r} = -mg\,dy$; therefore (6) can be written as

$$\tfrac{1}{2}mv^2 - \tfrac{1}{2}mv_0^2 = -mg \int_{y=R}^{y} dy,$$

where $y = R\cos\theta$. Thus, we obtain (4).

(ii) In general, T depends on v_0. In the limit $v_0 \to 0$ the integral in (2) diverges and $T \to \infty$, as one expects.

```
In[1]:= (* In this notebook u0 = v0/√Rg is a dimensionless initial velocity *)

      u0_min = 2 × 10^-5; u0_max = 2.0; u0_step = (u0_max - u0_min)/500;

      dat = Table[{u0, (2/u0) NIntegrate[ 1/√(1 + (4/u0^2) Sin[θ/2]^2), {θ, 0, π}]},
            {u0, u0_min, u0_max, u0_step}];

      ListLinePlot[dat, PlotRange → {0, 12}]
```

Question 7.14

For the particle of Question 7.13, express the equation of motion in plane polar coordinates (see equation (2) of Question 8.1). Then, use *Mathematica* to obtain numerical solutions for θ as a function of time for the initial conditions $\theta_o = 0$ and $v_o = \sqrt{Rg/50}$. Plot graphs of θ, $\dot\theta$, $\ddot\theta$ and N/mg versus t, up to $t = 3T$. Here, N is the normal force exerted by the wire on the particle and T is the period. Take $g = 9.8\,\mathrm{ms^{-2}}$ and $R = 0.10\,\mathrm{m}$.

Solution

The equation of motion is
$$m\ddot{\mathbf{r}} = m\mathbf{g} + \mathbf{N},$$
where $\mathbf{N} = N\hat{\mathbf{r}}$. In polar coordinates, and with θ measured from the vertical y-axis, $\mathbf{g} = -g(\hat{\mathbf{r}}\cos\theta + \hat{\boldsymbol{\theta}}\sin\theta)$. Here, r is constant and so the components of the equation of motion (see (2) of Question 8.1) are

$$\left.\begin{array}{l} mr\dot\theta^2 = mg\cos\theta - N \\ mr\ddot\theta = mg\sin\theta\,. \end{array}\right\} \quad (1)$$

(Note that the angle θ used in this problem is complementary to the angle θ in Question 8.1.) The following *Mathematica* notebook was used to solve $(1)_2$ for $\theta(t)$, and N was obtained by using this solution in $(1)_1$. The desired graphs are shown below. The period of the motion, obtained from (2) of Question 7.13, is 0.81 s.

```
In[1]:= g = 9.8; R = 0.1; v0 = √(Rg/50); θ0 = 0; θ0dot = v0/R;

        T = (2R/v0) NIntegrate[1/√(1 + (4Rg/v0²) Sin[θ/2]²), {θ, 0, π}];  (* T is the period *)

        tmax = 3 T;

        Sol = NDSolve[{-R θ''[t] + g Sin[θ[t]] == 0, θ[0] == θ0, θ'[0] == θ0dot},
              {θ[t], θ'[t], θ''[t]}, {t, 0, tmax}];

        Plot[Evaluate[θ[t] /. Sol], {t, 0, tmax}]

        Plot[Evaluate[θ'[t] /. Sol], {t, 0, tmax}]

        Plot[Evaluate[θ''[t] /. Sol], {t, 0, tmax}]

        Plot[(Evaluate[Cos[θ[t]] /. Sol] - (R/g) (Evaluate[θ'[t] /. Sol])²), {t, 0, tmax}]
```

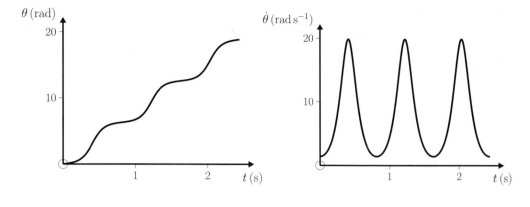

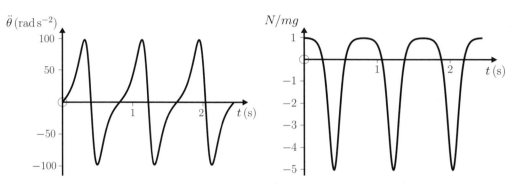

Question 7.15

A particle of mass m slides under gravity on a smooth wire lying in the vertical plane and having the shape $y = y(x)$.

(a) Show that the equation of motion can be written as

$$\left[1 + \left(\frac{dy}{dx}\right)^2\right]\ddot{x} + \left[g + \dot{x}^2 \frac{d^2y}{dx^2}\right]\frac{dy}{dx} = 0. \tag{1}$$

(b) Suppose $y(x) = \cosh x$ and the initial conditions are $x_o = 1.0\,\text{m}$, $\dot{x}_o = 0$ and $\dot{y}_o = 0$. Use *Mathematica* to find numerical solutions to (1) for $x(t)$ and $y(t)$. Plot graphs of $x(t)$ and $y(t)$; $\dot{x}(t)$ and $\dot{y}(t)$; $\ddot{x}(t)$ and $\ddot{y}(t)$; and $N(t)/mg$, where N is the normal force exerted by the wire on the particle. Take $0 \leq t \leq 2.5\,\text{s}$ and $g = 9.8\,\text{ms}^{-2}$.

(c) Repeat for $y(x) = \cos x$. Use the initial conditions $x_o = 0$, $\dot{x}_o = 2.0\,\text{ms}^{-1}$ and $\dot{y}_o = 0$. Take $0 \leq t \leq 4\,\text{s}$.

184 Solved Problems in Classical Mechanics

Solution

(a) In terms of the angle θ shown in the figure below, the equation of motion $m\ddot{\mathbf{r}} = m\mathbf{g} + \mathbf{N}$ has components $m\ddot{x} = -N\sin\theta$ and $m\ddot{y} = N\cos\theta - mg$. Eliminating θ from these equations and setting $\tan\theta = dy/dx$ gives $\ddot{x} + (g + \ddot{y})\,dy/dx = 0$. Now, $\dot{y} = \dot{x}\,dy/dx$ and so $\ddot{y} = \dot{x}^2\,d^2y/dx^2 + \ddot{x}\,dy/dx$. These results yield (1).

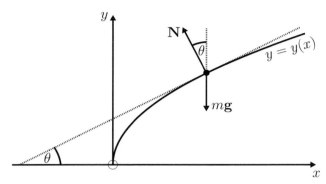

(b) The following *Mathematica* notebook was used to solve (1) for $x(t)$ and to calculate N when $y(x) = \cosh x$. The required graphs are shown below.

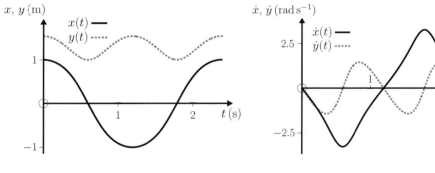

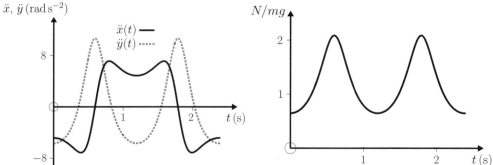

(c) Change `Cosh[x]` to `Cos[x]` in the *Mathematica* notebook below and use the given initial conditions. Note that here the x-motion is unbounded.

Motion in two and three dimensions 185

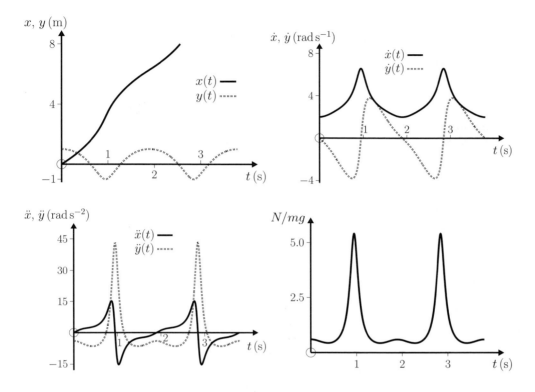

```
In[1]:= g = 9.8; x0 = 1.0; v0 = 0; t_max = 2.5; t_step = t_max/1000; f[x_] := Cosh[x]
        yVel[t_] := First[f'[x[t]] x'[t] /. Sol]
        yAcc[t_] := First[(f''[x[t]] x'[t]^2 + f'[x[t]] x''[t]) /. Sol]
        Force[t_] := First[1/g √((x''[t] /. Sol)^2 + (g + yAcc[t])^2)]
        Sol = NDSolve[{x''[t] (1 + f'[x[t]]^2) + (g + x'[t]^2 f''[x[t]]) f'[x[t]] == 0,
            x[0] == x0, x'[0] == v0}, {x[t], x'[t], x''[t]}, {t, 0, t_max}];
        xDisDat = Table[{t, First[x[t] /. Sol]}, {t, 0, t_max, t_step}];
        yDisDat = Table[{t, First[f[x[t]] /. Sol]}, {t, 0, t_max, t_step}];
        xVelDat = Table[{t, First[x'[t] /. Sol]}, {t, 0, t_max, t_step}];
        yVelDat = Table[{t, yVel[t]}, {t, 0, t_max, t_step}];
        xAccDat = Table[{t, First[x''[t] /. Sol]}, {t, 0, t_max, t_step}];
        yAccDat = Table[{t, yAcc[t]}, {t, 0, t_max, t_step}];
        ForcDat = Table[{t, Force[t]}, {t, 0, t_max, t_step}];
        ListLinePlot[{xDisDat, yDisDat}, PlotRange → All]
```

```
In[2]:= ListLinePlot[{xVelDat, yVelDat}, PlotRange → All]
        ListLinePlot[{xAccDat, yAccDat}, PlotRange → All]
        ListLinePlot[{ForcDat}, PlotRange → All]
```

Comment

Results for other shapes of wire can be obtained by changing $y(x)$ in the *Mathematica* notebook. Some interesting functions to try are $y = x^2$, $y = 1/x$, and $y = x^3$.

Question 7.16

A camel takes N steps while walking directly between two towns located on the same longitude, and whose latitudes differ by θ degrees. The camel takes S steps to cross a stadium. Express the Earth's circumference C (in units of stadia) in terms of N, θ, and S.

Solution

To circumnavigate the Earth would take $360/\theta$ journeys such as that completed by the camel. Therefore

$$C = \frac{360}{\theta} N \text{ camel steps} = \frac{360N}{\theta S} \text{ stadia.} \qquad (1)$$

Comments

(i) This method was used by Eratosthenes (a chief librarian in the famous library at Alexandria in the third century B. C.) to obtain a reasonable value for the size of the Earth. He knew that at noon on the summer solstice in Syene (the present Aswan), a vertical object cast no shadow on the ground. At the same date and time he found that the shadow of a vertical object in Alexandria subtended an angle of about $7\frac{1}{2}°$.[4] Because the sun's rays are nearly parallel, this is the angle θ in (1).

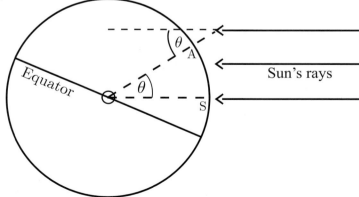

---

[4]   L. Hogben, *Science for the citizen*. London: George Allen and Unwin, 1956. Chap. 11.

It seems that Eratosthenes took the distance between Alexandria and Syene to be about 5200 stadia (this is the value of N/S in (1)), because he obtained for C approximately 250 000 stadia.[4] It is believed that this is within a few per cent of the modern value (according to Bertrand Russell,[5] the error was less than 1%).

(ii) This result, together with a later measurement of the gravitational constant G by Cavendish, made it possible to obtain a value for the mass of the Earth. These experiments by Eratosthenes and Cavendish were both selected in a recent poll of the 'most beautiful experiment in physics'.[6]

Question 7.17

A particle having mass m and charge q moves in a uniform magnetostatic field $\mathbf{B} = (0, 0, B)$ where B is a positive constant. The initial conditions at $t = 0$ are

$$\mathbf{r}_0 = 0 \quad \text{and} \quad \mathbf{v}_0 = \mathbf{u} = (u_1, u_2, u_3). \tag{1}$$

(a) Show that the trajectory is given by

$$\mathbf{r}(t) = \left(\frac{u_1}{\omega} \sin \omega t + \frac{u_2}{\omega}(1 - \cos \omega t),\ \frac{u_1}{\omega}(\cos \omega t - 1) + \frac{u_2}{\omega} \sin \omega t,\ u_3 t \right), \tag{2}$$

where

$$\omega = qB/m. \tag{3}$$

(Hint: Integrate the equation of motion once with respect to t and then decouple the resulting differential equations.)

(b) Make a convenient choice of axes, and then sketch this trajectory.
(c) Calculate the velocity of the particle and deduce that the speed is constant.

Solution

(a) First, integrate the equation of motion

$$m \frac{d\mathbf{v}}{dt} = q \frac{d\mathbf{r}}{dt} \times \mathbf{B} \tag{4}$$

with respect to t between $t = 0$ and t. Because $\mathbf{B}$ is constant we have

$$m \int_{\mathbf{u}}^{\mathbf{v}(t)} d\mathbf{v} = q \left(\int_{\mathbf{r}_0}^{\mathbf{r}} d\mathbf{r} \right) \times \mathbf{B}. \tag{5}$$

That is,

$$\mathbf{v}(t) = (q/m)(\mathbf{r} - \mathbf{r}_0) \times \mathbf{B} + \mathbf{u}. \tag{6}$$

---

[5] B. Russell, *History of western philosophy*, p. 225. London: George Allen and Unwin, 1962.
[6] R. P. Crease, "The most beautiful experiment," Physics World, vol. 15, pp. 19–20, September 2002.

With $\mathbf{B} = (0, 0, B)$ and for the initial conditions (1), the components of (6) are

$$\frac{dx}{dt} = \omega y + u_1, \qquad \frac{dy}{dt} = -\omega x + u_2, \qquad \frac{dz}{dt} = u_3. \qquad (7)$$

The first two equations in (7) can be decoupled by taking d/dt of $(7)_1$ and substituting $(7)_2$ in it. This yields

$$\frac{d^2 x}{dt^2} + \omega^2 x = \omega u_2. \qquad (8)$$

The general solution to (8) is the sum of the particular integral $x_p = u_2/\omega$ and the complementary function x_c. The latter is the general solution to (8) with $u_2 = 0$, and is given by the general solution of the equation of motion for a simple harmonic oscillator – see (20) of Question 4.1. Thus

$$x(t) = a \cos \omega t + b \sin \omega t + u_2/\omega, \qquad (9)$$

where the constants a and b are determined by the initial conditions: clearly, $(1)_1$ requires $a = -u_2/\omega$ and $(1)_2$ imposes $b = u_1/\omega$. Thus

$$x(t) = \frac{u_1}{\omega} \sin \omega t + \frac{u_2}{\omega}(1 - \cos \omega t). \qquad (10)$$

By substituting (10) in $(7)_1$ and solving for y we find that

$$y(t) = \frac{u_1}{\omega}(\cos \omega t - 1) + \frac{u_2}{\omega} \sin \omega t. \qquad (11)$$

Lastly, the solution to $(7)_3$ that is zero at $t = 0$ is

$$z(t) = u_3 t. \qquad (12)$$

Equations (10)–(12) yield the trajectory (2).

(b) The x- and y-components in (10) and (11) satisfy

$$\left(x - \frac{u_2}{\omega}\right)^2 + \left(y + \frac{u_1}{\omega}\right)^2 = \frac{u_1^2 + u_2^2}{\omega^2}, \qquad (13)$$

meaning that the projection of the motion onto the xy-plane is a circle of radius $\sqrt{u_1^2 + u_2^2}/\omega$ centred on $(u_2/\omega, -u_1/\omega)$. The period of this circular motion is $T = 2\pi/\omega$, and in this time the z-component (12) changes by $u_3 T$. Thus, the complete trajectory is a helix of constant pitch $D = u_3 T = 2\pi u_3/\omega$. To sketch this trajectory it is convenient to first rotate the coordinate system about the z-axis until the initial velocity vector $\mathbf{u}$ lies in the xz-plane (that is, $u_2 = 0$), and then shift the origin O by $-u_1/\omega$ along the y-axis. This simplifies (2) to

$$\mathbf{r}(t) = \left(\frac{u_1}{\omega} \sin \omega t, \; \frac{u_1}{\omega} \cos \omega t, \; u_3 t\right), \qquad (14)$$

which is a helix of radius u_1/ω, pitch $2\pi u_3/\omega$ and axis Oz, as depicted below. For this trajectory the initial conditions are $\mathbf{r}_0 = (0, \; u_1/\omega, \; 0)$ and $\mathbf{v}_0 = (u_1, \; 0, \; u_3)$. The figure below is for a particle of positive charge q and for $u_1, u_3 > 0$.

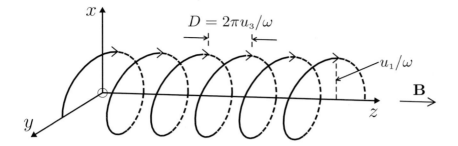

(c) The time derivative of (2) is the velocity

$$\mathbf{v}(t) = (u_1 \cos\omega t + u_2 \sin\omega t,\ -u_1 \sin\omega t + u_2 \cos\omega t,\ u_3). \tag{15}$$

The speed v is equal to the magnitude of (15), that is

$$v = \sqrt{u_1^2 + u_2^2 + u_3^2} = u. \tag{16}$$

Thus, the speed remains constant, equal to its initial value u.

Comments

(i) The motion is sensitive to an initial condition. If $u_3 = 0$ the trajectory is closed (bounded circular motion in a plane perpendicular to $\mathbf{B}$), whereas if $u_3 \neq 0$ the trajectory is open (unbounded motion along $\mathbf{B}$).

(ii) According to (16), the kinetic energy $K = \frac{1}{2}mv^2$ of the particle is a constant. This is to be expected because the magnetic force $\mathbf{F} = q\mathbf{v} \times \mathbf{B}$ is perpendicular to $\mathbf{v}$ and therefore the work–energy theorem $\dot{K} = \mathbf{F}\cdot\mathbf{v}$ (see Question 5.1) requires $\dot{K} = 0$.

(iii) According to electromagnetic theory, an accelerated charge radiates electromagnetic energy and the kinetic energy decreases during the motion. The analysis leading to (14) and (16) is approximate insofar as it neglects this energy loss. Usually this effect is very small – the fractional loss in energy per cycle is negligible, although it can become appreciable under certain conditions (see Question 7.21).

(iv) For a non-relativistic particle the angular frequency ω in (3) is independent of the speed v (and hence of the energy) of the particle. This property is the basis for the operation of a particle accelerator known as the cyclotron,[7] and ω is called the cyclotron frequency. For relativistic particles, ω depends on v (see Question 15.15), and a cyclotron cannot function properly. Consequently, the maximum energy produced by a cyclotron is limited to a few per cent of the rest-mass energy, and it is therefore not suitable for electrons, or for accelerating protons beyond about 20 MeV. A modification of the cyclotron that overcomes this energy limit is the frequency-modulated cyclotron (the synchrocyclotron).[7]

---

[7] See, for example, E. Persico, E. Ferrari, and S. E. Segre, *Principles of particle accelerators*. New York: Benjamin, 1968.

(v) The relevance of the theory of the simple harmonic oscillator to the classical problem (4), which is evident in (8) and (9), carries over to the quantum-mechanical problem;[8] thus the energy of bounded motion consists, in fact, of a set of discrete values
$$E = (n + \tfrac{1}{2})\hbar\omega \qquad (n = 0, 1, 2, \cdots), \qquad (17)$$
known as the Landau levels. (To specify the quantum-mechanical problem fully, one must select, in addition to the Hamiltonian operator H and the momentum-operator p_z, a third observable chosen from p_y (or p_x) and operators corresponding to the position (x_0, y_0) of the centre of the helical trajectory.[8] Related to this is the question of whether the angular momentum L_z is a constant of the motion.[8]) The fractional spacing of the levels (17) is $\Delta E/E = (n + \tfrac{1}{2})^{-1}$; for large n the spectrum forms a quasi-continuum and a classical description is possible.

Question 7.18

A particle of mass m and charge q is acted on by uniform magnetostatic and electrostatic fields that are perpendicular to each other:
$$\mathbf{B} = (0, 0, B) \quad \text{and} \quad \mathbf{E} = (0, E, 0), \qquad (1)$$
where B and E are positive constants. The initial conditions at time $t = 0$ are
$$\mathbf{r}_0 = 0 \quad \text{and} \quad \mathbf{v}_0 = \mathbf{u} = (u_1, u_2, u_3). \qquad (2)$$

(a) Show that the trajectory $\mathbf{r}(t)$ has components
$$x(t) = v_\mathrm{d} t + \frac{1}{\omega}(u_1 - v_\mathrm{d})\sin\omega t + \frac{u_2}{\omega}(1 - \cos\omega t) \qquad (3)$$
$$y(t) = \frac{1}{\omega}(u_1 - v_\mathrm{d})(\cos\omega t - 1) + \frac{u_2}{\omega}\sin\omega t \qquad (4)$$
$$z(t) = u_3 t, \qquad (5)$$
where $\omega = qB/m$ is the cyclotron frequency and $v_\mathrm{d} = E/B$.

(b) Sketch the trajectory for a particle that starts from rest.

(c) If $u_1 = v_\mathrm{d}$ and $u_3 = 0$, sketch the trajectories for $u_2 = \tfrac{3}{5}v_\mathrm{d}$, $u_2 = v_\mathrm{d}$, and $u_2 = 2v_\mathrm{d}$. Assume $q > 0$.

Solution

(a) This calculation is an extension of that in Question 7.17 to include the effect of an electrostatic field. We start by integrating the equation of motion
$$m\frac{d\mathbf{v}}{dt} = q\mathbf{E} + q\frac{d\mathbf{r}}{dt} \times \mathbf{B} \qquad (6)$$

---

[8] See, for example, O. L. de Lange and R. E. Raab, *Operator methods in quantum mechanics*. Oxford: Clarendon Press, 1991.

with respect to t between $t = 0$ and t. Because $\mathbf{B}$ and $\mathbf{E}$ are constants we see that

$$\mathbf{v}(t) = (q/m)\mathbf{E}\,t + (q/m)\,(\mathbf{r} - \mathbf{r}_0) \times \mathbf{B} + \mathbf{u}\,. \tag{7}$$

For the fields (1) and the initial conditions (2), the components of (7) are

$$\frac{dx}{dt} = \omega y + u_1\,, \qquad \frac{dy}{dt} = v_{\mathrm{d}}\omega t - \omega x + u_2\,, \qquad \frac{dz}{dt} = u_3\,. \tag{8}$$

We take d/dt of $(8)_2$ and substitute $(8)_1$ in it to obtain the decoupled equation

$$\frac{d^2 y}{dt^2} + \omega^2 y = \omega\,(v_{\mathrm{d}} - u_1)\,. \tag{9}$$

The general solution to (9) is

$$y(t) = a\cos\omega t + b\sin\omega t + (v_{\mathrm{d}} - u_1)/\omega\,, \tag{10}$$

where the constants a and b are fixed by the initial conditions (2) to be $a = (u_1 - v_{\mathrm{d}})/\omega$ and $b = u_2/\omega$. This proves (4). From $(8)_2$ and (4) we obtain (3), while (5) follows from an integration of $(8)_3$. We remark that if $\mathbf{E} = 0$ then (3)–(5) reduce to the trajectory (2) in Question 7.17.

(b) If the particle is initially at rest ($\mathbf{u}=0$), (3)–(5) become

$$x(t) = (v_{\mathrm{d}}/\omega)(\omega t - \sin\omega t)\,, \qquad y(t) = (v_{\mathrm{d}}/\omega)(1 - \cos\omega t)\,, \qquad z(t) = 0\,, \tag{11}$$

which represent a cycloid in the xy-plane:

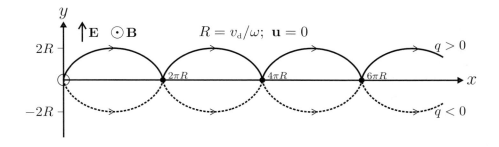

(c) If $u_1 = v_{\mathrm{d}}$ and $u_3 = 0$, (3)–(5) become

$$x(t) = \frac{1}{\omega}[v_{\mathrm{d}}\omega t + u_2(1 - \cos\omega t)]\,, \qquad y(t) = \frac{u_2}{\omega}\sin\omega t\,, \qquad z(t) = 0\,. \tag{12}$$

The curves represented by (12) depend on the relative values of u_2 and v_{d}:

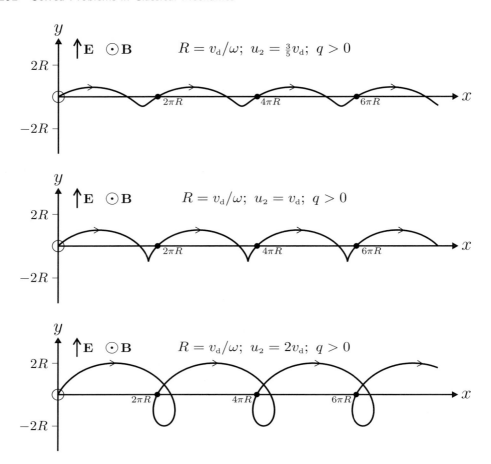

Comments

(i) Starting with (3) and (4) we can readily show that

$$(x - v_\mathrm{d} t - u_2/\omega)^2 + (y - \{v_\mathrm{d} - u_1\}/\omega)^2 = \{(u_1 - v_\mathrm{d})^2 + u_2^2\}/\omega^2, \quad (13)$$

meaning that the projection of the motion onto the xy-plane is a circle with a constant velocity $v_\mathrm{d}\hat{\mathbf{x}} = E\hat{\mathbf{x}}/B$ superimposed on it: to an observer travelling with velocity $v_\mathrm{d}\hat{\mathbf{x}}$, and for $u_3 = 0$, the particle moves in a circle of radius $\sqrt{(u_1 - v_\mathrm{d})^2 + u_2^2}/\omega$ centred at $(u_2/\omega, (v_\mathrm{d} - u_1)/\omega)$.

(ii) Thus, a particle projected in the xy-plane remains in this plane and drifts with velocity v_d along the direction of $\mathbf{E} \times \mathbf{B}$; this behaviour is known as $\mathbf{E} \times \mathbf{B}$ drift. The direction of the drift is independent of the sign of the charge q: a particle starting from rest drifts along $\mathbf{E} \times \mathbf{B}$ irrespective of the sign of q – see (11) and the figure following it. (More generally, when $q \to -q$ and $u_2 \to -u_2$ in (3) and (4), then $x \to x$ and $y \to -y$.)

(iii) The trajectory described by (3) and (4) is a family of curves known as a trochoid. These are the curves traced by a point P that is fixed on a disc that rolls along a line. (P may be outside the disc, but it is fixed relative to the disc.) If P lies inside the disc, the curve is known as a curtate cycloid (see the figure for $u_2 = \tfrac{3}{5}v_\mathrm{d}$); if P is on the edge of the disc, the curve is a cycloid (see the figure for $u_2 = v_\mathrm{d}$); if P lies outside the disc, the curve is a prolate cycloid (see the figure for $u_2 = 2v_\mathrm{d}$).

(iv) According to (11), for a particle that starts from rest at the origin, the kinetic energy
$$K = \tfrac{1}{2}m(\dot{x}^2 + \dot{y}^2) = mv_\mathrm{d}^2(1 - \cos\omega t) \tag{14}$$
oscillates between 0 and $2mv_\mathrm{d}^2$, while the electric potential energy $V = -qEy = -K$ oscillates between 0 and $-2mv_\mathrm{d}^2$. The sum $K+V$ maintains a constant value of zero.

(v) The results (3)–(5) are readily extended to the case where $\mathbf{E}$ has also a constant component along $\mathbf{B}$, that is $\mathbf{E} = (0, E, E_\parallel)$. The effect is to replace (5) by
$$z(t) = u_3 t + qE_\parallel t^2 / 2m, \tag{15}$$
and thus there will be parabolic motion along the z-axis superimposed on the motions discussed above.

Question 7.19

(a) A particle of mass m and charge q moving with constant velocity $v_0 \hat{\mathbf{x}}$ (where $v_0 > 0$) enters a region $0 \leq x \leq L$ where there is a uniform electrostatic field $\mathbf{E} = E\hat{\mathbf{y}}$. Neglecting the effect of gravity, show that the position at which the particle strikes a vertical screen placed at $x = L + D$ is
$$y_E = \frac{qEL}{mv_0^2}\left(\tfrac{1}{2}L + D\right). \tag{1}$$

(b) If the electric field is replaced by a weak, uniform magnetostatic field $\mathbf{B} = B\hat{\mathbf{z}}$, show that the charge strikes the screen at
$$y_B = -\frac{qBL}{mv_0}\left(\tfrac{1}{2}L + D\right). \tag{2}$$

Solution

(a) Along OP_1 in the diagram below, the electric field acts on the particle and its coordinates are given by (see Question 3.1)
$$x = v_0 t, \qquad y = \frac{qE}{2m}t^2. \tag{3}$$
Therefore, the portion OP_1 of the trajectory is parabolic:
$$y = \frac{qE}{2mv_0^2}x^2 \qquad (0 \leq x \leq L). \tag{4}$$

194 Solved Problems in Classical Mechanics

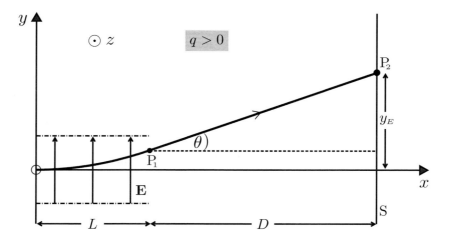

For $x > L$ the force on the particle is zero and therefore the portion $P_1 P_2$ of the trajectory is a straight line with slope

$$\tan\theta = \left(\frac{dy}{dx}\right)_{x=L} = \frac{qEL}{mv_0^2}. \tag{5}$$

From (4) and (5) it follows that the equation of $P_1 P_2$ is given by

$$y = \frac{qEL^2}{2mv_0^2} + \frac{qEL}{mv_0^2}(x - L). \tag{6}$$

Setting $x = L+D$ in (6) we obtain (1). (The trajectory sketched in the above figure is for a positively charged particle; for a negative charge the sense of deflection is opposite – that is, P_2 is below the x-axis.)

(b)

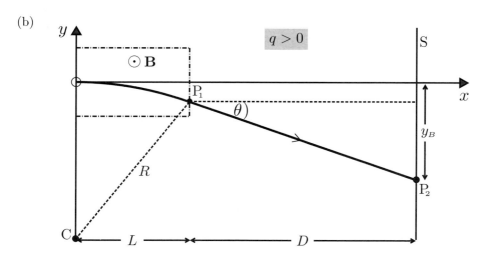

In the magnetic field, the particle moves in a circle of radius

$$R = mv_0/qB \qquad (7)$$

(see Question 7.17) and the equation of the portion OP_1 of the trajectory is

$$x^2 + (y + R)^2 = R^2 \qquad (0 \leq x \leq L). \qquad (8)$$

For weak fields, R is large compared to y, and so in (8) we can neglect the term in y^2. Then, (7) and (8) yield

$$y = -\frac{qB}{2mv_0} x^2 \qquad (0 \leq x \leq L). \qquad (9)$$

For the segment P_1P_2 the slope is $\tan\theta = \left(\dfrac{dy}{dx}\right)_{x=L} = \dfrac{-qBL}{mv_0}$, and therefore

$$y = -\frac{qBL^2}{2mv_0} - \frac{qBL}{mv_0}(x - L). \qquad (10)$$

Setting $x = L + D$ in (10) gives (2).

Comments

(i) Equations (1) and (2) provide the basis for interpreting a famous experiment performed in 1897 by J. J. Thomson to measure the charge-to-mass ratio q/m of the electron. The lengths L and D, the fields E and B, and the deflections y_E and y_B are all accessible to measurement, and so values for the two unknowns v_0 and q/m can be extracted from (1) and (2). Thomson's procedure differed slightly from this in that he measured v_0 by using the B-field to cancel exactly the deflection due to the E-field. That is, $y_E + y_B = 0$ and therefore $v_0 = E/B$. Then, measurements with the E-field alone yielded q/m from (1).

(ii) By 1899, Thomson had measured also the electronic charge q and inferred that the mass m is about a thousand times less than that of a hydrogen atom. In this way Thomson discovered the electron and concluded that atoms are routinely split by electrification.[9]

Question 7.20

The single particle of Question 7.19 is now replaced by a beam of non-interacting particles all having the same mass and charge but travelling with a range of initial velocities $v_0\hat{\mathbf{x}}$. Also, the magnetic field is applied parallel (or anti-parallel) to the electric field (that is, $\mathbf{B} = B\hat{\mathbf{y}}$). Show that the motion is now three-dimensional and that the coordinates at which the particles strike the screen must lie on a parabola

$$y = \frac{m}{q} \frac{E}{B^2 L(\frac{1}{2}L + D)} z^2. \qquad (1)$$

---

[9] J. F. Mulligan, "The personal and professional interactions of J. J. Thomson and Arthur Schuster," American Journal of Physics, vol. 65, pp. 954 – 963, 1997.

Solution

The motion is three-dimensional because **E** produces a y-deflection and **B** produces a z-deflection. The y-deflection on the screen due to **E** is given by (1) of Question 7.19:

$$y = \frac{qEL}{mv_0^2}\left(\tfrac{1}{2}L + D\right). \tag{2}$$

By a similar calculation to (b) of Question 7.19 we see that a magnetic field $\mathbf{B} = B\hat{\mathbf{y}}$ produces a z-deflection on the screen given by

$$z = \frac{qBL}{mv_0}\left(\tfrac{1}{2}L + D\right). \tag{3}$$

By taking the ratio of (2) and (3)² we eliminate v_0 and obtain (1).

Comments

(i) According to (3), particles with different initial velocity v_0 strike the screen at different points on the same parabolic curve (1):

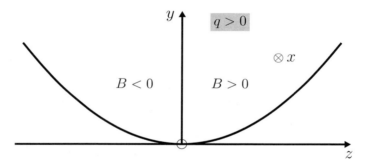

(ii) Thomson used this modification of his experiment to measure also the charge-to-mass ratio q/m of positive ions, finding that the ionic masses are much larger than the electron mass. He also found that certain chemically pure gases produced more than one parabola, and therefore had more than one value of q/m, thus heralding the discovery of isotopes.

(iii) Variations of Thomson's cathode ray tubes found widespread application in devices such as oscilloscopes, television sets and computer monitors. It is only recently that they have been surpassed by the less bulky liquid crystal displays and plasma screens.

Question 7.21

A particle of charge q and rest mass m moves in a circle of radius R in a uniform magnetic field **B**. It has kinetic energy‡

---
‡This question involves some results from the theory of special relativity, see Chapter 15.

$$K = (\gamma - 1)mc^2, \tag{1}$$

and the power it radiates is given by[7]

$$P = \frac{1}{4\pi\epsilon_0} \frac{2cq^2\beta^4\gamma^4}{3R^2}. \tag{2}$$

Here, c is the speed of light in vacuum,

$$\beta = v/c, \qquad \gamma = (1-\beta^2)^{-1/2}. \tag{3}$$

Calculate the fractional loss in kinetic energy per cycle, $\Delta K/K$:

(a) for a proton with $K = 0.08$ MeV and $R = 3.2$ cm;
(b) for a deuteron with $K = 24$ MeV and $R = 76$ cm;
(c) for a 50-GeV proton when $R = 4.3$ km;
(d) for a 50-GeV electron when $R = 4.3$ km.

Use the rest mass energies: $mc^2 = 938$ MeV (proton), 1877 MeV (deuteron), 0.51 MeV (electron). Also, $1\,\text{eV} \approx 1.6 \times 10^{-19}$ J, 1 MeV= 10^6 eV and 1 GeV= 10^9 eV.

Solution

The energy radiated in one cycle is $\Delta K = PT$, where $T = 2\pi R/v = 2\pi R/\beta c$ is the period. It therefore follows from (1) and (2) that the fractional loss in energy per cycle is given by

$$\frac{\Delta K}{K} = \frac{1}{4\pi\epsilon_0} \frac{4\pi q^2 \beta^3 \gamma^4}{3KR}. \tag{4}$$

This result applies at all particle velocities less than c.

(a) From (1), $\gamma - 1 = K/mc^2 = 0.08/938 = 8.5 \times 10^{-5}$. From (3)$_2$, $\beta = 1.3 \times 10^{-2}$. Then (4) gives

$$\frac{\Delta K}{K} = 9 \times 10^9 \frac{4\pi(1.6 \times 10^{-19})^2 \times (1.3 \times 10^{-2})^3 \times 1^4}{3 \times (0.08 \times 1.6 \times 10^{-13}) \times 3.2 \times 10^{-2}} = 5.2 \times 10^{-18}. \tag{5}$$

(b) It follows from (1) and (3) that, $\gamma - 1 = 24/1877 = 1.3 \times 10^{-2}$ and $\beta = 0.16$. Then (4) gives

$$\frac{\Delta K}{K} = 9 \times 10^9 \frac{4\pi(1.6 \times 10^{-19})^2 \times (0.16)^3 \times (1.013)^4}{3 \times (24 \times 1.6 \times 10^{-13}) \times 0.76} = 1.4 \times 10^{-18}. \tag{6}$$

(c) Here, $\gamma - 1 = 50\,000/938 = 53.3$ and $\beta \approx 1$. Then

$$\frac{\Delta K}{K} = 9 \times 10^9 \frac{4\pi(1.6 \times 10^{-19})^2 \times 1^3 \times (54.3)^4}{3 \times (50 \times 10^3 \times 1.6 \times 10^{-13}) \times 4300} = 2.4 \times 10^{-16}. \tag{7}$$

(d) From (1) and (3), $\gamma - 1 = 50\,000/0.51 = 9.8 \times 10^4$ and $\beta \approx 1$. Then

$$\frac{\Delta K}{K} = 9 \times 10^9 \frac{4\pi(1.6 \times 10^{-19})^2 \times 1^3 \times (9.8 \times 10^4)^4}{3 \times (50 \times 10^3 \times 1.6 \times 10^{-13}) \times 4300} = 2.6 \times 10^{-3}. \tag{8}$$

Comments

(i) The data for the protons in (a) above are for the first ($2\frac{1}{2}$ inch diameter) cyclotron that was constructed by Lawrence and Livingstone in Berkeley in 1932. The deuterons in (b) were produced in a later (1946) cyclotron that had a diameter of 60 inches. The relativistic protons and electrons considered in (c) and (d) are for the LEP electron synchrotron at CERN.

(ii) In most instances the fractional energy loss from radiation is small and can be neglected, as we did in Question 7.17.

(iii) The large increase in energy loss for the relativistic electrons in (d) compared with the relativistic protons in (c) is due to the large γ-factor of the electrons, see (7) and (8).

(iv) The above is for a single particle. For a circulating current comprising N particles there can be interesting 'coherence' effects. For example, if the particles form a 'bunch' whose dimensions are small compared to the main wavelength of the radiation, then this bunch behaves like a particle of charge Nq, and in (4) the factor q^2 is replaced by $(Nq)^2$: the energy loss varies as N^2 rather than N.[7]

(v) By contrast with this, one can think of a ring of superconducting material held at a temperature below its critical temperature, and in which a circulating electric current has been induced. Sensitive experiments (in which the magnetic dipole moment of the current loop is accurately monitored) reveal no measurable decrease in current over long periods of time. This phenomenon is quantum mechanical: interaction between the electrons has produced a macroscopic quantum state consisting of paired electrons, and the magnetic flux through the ring is constant and is quantized in units of $h/2e$ (the flux quantum).[10]

Question 7.22

Consider a classical model of the hydrogen atom in which an electron of mass m and charge q moves initially in a circular orbit of radius r_0 about a proton. This orbit is unstable because the accelerated electron should radiate electromagnetic energy and spiral towards the proton. By using the non-relativistic form of (2) of Question 7.21 for the rate at which the electron loses energy, show that the lifetime of hydrogen in this classical model is given approximately by

$$\tau = \frac{m^2 c^3 r_0^3}{4(q^2/4\pi\epsilon_0)^2}. \tag{1}$$

Solution

Here, we consider only the motion of the electron since it is about 1800 times lighter than the proton. For non-relativistic motion ($v \ll c$) in a circular orbit, the equation of motion of the electron is

---

[10] See, for example, M. Tinkham, *Introduction to superconductivity*. New York: McGraw-Hill, 1975.

$$ma = m\frac{v^2}{r} = \frac{1}{4\pi\epsilon_0}\frac{q^2}{r^2}. \tag{2}$$

The energy of the electron is the sum of its kinetic and potential energies:

$$E = \tfrac{1}{2}mv^2 - \frac{1}{4\pi\epsilon_0}\frac{q^2}{r} = -\frac{1}{4\pi\epsilon_0}\frac{q^2}{2r}. \tag{3}$$

The rate at which the orbit decays due to radiation can be written

$$\frac{dr}{dt} = \frac{dE/dt}{dE/dr}. \tag{4}$$

For non-relativistic motion, (2) of Question 7.21 gives for the rate of change of the electron's energy

$$\frac{dE}{dt} = -\frac{1}{4\pi\epsilon_0}\frac{2q^2 a^2}{3c^3}, \tag{5}$$

while from (3) we have

$$\frac{dE}{dr} = \frac{1}{4\pi\epsilon_0}\frac{q^2}{2r^2}. \tag{6}$$

Equations (4)–(6) show that

$$\frac{dr}{dt} = -\left(\frac{q^2}{4\pi\epsilon_0}\right)^2 \frac{4}{3m^2 c^3 r^2}, \tag{7}$$

and therefore

$$\int_0^\tau dt = -\frac{3m^2 c^3}{4(q^2/4\pi\epsilon_0)^2}\int_{r_0}^0 r^2 dr, \tag{8}$$

which is (1).

Comments

(i) Using (2), one can readily check that the motion of the electron is non-relativistic for most of its journey toward the proton. Also, according to (7), the fractional change in r per orbit is small, and so the actual (spiral) trajectory is reasonably approximated by circles.

(ii) Equation (1) can be expressed as

$$\tau = \frac{r_0^3}{4r_e^2 c}, \tag{9}$$

where r_e is the classical radius of the electron, defined by $mc^2 = q^2/4\pi\epsilon_0 r_e$, and having the approximate value 2.8×10^{-15} m. If we take the diameter of the initial electron orbit to be 0.5×10^{-10} m, then (9) gives a classical lifetime for the hydrogen atom of

$$\tau = \frac{(0.5 \times 10^{-10})^3}{4 \times (2.8 \times 10^{-15})^2 \times 3.0 \times 10^8} = 1.3 \times 10^{-11}\,\text{s}. \tag{10}$$

(iii) This instability of the classical hydrogen atom was but one of the serious difficulties facing classical physics during several years spanning 1900. Other 'clouds' were the 'ultraviolet catastrophe' of blackbody radiation, the photoelectric effect, the specific heats of solids at low temperature, the specific heat of a diatomic gas, and the Gibbs paradox for the extensive property of the entropy of an ideal gas. The resolution of these, and other problems, was achieved during the first quarter of the twentieth century, and it involved the creation of a new physical theory known as quantum mechanics. Today, it is recognized that classical physics applies to a large, but nevertheless restricted, domain of phenomena. There is as yet no experimental evidence that indicates any breakdown of quantum physics. The development of this theory is one of the greatest achievements of the human intellect, and its epic history is recounted in detail in the monumental set of six volumes by Mehra and Rechenberg.[11]

(iv) The first use of quantum ideas in the resolution of the hydrogen-atom problem was made by Bohr in 1913. In a drastic departure from classical theory he proposed that the angular momentum of the atomic electron can have only a discrete set of values $L = mvr = nh/2\pi$, where $n = 1, 2, \cdots$ and h is Planck's constant. He, nevertheless, retained the classical equation of motion[‡] $mv^2/r = e^2/4\pi\epsilon_0 r^2$ for a circular electronic orbit, and concluded from these two equations that the electronic orbital radii can have just a discrete set of values $r_n = \epsilon_0 h^2 n^2/\pi m e^2$, the so-called Bohr radii. Correspondingly, the electronic energy $E = \frac{1}{2}mv^2 - e^2/4\pi\epsilon_0 r$ is also restricted to a discrete set of values, the Bohr energy levels

$$E_n = -\frac{me^4}{8\epsilon_0^2 h^2} \frac{1}{n^2}. \qquad (11)$$

He then made a connection with experiment by proposing that the spectral lines of hydrogen are associated with 'quantum jumps' of the electron between Bohr orbits differing in energy by ΔE, according to $\Delta E = h\nu$ where, following Planck and Einstein, $h\nu$ is the energy of an emitted or absorbed quantum (photon) of frequency ν. In this way, Bohr was able to account for the observed spectrum of hydrogen (but not its fine structure). In particular, he obtained from (11) the expression

$$R_{\text{H}} = \frac{me^4}{8\epsilon_0^2 h^3 c} = 1.097 \times 10^7 \, \text{m}^{-1} \qquad (12)$$

for the Rydberg constant, which sets the scale of the spectrum. Despite the fact that Bohr's theory is a curious mixture of classical and quantum concepts, it made a strong impact: the good agreement he found between the theoretical expression (12) for R_{H} involving Planck's constant and the measured value was regarded as a compelling indication that a theory of quantum mechanics should exist. The realization of this theory took another 12 years.[11]

‡Here, and in what follows, we have expressed Bohr's results in terms of SI units.

---

[11]   J. Mehra and H. Rechenberg, *The historical development of quantum theory*. New York: Springer, 1982–2001. Vols. 1-6.

(v) After the formulation of quantum mechanics in 1925, the earliest applications were to the hydrogen-atom problem: independent calculations by Pauli (who used the matrix mechanics of Heisenberg, Born and Jordan) and Schrödinger (who used his wave mechanics) showed how the new theory gave the Bohr energy levels for a non-relativistic hydrogen atom.[11] This success was regarded as an important initial test of the theory, which was soon extended to the relativistic atom (see Question 15.16).

(vi) In modern particle physics the stability of atoms such as hydrogen is viewed in terms of the classification of particles into baryons (e.g. the proton) and leptons (e.g. the electron), and the conservation laws of baryon number B and lepton number L. Accordingly, a proton ($B = 1$, $L = 0$) cannot decay into a positron[†] ($B = 0$, $L = -1$) and a photon ($B = 0$, $L = 0$) because this would violate conservation of baryon and lepton numbers. (The current experimental lower limit for the lifetime of a proton is $\sim 10^{36}$ yr.) Similarly, a hydrogen atom ($B = 1$, $L = 1$) cannot decay into photons. However, this decay is not prohibited for a positronium atom (a bound state of a positron and an electron), where $B = 0$ and $L = 0$. In fact, the longest lifetime of the ground state of positronium is $\sim 10^{-7}$ s. Of course, this lifetime is determined by quantum mechanics and not by the classical result (1).

(vii) The developments mentioned above provide a good illustration of Thomas Kuhn's famous description of the nature of scientific progress, as a cyclic passage through various stages:

$$\text{regular science} \;\to\; \text{crisis} \;\to\; \text{revolution} \;\to\; \text{regular science}.$$

In regular science the consequences and applications of a new theory or model are developed and tested. Eventually, a crisis develops when experimental results are obtained that are at variance with this model (for example, the 'clouds' referred to earlier). This precipitates a revolution when fundamental changes to the science are implemented, and thereafter regular science resumes.[12]

Question 7.23

A particle of mass m and charge q is subject to a non-uniform magnetostatic field

$$\mathbf{B} = (0,\; -\alpha y,\; 1 + \alpha z)B\,, \tag{1}$$

where α and B are positive constants.

(a) Show that the components of the equation of motion can be expressed as

$$\frac{dX}{d\tau} - (1 + Z)Y = D \tag{2}$$

[†]i.e. an anti-electron.

[12] T. S. Kuhn, *The structure of scientific revolutions*. Chicago: Chicago University Press, 1970.

$$\frac{d^2Y}{d\tau^2} + (1+Z)^2 Y + D(1+Z) = 0 \qquad (3)$$

$$\frac{d^2Z}{d\tau^2} + (1+Z)Y^2 + DY = 0. \qquad (4)$$

Here, D is a constant which is determined below and

$$(X, Y, Z) = (\alpha x, \alpha y, \alpha z) \quad \text{and} \quad \tau = qBt/m = \omega t \qquad (5)$$

are dimensionless coordinates and time.

(b) Consider a coordinate system in which the initial conditions at $t = 0$ are

$$(X, Y, Z)_0 = (0, Y_0, 0) \quad \text{and} \quad \mathbf{V}_0 = (U_1, 0, U_3), \qquad (6)$$

where $U_i = \alpha u_i / \omega$ are dimensionless initial velocities. Take

$$Y_0 = 0.66, \qquad U_1 = 0.02, \qquad U_3 = 0.08, \qquad (7)$$

and use *Mathematica* to solve the coupled, non-linear equations (2)–(4) for $0 \leq \tau \leq 140$. (Note that according to (2) and (6) the constant D is fixed in terms of initial conditions by $D = U_1 - Y_0 = -0.64$.) Present the results for $X(\tau)$, $Y(\tau)$ and $Z(\tau)$ graphically, and also plot graphs of $Y(Z)$ and $Y(X)$.

(c) Describe the motion represented by these graphs.

Solution

We remark that the field (1) satisfies $\nabla \cdot \mathbf{B} = 0$ (as any magnetic field must) and $\nabla \times \mathbf{B} = 0$ (which is required of static fields in source-free regions). Note also that α^{-1} is a length scale.

(a) For the field (1), the equation of motion $m\ddot{\mathbf{r}} = q\dot{\mathbf{r}} \times \mathbf{B}$ has components

$$\ddot{x} = \omega \left\{ \dot{y} + \alpha \frac{d}{dt}(yz) \right\} \qquad (8)$$

$$\ddot{y} = -\omega(1 + \alpha z)\dot{x} \qquad (9)$$

$$\ddot{z} = -\omega \alpha y \dot{x}, \qquad (10)$$

where $\omega = qB/m$ is the cyclotron frequency for the uniform field $(0, 0, B)$ corresponding to $\alpha = 0$. Integration of (8) gives

$$\dot{x} = \omega(1 + \alpha z)y + D_0, \qquad (11)$$

where D_0 is a constant of integration. By substituting (11) in (9) and (10) we obtain two coupled equations involving y and z only:

$$\ddot{y} = -\omega^2(1 + \alpha z)^2 y - \omega D_0 (1 + \alpha z) \qquad (12)$$

$$\ddot{z} = -\omega^2 \alpha (1 + \alpha z) y^2 - \omega D_0 \alpha y. \qquad (13)$$

If we multiply (11) by α/ω and (12) and (13) by α/ω^2, and use (5), we obtain the dimensionless forms (2)–(4) with $D = \alpha D_0 / \omega$.

(b) The *Mathematica* notebook used to solve (2)–(4) for $X(\tau)$, $Y(\tau)$, and $Z(\tau)$, and the graphs obtained from it, are shown below.

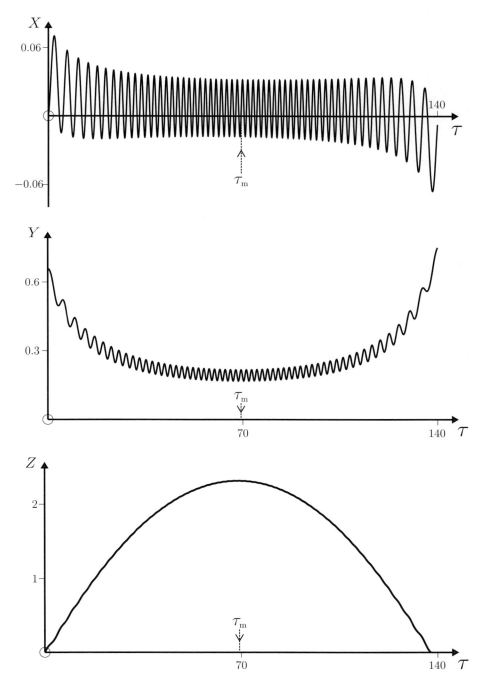

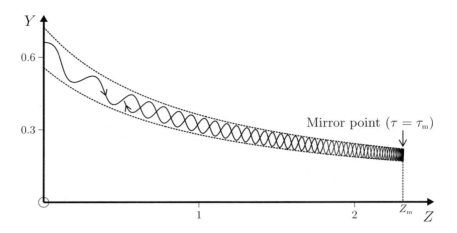

(c) Three features are evident in these graphs:

1. Just as in the case of motion in a uniform field (see Question 7.17), the particle spirals about the field lines (these lines are discussed in Question 5.26).

2. There is a turning point at the instant $\tau = \tau_m$ in the motion along the Z-axis: for $\tau < \tau_m$ the particle spirals in the positive Z-direction; for $\tau > \tau_m$ it spirals back in the negative Z-direction; and at $\tau = \tau_m = 69.35$ the component $\dot{Z} = 0$. The spatial point at which $Z = Z_m = 2.316$ is known as the mirror point. Thus, when charged particles spiral into regions of increasing magnetic field (see Question 5.26) their forward motion decreases until they are eventually reflected.

3. The spiralling and forward (or backward) motions are accompanied by drift in the X- and Y-directions. For $\tau < \tau_m (> \tau_m)$ the particle drifts in the negative (positive) Y-direction. The drift along the X-axis is slower than this and is in the negative X-direction. Close to the mirror point the drift velocities are low.

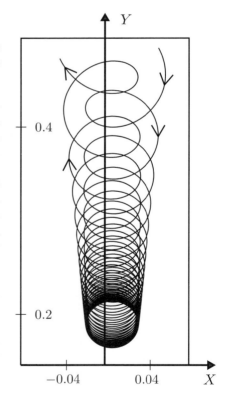

Comments

(i) The above calculation neglects radiation by the particle (see Question 7.21). In this approximation the kinetic energy is constant (because the field does no work

on the particle), and so
$$\dot{X}^2 + \dot{Y}^2 + \dot{Z}^2 = U_1^2 + U_3^2, \tag{14}$$
where a dot signifies differentiation with respect to τ. This can be expressed in two-dimensional form by using (2) to eliminate $\dot{X}$ in favour of Y and Z:
$$\dot{Y}^2 + \dot{Z}^2 = U_1^2 + U_3^2 - \{D + (1+Z)Y\}^2. \tag{15}$$
Because $\dot{Y}^2 + \dot{Z}^2$ cannot be negative, (15) shows that there are two boundary curves to the graph of $Y(Z)$, obtained by setting the right-hand side of (15) equal to zero. That is,
$$Y = \frac{-D \pm \sqrt{U_1^2 + U_3^2}}{1+Z} = \frac{0.64 \pm 0.0825}{1+Z}. \tag{16}$$
These boundaries – that are indicated by the two dotted curves in the above plot of $Y(Z)$ – are, in fact, magnetic field lines (see (3) in Question 5.26).

(ii) The magnetic field in this question is an example of a 'mirror field', characterized by the property that charged particles are reflected back from regions where the field is large. Magnetic mirrors find application in, *inter alia*, fusion devices that are used to confine charged particles.

(iii) An important example of a magnetic mirror in nature is the magnetic field around the Earth. Charged particles from the solar wind (usually electrons, but also protons and ions) that become trapped in the Earth's field perform three distinct motions:

☞ they spiral around field lines with a period of order milliseconds;

☞ they move along field lines, oscillating between the North and South poles with a period of order seconds (this is the mirror effect); and

☞ they also perform a transverse drift around the Earth with a period of order hours.

(iv) The regions of space around the Earth populated by these charged particles are called the Van Allen radiation belts. They are responsible for the polar aurora when the trapped particles interact with atmospheric oxygen and nitrogen, causing these molecules to fluoresce.

```
In[1]:= X0 = 0; Y0 = 0.66; Z0 = 0;
       U1 = 0.02; U2 = 0; U3 = 0.08; τmax = 140;
       𝒟 = U1 - Y0;  (* Here 𝒟 represents the constant D used in the Question *)
       f1[τ_] := (-𝒟 - √(U1² + U3²))/(1.0 + Z[τ]);  f2[τ_] := (-𝒟 + √(U1² + U3²))/(1.0 + Z[τ]);
       Sol = NDSolve[{X'[τ] - ((1 + Z[τ]) Y[τ]) - 𝒟 == 0,
              Y''[τ] + (1 + Z[τ])² Y[τ] + 𝒟(1 + Z[τ]) == 0,
              Z''[τ] + (1 + Z[τ]) Y[τ]² + 𝒟 Y[τ] == 0,
              X[0] == X0, Y'[0] == U2, Y[0] == Y0, Z'[0] == U3, Z[0] == Z0},
              {X[τ], Y[τ], Z[τ], X'[τ], Y'[τ], Z'[τ]}, {τ, 0, τmax}, MaxSteps → 100000];
```

```
In[2]:= Plot[X[τ]/.Sol, {τ, 0, τmax}, PlotRange → {{0, τmax}, {-0.07, 0.07}}]
       Plot[Y[τ]/.Sol, {τ, 0, τmax}, PlotRange → {{0, τmax}, {0, 0.8}}]
       Plot[Z[τ]/.Sol, {τ, 0, τmax}, PlotRange → {{0, τmax}, {0, 2.5}}]
       ParametricPlot[{{{Z[τ], Y[τ]}/.Sol}, {{Z[τ], f1[τ]}/.Sol},
           {{Z[τ], f2[τ]}/.Sol}}, {τ, 0, τmax}, PlotRange → {{0, 2.5}, {0, 0.8}}]
```

Question 7.24

Consider the previous question for motion of a charged particle in a non-uniform magnetostatic field $\mathbf{B} = (0, -\alpha y, 1+\alpha z)B$ in the limit of small departure from the uniform-field problem. That is, assume

$$\alpha \ll 1 \quad \text{and} \quad y_0 \approx u_1/\omega = r_{\text{L}}; \quad \text{i.e.} \quad Y_0 \approx \alpha r_{\text{L}} = U_1, \tag{1}$$

where the Larmor radius r_{L} is the orbital radius in a uniform field.

(a) Show that the dimensionless Z-coordinate of the motion is given by a parabolic approximation

$$Z(\tau) = U_3 \tau - \tfrac{1}{4} U_1^2 \tau^2, \tag{2}$$

where the notation is that of the previous question.

(b) Hence, obtain approximate expressions for the time τ_{m} to reach the mirror point and the coordinate Z_{m} of this point.

(c) Compare the approximation (2) with the result of a numerical calculation (using the above *Mathematica* notebook) for

$$U_1 = 0.02, \qquad U_3 = 0.08, \qquad Y_0 = 2U_1 = 0.04, \tag{3}$$

on a plot of $Z(\tau)$ for $0 \leq \tau \leq 800$.

Solution

(a) When (1) is satisfied we can approximate x and y on the right-hand side of (10) in Question 7.23 by their expressions for a uniform field $(0, 0, B)$, namely

$$x(t) = (u_1/\omega) \sin \omega t, \qquad y(t) = (u_1/\omega) \cos \omega t + c, \tag{4}$$

where c is a constant (see Question 7.17). Then, in dimensionless form we have for (10):

$$\frac{d^2 Z}{d\tau^2} = -Y \frac{dX}{d\tau} = -(U_1 \cos \tau + C) U_1 \cos \tau, \tag{5}$$

where $C = \alpha c$. Therefore,

$$\dot{Z}(\tau) = U_3 - \tfrac{1}{2} U_1^2 \tau - \tfrac{1}{4} U_1^2 \sin 2\tau - C U_1 \sin \tau. \tag{6}$$

The last two terms in (6) are rapidly oscillating contributions about an average value

$$\dot{Z}(\tau) = U_3 - \tfrac{1}{2} U_1^2 \tau, \tag{7}$$

and consequently we have the parabolic approximation (2) for $Z(\tau)$.

(b) At the mirror point $\dot{Z} = 0$ and therefore

$$\tau_m = 2U_3/U_1^2 \quad \text{and} \quad Z_m = (U_3/U_1)^2. \tag{8}$$

(c) The parabolic approximation (2) with initial conditions (3) is compared with the result of the numerical analysis in the following figure and table.

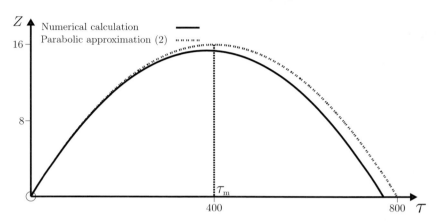

	τ_m	Z_m
Numerical calculation	384.9	15.4
Equation (8)	400.0	16.0

Improved agreement is obtained with decreasing Y_0 (see below).

Comments

(i) The following graph of $Z_m(\alpha)$ (for $U_1 = 2\alpha$, $U_3 = 8\alpha$ and $Y_0 = 6\alpha$) shows the agreement between the numerical values and the approximation $(8)_2$ at low α, and the increasing disparity as α increases.

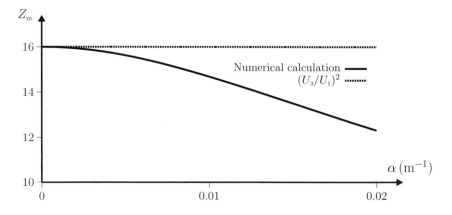

(ii) The final graph shows Z_m as a function of $y_0/r_L = Y_0/\alpha r_L$ for $\alpha = 0.001\,\text{m}^{-1}$, with $U_1 = 0.02$ and $U_3 = 0.08$. Again, we see the expected agreement with the approximation $(8)_2$ for low values of y_0/r_L (where (4) is a good approximation), and increasing disagreement at higher values of y_0/r_L (where (4) becomes a poor approximation).

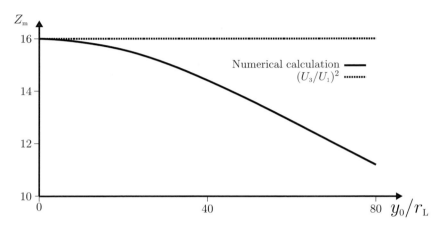

Question 7.25

A particle of mass m moves in the xy-plane, acted on by a linear restoring force $\mathbf{F} = -k\mathbf{r} = -k(x, y)$, where k is a positive constant. Determine and sketch the trajectory if the initial conditions at $t = 0$ are

$$\mathbf{r}_0 = (x_0, 0) \quad \text{and} \quad \mathbf{v}_0 = (0, v_0). \tag{1}$$

Solution

The equation of motion

$$m\frac{d^2\mathbf{r}}{dt^2} = -k\mathbf{r} \tag{2}$$

yields the two oscillator equations

$$\frac{d^2x}{dt^2} + \omega^2 x = 0, \qquad \frac{d^2y}{dt^2} + \omega^2 y = 0, \tag{3}$$

where $\omega^2 = k/m > 0$. The solutions to (3) that satisfy the initial conditions (1) are clearly

$$x(t) = x_0 \cos\omega t, \qquad y(t) = (v_0/\omega)\sin\omega t. \tag{4}$$

Therefore, the trajectory is the ellipse

$$\frac{x^2}{x_0^2} + \frac{y^2}{(v_0/\omega)^2} = 1, \tag{5}$$

with axes of length $2x_0$ and $2v_0/\omega$. The centre of the ellipse is at the centre of force O. There are two special cases: if $v_0 = \omega x_0$ the trajectory is a circle of radius x_0, and if $v_0 = 0$ the particle moves in a straight line between $-x_0$ and x_0 along the x-axis.

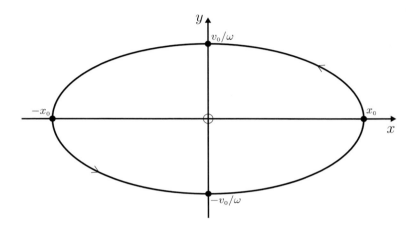

Comments

(i) This system is known as an isotropic, two-dimensional harmonic oscillator.
(ii) Use of the initial conditions (1) is equivalent to orienting the coordinate axes in a convenient way, namely with the x-axis directed to a turning point of the motion (see also Question 8.10).
(iii) We could have used the initial conditions $\mathbf{r}_0 = (x_0, y_0)$ and $\mathbf{v}_0 = (u_1, u_2)$. For these the solutions to (4) are

$$x(t) = A_1 \cos(\omega t + \phi_1), \qquad y(t) = A_2 \cos(\omega t + \phi_2), \tag{6}$$

where

$$A_1 = \sqrt{x_0^2 + (u_1/\omega)^2}, \qquad \tan\phi_1 = -u_1/\omega x_0, \tag{7}$$

and similarly for A_2 and ϕ_2 (see Question 4.1). However, this does not produce any additional types of trajectory: a short calculation starting with (6) shows that

$$\frac{x^2}{A_1^2} + \frac{y^2}{A_2^2} - \frac{2xy}{A_1 A_2}\cos(\phi_2 - \phi_1) = \sin^2(\phi_2 - \phi_1). \tag{8}$$

If $\phi_2 - \phi_1 = \frac{1}{2}\pi$ then (8) reduces to an ellipse in the form (5). If $\phi_2 - \phi_1 = 0$ or π then (8) reduces to a straight line $y = \pm(A_2/A_1)x$, while for all other values of $\phi_2 - \phi_1$ it is an ellipse with its axes rotated in the xy-plane.

(iv) For an isotropic, three-dimensional harmonic oscillator $\mathbf{F} = -k(x, y, z)$, and in addition to (6) we have

$$z(t) = A_3 \cos(\omega t + \phi_3). \tag{9}$$

Equations (6) and (9) imply a linear relationship
$$z = ax + by, \qquad (10)$$
which is the equation of a plane. That the motion is confined to a plane is not surprising because the central force $-k\mathbf{r}$ conserves the angular momentum of the particle, and therefore the motion is restricted to the plane defined by the initial vectors $\mathbf{r}_0$ and $\mathbf{p}_0 = m\mathbf{v}_0$ (see Question 6.15). If we orient the x- and y-axes in this plane then the solutions are (6) and $z = 0$: the trajectory is either an ellipse, a circle or a straight line.

(v) It is also interesting to solve and analyze the isotropic, three-dimensional harmonic oscillator using plane polar coordinates (see Chapters 8 and 9).

Question 7.26

Consider an anisotropic, two-dimensional harmonic oscillator with $\mathbf{F} = -(k_1 x, k_2 y)$, where the k_i are positive constants. Obtain the trajectory and deduce the condition for closed orbits.

Solution

The components of the equation of motion are
$$\frac{d^2 x}{dt^2} + \omega_1^2 x = 0, \qquad \frac{d^2 y}{dt^2} + \omega_2^2 y = 0, \qquad (1)$$
where $\omega_i = \sqrt{k_i/m}$. The general solutions are
$$x(t) = A_1 \cos(\omega_1 t + \phi_1), \qquad y(t) = A_2 \cos(\omega_2 t + \phi_2), \qquad (2)$$
where the constants A_i and ϕ_i depend on the initial conditions $\mathbf{r}_0$ and $\mathbf{v}_0$ in the usual manner (see Question 4.1). The nature of the trajectory described by (2) depends crucially on the ratio ω_2/ω_1 of the angular frequencies. If this ratio is a rational number n_2/n_1, where n_1 and n_2 are integers, then
$$\frac{n_2}{n_1} = \frac{\omega_2}{\omega_1} = \frac{T_1}{T_2}, \qquad (3)$$
where T_1 and T_2 are the periods of the x- and y-oscillations. It follows that the orbit is closed because $n_1 T_1 = n_2 T_2$, meaning that when the particle has performed n_1 complete oscillations in the x-direction it has also performed n_2 complete oscillations in the y-direction, and has therefore returned to its starting point $\mathbf{r}_0$ with the same initial velocity $\mathbf{v}_0$. The motion is periodic.

If ω_2/ω_1 is an irrational number, the orbit is open: the particle never regains the initial conditions $\mathbf{r}_0$, $\mathbf{v}_0$ (it never passes twice through the same point with the same velocity). It can be shown that in this case the trajectory 'fills' the rectangle $2A_1 \times 2A_2$ in the xy-plane in the sense that given sufficient time, the particle will pass arbitrarily close to each point in this rectangle.

Comments

(i) This example illustrates again that the dynamics can be sensitive to even very small changes in a parameter – in this case a ratio of frequencies (see also Questions 2.1, 2.2, 3.3, 7.12, and Chapter 13).

(ii) The trajectory (2) is illustrated in the following two figures that are for $A_2/A_1 = 1$, $\phi_2 = \phi_1 = 0$, $\omega_2/\omega_1 = 3$ and $A_2/A_1 = 1$, $\phi_2 = \phi_1 = 0$, $\omega_2/\omega_1 = \pi$. The first trajectory is closed and periodic; the second is open.

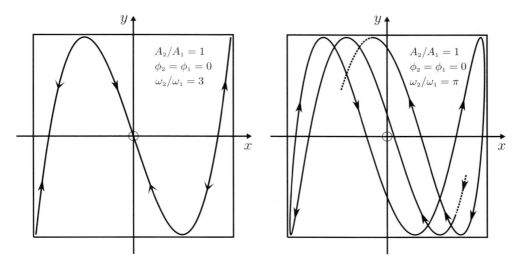

(iii) In general, the shape of the curve is sensitive to the phase difference $\delta = \phi_2 - \phi_1$, as illustrated in the following so-called Lissajous figures for various values of δ. Here, $A_2/A_1 = 1$ and $\omega_2/\omega_1 = 2$:

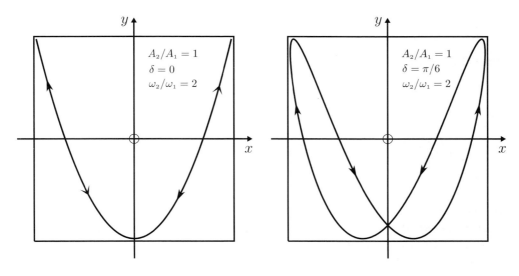

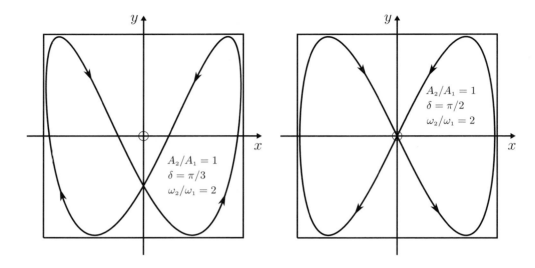

(iv) The above analysis is readily extended to an anisotropic, three-dimensional harmonic oscillator with $\mathbf{F} = -(k_1 x, \, k_2 y, \, k_3 z)$. Then, in addition to (2) we have

$$z(t) = A_3 \cos(\omega_3 t + \phi_3), \qquad (4)$$

where $\omega_3 = \sqrt{k_3/m}$. The condition for closed orbits is now

$$\omega_2/\omega_1 = n_2/n_1 \quad \text{and} \quad \omega_3/\omega_2 = n_3/n_2, \qquad (5)$$

because then $n_1 T_1 = n_2 T_2 = n_3 T_3$, meaning that n_1 oscillations in the x-direction, n_2 oscillations in the y-direction, and n_3 oscillations in the z-direction are completed in the same time, and the particle returns to its starting position $\mathbf{r}_0$ with the same initial velocity $\mathbf{v}_0$. Otherwise the orbit is open, and the trajectory fills the rectangular box $2A_1 \times 2A_2 \times 2A_3$ in the sense mentioned above.

Question 7.27

A particle of mass m is acted on by an inverse-square force (the Coulomb problem)

$$\mathbf{F} = -k \frac{\mathbf{r}}{r^3}, \qquad (1)$$

where the constant k can be positive or negative. The particle passes the point $\mathbf{r}_0$ with velocity $\mathbf{v}_0$, where

$$\mathbf{r}_0 = (x_0, \, y_0, \, 0) \quad \text{and} \quad \mathbf{v}_0 = (u_1, \, u_2, \, 0). \qquad (2)$$

(a) Argue that the motion is confined to the xy-plane, and show that the Cartesian form of the trajectory is

$$A_1 x + A_2 y + (L^2/mk) = (x^2 + y^2)^{1/2}. \qquad (3)$$

Here, L is the magnitude of the angular momentum $\mathbf{L} = \mathbf{r} \times \mathbf{p} = (0, 0, L)$, and A_1 and A_2 are constants such that

$$A_1^2 + A_2^2 = 1 + (2L^2 E/mk^2), \tag{4}$$

where $E = \frac{1}{2}mv^2 - k/r$ is the energy. (Hint: Use conservation of angular momentum and the kinematic identities

$$\frac{1}{y}\frac{d}{dt}\frac{x}{r} = -\frac{1}{x}\frac{d}{dt}\frac{y}{r} = \frac{1}{r^3}\left(y\frac{dx}{dt} - x\frac{dy}{dt}\right), \quad \text{where } r = \sqrt{x^2 + y^2}.) \tag{5}$$

(b) The task of interpreting (3) is simplified if we choose the x-axis to point along the vector $\mathbf{A} = (A_1, A_2, 0)$. In such a system $A_2 = 0$. Hence, obtain from (3) the trajectory in the form

$$y^2 = (2L^2/mk)(\pm x + L^2/2mk) \tag{6}$$

if $E = 0$, and

$$\frac{\left(x + \frac{k}{2E}\sqrt{1 + \frac{2L^2 E}{mk^2}}\right)^2}{(k/2E)^2} - \frac{y^2}{L^2/2mE} = 1 \tag{7}$$

if $E \neq 0$.

(c) Interpret and sketch the possible trajectories represented by (6) and (7).

Solution

(a) Because the force (1) is central (always directed toward or away from the origin), the motion is confined to the plane defined by the position and velocity vectors $\mathbf{r}$ and $\mathbf{v}$ at any instant: so (2) means that the motion is in the xy-plane. The components of the equation of motion in this system are

$$m\frac{d^2 x}{dt^2} = -k\frac{x}{r^3}, \qquad m\frac{d^2 y}{dt^2} = -k\frac{y}{r^3}, \tag{8}$$

where $r = \sqrt{x^2 + y^2}$. It follows that $y\ddot{x} = x\ddot{y}$, which implies conservation of angular momentum $\mathbf{L} = \mathbf{r} \times \mathbf{p}$ in this system:

$$x\frac{dy}{dt} - y\frac{dx}{dt} = \frac{L}{m}. \tag{9}$$

From (5), (8) and (9) we have

$$\frac{d}{dt}\frac{x}{r} = \frac{L}{k}\frac{d^2 y}{dt^2}, \qquad \frac{d}{dt}\frac{y}{r} = -\frac{L}{k}\frac{d^2 x}{dt^2}. \tag{10}$$

Integration of (10) gives the pair of first-order, coupled equations

$$\frac{x}{r} = \frac{L}{k}\frac{dy}{dt} + A_1, \qquad \frac{y}{r} = -\frac{L}{k}\frac{dx}{dt} + A_2, \tag{11}$$

where A_1 and A_2 are constants. The trajectory (3) is obtained after using (11) to eliminate dx/dt and dy/dt from (9). It also follows from (11) and (9) that

$$A_1^2 + A_2^2 = \left(\frac{x}{r} - \frac{L}{k}\dot{y}\right)^2 + \left(\frac{y}{r} + \frac{L}{k}\dot{x}\right)^2$$

$$= \frac{x^2 + y^2}{r^2} + \left(\frac{L}{k}\right)^2(\dot{x}^2 + \dot{y}^2) + \frac{2L}{kr}(y\dot{x} - x\dot{y})$$

$$= 1 + (L/k)^2\{\dot{x}^2 + \dot{y}^2 - 2k/mr\},$$

which is (4) because the quantity in braces is $(2/m)$ times the energy E.

(b) By squaring both sides of (3), with $A_2 = 0$, and rearranging terms we have

$$(1 - A_1^2)x^2 - 2(L^2/mk)A_1 x + y^2 = (L^2/mk)^2, \tag{12}$$

where, according to (4),

$$A_1^2 = 1 + 2L^2 E/mk^2. \tag{13}$$

When $E = 0$, $A_1 = \pm 1$ and (12) reduces to (6). When $E \neq 0$ we divide (12) by $1 - A_1^2 = -2L^2 E/mk^2$ and complete the square in x to obtain (7).

(c) To interpret these results we consider separately attractive and repulsive forces.

☞ **Attractive force $k > 0$**

The energy E can be zero, negative or positive. For $E = 0$ the trajectories (6) are parabolas. For $E \neq 0$ the trajectory is given by (7), and to interpret it we recall that the curve

$$X^2/a^2 - Y^2/b^2 = 1 \qquad (a^2, b^2 > 0) \tag{14}$$

is a hyperbola in the XY-plane, while the curve

$$X^2/a^2 + Y^2/b^2 = 1 \qquad (a^2, b^2 > 0) \tag{15}$$

is an ellipse. If $a > b (> 0)$ then a is the length of the semi-major axis, b is the length of the semi-minor axis of the ellipse. Also, the eccentricity is

$$e = \sqrt{1 + b^2/a^2} \tag{16}$$

for a hyperbola and

$$e = \sqrt{1 - b^2/a^2} \tag{17}$$

for an ellipse. Therefore, by inspection of (7), we conclude that for $E > 0$ the trajectory is a hyperbola with $a^2 = (k/2E)^2$, $b^2 = L^2/2mE$, and for $E < 0$ the trajectory is an ellipse with $a^2 = (k/2E)^2$ and $b^2 = -L^2/2mE$. In both cases the eccentricities (16) and (17) are

$$e = \sqrt{1 + (2L^2 E/mk^2)} = |A_1|. \tag{18}$$

Equation (7) also shows that the centre of force $x = 0$, $y = 0$ is not at the centre $X = 0$, $Y = 0$ of the ellipse, but at

$$X = \frac{k}{2E}\sqrt{1 + \frac{2L^2 E}{mk^2}} = ea, \qquad Y = 0, \tag{19}$$

which is the position of one of the foci of the ellipse. Similarly, for the hyperbolic and parabolic orbits the centre of force is at a focus.

☞ **Repulsive force $k < 0$**

Here, the energy E in (7) is necessarily positive, and consequently only hyperbolic orbits occur.

Note that with the above choice of axes, a point at which the trajectory cuts the x-axis is a turning point (a minimum or maximum) of the motion: this is evident from $(11)_2$, where $A_2 = 0$ and $y = 0$ imply $\dot{x} = 0$. Diagrams of the various orbits are given in Question 8.9.

Comments

(i) This problem is usually solved in terms of plane polar coordinates (see Question 8.9), although the above solution based on Cartesian coordinates is, if anything, more direct. Of course, there are advantages to using polar coordinates: they are part of a general approach to solving and analyzing dynamics in a spherically symmetric potential, and examples are given in Chapter 8.

(ii) The Cartesian trajectory (3) can be written in terms of polar coordinates (r, θ) by setting $x = r\cos\theta$, $y = r\sin\theta$ and $A_1 = A\cos\theta_0$, $A_2 = A\sin\theta_0$. Then, (3) and (4) give the polar equation of a conic section (ellipse, hyperbola or parabola):

$$r(\theta) = \frac{L^2/mk}{1 - \sqrt{1 + (2L^2 E/mk^2)}\,\cos(\theta - \theta_0)}. \tag{20}$$

The choice $\theta_0 = 0$ aligns the x-axis with $\mathbf{A}$, making (20) equivalent to (6) and (7).

(iii) An essential feature of the above solution in Cartesian coordinates is the use of the two constants of integration A_1 and A_2 that appear in (11). The existence of these constants is associated with a 'hidden symmetry' possessed by the inverse-square force. This topic has been the subject of extensive study since the seventeenth century (see Chapter 9).

(iv) The solution to the problem of motion in an inverse-square force and its application to planetary motion (the Kepler problem) was the first major accomplishment of Newtonian mechanics, and it established the connection between Kepler's laws of motion and the law of universal gravitation. Other applications followed, such as to Rutherford scattering.

(v) In Newtonian dynamics the trajectory always involves (at least) the initial conditions $\mathbf{r}_0$ and $\mathbf{v}_0$, and the reader may wonder where these occur in (6) and (7). The answer is that they enter through the energy $E = \frac{1}{2}mv_0^2 - k/r_0$ and the angular momentum $\mathbf{L} = m\mathbf{r}_0 \times \mathbf{v}_0$, which are both conserved quantities.

8
Spherically symmetric potentials

The examples in this chapter deal with motion of a particle in a spherically symmetric potential $V(r)$, the corresponding force

$$\mathbf{F} = -\frac{dV}{dr}\hat{\mathbf{r}}$$

being central and isotropic. We have already seen in Questions 5.3 and 6.15 that for such motion the energy E and angular momentum $\mathbf{L}$ of the particle are conserved. The conservation of $\mathbf{L}$ has the immediate consequence that for $\mathbf{L} \neq 0$ the motion is confined to the plane defined by the initial position and momentum vectors $\mathbf{r}_0$ and $\mathbf{p}_0 = m\mathbf{v}_0$. The angular momentum is perpendicular to this plane, and $\mathbf{r}$, $\mathbf{p}$ and $\mathbf{L} = \mathbf{r} \times \mathbf{p}$ form a right-handed set. Usually, the motion is analyzed in terms of plane polar coordinates (r, θ) of an inertial frame, although we could also use other coordinates such as two-dimensional Cartesian coordinates (x, y), as in Questions 7.25–7.27.

These two sets of coordinates and their unit vectors are depicted below. Clearly,

$$x = r\cos\theta, \qquad y = r\sin\theta.$$

The inverse relations are

$$r = \sqrt{x^2 + y^2}, \qquad \theta = \tan^{-1}\left(\frac{y}{x}\right).$$

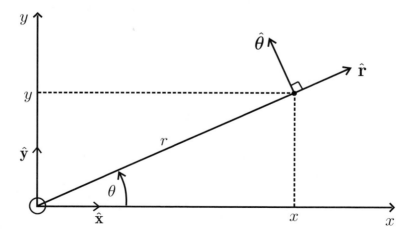

Question 8.1

Prove that in plane polar coordinates (r, θ) the velocity and acceleration vectors are given by

$$\mathbf{v} = \dot{r}\hat{\mathbf{r}} + r\dot{\theta}\hat{\boldsymbol{\theta}} \tag{1}$$

$$\mathbf{a} = (\ddot{r} - r\dot{\theta}^2)\hat{\mathbf{r}} + (r\ddot{\theta} + 2\dot{r}\dot{\theta})\hat{\boldsymbol{\theta}}. \tag{2}$$

Solution

It is clear that the two unit vectors $\hat{\mathbf{r}}$ and $\hat{\boldsymbol{\theta}}$ shown in the figure above are time dependent. To calculate their rates of change, we first resolve them into x- and y-components:

$$\hat{\mathbf{r}} = \hat{\mathbf{x}}\cos\theta + \hat{\mathbf{y}}\sin\theta, \qquad \hat{\boldsymbol{\theta}} = -\hat{\mathbf{x}}\sin\theta + \hat{\mathbf{y}}\cos\theta. \tag{3}$$

Then

$$\frac{d\hat{\mathbf{r}}}{dt} = \dot{\theta}(-\hat{\mathbf{x}}\sin\theta + \hat{\mathbf{y}}\cos\theta) = \dot{\theta}\hat{\boldsymbol{\theta}} \tag{4}$$

$$\frac{d\hat{\boldsymbol{\theta}}}{dt} = -\dot{\theta}(\hat{\mathbf{x}}\cos\theta + \hat{\mathbf{y}}\sin\theta) = -\dot{\theta}\hat{\mathbf{r}}. \tag{5}$$

(In obtaining (4) and (5) from (3) we have used the fact that $d\hat{\mathbf{x}}/dt$ and $d\hat{\mathbf{y}}/dt$ are both zero.) Now

$$\mathbf{r} = r\hat{\mathbf{r}}, \tag{6}$$

and hence the velocity is

$$\mathbf{v} = \frac{dr}{dt}\hat{\mathbf{r}} + r\frac{d\hat{\mathbf{r}}}{dt}. \tag{7}$$

Equations (4) and (7) yield (1). The acceleration is the derivative of (1):

$$\mathbf{a} = \ddot{r}\hat{\mathbf{r}} + \dot{r}\frac{d\hat{\mathbf{r}}}{dt} + \dot{r}\dot{\theta}\hat{\boldsymbol{\theta}} + r\ddot{\theta}\hat{\boldsymbol{\theta}} + r\dot{\theta}\frac{d\hat{\boldsymbol{\theta}}}{dt}. \tag{8}$$

If we substitute (4) and (5) into (8), we obtain (2).

Comments

(i) Equations (1) and (2) express the velocity and acceleration in terms of radial and transverse components (that is, components along $\hat{\mathbf{r}}$ and $\hat{\boldsymbol{\theta}}$). The usefulness of these results will become apparent in the questions below.

(ii) For motion in a circle, r is constant and (1) becomes

$$\mathbf{v} = r\dot{\theta}\hat{\boldsymbol{\theta}}; \tag{9}$$

the velocity is transverse. If the angular speed $\dot{\theta}$ is also constant then (2) simplifies to

$$\mathbf{a} = -r\dot{\theta}^2\hat{\mathbf{r}} = -v^2\hat{\mathbf{r}}/r; \tag{10}$$

the acceleration is radial (directed towards the centre of the circle).

Question 8.2

Show that the angular momentum $\mathbf{L}$ and the energy E of a particle of mass m moving in a potential $V(r)$ can be expressed in terms of plane polar coordinates as

(a) $\quad \mathbf{L} = mr^2 \dot\theta \hat{\mathbf{z}},$ (1)

(b) $\quad E = \tfrac{1}{2}m\dot r^2 + \tfrac{1}{2}mr^2\dot\theta^2 + V(r).$ (2)

Here, $\hat{\mathbf{z}} = \hat{\mathbf{r}} \times \hat{\boldsymbol\theta}$ is a unit vector perpendicular to the plane of the trajectory.

Solution

(a) $\mathbf{L} = \mathbf{r} \times \mathbf{p} = m\mathbf{r} \times \mathbf{v} = mr\hat{\mathbf{r}} \times (\dot r \hat{\mathbf{r}} + r\dot\theta \hat{\boldsymbol\theta}),$ (3)

where in the last step we have used (1) and (6) of the previous question. Now $\hat{\mathbf{r}} \times \hat{\mathbf{r}} = 0$, and hence (3) reduces to (1).

(b) $E = \tfrac{1}{2}mv^2 + V(r),$ (4)

where
$$v^2 \equiv \mathbf{v} \cdot \mathbf{v} = (\dot r \hat{\mathbf{r}} + r\dot\theta \hat{\boldsymbol\theta}) \cdot (\dot r \hat{\mathbf{r}} + r\dot\theta \hat{\boldsymbol\theta}) = \dot r^2 + r^2 \dot\theta^2. \quad (5)$$

From (4) and (5) we obtain (2).

Comments

(i) We can use (1) to eliminate $\dot\theta$ from (2). Then

$$E = \tfrac{1}{2}m\dot r^2 + V(r) + \frac{L^2}{2mr^2}, \quad (6)$$

where
$$L = mr^2\dot\theta \quad (7)$$

is the component of L along $\hat{\mathbf{z}}$. Equation (6) is a very interesting result that is often written as

$$E = \tfrac{1}{2}m\dot r^2 + V_{\mathrm{e}}(r), \quad (8)$$

where
$$V_{\mathrm{e}}(r) = V(r) + \frac{L^2}{2mr^2}. \quad (9)$$

Equation (8) is just the expression for the energy of a particle of mass m moving in one dimension (along $\hat{\mathbf{r}}$) in an 'effective potential' $V_{\mathrm{e}}(r)$. Evidently, we are now viewing the motion from a frame that rotates in such a way that the angular position of the particle is fixed and the motion is purely radial. Such a frame is clearly non-inertial: in it there is an effective force

$$\mathbf{F}_{\mathrm{e}} = -\frac{dV_{\mathrm{e}}(r)}{dr}\hat{\mathbf{r}} = -\frac{dV(r)}{dr}\hat{\mathbf{r}} + \frac{L^2}{mr^3}\hat{\mathbf{r}}, \quad (10)$$

which consists of the physical force $(-dV/dr)\hat{\mathbf{r}}$, experienced in an inertial frame, plus a non-inertial (centrifugal) force

$$\mathbf{F}_{cf} = \frac{L^2}{mr^3}\hat{\mathbf{r}} \qquad (11)$$

associated with the rotation (see Chapters 1 and 14).

(ii) Equations (8) and (9) enable us to describe the motion in terms of energy diagrams, just as we did for the one-dimensional problems in Questions 5.12–5.15. For examples based on (8) and (9), see Questions 8.4 to 8.6.

(iii) In principle, (8) provides the solution $r(t)$ for the radial motion in the inverse form

$$t(r) = \int_{r_0}^{r} \sqrt{\frac{m}{2\{E - V_e(r)\}}}\, dr, \qquad (12)$$

where $r_0 = r(0)$ and we have assumed a positive root. The solution (12) is analogous to the solution for motion in a one-dimensional potential $V(x)$ – see (1) of Question 5.16. (In both cases the possibility of a negative root is related to time reversal $t \to -t$: if $r(t)$ is a solution then $r(-t)$ is also a solution.) Inversion of (12) gives the radial position $r(t)$. Using this in (7) we can express the angular position as

$$\theta(t) = \frac{L}{m}\int_{0}^{t} \frac{dt}{r^2(t)} + \theta_0. \qquad (13)$$

The solutions (12) and (13) specify completely the time-dependent orbit.

(iv) These time-dependent solutions can be rather complicated (as in Questions 8.18 and 8.19) and it is sometimes preferable to work instead with the geometric orbit $r(\theta)$. The next question shows how this is done.

Question 8.3

Starting with (8) of Question 8.2, and by means of a change of variable from dr/dt to $dr/d\theta$, obtain the formal solution for the geometric orbit in inverse form:

$$\theta(r) = \theta(r_0) + \frac{L}{\sqrt{2m}}\int_{r_0}^{r} \frac{dr}{r^2\sqrt{E - V_e(r)}}. \qquad (1)$$

Solution

Begin by writing

$$\frac{dr}{dt} = \frac{d\theta}{dt}\frac{dr}{d\theta} = \frac{L}{mr^2}\frac{dr}{d\theta}, \qquad (2)$$

where in the last step we have used (7) of Question 8.2. Thus, (8) of Question 8.2 becomes

$$E = \frac{L^2}{2mr^4}\left(\frac{dr}{d\theta}\right)^2 + V_e(r), \qquad (3)$$

and hence

$$d\theta = \frac{L}{\sqrt{2m}}\frac{dr}{r^2\sqrt{E - V_e(r)}}. \qquad (4)$$

Integration of (4) between $r = r_0$ and r, and $\theta = \theta(r_0)$ and $\theta(r)$, yields (1).

Comments

(i) In applications of (1) it is convenient to write the definite integral as an indefinite integral by incorporating the contribution from the lower limit r_0 into $\theta(r_0)$; the resulting combination we denote by θ_0. Thus

$$\theta(r) = \theta_0 + \frac{L}{\sqrt{2m}} \int \frac{dr}{r^2 \sqrt{E - V_e(r)}}. \tag{5}$$

(ii) There are three force fields for which (5) yields simple, invertible results, namely the inverse-square, the linear, and the inverse-cube forces (see Questions 8.9–8.11). Some other power-law forces are integrable in terms of elliptic functions.[1]

Question 8.4

A particle of mass m moves in an isotropic oscillator potential $V = \frac{1}{2}kr^2$ (k is a positive constant).

(a) Sketch the graph of the effective potential $V_e(r)$ versus r.
(b) Use this graph to discuss the possible motions of the particle.

Solution

(a) From (9) of Question 8.2 we have

$$V_e(r) = \tfrac{1}{2}kr^2 + \frac{L^2}{2mr^2} \qquad (r \geq 0). \tag{1}$$

It is a simple matter to picture this function. When $r \to 0_+$ the second term in (1) dominates and $V_e \to \infty$. When $r \to \infty$, the first term in (1) dominates and $V_e \to \infty$. There is one turning point: $dV_e/dr = 0$ when $kr - L^2/mr^3 = 0$, and so the turning point is at

$$r_0 = (L^2/mk)^{1/4}. \tag{2}$$

The value

$$V_e(r_0) = \sqrt{kL^2/m} \tag{3}$$

is a minimum. Thus, we have the following graph:

---

[1] See, for example, H. Goldstein, *Classical mechanics*. Reading: Addison-Wesley, 2nd edn, 1980.

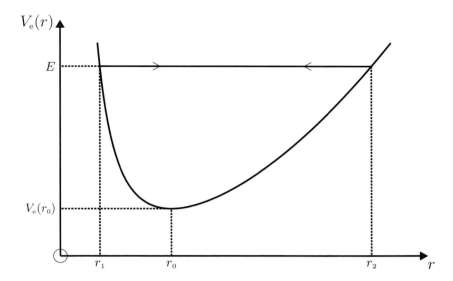

(b) To interpret an energy diagram such as the one above, it is necessary to distinguish between motion relative to the inertial frame S (with polar coordinates (r, θ)) and motion relative to a non-inertial frame S' that is rotating with angular velocity $\dot\theta \hat{\mathbf{z}}$ with respect to S. In S' the motion is one-dimensional – it is specified by $r(t)$ alone – while in S the motion is specified by both $r(t)$ and $\theta(t)$. An energy diagram allows us to describe the motion in S' in the same manner as for one-dimensional motion in a potential $V(x)$ (see Questions 5.12–5.15). Note that it is impossible for E to be less than $V_e(r)$ because, according to (8) of Question 8.2, that would mean the radial speed $\dot r$ is imaginary. We can now interpret the above energy diagram, where we have drawn a horizontal line to indicate a constant value of the energy $E \; (\geq \sqrt{kL^2/m}\,)$. We see immediately that the motion in S' is simply an oscillation between the classical turning points r_1 and r_2 that are the positive roots of the equation

$$V_e = \tfrac{1}{2}kr^2 + \frac{L^2}{2mr^2} = E\,; \qquad (4)$$

that is,

$$r_1 = \sqrt{\frac{E}{k}}\sqrt{1 - \sqrt{1 - \frac{kL^2}{mE^2}}}\,, \qquad r_2 = \sqrt{\frac{E}{k}}\sqrt{1 + \sqrt{1 - \frac{kL^2}{mE^2}}}\,. \qquad (5)$$

Thus the particle is always bound by the potential. The simplest motion occurs when $E = \sqrt{kL^2/m}$: the particle is at rest in S', and the trajectory in S is therefore a circle of radius $r_0 = (L^2/mk)^{1/4}$. The fact that $V_e(r)$ has a minimum at r_0 means that in S' the particle is in stable equilibrium at r_0, and that the circular orbit in S is stable. To determine the nature of the orbit for $E > \sqrt{kL^2/m}$ requires a detailed calculation: this shows that the orbit is an ellipse with centre of force at the origin, and with minor and major axes equal to $2r_1$ and $2r_2$, respectively (see Questions 7.25 and 8.10).

Comment

The term $L^2/2mr^2$ in the effective potential (1) plays the role of a 'centrifugal barrier': the associated repulsive centrifugal force $L^2\hat{\mathbf{r}}/mr^3$ prevents the particle from falling into the centre of force under the attraction of the linear restoring force $-k\mathbf{r}$ associated with the term $\frac{1}{2}kr^2$ in (1).

Question 8.5

A particle of mass m moves in an attractive Coulomb potential $V = -k/r$ (k is a positive constant).

(a) Sketch the graph of the effective potential $V_e(r)$ versus r.
(b) Use this graph to discuss the possible motions of the particle.

Solution

(a) From (9) of Question 8.2 we have

$$V_e(r) = -\frac{k}{r} + \frac{L^2}{2mr^2} \qquad (r \geq 0). \qquad (1)$$

When $r \to 0_+$, the second term in (1) dominates and $V_e \to \infty$. When $r \to \infty$, the first term in (1) dominates and $V_e \to 0_-$. There is one turning point at a finite value of r: $dV_e/dr = 0$ when $(k/r^2) - L^2/mr^3 = 0$, and therefore the turning point is at

$$r_o = L^2/mk. \qquad (2)$$

The value

$$V_e(r_o) = -mk^2/2L^2 \qquad (3)$$

is a minimum. Thus, we have the following graph:

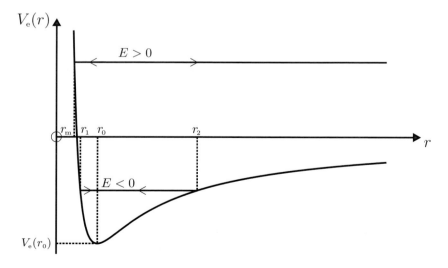

(b) On the graph we have also drawn horizontal lines to indicate constant positive and negative energies. We see that if $E \geq 0$ the motion is unbounded. In the rotating frame S′ in which the angular position of the particle is fixed (see Question 8.2), a particle moving toward the centre of force comes to rest at a distance r_m where $V(r_m) = E$. After that, the particle moves to the right, attaining an asymptotic velocity $\mathbf{v}_\infty = \sqrt{2E/m}\,\hat{\mathbf{r}}$ at large r. If $E < 0$ the motion is bounded: in S′ the particle oscillates between classical turning points r_1 and r_2 that are roots of the equation

$$V_e = -\frac{k}{r} + \frac{L^2}{2mr^2} = E\,; \tag{4}$$

that is,

$$r_1 = -\frac{k}{2E}\left(1 - \sqrt{1 + \frac{2EL^2}{mk^2}}\right), \qquad r_2 = -\frac{k}{2E}\left(1 + \sqrt{1 + \frac{2EL^2}{mk^2}}\right). \tag{5}$$

A particle with energy $E = -mk^2/2L^2$ moves in a stable circular orbit of radius $r_0 = L^2/mk$. To determine the nature of the orbit for $E > -mk^2/2L^2$ requires a detailed calculation. This shows that the orbit is an ellipse if $E < 0$, a parabola if $E = 0$, and a hyperbola if $E > 0$ (see Question 8.9).

Comments

(i) Note again the role of the centrifugal barrier associated with the second term in (1), which prevents the particle from falling into the centre of force under the attraction of the inverse-square force $-k\hat{\mathbf{r}}/r^2$ associated with the Coulomb potential in (1).

(ii) The following energy diagram is for a repulsive Coulomb potential (that is, for $k < 0$ in (1)). The motion is always unbounded: an incoming particle is scattered by the potential. The distance of closest approach, r_1, decreases as E increases.

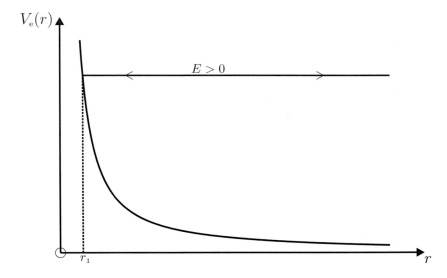

Question 8.6

A particle of mass m moves in an attractive Yukawa potential $V(r) = -(k/r)e^{-r/\lambda}$, where k and λ are positive constants. Analyze the effective potential $V_e(r)$ and plot the four distinct types of graph of $V_e(r)$ versus r. Use these to discuss the possible motions of the particle: in particular, show that there exists a critical value $\lambda_c = 1.19053 L^2/mk$, above which the potential can bind the particle.

Solution

The constant λ is the range of the potential. From Question 8.2 we have

$$V_e(r) = -\frac{k}{r} e^{-r/\lambda} + \frac{L^2}{2mr^2} \qquad (r \geq 0). \tag{1}$$

When $r \to 0$, $V_e \to \infty$ and when $r \to \infty$, $V_e \to 0_+$, because in both limits the second term in (1) dominates. In its dependence on the range λ, the function (1) has a rather rich behaviour. To analyze this it is convenient to use the dimensionless quantity $u = r/\lambda$, in terms of which

$$V_e(u) = \frac{k}{\lambda}\left(-\frac{1}{u}e^{-u} + \frac{1}{2\alpha u^2}\right), \tag{2}$$

where

$$\alpha = (mk/L^2)\lambda \tag{3}$$

is a dimensionless, positive quantity. The condition $dV_e/du = 0$ for a turning point(s) can be written

$$G(u) = e^u - \alpha(u^2 + u) = 0 \qquad (u \geq 0). \tag{4}$$

The function $G(u)$ has two real roots if α exceeds a value α_c (which is calculated below). These roots become coincident at $\alpha = \alpha_c$, and below α_c there are no real, positive roots:

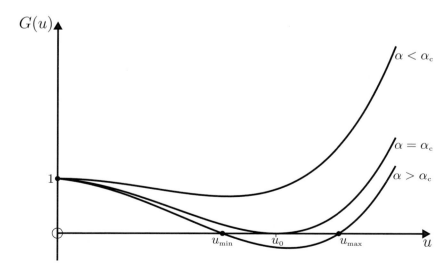

The value of α_c is calculated by solving the simultaneous equations $G = 0$ and $dG/du = e^u - \alpha(2u+1) = 0$. This yields

$$u_0 = \tfrac{1}{2}(1 + \sqrt{5}) = 1.61803 \tag{5}$$

$$\alpha_c = (2 + \sqrt{5})^{-1} e^{(1+\sqrt{5})/2} = 1.19053 . \tag{6}$$

A numerical evaluation[‡] of the two roots $u_{\min}(\alpha)$ and $u_{\max}(\alpha)$ of (4), where $V_e(u)$ has a turning point, provides the following graph:

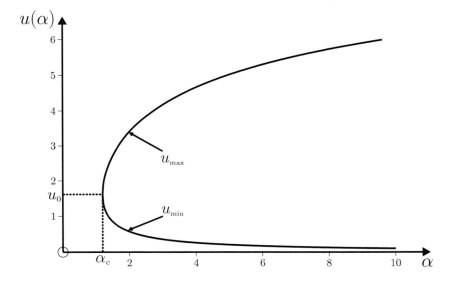

There are four distinct graphs of $V_e(u)$, corresponding to various values of α (that is, of the range λ) in (2):

$\alpha < \alpha_c$

$V_e(u)$ has no turning points: it is a positive, monotonic decreasing function. The motion is unbounded – an incoming particle is scattered by the potential:

[‡]The *Mathematica* notebook is:

```
In[1]:= αmini = 1.19053; αmaxi = 10.0; αstep = (αmaxi - αmini)/500;

LowerRoot = Table[{α, x/.FindRoot[e^x - α (x^2 + x) == 0,
   {x, 0.001, 1.19503}]}, {α, αmini, αmaxi, αstep}];

UpperRoot = Table[{α, x/.FindRoot[e^x - α (x^2 + x) == 0,
   {x, 11.9503}]}, {α, αmini, αmaxi, αstep}];

ListPlot[Join[LowerRoot, UpperRoot]]
```

226 *Solved Problems in Classical Mechanics*

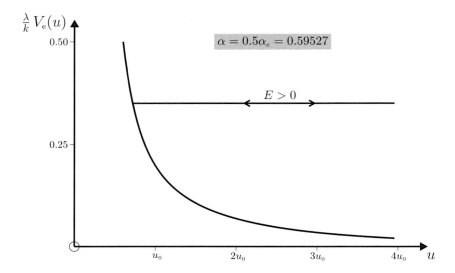

Here, and in the following graphs, we plot $V_e(u)$ in units of k/λ – see (2).

$\boxed{\alpha = \alpha_c}$

A point of inflection with a horizontal tangent appears at u_0. Consequently, there is an unstable circular orbit of radius u_0 for a particle with energy[†] $E = V_e(u_0) = 0.03787 k/\lambda$. Otherwise, the motion is unbounded.

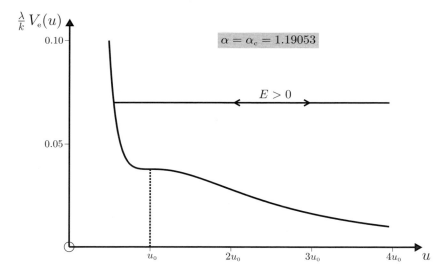

[†]See (8) of Question 8.2.

$1.19053 < \alpha < 1.35914$

For $\alpha > \alpha_c = 1.19053$ the function $V_e(u)$ has a minimum at $u_{\min}$ and a maximum at $u_{\max}$. Also, $V_e(u_{\min}) > 0$ if α is not too large: the value of α below which $V_e(u_{\min}) > 0$ is calculated by solving $V_e(u) = 0$ and $G(u) = 0$ simultaneously. According to (2) and (4), this yields

$$u_{\min} = 1, \qquad \alpha = 1.35914. \tag{7}$$

The next graph is for a value of α less than $(7)_2$, namely $\alpha = 1.1\alpha_c = 1.30958$.

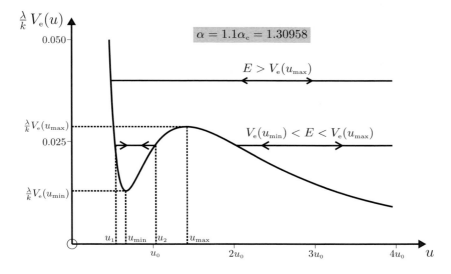

From this graph we see that there is a stable circular orbit of radius $u_{\min}$ for a particle with energy $E = V_e(u_{\min})$, and an unstable circular orbit of radius $u_{\max}$ for a particle with energy $E = V_e(u_{\max})$. For $\alpha = 1.1\alpha_c$ the roots of (4) are $u_{\min} = 1.07316$ and $u_{\max} = 2.30118$. Then, $V_e(u_{\min}) = 0.01290\,k/\lambda$ and $V_e(u_{\max}) = 0.02858\,k/\lambda$. A particle with energy E such that $V_e(u_{\min}) < E < V_e(u_{\max})$, and that is located inside the well, performs a bounded motion: its dimensionless radial coordinate oscillates between the classical turning points u_1 and u_2, which are the roots of $V_e(u) = E$. All other motions are unbounded – an incoming particle is scattered by the potential.

$\alpha > 1.35914$

Here, $V_e(u_{\min}) < 0$. The possible motions of the particle are the same as in the previous case, with the additional feature that now E can be negative. The following graph is for $\alpha = 1.2\alpha_c = 1.42864$.

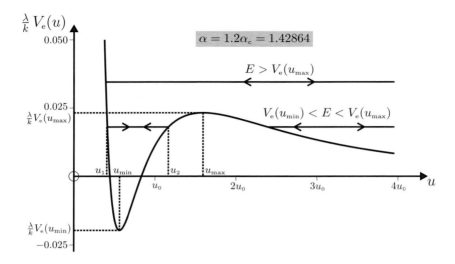

We see from the above discussion that a Yukawa potential can bind the particle if the range of the potential λ is larger than

$$\lambda_c = 1.19053 \frac{L^2}{mk}. \tag{8}$$

Comments

(i) For uniform motion in a circular orbit of radius r the acceleration is $-v^2\hat{\mathbf{r}}/r$ (see Question 8.1), and the equation of motion can be written

$$m\frac{v^2}{r} = \frac{dV}{dr} = k\left(\frac{1}{r^2} + \frac{1}{\lambda r}\right)e^{-r/\lambda}. \tag{9}$$

Thus

$$v = \sqrt{\frac{k(1+r/\lambda)}{mr}}\, e^{-r/2\lambda}, \tag{10}$$

and the period $T = 2\pi r/v$ can be expressed in terms of r:

$$T = 2\pi\sqrt{\frac{mr^3}{k(1+r/\lambda)}}\, e^{r/2\lambda}. \tag{11}$$

(ii) We have seen that a barrier creates a classically forbidden region where $E < V$. However, in quantum mechanics a particle with energy $E < V$ can tunnel through a barrier of finite width. Thus, in the third case above, tunnelling of a bound particle can occur. In the last case considered, tunnelling can occur if $E > 0$; but if $E < 0$ the barrier width is infinite (see the last diagram) and tunnelling cannot occur.

Question 8.7

Show that when the equation of motion for a particle of mass m moving in a central, isotropic field, namely

$$m\ddot{\mathbf{r}} = F(r)\hat{\mathbf{r}}, \qquad (1)$$

is expressed in terms of plane polar coordinates (r, θ), it yields

$$m(\ddot{r} - r\dot{\theta}^2) = F(r) \qquad (2)$$

$$\frac{d}{dt}(mr^2\dot{\theta}) = 0. \qquad (3)$$

Solution

The acceleration $\mathbf{a} = \ddot{\mathbf{r}}$ is given in plane polar coordinates by (2) of Question 8.1: thus, (1) can be written

$$m(\ddot{r} - r\dot{\theta}^2)\hat{\mathbf{r}} + m(r\ddot{\theta} + 2\dot{r}\dot{\theta})\hat{\boldsymbol{\theta}} = F(r)\hat{\mathbf{r}}. \qquad (4)$$

Now, if $c_1\hat{\mathbf{r}} + c_2\hat{\boldsymbol{\theta}} = 0$ then $c_1 = c_2 = 0$. Therefore (4) yields (2) and (3).

Comments

(i) Equation (3) expresses conservation of the angular momentum $\mathbf{L} = L\hat{\mathbf{z}}$, where $L = mr^2\dot{\theta}$ (see (1) of Question 8.2).

(ii) We can express (2) in terms of r alone by replacing $\dot{\theta}$ with L/mr^2:

$$m\ddot{r} - \frac{L^2}{mr^3} = F(r). \qquad (5)$$

By solving this second-order differential equation we can determine $r(t)$. This procedure is equivalent to evaluating the integral (12) of Question 8.2 to obtain the solution in the inverse form $t(r)$; which of these two approaches one uses is a matter of convenience (see below).

(iii) Note that, as one expects, (5) implies that the energy $E = \tfrac{1}{2}m\dot{r}^2 + V_e(r)$ is constant (see (10) of Question 8.2).

Question 8.8

Prove that (5) of Question 8.7 can be written as

$$\frac{d^2u}{d\theta^2} + u = -\frac{m}{L^2}\frac{1}{u^2}F\left(\frac{1}{u}\right), \qquad (1)$$

where

$$u = \frac{1}{r}. \qquad (2)$$

230 Solved Problems in Classical Mechanics

Solution

We write
$$\frac{dr}{dt} = \frac{d}{dt}\left(\frac{1}{u}\right) = -\frac{1}{u^2}\frac{du}{dt} = -\frac{1}{u^2}\frac{d\theta}{dt}\frac{du}{d\theta} = -\frac{L}{m}\frac{du}{d\theta}, \tag{3}$$

where the result $\dot\theta = L/mr^2 = Lu^2/m$ has been used in the last step. Then
$$\frac{d^2r}{dt^2} = -\frac{L}{m}\frac{d\theta}{dt}\frac{d}{d\theta}\frac{du}{d\theta} = -\left(\frac{L}{m}\right)^2 u^2 \frac{d^2u}{d\theta^2}. \tag{4}$$

Use of (2) and (4) in (5) of Question 8.7 yields (1).

Comment

Equation (1) can be used in two ways. First, to solve the so-called direct problem: given the force $F(r)$, calculate the trajectory $r(\theta)$. (This procedure is equivalent to evaluating the integral (5) of Question 8.3 to obtain the solution $\theta(r)$.) Secondly, to solve the so-called inverse problem: given the trajectory $r(\theta)$, calculate the force $F(r)$. These applications are illustrated in the examples below.

Question 8.9

Determine the geometric form of the trajectory $r(\theta)$ for a particle of mass m moving in a Coulomb potential $V(r) = -k/r$:

(a) by evaluating the integral (5) of Question 8.3;
(b) by solving the differential equation (1) of Question 8.8.

Solution

(a) In (5) of Question 8.3 we substitute $V_e = -k/r + L^2/2mr^2$ and make the change of variable $r = 1/u$. Then
$$\theta = \theta_0 - \int \frac{du}{\sqrt{-u^2 + (2mk/L^2)u + 2mE/L^2}}. \tag{1}$$
Now
$$\int \frac{dx}{\sqrt{Ax^2 + Bx + C}} = \frac{1}{\sqrt{-A}} \cos^{-1}\frac{2Ax + B}{\sqrt{B^2 - 4AC}} \qquad (A < 0). \tag{2}$$

It follows from (1) and (2) that
$$\theta = \theta_0 - \cos^{-1}\left(\frac{1 - r_0 u}{e}\right), \tag{3}$$

where[‡]

[‡]Use of the symbol e should not be confused with the base of the natural logarithm. Which is intended should be clear from the context.

$$r_o = \frac{L^2}{mk}, \qquad e = \sqrt{1 + \frac{2L^2 E}{mk^2}}. \tag{4}$$

Equation (3) can be inverted to obtain $u(\theta)$ and hence $r(\theta) = 1/u(\theta)$:

$$r(\theta) = \frac{r_o}{1 - e\cos(\theta - \theta_o)}. \tag{5}$$

(b) With $F = -k/r^2 = -ku^2$, (1) of Question 8.8 is

$$\frac{d^2 u}{d\theta^2} + u = \frac{mk}{L^2}. \tag{6}$$

If we write $u = w + mk/L^2$ then (6) becomes

$$\frac{d^2 w}{d\theta^2} + w = 0, \tag{7}$$

and hence $w = C\cos(\theta - \theta_o)$ where C and θ_o are arbitrary constants. Thus

$$r(\theta) = \frac{1}{mk/L^2 + C\cos(\theta - \theta_o)}. \tag{8}$$

The constant C can be determined by using (8) to evaluate the energy given by (3) of Question 8.3. A short calculation shows that

$$E = \frac{L^2 C^2}{2m} - \frac{mk^2}{2L^2}, \tag{9}$$

and hence

$$C = \pm \frac{mk}{L^2}\sqrt{1 + \frac{2EL^2}{mk^2}} = \pm e/r_o, \tag{10}$$

according to (4) above. If we choose the negative sign in (10) then (8) is identical to (5). (Choice of the positive sign in (10) is equivalent to a rotation of axes by π – see below.) In interpreting the above results it should be kept in mind that for an attractive (repulsive) potential $k > 0$ (< 0).

Comments

(i) Readers who are familiar with the theory of conic sections will recognize (5) as the general polar equation of a conic with eccentricity e.
(ii) For readers who are unfamiliar with this theory, it may be helpful to convert (5) into its various Cartesian forms in order to obtain the possible trajectories. We first choose the arbitrary constant $\theta_o = 0$. (This is equivalent to orienting the axes in a convenient manner.) Then

$$r(\theta) = \frac{r_o}{1 - e\cos\theta}. \tag{11}$$

Now choose the Cartesian axes shown, with origin O displaced from the centre of force F by an amount ea along the x-axis, where a is to be determined. Then $x = r\cos\theta - ea$ and $y = r\sin\theta$. Note that if $a > 0$ (< 0) then the origin is to the right (left) of F.

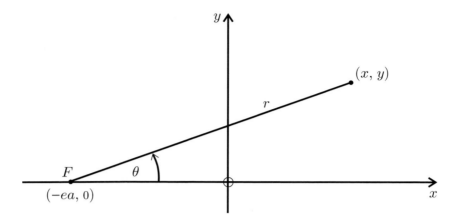

We express (11) in terms of x and y by setting $r\cos\theta = x + ea$ and $r = \sqrt{(x+ea)^2 + y^2}$. A short calculation shows that if $e \neq 1$, then the choice

$$a = \frac{r_0}{1-e^2}, \tag{12}$$

yields a simple quadratic form:

$$x^2 + \frac{1}{1-e^2} y^2 = \frac{r_0^2}{(1-e^2)^2}. \tag{13}$$

If $e = 1$ then the choice $a = -\frac{1}{2}r_0$ gives

$$y^2 = 2r_0 x. \tag{14}$$

Equation (14) represents a parabola. To interpret (13) we must consider whether e is less than or greater than one (that is, whether $E < 0$ or $E > 0$, see (4)). If $e < 1$, we can write (13) as

$$\frac{x^2}{a^2} + \frac{y^2}{b^2} = 1, \tag{15}$$

where

$$b^2 = \frac{r_0^2}{1-e^2} \tag{16}$$

is positive, and hence b is real.

If $e > 1$, we write (13) as

$$\frac{x^2}{a^2} - \frac{y^2}{b^2} = 1, \tag{17}$$

where

$$b^2 = \frac{r_0^2}{e^2 - 1}. \tag{18}$$

Again, b is real. We recognize (15) as the equation of an ellipse and (17) as that of a hyperbola.

(iii) Consider an attractive potential ($k > 0$). According to (4), $r_0 > 0$. From (11), or equivalently (14), (15) and (17), we have three distinct types of trajectory (it is useful to compare the following analysis with the discussion of the energy diagram in Question 8.5):

$e < 1$ (that is, $-mk^2/2L^2 < E < 0$)

Here, a in (12) is positive and (15) yields the trajectory

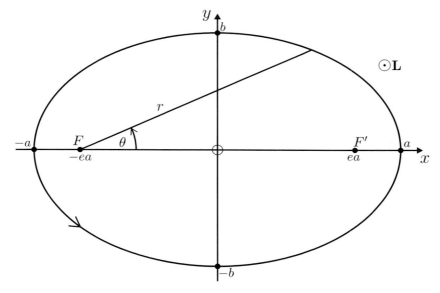

The points F and F' are the foci of the ellipse (the sum of the distances of the particle from the two foci is a constant). The centre of force is at a focus (F), in accordance with Kepler's first law. (For a trajectory $r = r_0/(1 + e\cos\theta)$ the centre of force is at F'.) If $\theta_0 \neq 0$ in (5) then the ellipse is rotated with respect to the x- and y-axes. The minimum and maximum values of r are

$$\left. \begin{array}{ll} r_{\min} = \dfrac{r_0}{1+e} = (1-e)a & \text{when } \theta = \pi, \\[2mm] r_{\max} = \dfrac{r_0}{1-e} = (1+e)a & \text{when } \theta = 0. \end{array} \right\} \qquad (19)$$

The major axis of the ellipse has length $2a$, and the minor axis has length $2b$. From (4) and (12) we have

$$E = -\frac{k}{2a}, \qquad (20)$$

and therefore all orbits with the same major axis have the same energy. This degeneracy is associated with the extra symmetry possessed by the Coulomb problem (see the comments for Question 9.9). Note that when $E = -mk^2/2L^2$

we have $e=0$: the foci F and F' coincide with the origin O and the trajectory is a circle of radius r_0, as is immediately evident from (5).

$e > 1$ (that is, $E > 0$)

Equation (17) represents a hyperbola comprising two branches:

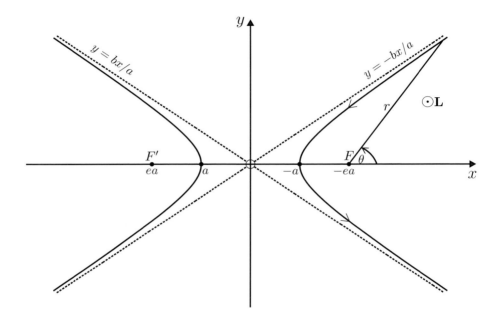

Note that here a, given by (12), is negative. Again there are two foci, F and F': each branch is a curve traced by a point that moves in such a way that the difference of its distances from F and F' is constant. The dotted lines are the asymptotes. If the centre of an attractive force is at F then the trajectory followed is the right-hand branch. The distance of closest approach to F is

$$r_{\min} = \frac{r_0}{1+e} = (e-1)|a| \qquad \text{when } \theta = \pi. \qquad (21)$$

For a repulsive potential ($k < 0$) the energy E is necessarily positive, and therefore $e > 1$. The trajectory is a hyperbola (following the right-hand branch if the centre of force is at the left-hand focus F').

$e = 1$ (that is, $E = 0$)

Equation (14) represents a parabola. The centre of an attractive force is at the focus F.

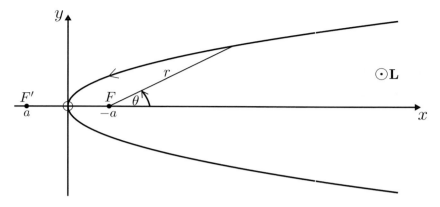

(iv) The question answered above, namely given the potential (or force), what is the trajectory, is known as the direct problem (sometimes it is referred to in the literature as the inverse problem). The direct problem for an inverse-square force was first solved by Hermann in 1710.[2]

Question 8.10

Determine the trajectory $r(\theta)$ for a particle of mass m moving in an oscillator potential $V(r) = \frac{1}{2}kr^2$ by evaluating the integral (5) of Question 8.3.

Solution

In (5) of Question 8.3 we substitute $V_e = \frac{1}{2}kr^2 + L^2/2mr^2$ and make the change of variable $r = \sqrt{u}$. Then

$$\theta = \theta_0 + \frac{L}{2}\int \frac{du}{u\sqrt{(-mku^2 + 2mEu - L^2)}}. \tag{1}$$

Now

$$\int \frac{dx}{x\sqrt{Ax^2 + Bx + C}} = \frac{1}{\sqrt{-C}}\sin^{-1}\frac{Bx + 2C}{x\sqrt{B^2 - 4AC}} \quad (C < 0). \tag{2}$$

From (1) and (2) we find

$$\theta = \theta_0 + \tfrac{1}{2}\sin^{-1}\frac{u - r_0^2}{eu}, \tag{3}$$

where

$$r_0 = \frac{L}{\sqrt{mE}}, \qquad e = \sqrt{1 - \frac{kL^2}{mE^2}}. \tag{4}$$

Note that $E \geq \sqrt{kL^2/m}$ (see Question 8.4) and so e is real. Equation (3) can be inverted to obtain $u(\theta)$ and hence $r^2(\theta) = u(\theta)$:

---

[2] O. Volk, "Miscellanea from the history of celestial mechanics," Celestial Mechanics, vol. 14, pp. 365–382, 1976.

$$r^2(\theta) = \frac{r_0^2}{1 - e\sin 2(\theta - \theta_0)}. \tag{5}$$

Comments

(i) To interpret the trajectory (5) we convert to Cartesian coordinates, placing the centre of force at the origin O. Then $x = r\cos\theta$ and $y = r\sin\theta$. We also make a convenient choice of $\theta_0 = -\tfrac{1}{4}\pi$; then $\sin 2(\theta - \theta_0) = \cos 2\theta = \cos^2\theta - \sin^2\theta$ and (5) can be expressed as

$$\frac{x^2}{a^2} + \frac{y^2}{b^2} = 1, \tag{6}$$

where

$$a = \frac{r_0}{\sqrt{1-e}}, \qquad b = \frac{r_0}{\sqrt{1+e}}. \tag{7}$$

The orbit is an ellipse with major axis equal to $2a$, minor axis equal to $2b$, and centre of force at the origin, as shown in the diagram below.

(ii) This result can be obtained more simply by solving the equation of motion in Cartesian coordinates (see Question 7.25).

(iii) Solution by means of the differential equation (1) of Question 8.8, which for this problem reads

$$\frac{d^2u}{d\theta^2} + u = \frac{mk}{L^2}\frac{1}{u^3}, \tag{8}$$

is more difficult than either of the above two methods; but the reader can verify that (5) does satisfy (8). In the next question it is simpler to use the differential equation.

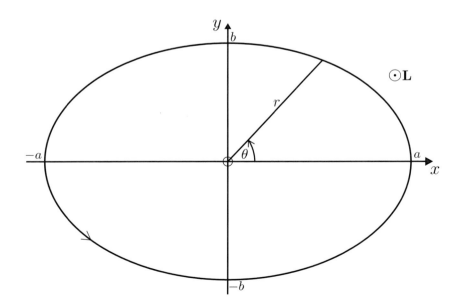

Question 8.11

Determine the trajectory $r(\theta)$ for a particle of mass m moving in an attractive potential $V(r) = -k/2r^2$ (where k is a positive constant) by solving the differential equation (1) of Question 8.8.

Solution

The force corresponding to the above potential is

$$\mathbf{F} = -\frac{dV}{dr}\hat{\mathbf{r}} = -\frac{k}{r^3}\hat{\mathbf{r}}, \tag{1}$$

and therefore (1) of Question 8.8 becomes

$$\frac{d^2u}{d\theta^2} + \left(1 - \frac{mk}{L^2}\right)u = 0. \tag{2}$$

The form of the general solution to (2) depends on whether the quantity in brackets is positive, negative or zero. These three cases are considered below.

1. $L > \sqrt{mk}$

Here
$$u(\theta) = \frac{1}{r(\theta)} = C\cos\alpha(\theta - \theta_0), \tag{3}$$

where C and θ_0 are constants, and

$$\alpha = \sqrt{1 - \frac{mk}{L^2}} \tag{4}$$

is real. The constant C can be evaluated by using the expression for the energy given in (3) of Question 8.3:

$$E = \frac{L^2}{2mr^4}\left(\frac{dr}{d\theta}\right)^2 + \frac{L^2}{2m}\left(1 - \frac{mk}{L^2}\right)\frac{1}{r^2}. \tag{5}$$

From (3) and (5) we find that

$$C = \sqrt{2mE/\alpha^2 L^2}. \tag{6}$$

Note that C is real because, according to (5), $E > 0$.

2. $L < \sqrt{mk}$

Here
$$u(\theta) = \frac{1}{r(\theta)} = \tfrac{1}{2}(Ce^{\beta\theta} + De^{-\beta\theta}), \qquad (7)$$

where C and D are constants, and
$$\beta = \sqrt{\frac{mk}{L^2} - 1} \qquad (8)$$

is real. From (7) and (5) we obtain
$$E = -\beta^2 \frac{L^2}{2m} CD. \qquad (9)$$

There are three cases to consider, namely whether C and D have the same or opposite sign, or whether one of them is zero (that is, whether E is negative, positive, or zero).

☞ $E < 0$

Write $C = Be^{-\beta\theta_o}$, $D = Be^{\beta\theta_o}$, where B and θ_o are constants. Then, (7) becomes
$$\frac{1}{r(\theta)} = B\cosh\beta(\theta - \theta_o), \qquad (10)$$

where, according to (9),
$$B = \sqrt{-2mE/\beta^2 L^2}. \qquad (11)$$

☞ $E > 0$

Write $C = Be^{-\beta\theta_o}$, $D = -Be^{\beta\theta_o}$ to obtain
$$\frac{1}{r(\theta)} = B\sinh\beta(\theta - \theta_o), \qquad (12)$$

where
$$B = \sqrt{2mE/\beta^2 L^2}. \qquad (13)$$

☞ $E = 0$

Either D or C in (9) is zero, and from (7) the corresponding solutions are
$$\frac{1}{r(\theta)} = Be^{\beta\theta}, \qquad (14)$$

$$\frac{1}{r(\theta)} = Be^{-\beta\theta}, \qquad (15)$$

where B is a constant.

3. $L = \sqrt{mk}$

Here
$$u(\theta) = \frac{1}{r(\theta)} = C\theta + D, \qquad (16)$$
where C and D are constants. From (16) and (5) we find
$$C = \sqrt{2mE/L^2}. \qquad (17)$$

When $E = 0$, we see from (16) and (17) that $r = 1/D$ and the orbit is circular. In a frame rotating with angular velocity $d\theta/dt$, the particle is at rest in a state of neutral equilibrium (neutral because $V_e = 0$ is constant when $L = \sqrt{mk}$).

Comments

(i) The trajectories of a particle moving in an inverse-cube force were first obtained by Johann Bernoulli in 1710.[2]
(ii) The following diagrams depict the trajectories obtained above.

1. $L > \sqrt{mk}$

From (3) and (6) above, $r(\theta) = \left[\sqrt{2mE/\alpha^2 L^2} \cos\alpha(\theta - \theta_0)\right]^{-1}$ where $E > 0$. The particle starts at θ_0, and escapes to infinity. The number of spirals increases with decreasing α. In calculating the trajectory below, we have taken $\alpha = 1/7$ and $\theta_0 = \pi/3$ and set the quantity under the square root equal to one. The dotted line shows the value of θ at which r becomes infinite, that is $\theta = \theta_0 + \pi/2\alpha$.

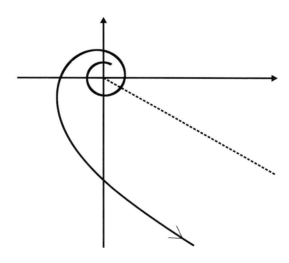

2. $L < \sqrt{mk}$

☞ $E < 0$

From (10) and (11) above, $r(\theta) = \left[\sqrt{-2mE/\beta^2 L^2}\cosh\beta(\theta-\theta_0)\right]^{-1}$ for $E < 0$. The particle starts at θ_0, and spirals into the centre of force at the origin. In calculating the trajectory below, we have taken $\beta = 1/7$ and $\theta_0 = \frac{1}{6}\pi$, and set the quantity under the square root equal to one.

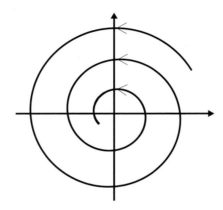

☞ $E > 0$

From (12) and (13) above, $r(\theta) = \left[\sqrt{2mE/\beta^2 L^2}\sinh\beta(\theta-\theta_0)\right]^{-1}$ for $E > 0$. The particle starts at $\theta > \theta_0$, and escapes to infinity if $\dot{r}_0 > 0$ (radial motion is outward). In calculating the trajectory below, the fraction under the square root is set equal to one, $\beta = 1/100$, an initial value of $\theta = 3\pi$ was used and $\theta_0 = \frac{1}{6}\pi$. The dotted line shows the value of θ at which r becomes infinite, that is, $\theta = \theta_0$. The number of spirals increases if the initial value of θ is increased (i.e. if the particle starts closer to the centre of force).

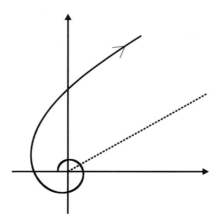

If $\dot{r}_0 < 0$ (radial motion is inward), the particle spirals into the centre of force. In calculating the trajectory below, the quantity under the square root is set equal to one, $\beta = 1/100$, an initial value of $\theta = 9\pi/4$ was used and $\theta_0 = \frac{1}{6}\pi$.

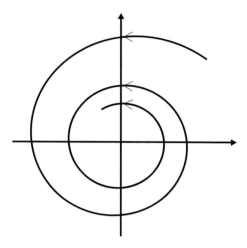

☞ $E = 0$

The trajectories are given by (14) and (15). The particle either spirals into the centre of force or out to infinity.

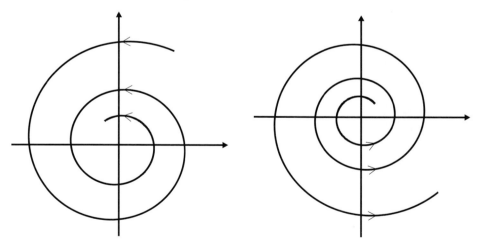

3. $L = \sqrt{mk}$

Here, $r(\theta) = \left[\sqrt{2mE/L^2}\,\theta + D\right]^{-1}$ where $E > 0$. The particle spirals into the centre of force if $\dot{r}_0 < 0$. In calculating the trajectory below, the quantity under the square root is set equal to one, $D = 1$ and $\frac{1}{6}\pi \leq \theta \leq 4\frac{3}{4}\pi$.

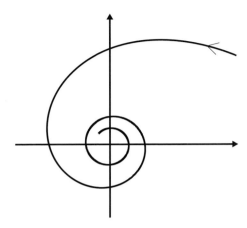

If $\dot{r}_0 > 0$, the particle escapes to infinity. The dotted line shows the value of θ at which r becomes infinite, that is, $\theta = -D\sqrt{L^2/2mE}$. In calculating the trajectory below, the quantity under the square root is set equal to one, $D = 1$ and $-1 < \theta \leq \frac{3}{4}\pi$. Note that when $E = 0$ the trajectory is a circular orbit. Apart from these circular orbits, none of the other trajectories is closed.

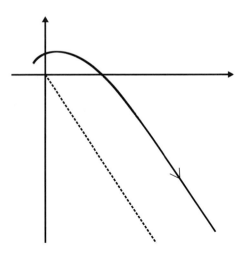

(iii) For the effective potential of an inverse-cube force

$$V_e(r) = \frac{L^2}{2m}\left(1 - \frac{mk}{L^2}\right)\frac{1}{r^2}, \qquad (18)$$

and there are three possible graphs:

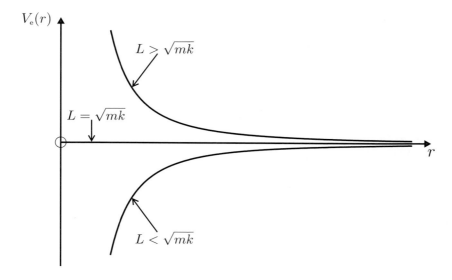

It is a useful exercise for the reader to reconcile the various trajectories depicted above with these energy diagrams. For example, it is clear from the energy diagrams that for $L > \sqrt{mk}$ the motion is unbounded (the particle eventually escapes to infinity); the centrifugal barrier is strong enough to prevent the particle from falling into the centre of force. It is also clear that for $L < \sqrt{mk}$ the motion is bounded for $E < 0$, and either bounded or unbounded for $E \geq 0$, depending on the sign of the radial velocity $\dot{r}$. The centrifugal barrier is now too weak to prevent an incoming particle from falling into the centre of force.

Question 8.12

Extend the calculation of Question 8.9(a) to obtain the bounded trajectories $r(\theta)$ for motion in a perturbed, attractive Coulomb potential.

$$V(r) = -\frac{k}{r} - \frac{\alpha}{r^2}, \tag{1}$$

where $k\ (> 0)$ and α are constants.

Solution

The effective potential is

$$V_{\rm e}(r) = -\frac{k}{r} + \frac{L^2 - 2m\alpha}{2mr^2}. \tag{2}$$

It follows from (2) that for bounded motion we require

$$\alpha < \frac{L^2}{2m} \quad \text{and} \quad -\frac{mk^2}{2(L^2 - 2m\alpha)} \leq E < 0.$$

(See the first energy diagram in Question 8.5.) Thus, proceeding as in Question 8.9(a), we have

$$\theta = \theta_0 - \frac{L}{\sqrt{2m}} \int \frac{du}{\sqrt{-(\beta^2 L^2/2m)u^2 + ku + E}}$$

$$= \theta_0 - \frac{1}{\beta} \cos^{-1}\left\{\frac{-(\beta^2 L^2/m)u + k}{\sqrt{k^2 + (2\beta^2 L^2 E/m)}}\right\}, \quad (3)$$

where $u = 1/r$ and $\beta^2 = 1 - 2m\alpha/L^2 > 0$. Hence

$$r(\theta) = \frac{r_0}{1 - e\cos\beta(\theta - \theta_0)}, \quad (4)$$

where

$$r_0 = \beta^2 \frac{L^2}{mk}, \qquad e = \sqrt{1 + \beta^2 \frac{2L^2 E}{mk^2}}. \quad (5)$$

Comments

(i) We see that the effect of the perturbation $-\alpha/r^2$ is contained in the parameter β in (4) and (5). When $\alpha = 0$ (i.e. $\beta = 1$), (4) reduces to the trajectory in a Coulomb potential (see Question 8.9).

(ii) For bounded motion we saw above that $\alpha < L^2/2m$ and $-mk^2/2\beta^2 L^2 \leq E < 0$. Therefore, $0 \leq e < 1$. For $e = 0$ the orbit is circular (and stable). If $e \neq 0$, it is evident from (4) that for rational values of β the orbit is closed (the particle eventually returns to its starting point), while for irrational values of β the orbit is open (the particle never returns to its starting point).

(iii) For $0 < e < 1$ the orbit can precess. This is illustrated in the two figures below, (a) for a rational $\beta = 4/5$ and (b) for an irrational $\beta = (2+\pi)/2\pi$. In (a) the orbit

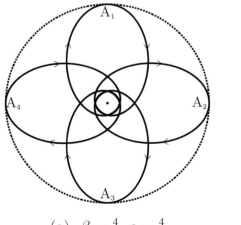
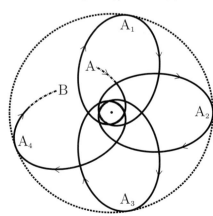

(a) $\beta = \frac{4}{5}$, $e = \frac{4}{5}$ (b) $\beta = \frac{2+\pi}{2\pi}$, $e = \frac{4}{5}$

is closed, while in (b) it is open. The points A_1, A_2, A_3 and A_4 are successive aphelia. In (b) a portion of the trajectory is shown, between points A and B; the actual trajectory fills the entire annulus between the turning points $r_{\min}$ and $r_{\max}$. In these figures, the dotted circle shows the maximum value $r_0/(1-e)$ of r. The perturbing potential is attractive because $\beta < 1$.

(iv) If $\beta = \text{(integer)}^{-1}$ one finds that the position of the aphelion (the furthest distance from the force centre) is fixed on the outer circle:

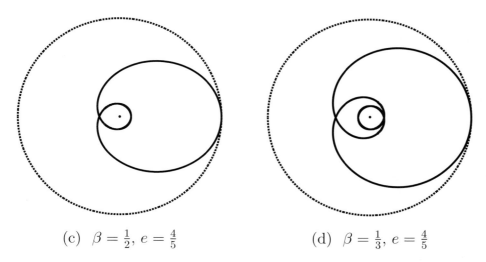

(c) $\beta = \frac{1}{2}$, $e = \frac{4}{5}$

(d) $\beta = \frac{1}{3}$, $e = \frac{4}{5}$

(v) For a repulsive perturbing potential $\beta > 1$. Typical trajectories are shown below: (e) for a rational $\beta = 3$ that produces a closed orbit, and (f) for an irrational $\beta = (3+8\pi)/3\pi$ where the orbit is open (a finite portion of the trajectory between initial and final points A and B is shown).

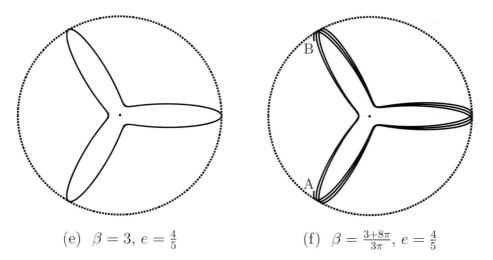

(e) $\beta = 3$, $e = \frac{4}{5}$

(f) $\beta = \frac{3+8\pi}{3\pi}$, $e = \frac{4}{5}$

The topic of closed and open orbits is encountered again in Questions 8.15 and 8.16, and also in Chapter 9, where it is related to the concept of hidden symmetry and the breaking of this symmetry.

(vi) The next question deals with the so-called inverse problem, namely given the trajectory $r(\theta)$ find the force $\mathbf{F}(\mathbf{r})$.

Question 8.13

For each of the following trajectories $r(\theta)$, determine the corresponding force $\mathbf{F} = F(r)\hat{\mathbf{r}}$.

(a) $\quad r(\theta) = \dfrac{r_0}{1 - e\cos\theta}$, $\hspace{2cm}$ (1)

(b) $\quad r(\theta) = \dfrac{r_0^2}{1 - e\sin 2\theta}$, $\hspace{2cm}$ (2)

(c) $\quad r(\theta) = Ae^{\beta\theta}$, $\hspace{2cm}$ (3)

(d) $\quad r(\theta) = (A\theta + B)^{-1}$, $\hspace{2cm}$ (4)

where r_0, e, β, A and B are constants.

Solution

To obtain the force from the trajectory we use the equation of motion in the form (1) of Question 8.8:

$$F\left(\frac{1}{u}\right) = -\frac{L^2 u^2}{m}\left(\frac{d^2 u}{d\theta^2} + u\right), \hspace{1cm} (5)$$

where $u = 1/r$. Note that in any application of (5) the result obtained for F must, by definition, be independent of the angular position θ.

(a) From (1) and (5) we have

$$F\left(\frac{1}{u}\right) = -\frac{L^2 u^2}{m}\left\{\frac{e}{r_0}\cos\theta + \frac{1}{r_0}(1 - e\cos\theta)\right\} = -\frac{L^2 u^2}{mr_0}.$$

So

$$\mathbf{F} = -\left(\frac{L^2}{mr_0}\right)\frac{1}{r^2}\hat{\mathbf{r}}, \hspace{1cm} (6)$$

which is an attractive inverse-square force directed towards a focus of the conic section (1) (see Question 8.9).

(b) Here, $u = \sqrt{1 - e\sin 2\theta}/r_0$, and a short calculation shows that

$$\frac{d^2 u}{d\theta^2} + u = \frac{1 - e^2}{r_0^4 u^3}. \hspace{1cm} (7)$$

From (5) and (7) we have

$$F\left(\frac{1}{u}\right) = -\frac{(1-e^2)L^2}{mr_0^4 u}, \tag{8}$$

and hence

$$\mathbf{F} = -\frac{(1-e^2)L^2}{mr_0^4} r\hat{\mathbf{r}}, \tag{9}$$

which is an attractive, linear (Hooke's-law) force directed toward the centre of the ellipse (2) (see also Question 8.10).

(c) From (3) and (5) we have

$$F\left(\frac{1}{u}\right) = -\frac{L^2 u^2}{m}(\beta^2 + 1)u. \tag{10}$$

Thus

$$\mathbf{F} = -\frac{L^2(\beta^2+1)}{m}\frac{1}{r^3}\hat{\mathbf{r}}, \tag{11}$$

which is an attractive, inverse-cube force.

(d) From (4) and (5) we have

$$F\left(\frac{1}{u}\right) = -\frac{L^2 u^2}{m}u, \tag{12}$$

and therefore

$$\mathbf{F} = -\frac{L^2}{m}\frac{1}{r^3}\hat{\mathbf{r}}, \tag{13}$$

which is also an attractive, inverse-cube force.

Comments

(i) The direct problems for the above three forces were solved in Questions 8.9–8.11. The solution of an inverse problem involves differentiation, and it is invariably easier than the solution of the corresponding direct problem, which involves integration.

(ii) The inverse problems leading to the inverse-square and linear central forces (6) and (9) were first solved by Newton.[3] Thus, the result that elliptical motion can occur either in an inverse-square force field directed to one of the foci of the ellipse, or in a linear force field directed to the centre of the ellipse, is due to Newton.

Question 8.14

A particle of mass m moves in a central, isotropic force field $F(r)\hat{\mathbf{r}}$.

---

[3] S. Chandrasekhar, *Newton's Principia for the common reader*, Chaps. 5 and 6. Oxford: Clarendon Press, 1995.

(a) Show that the condition for a circular orbit of radius r_0 is
$$F(r_0) = -L^2/mr_0^3, \tag{1}$$
where L is the angular momentum of the particle.

(b) Prove that this orbit is stable provided
$$\left.\frac{dF}{dr}\right|_{r=r_0} < -\frac{3}{r_0} F(r_0). \tag{2}$$

(c) Hence determine the values of the constant n for which the power-law force $F(r) = -k/r^n$ has stable circular orbits.

Solution

(a) The radial part of the equation of motion is (see Question 8.7)
$$m\ddot{r} - mr\dot\theta^2 = F(r). \tag{3}$$

For a circular orbit $r = r_0$ is a constant, and (3) becomes $-mr_0\dot\theta^2 = F(r_0)$, which is (1) because $L = mr_0^2\dot\theta$ (see Question 8.2). The same result follows by setting the effective force (10) of Question 8.2 equal to zero.

(b) For a stable circular orbit the effective potential (see Question 8.2)
$$V_e = V(r) + L^2/2mr^2 \tag{4}$$
must have a minimum at $r = r_0$. That is,
$$\frac{d^2}{dr^2}\left(V(r) + \frac{L^2}{2mr^2}\right) > 0 \quad \text{at } r = r_0. \tag{5}$$

This is (2) because $dV/dr = -F(r)$, and $F(r_0)$ is given by (1).

(c) If the force is repulsive ($F > 0$ and hence $k < 0$), (1) cannot be satisfied: no circular orbits – for that matter, no bounded orbits – are possible in a repulsive power-law force field. We need consider only an attractive force, $k > 0$. Now (2) requires
$$nk/r_0^{n+1} < 3k/r_0^{n+1}. \tag{6}$$
Because k and r_0 are positive, this means that $n < 3$ for stable circular orbits.

Comment

In this connection it is interesting to sketch the curves of the effective potential (4) for the power-law force $F = -k/r^n$:
$$\left.\begin{aligned} V_e(r) &= -\frac{k}{n-1}\frac{1}{r^{n-1}} + \frac{L^2}{2mr^2} \quad (n \neq 1) \\ &= k\ln r + \frac{L^2}{2mr^2} \quad (n = 1). \end{aligned}\right\} \tag{7}$$

So, for $k > 0$ we have the following energy diagrams that demonstrate the stability of circular orbits for $n < 3$. Note that for $n \leq 1$ all trajectories are bounded, whereas for $1 < n < 3$ both unbounded and bounded trajectories occur. If $n > 3$, circular orbits are possible ((1) can be satisfied), but they are unstable (the inequality (2) cannot be satisfied). The energy diagram for $n = 3$, given on page 243, shows that the stability of a circular orbit is neutral in this case.

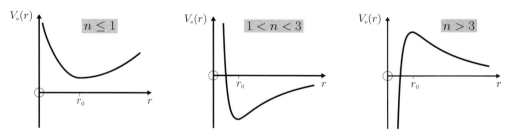

Question 8.15

(a) Let T_r be the period of radial, bounded motion of a particle with mass m and energy E in an effective potential $V_e(r)$. Show that

$$T_r = \int_{r_1}^{r_2} \sqrt{\frac{2m}{\{E - V_e(r)\}}}\, dr \,, \tag{1}$$

where r_1 and r_2 are the classical turning points. (Hint: Refer to Question 8.2.)

(b) Let T_θ be the period of the angular motion of the above particle. Show that

$$T_\theta = \frac{m}{L}\int_0^{2\pi} r^2(\theta)\, d\theta \,. \tag{2}$$

(c) Evaluate T_r and T_θ for an attractive Coulomb potential $V = -k/r$.
(d) Evaluate T_r and T_θ for an isotropic oscillator potential $V = \frac{1}{2}kr^2$.

Solution

(a) The radial period T_r is the period of one-dimensional motion in the rotating frame S' referred to in Question 8.2 (that is, the period of one-dimensional bounded motion in the effective potential $V_e(r)$). T_r is equal to the time taken for the motion $r_1 \to r_2 \to r_1$, which is twice the time taken to go from $r_1 \to r_2$. Thus, (1) follows directly from (12) of Question 8.2.

(b) From (7) of Question 8.2,

$$dt = (m/L)r^2 d\theta \,. \tag{3}$$

The angular period T_θ is the time taken for the angular position to change by 2π, and therefore integration of the right-hand side of (3) between $\theta = 0$ and $\theta = 2\pi$ yields (2).

(c) For a Coulomb potential $V = -k/r$, (1) gives

$$T_r = \int_{r_1}^{r_2} \sqrt{\frac{2m}{\{E + k/r - L^2/2mr^2\}}}\, dr$$

$$= 2m \int_{r_1}^{r_2} \frac{r\, dr}{\sqrt{2mEr^2 + 2mkr - L^2}}, \qquad (4)$$

where $E < 0$ for bounded motion. Now

$$\int \frac{r\, dr}{\sqrt{Ar^2 + Br + C}} = \frac{\sqrt{Ar^2 + Br + C}}{A} + \frac{B}{2A\sqrt{-A}} \sin^{-1}\left\{\frac{2Ar + B}{\sqrt{B^2 - 4AC}}\right\}, \qquad (5)$$

for $A < 0$. Thus

$$\int_{r_1}^{r_2} \frac{r\, dr}{\sqrt{2mEr^2 + 2mkr - L^2}} = \frac{k}{2E\sqrt{-2mE}}\left\{\sin^{-1}(1) - \sin^{-1}(-1)\right\}$$

$$= \frac{\pi k}{2E\sqrt{-2mE}}, \qquad (6)$$

because for $r = r_1$ and $r = r_2$ the quantity $2mEr^2 + 2mkr - L^2 = 0$ and hence $4mEr + 2mk = \pm\sqrt{4m^2k^2 + 8mL^2E}$. From (4) and (6) we have

$$T_r = \pi k \sqrt{\frac{-m}{2E^3}}. \qquad (7)$$

To calculate the angular period T_θ for a Coulomb potential, use the trajectory $r(\theta)$ given by (11) of Question 8.9. Then, (2) gives

$$T_\theta = \frac{L^3}{mk^2} \int_0^{2\pi} \frac{d\theta}{(1 - e\cos\theta)^2}, \qquad (8)$$

where

$$e = \sqrt{1 + \frac{2L^2 E}{mk^2}} < 1. \qquad (9)$$

Now

$$\int \frac{d\theta}{(1 - e\cos\theta)^2} = \frac{e}{1 - e^2}\frac{\sin\theta}{1 - e\cos\theta} + \frac{2}{(1 - e^2)^{3/2}} \tan^{-1}\left[\frac{\sqrt{1 - e^2}\tan\tfrac{1}{2}\theta}{1 - e}\right] \qquad (10)$$

for $e^2 < 1$. Thus

$$\int_0^{2\pi} \frac{d\theta}{(1 - e\cos\theta)^2} = 2\int_0^{\pi} \frac{d\theta}{(1 - e\cos\theta)^2} = \frac{4\tan^{-1}\infty}{(1 - e^2)^{3/2}}$$

$$= \frac{2\pi}{(1 - e^2)^{3/2}}. \qquad (11)$$

From (8), (9) and (11) we have

$$T_\theta = \frac{L^3}{mk^2} 2\pi \left(-\frac{mk^2}{2L^2 E}\right)^{3/2} = \pi k \sqrt{\frac{-m}{2E^3}}. \qquad (12)$$

(d) For the oscillator potential $V = \frac{1}{2}kr^2$, (1) gives

$$T_r = \int_{r_1}^{r_2} \sqrt{\frac{2m}{\{E - \frac{1}{2}kr^2 - L^2/2mr^2\}}}\, dr. \tag{13}$$

The substitution $u = r^2$ converts this to

$$T_r = m\int_{u_1}^{u_2} \frac{du}{\sqrt{-mku^2 + 2mEu - L^2}}. \tag{14}$$

By definition u_1 and u_2 are the roots of the quadratic function in the denominator of (14). Thus, according to (2) of Question 8.9, the integral in (14) is equal to $\{\cos^{-1}(-1) - \cos^{-1}(1)\}/\sqrt{mk} = \pi/\sqrt{mk}$, and therefore

$$T_r = \pi\sqrt{\frac{m}{k}}. \tag{15}$$

To calculate T_θ for the oscillator, use the trajectory $r(\theta)$ given by (5) of Question 8.10 with $\theta_0 = \frac{1}{4}\pi$. Then, (2) and the symmetry of the trajectory yield

$$T_\theta = \frac{L}{E} 4\int_0^{\pi/2} \frac{d\theta}{1 + e\cos 2\theta}, \tag{16}$$

where $e = \sqrt{1 - kL^2/mE^2}$. Now

$$\int \frac{d\alpha}{1 + e\cos\alpha} = \frac{2}{\sqrt{1 - e^2}} \tan^{-1}\left[\frac{\sqrt{1 - e^2}\tan\frac{1}{2}\alpha}{1 + e}\right]. \tag{17}$$

Consequently, the integral in (16) is equal to $\frac{1}{2}\pi/\sqrt{1 - e^2}$, and so

$$T_\theta = \frac{L}{E} 2\pi\sqrt{\frac{mE^2}{kL^2}} = 2\pi\sqrt{\frac{m}{k}}. \tag{18}$$

Comments

(i) For the Coulomb potential the dependence on the angular momentum L cancels in the above calculations of T_r and T_θ, while for the oscillator the dependences on both L and E cancel (see, for example, the steps from (8) to (12), and from (16) to (18)).

(ii) For the Coulomb potential, (7) and (12) show that the angular and radial periods are equal, $T_\theta = T_r$ – the motion is singly periodic. For the oscillator, (15) and (18) show that $T_\theta = 2T_r$ – the motion is doubly periodic. (The difference arises because the centre of force for the oscillator is at the centre of the ellipse and consequently a radial cycle $r_1 \to r_2 \to r_1$ is completed in half an angular cycle.) It follows that *all* bounded orbits of these potentials are closed. In fact, these are the only potentials that have this remarkable property, a result that is known as Bertrand's theorem (see Question 9.12). The special feature that sets the non-relativistic Coulomb and oscillator problems apart is the 'hidden' symmetry they possess (see Chapter 9).

(iii) In general, bounded motion in a potential $V(r)$ is doubly periodic, with different periods T_θ and T_r. The ratio T_θ/T_r enables us to distinguish between closed and open orbits.

(iv) If T_θ/T_r is a rational number, that is

$$T_\theta/T_r = p/q, \tag{19}$$

where p and q are integers, then after p cycles of the radial motion the particle will have executed q complete revolutions – it will have returned to its initial position. The orbit is closed. For the Coulomb potential $T_\theta/T_r = 1$ and the orbit closes after just one radial cycle and one revolution (θ changes by 2π during the motion $r_1 \to r_2 \to r_1$). Clearly, the same remark applies to the circular orbits of any potential. If T_r is a little larger than T_θ then during one radial cycle the angular position θ will change by a little more than 2π. The resulting motion, which is known as precession, is depicted below. Here, T_θ is the time taken to go from A to B, and T_r is the time taken to go from A to C. It takes a large number of revolutions for the orbit to close (for example, 100 if $p/q = 99/100$).

(v) If T_θ/T_r is not a rational number then the orbit will not close – it is an open orbit. All points in the classically accessible region $r_1 \leq r \leq r_2$ and $0 \leq \theta \leq 2\pi$ (the region between the circles of radii r_1 and r_2 in the figure) will eventually be reached by the particle.

(vi) Equation (7) and the relation $E = -k/2a$ (see Question 8.9) show that the period T for the Coulomb problem is related to the length $2a$ of the major axis by

$$T = 2\pi\sqrt{ma^3/k}. \tag{20}$$

This is Kepler's third law (see also Question 10.11).

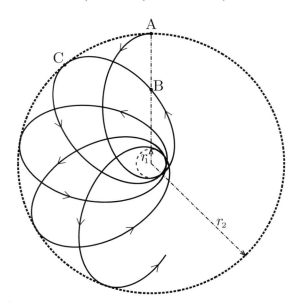

Question 8.16

Consider bounded motion in the perturbed, attractive Coulomb potential

$$V(r) = -\frac{k}{r} - \frac{\alpha}{r^2}, \tag{1}$$

where $k\ (>0)$ and α are constants. Let $\beta = \sqrt{1 - 2m\alpha/L^2}$. Using the definite integrals provided in the solution to Question 8.15, prove that if β is rational (i.e. $\beta = p/q$, where p and q are integers) then the ratio of the angular and radial periods is given by

$$\frac{T_\theta}{T_r} = \beta. \tag{2}$$

(Hint: For the angular period T_θ consider q complete cycles.)

Solution

The effective potential corresponding to (1) is

$$V_e(r) = -\frac{k}{r} + \frac{\beta^2 L^2}{2mr^2}. \tag{3}$$

This potential follows from the unperturbed effective potential by making the replacement $L \to \beta L$. Thus, the radial period T_r for the perturbed motion can be obtained by making this replacement in (7) of Question 8.15. But the latter result is independent of L and therefore the radial period is unaffected by the perturbation: from (7)

$$T_r = \pi k \sqrt{\frac{-m}{2E^3}}. \tag{4}$$

(If the perturbation decreases (increases) the integrand in the general formula (1) of Question 8.15, then for a given E it increases (decreases) the range of integration $[r_1, r_2]$; these two effects compensate, leaving T_r unchanged.)

For the angular period T_θ we use (2) of Question 8.15, and the trajectory

$$r(\theta) = \frac{\beta^2 L^2}{mk(1 - e \cos \beta\theta)}, \quad \text{with} \quad e = \sqrt{1 + \beta^2 \frac{2L^2 E}{mk^2}}, \tag{5}$$

(see Question 8.12). Then

$$T_\theta = \frac{\beta^4 L^3}{mk^2} \int_0^{2\pi} \frac{d\theta}{(1 - e \cos \beta\theta)^2}. \tag{6}$$

In (6) we make the substitution $u = \beta\theta$ and consider q complete angular cycles. Then

$$qT_\theta = \frac{\beta^3 L^3}{mk^2} \int_0^{2\pi q\beta} \frac{du}{(1 - e \cos u)^2}. \tag{7}$$

For rational β ($= p/q$), (7) becomes

$$qT_\theta = p\frac{\beta^3 L^3}{mk^2}\int_0^{2\pi}\frac{du}{(1 - e\cos u)^2} = p\pi k\sqrt{\frac{-m}{2E^3}}. \tag{8}$$

Here, we have used (11) of Question 8.15 and $(5)_2$. According to (4) and (8)

$$T_\theta = \beta T_r. \tag{9}$$

Comments

(i) It follows from (2) that the trajectory is closed if β is rational. This conclusion is consistent with the trajectories calculated in Question 8.12.

(ii) The above example illustrates how a perturbation destroys a special property of motion in a Coulomb potential (namely, that all bounded orbits are closed and singly periodic): the bounded trajectories of the perturbed potential (1) are closed only for discrete values of β, and hence of the angular momentum

$$L = \sqrt{\frac{2m\alpha}{1 - (p/q)^2}}, \tag{10}$$

where p and q are integers (see also Question 8.12).

(iii) It is straightforward to show that a similar result holds for the perturbed oscillator potential $V(r) = \tfrac{1}{2}kr^2 - \alpha/r^2$, where the discrete values that produce closed orbits are given by

$$L = \sqrt{\frac{2m\alpha}{1 - (p/2q)^2}}. \tag{11}$$

In both instances the perturbed closed orbits are doubly periodic.

Question 8.17

Determine the time-dependent trajectory $\bigl(r(t),\,\theta(t)\bigr)$ for motion in a spiral orbit

$$r(\theta) = Ae^{\beta\theta}, \tag{1}$$

where A and β are constants. (Hint: Start with (7) of Question 8.2.)

Solution

For the given orbit, and by integrating (7) of Question 8.2, we have

$$t = \frac{mA^2}{L}\int_{\theta_0}^{\theta(t)} e^{2\beta\theta}\,d\theta = \frac{mA^2}{2\beta L}\left(e^{2\beta\theta(t)} - e^{2\beta\theta_0}\right), \tag{2}$$

where $\theta_0 = \theta(0)$. Inversion of (2) gives

$$\theta(t) = \frac{1}{2\beta} \ln\left(\frac{2\beta L}{mA^2} t + e^{2\beta\theta_0}\right). \qquad (3)$$

From (1) and (3) we obtain

$$r(t) = A\sqrt{\frac{2\beta L}{mA^2} t + e^{2\beta\theta_0}}. \qquad (4)$$

Comments

(i) It is a good exercise for the reader to reconcile (3) and (4) with the relevant polar plots in Question 8.11.
(ii) In certain cases it is not feasible to invert the equations for $t(\theta)$ and $t(r)$. The next question is an important example of this type.

Question 8.18

Consider the elliptical orbits of a Coulomb potential described in Question 8.9.

(a) Given the initial conditions

$$\theta = 0 \quad \text{and} \quad r = r_{\min} = (1-e)a \qquad (1)$$

at $t = 0$, show that the time-dependent trajectory $(r(t), \theta(t))$ is given in inverse form for the first half-cycle, $0 \leq t \leq \frac{1}{2}T$, by

$$t(u) = \frac{T}{4}\left[1 - \frac{2}{\pi}\sqrt{-u^2 + 2u + e^2 - 1} - \frac{2}{\pi}\sin^{-1}\left(\frac{1-u}{e}\right)\right], \qquad (2)$$

$$t(\theta) = \frac{T}{4}\left[1 - \frac{2}{\pi} e \sqrt{1-e^2}\,\frac{\sin\theta}{1+e\cos\theta} - \frac{2}{\pi}\sin^{-1}\left(\frac{e+\cos\theta}{1+e\cos\theta}\right)\right], \qquad (3)$$

for $(1-e) \leq u \leq (1+e)$ and $0 \leq \theta \leq \pi$. Here, $u = r/a$, T is the period, e is the eccentricity, and $2a$ is the length of the major axis of the ellipse. (Hint: Start with (12) of Question 8.2, and express $t(r)$ in terms of a, e and T by using the relations (see Questions 8.9 and 8.15)

$$E = -k/2a, \qquad L^2 = (mk^2/2E)(e^2 - 1), \qquad T = 2\pi\sqrt{ma^3/k}.\big) \qquad (4)$$

(b) Show that for the second half-cycle, $\frac{1}{2}T \leq t \leq T$,

$$t(u) = T - (2) \quad \text{and} \quad t(\theta) = T - (3), \qquad (5)$$

where (2) and (3) denote the right-hand sides of (2) and (3), respectively. Make parametric plots of $u(t/T)$ and $\theta(t/T)$ versus t/T for $e = 0.1$, $e = 0.5$ and $e = 0.9$. For $u(t/T)$ show the first three cycles, and for $\theta(t/T)$ show the first cycle.

Solution

(a) For the Coulomb potential $V = -k/r$, (12) of Question 8.2 yields

$$t(r) = \sqrt{\frac{m}{2}} \int_{r_0}^{r} \frac{dr}{\sqrt{E + k/r - L^2/2mr^2}}, \qquad (6)$$

where the initial condition $(1)_2$ requires $r_0 = (1-e)a$. In (6) we make the change of variable $u = r/a$, and use (4) to express the result in terms of a, e and T:

$$t(u) = \frac{T}{2\pi} \int_{1-e}^{u} \frac{u\,du}{\sqrt{-u^2 + 2u + e^2 - 1}}. \qquad (7)$$

According to (5) of Question 8.15,

$$\int \frac{u\,du}{\sqrt{-u^2 + 2u + e^2 - 1}} = -\sqrt{-u^2 + 2u + e^2 - 1} - \sin^{-1}\left(\frac{1-u}{e}\right), \qquad (8)$$

a result that can easily be checked by differentiation. From (7) and (8) we find that $t(u)$ is given by (2). The calculation of $t(\theta)$ in (3) is similar. The polar equation that is consistent with the initial conditions (1) is obtained by choosing $\theta_0 = \pi$ in (5) of Question 8.9; that is,

$$r(\theta) = \frac{a(1-e^2)}{(1+e\cos\theta)}.$$

After substituting this in (3) of Question 8.15, we use the integral[‡]

$$\int \frac{d\theta}{(1+e\cos\theta)^2} = -\frac{e}{1-e^2}\frac{\sin\theta}{1+e\cos\theta} - \frac{1}{(1-e^2)^{3/2}}\sin^{-1}\left[\frac{e+\cos\theta}{1+e\cos\theta}\right], \qquad (9)$$

where $0 \leq \theta \leq \pi$ and $e < 1$. This yields (3).

(b) The second half-cycle starts at $t = \tfrac{1}{2}T$ and $r = r_{\max} = (1+e)a$. Thus, for $\tfrac{1}{2}T \leq t \leq T$ we have, instead of (7),

$$t(r) = \tfrac{1}{2}T - \frac{T}{2\pi}\int_{1+e}^{u}\frac{u\,du}{\sqrt{-u^2+2u+e^2-1}}. \qquad (10)$$

(The negative sign before the integral in (10) takes account of the change in the direction of motion during the second half of the cycle.) Equations (8) and (10) lead to $(5)_1$. A similar calculation yields $(5)_2$. By continuing in this manner one can construct the various half-cycles of the motion. The required plots are shown below. We remark that if the apparent cusp-like structures at $t = T, 2T, 3T \cdots$ in the first graph are examined on a finer scale, one finds that the slope is always continuous and equal to zero there. General formulas for $t(u)$ and $t(\theta)$ in the various half-cycles are given in (11) and (12).

[‡]The validity of (9) can be checked by differentiation. It is equivalent to (10) of Question 8.15 with e replaced by $-e$.

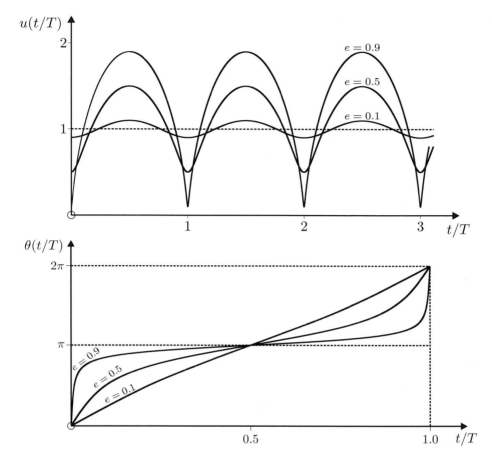

Comments

(i) The solutions (2) and (3) have the simple feature that the parameters m, k, L and E of the problem influence the dimensionless radial position $u(t/T)$ and the angular position $\theta(t/T)$ only through the eccentricity e.

(ii) The substitution $u = (1-e^2)/(1+e\cos\theta)$ changes (2) into (3) as required (the solution to $t(u) = t(\theta)$ is the polar equation for u).

(iii) It is straightforward to generalize (2) and (3) to the nth half-period (of duration $\frac{1}{2}T$), and the results are

$$\frac{t(u)}{T} = \frac{(2n-1)}{4} + \frac{(-1)^n}{2\pi}\left[\sqrt{-u^2+2u+e^2-1} + \sin^{-1}\left(\frac{1-u}{e}\right)\right], \quad (11)$$

$$\frac{t(\theta)}{T} = \frac{(2n-1)}{4} + \frac{(-1)^n}{2\pi}\left[e\sqrt{1-e^2}\frac{|\sin\theta|}{1+e\cos\theta} + \sin^{-1}\left(\frac{e+\cos\theta}{1+e\cos\theta}\right)\right], \quad (12)$$

for $n = 1, 2, \cdots$ and $(1-e) \leq u \leq (1+e)$ and $(n-1)\pi \leq \theta \leq n\pi$. We remark that the modulus sign in (12) is essential to obtain the correct results for even n.

(iii) The task of determining the time-dependent trajectories for the Kepler problem was one of the first problems considered at the dawn of Newtonian mechanics.[1,2] The historical approach differs from the 'head-on' method that we used to obtain (2) and (3) in that it is based on the introduction of an intermediate variable, an angle ψ known as the 'eccentric anomaly'. In terms of ψ, the coordinates of the particle relative to the centre of the ellipse are $x = a\cos\psi$, $y = b\sin\psi$. For the polar equation (9), the focus is at $F' = (ea, 0)$. Thus, the x-coordinate relative to F' is $x' = x - ea$ (see Question 8.9), and so the distance from F' is

$$r = \sqrt{x'^2 + y^2} = a(1 - e\cos\psi). \tag{13}$$

During a complete cycle from perigee to apogee to perigee, ψ changes from 0 to π to 2π. The change of variable (13) simplifies (7) to

$$t = \frac{T}{2\pi}\int_0^{\psi}(1 - e\cos\psi)\,d\psi.$$

Hence

$$\omega t = \psi - e\sin\psi, \tag{14}$$

where $\omega = 2\pi/T$. In addition, there is a relation between the angular positions θ and ψ that follows by equating $r(\theta)$ and (13), namely

$$\tan\frac{\theta}{2} = \sqrt{\frac{1+e}{1-e}}\tan\frac{\psi}{2}. \tag{15}$$

The transcendental equation (14) is known as Kepler's equation. It was first derived by Newton,[3] who also deduced its form for hyperbolic trajectories.

(iv) In the above, the centre of force is located at the right-hand focus of the ellipse. In problems where the centre of force is at the left-hand focus, one need simply replace e with $-e$ in (13)–(15).

(v) Equations (13)–(15) are the desired solution in terms of the eccentric anomaly. To find the position $(r(t), \theta(t))$ on the orbit at time t, they are used as follows. First, the eccentric anomaly is calculated from (14), and then $r(t)$ and $\theta(t)$ are obtained from (13) and (15). For examples, see Questions 10.9 and 11.6.

(vi) Because of the importance of obtaining accurate numerical solutions for $\mathbf{r}(t)$, the Kepler problem has attracted considerable attention since the seventeenth century, when Newton described an iterative procedure for obtaining approximate solutions.[3]

Question 8.19

Extend the calculation of Question 8.18 to obtain the time-dependent trajectory $(r(t), \theta(t))$ in inverse form for hyperbolic orbits of an attractive Coulomb potential.

Solution

For the hyperbolic orbits $E = k/2|a|$, $e > 1$, and $r_{\min} = (e-1)|a|$ (see Question 8.9). (In what follows we omit the modulus sign on a.) Consequently, (6) and (4)$_2$ of Question 8.18 yield

$$t(u) = \sqrt{\frac{ma^3}{k}} \int_{e-1}^{u} \frac{u\,du}{\sqrt{u^2 + 2u + 1 - e^2}}, \qquad (1)$$

where $u = r/a$ ($\geq e - 1$). The integral

$$\int \frac{u\,du}{\sqrt{u^2 + 2u + 1 - e^2}} = \sqrt{u^2 + 2u + 1 - e^2} - \ln\left(\sqrt{u^2 + 2u + 1 - e^2} + u + 1\right) \quad (2)$$

($e > 1$) can readily be checked by differentiation. Therefore, (1) gives

$$t(u) = \tau\left[\sqrt{u^2 + 2u + 1 - e^2} - \ln\frac{1}{e}\left(\sqrt{u^2 + 2u + 1 - e^2} + u + 1\right)\right], \qquad (3)$$

for $u \geq e - 1$. Here, τ is a time scale for the motion: $\tau = \sqrt{ma^3/k} = a/v_\infty$, where v_∞ is the asymptotic velocity. The calculation of $t(\theta)$ is similar. By substituting the polar equation $u = (e^2 - 1)/(1 + e\cos\theta)$ in (3) of Question 8.15 and using the integral

$$\int \frac{d\theta}{(1 + e\cos\theta)^2} = \frac{e}{1 - e^2}\frac{\sin\theta}{1 + e\cos\theta} - \frac{1}{(e^2 - 1)^{3/2}}\ln\left[\frac{\sqrt{e^2 - 1}\sin\theta + \cos\theta + e}{1 + e\cos\theta}\right] \quad (4)$$

($e > 1$) and the initial condition $\theta = 0$ at $t = 0$, we obtain

$$t(\theta) = \tau\left[e\sqrt{e^2 - 1}\frac{\sin\theta}{1 + e\cos\theta} - \ln\left(\frac{\sqrt{e^2 - 1}\sin\theta + \cos\theta + e}{1 + e\cos\theta}\right)\right], \qquad (5)$$

for $\theta \leq \cos^{-1}(-1/e)$.

Comment

Graphs of $u(t/\tau)$ and $\theta(t/\tau)$ versus t/τ obtained from (3) and (4) for three values of the eccentricity e are shown below. The dotted lines are the asymptotes $u = t/\tau$ and $\theta = \cos^{-1}(-1/e)$. We remark that curves of $u(t/\tau)$ for different values of e can cross, as illustrated in the first figure.

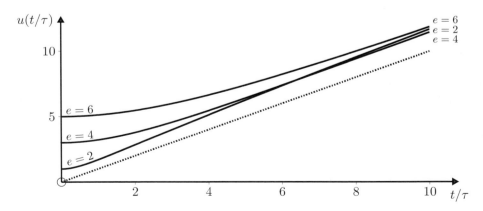

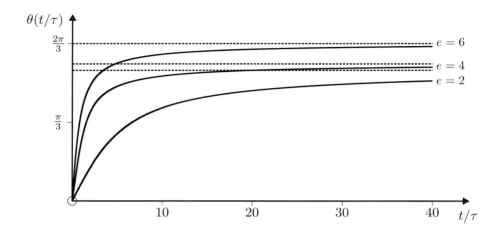

Question 8.20

At time $t = 0$ a rocket of mass m is at a distance of two Earth radii ($2R_e$) from the centre of the Earth and is travelling perpendicular to the equatorial plane with speed u. The mass of the Earth $M_e \gg m$. Suppose that

$$u = \alpha v_\odot, \quad \text{where} \quad v_\odot = \sqrt{GM_e/2R_e} \tag{1}$$

and α is a positive number.

(a) Show that a circular orbit requires $\alpha = 1$.
(b) Show that the escape velocity of the rocket corresponds to $\alpha = \sqrt{2}$.
(c) Show that for the rocket to strike a point on the Earth's surface at angular position λ (measured from the Equator) requires

$$\alpha = \sqrt{\frac{1 - \cos\lambda}{2 - \cos\lambda}}. \tag{2}$$

(Neglect atmospheric drag when the rocket enters the Earth's atmosphere.)

(d) Use *Mathematica* to find the time taken by the rocket to reach the Arctic Circle ($\lambda = 66\frac{1}{2}°$). Then, plot the trajectory for $\alpha = 0.613$, $\sqrt{2/3}$, 0.9, 1.0, and 1.1.

Solution

Since $M_e \gg m$, it is a good approximation to take the centre of mass of the Earth–rocket system at the centre of the Earth (see Chapter 10).

(a) Then, the equation of motion of the rocket for a circular orbit of radius $2R_e$ is

$$m\frac{u^2}{2R_e} = \frac{GM_e m}{(2R_e)^2}, \tag{3}$$

and therefore $u = v_\odot$.

(b) Conservation of energy requires

$$\tfrac{1}{2}mu^2 - \frac{GM_e m}{2R_e} = \tfrac{1}{2}mv_\infty^2, \qquad (4)$$

where v_∞ is the speed of the rocket at $r = \infty$. By definition, the escape velocity is the value of u corresponding to $v_\infty = 0$: that is, $u = \sqrt{2}\,v_\odot$.

(c) In the diagram below, the circle represents the surface of the Earth. The trajectory AB of the rocket is part of an ellipse with one focus at the centre O of the Earth.

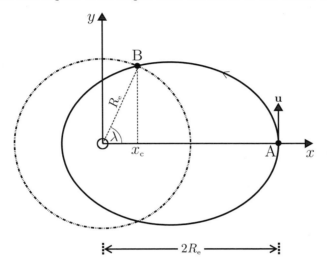

The point of intersection B between the circle and the ellipse is determined by the simultaneous equations

$$x^2 + y^2 = R_e^2 \quad \text{and} \quad \left(\frac{x - ea}{a}\right)^2 + \left(\frac{y}{b}\right)^2 = 1, \qquad (5)$$

where $b = a\sqrt{1 - e^2}$ (see Question 8.9). The positive solution to (5) for x is

$$x_c = \frac{-a(1 - e^2) + R_e}{e}. \qquad (6)$$

Also, $a(1 + e) = 2R_e$ (see Question 8.9) and $x_c = R_e \cos\lambda$. If we use these two equations to eliminate e and x_c from (6) we find that

$$a = \left(\frac{2 - \cos\lambda}{3 - \cos\lambda}\right) 2R_e. \qquad (7)$$

Now, the energy of the rocket is

$$E = \tfrac{1}{2}mu^2 - \frac{GM_e m}{2R_e} = -\frac{GM_e m}{2R_e}(1 - \tfrac{1}{2}\alpha^2). \qquad (8)$$

From (7) and (8), and (20) of Question 8.9 (with $k = GM_e m$), we obtain the desired result (2). Note that $\lambda = 180°$ requires $\alpha = \sqrt{2/3}$, and $\lambda = 66\frac{1}{2}°$ requires $\alpha = 0.613$.

(d) The *Mathematica* notebook given below was used to solve the equation of motion $m\ddot{\mathbf{r}} = -\dfrac{GM_e m}{r^3}\mathbf{r}$ for the trajectory $(x(t), y(t))$. This yields the trajectories shown in the figure. With $\alpha = 0.613$ we find that the rocket reaches the ground after $t = 2715\,\text{s} \approx 45\,\text{min}$.

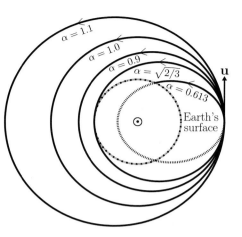

Comment

If $\alpha < \sqrt{2}$, (8) shows that $E < 0$. The trajectory is an ellipse if $\sqrt{2/3} < \alpha < \sqrt{2}$, and part of an ellipse if $\alpha \leq \sqrt{2/3}$ (see (2) with $\lambda \leq \pi$). The left-hand focus is at O if $\alpha < 1$; the right-hand focus is at O if $\alpha > 1$; and the orbit is a circle if $\alpha = 1$. For $\alpha > \sqrt{2}$, $E > 0$ and the trajectory is a hyperbola. If $\alpha = \sqrt{2}$ then $E = 0$ and the trajectory is a parabola.

```mathematica
In[1]:= M_e = 5.99 × 10^24; R_e = 6.37 × 10^6; G = 6.67 × 10^-11;

r_0 = 2 R_e; V_circ = Sqrt[G M_e / r_0]; λ = 66.5 π/180;

α = Sqrt[(1 - Cos[λ])/(2 - Cos[λ])]; V = α V_circ; a = r_0/(2 - α^2); T = Sqrt[4 π^2 a^3 / (G M_e)]; tmax = T;

r[t_] := {x[t], y[t]};

x0 = r_0; vx0 = 0; y0 = 0; vy0 = V;

EqnMotion = Thread[r''[t] + (G M_e r[t])/Dot[r[t], r[t]]^(3/2) == 0];

InitCon = Join[Thread[r[0] == {x0, y0}], Thread[r'[0] == {vx0, vy0}]];

Sol = NDSolve[Join[EqnMotion, InitCon], {x[t], y[t]}, {t, 0, tmax}];

ParametricPlot[{{Evaluate[{x[t]/R_e, y[t]/R_e}] /. Sol}}, {t, 0, tmax}]

Sol = FindRoot[x[t] == R_e Cos[λ] /. Sol, {t, T/4}]
```

9

The Coulomb and oscillator problems

The Coulomb (or Kepler) and oscillator problems play a special role in both classical and quantum physics. Despite their simplicity, these two problems have important applications and they possess interesting theoretical properties; not surprisingly, a large number of papers have been devoted to them since the seventeenth century. The following questions illustrate some of their special properties (extra constants of the motion, hidden symmetries, closed orbits and transformations).

Question 9.1

Consider the equation of motion for a particle of mass m in a Coulomb potential:

$$\frac{d\mathbf{p}}{dt} = -\frac{k}{r^2}\hat{\mathbf{r}}. \tag{1}$$

(a) Cast (1) into the form of a conservation equation

$$\frac{d\mathbf{B}}{dt} = 0, \tag{2}$$

where the conserved vector $\mathbf{B}$ is given by

$$\mathbf{B} = \frac{L}{mk}\mathbf{p} - \hat{\mathbf{L}} \times \hat{\mathbf{r}}. \tag{3}$$

(Hint: Use the relation $d\hat{\boldsymbol{\theta}}/dt = -(d\theta/dt)\hat{\mathbf{r}}$ for the unit vectors of plane polar coordinates – see Question 8.1.)

(b) Use $\mathbf{B}$ and $\mathbf{L}$ to construct a second conserved vector $\mathbf{A}$, and show that it can be expressed as

$$\mathbf{A} = \hat{\mathbf{r}} + \frac{1}{mk}\mathbf{L} \times \mathbf{p}. \tag{4}$$

Solution

(a) The relation (see Question 8.1)

$$\frac{d\hat{\boldsymbol{\theta}}}{dt} = -\frac{d\theta}{dt}\hat{\mathbf{r}} = -\frac{L}{mr^2}\hat{\mathbf{r}} \qquad (5)$$

enables us to write (1) as

$$\frac{d}{dt}\left(\frac{L}{mk}\mathbf{p} - \hat{\boldsymbol{\theta}}\right) = 0. \qquad (6)$$

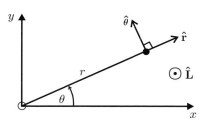

This is (2) with

$$\mathbf{B} = \frac{L}{mk}\mathbf{p} - \hat{\boldsymbol{\theta}}. \qquad (7)$$

Because $\hat{\boldsymbol{\theta}} = \hat{\mathbf{L}} \times \hat{\mathbf{r}}$ (see figure), this can also be written as (3).

(b) By taking the vector product of $\hat{\mathbf{L}}$ with $\mathbf{B}$ we can create a second conserved vector that is orthogonal to both $\mathbf{B}$ and $\mathbf{L}$:

$$\mathbf{A} = \hat{\mathbf{L}} \times \mathbf{B}. \qquad (8)$$

From (7) and (8), and with $\hat{\mathbf{L}} \times \hat{\boldsymbol{\theta}} = -\hat{\mathbf{r}}$ (see figure), we obtain (4).

Comments

(i) The conserved vectors $\mathbf{A}$ and $\mathbf{B}$ are dimensionless and they lie in the plane of the trajectory (they are perpendicular to $\mathbf{L}$).
(ii) These two vectors have been known and studied for a long time.[1,2] We will refer to them as the Laplace vector $\mathbf{A}$ and the Hamilton vector $\mathbf{B}$. (In the literature, $\mathbf{A}$ is also known as the Runge–Lenz vector, or the Hermann–Bernoulli–Laplace vector, or the second Laplace vector, and $\mathbf{B}$ is also called the first Laplace vector.[1,2])
(iii) A conserved quantity like $\mathbf{A}$ or $\mathbf{B}$ is referred to as a constant of the motion. Thus we have found that the Coulomb problem possesses a scalar constant of the motion E and three vector constants of the motion $\mathbf{L}$, $\mathbf{A}$ and $\mathbf{B}$; that is, ten constants in all. Of course, not all of these can be independent: the number of independent constants is determined in the next question.
(iv) The vectors $\mathbf{A}$ and $\mathbf{B}$ have interesting properties and applications, some of which are illustrated in the following three questions.

---

[1] H. Goldstein, "More on the prehistory of the Laplace or Runge-Lenz vector," American Journal of Physics, vol. 44, pp. 1123–1124, 1976.
[2] O. Volk, "Miscellanea from the history of celestial mechanics," Celestial Mechanics, vol. 14, pp. 365–382, 1976.

Question 9.2

Prove that the magnitudes of the Laplace and Hamilton vectors for the Coulomb problem satisfy

$$A^2 = B^2 = 1 + \frac{2L^2 E}{mk^2}, \tag{1}$$

where E is the energy

$$E = \frac{1}{2}m\left(\frac{dr}{dt}\right)^2 + \frac{L^2}{2mr^2} - \frac{k}{r}. \tag{2}$$

Solution

In the construction $\mathbf{A} = \hat{\mathbf{L}} \times \mathbf{B}$, the vector $\mathbf{B}$ is perpendicular to $\hat{\mathbf{L}}$ (see Question 9.1). It follows immediately that $A = B$. We evaluate B^2 by first expressing $\mathbf{B}$ in terms of radial and transverse components. From (7) of Question 9.1, and using

$$\mathbf{p} = m\dot{r}\hat{\mathbf{r}} + mr\dot{\theta}\hat{\boldsymbol{\theta}} = m\dot{r}\hat{\mathbf{r}} + (L/r)\hat{\boldsymbol{\theta}}, \tag{3}$$

(see Question 8.1) we have

$$\mathbf{B} = (L/k)\dot{r}\hat{\mathbf{r}} + \{(L^2/mkr) - 1\}\hat{\boldsymbol{\theta}}. \tag{4}$$

Now, $B^2 = \mathbf{B} \cdot \mathbf{B}$ and $\hat{\mathbf{r}} \cdot \hat{\boldsymbol{\theta}} = 0$, while $\hat{\mathbf{r}} \cdot \hat{\mathbf{r}} = \hat{\boldsymbol{\theta}} \cdot \hat{\boldsymbol{\theta}} = 1$. Thus, the scalar product of (4) with itself yields

$$B^2 = 1 + (L^2/k^2)\dot{r}^2 + L^4/m^2k^2r^2 - 2L^2/mkr, \tag{5}$$

which is (1) with E given by (2).

Comments

(i) The energy E is not an independent constant of the motion, being related to the magnitudes of the Laplace and Hamilton vectors by (1).
(ii) We can now specify the number of independent constants of the motion for the Coulomb problem. They are the six components of $\mathbf{L}$ and $\mathbf{A}$ (or $\mathbf{B}$) minus one because of the orthogonality condition

$$\mathbf{L} \cdot \mathbf{A} = 0. \tag{6}$$

That is, a total of five, which is the maximum number of independent constants of the motion that allows a continuous trajectory in phase space. It seems that this property of the Coulomb problem was first proved by Laplace.[1,2]

Question 9.3

Use the Laplace vector

$$\mathbf{A} = \hat{\mathbf{r}} + \frac{1}{mk}\mathbf{L} \times \mathbf{p} \tag{1}$$

to determine the polar equation $r = r(\theta)$ of the trajectory for the Coulomb problem.

Solution

The scalar product of **r** with (1) is

$$\mathbf{r} \cdot \mathbf{A} = r - \frac{1}{mk} L^2. \tag{2}$$

(Here, we have used the identity $\mathbf{r} \cdot (\mathbf{L} \times \mathbf{p}) = -\mathbf{L} \cdot (\mathbf{r} \times \mathbf{p}) = -L^2$.) But

$$\mathbf{r} \cdot \mathbf{A} = rA \cos\theta, \tag{3}$$

where θ is the angle between **r** and **A**. From (2) and (3) we obtain the polar equation of the trajectory

$$r(\theta) = \frac{r_0}{1 - e\cos\theta}, \tag{4}$$

where

$$r_0 = L^2/mk, \qquad e = A. \tag{5}$$

Comments

(i) Equation (4) is the polar equation of a conic section with *semilatus rectum* r_0 and eccentricity e. With A given by (1) of Question 9.2, the solution (4) is the same as that obtained by solving the equation of motion (see Question 8.9).

(ii) We see from (3) and (4) that the vector **A** points along the symmetry axis of the conic section (the line $\theta = 0$), and it is directed from a focus F away from the point of closest approach. This is illustrated for an elliptical orbit ($E < 0$) in the figure below. In the case of planetary motion, **A** is directed towards the aphelion. While **A** is along the major axis of the ellipse, **B** is parallel to the minor axis.

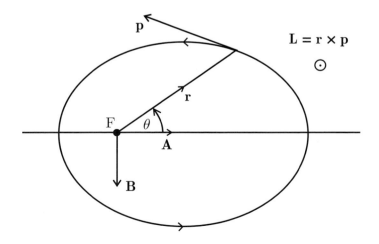

(iii) The same trajectory is obtained using the Hamilton vector **B**: simply take the scalar product of $\hat{\boldsymbol{\theta}}$ with (4) of Question 9.2,

$$\hat{\boldsymbol{\theta}} \cdot \mathbf{B} = \frac{L^2}{mkr} - 1, \tag{6}$$

and note that from the figure

$$\hat{\boldsymbol{\theta}} \cdot \mathbf{B} = B\cos(\pi - \theta) = -B\cos\theta. \tag{7}$$

Equations (6) and (7) yield (4) with $e = B = A$.

(iv) Thus, either the Laplace vector or the Hamilton vector can be used to obtain the trajectory $r = r(\theta)$ without solving a differential equation or performing any integration.

Question 9.4

(a) Use the Hamilton vector in the form

$$\mathbf{B} = \frac{L}{mk}\mathbf{p} - \hat{\boldsymbol{\theta}} \tag{1}$$

(see (7) of Question 9.1) to show that in momentum space the trajectories of the Coulomb problem are circular if $L \neq 0$.

(b) Sketch the trajectory in momentum space corresponding to each trajectory in configuration (coordinate) space.

Solution

(a) According to (1),

$$\mathbf{p} - mkL^{-1}\mathbf{B} = mkL^{-1}\hat{\boldsymbol{\theta}}. \tag{2}$$

Choose Cartesian axes such that $\mathbf{A} = (A, 0, 0)$ and $\mathbf{B} = (0, B, 0)$, and recall that $B = A$ (see Question 9.2). Then, (2) yields

$$p_x^2 + (p_y - mkL^{-1}A)^2 = (mkL^{-1})^2. \tag{3}$$

Thus, the trajectory in momentum space (the so-called hodograph) is a circle of radius mkL^{-1} with centre at $(0, mkL^{-1}A, 0)$.

(b) In each of the diagrams below, **B** points from the origin O in momentum space to the centre C of the hodograph. There are four cases:

$E = -mk^2/2L^2$, $A = 0$ (circular orbit in configuration space)

The centre of the hodograph is at the origin O of coordinates in momentum space. The magnitude p of the momentum is a constant, mk/L.

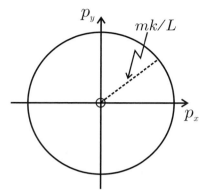

$-mk^2/2L^2 < E < 0,\ 0 < A < 1$ (elliptical orbit in configuration space)

The origin O is inside the hodograph.

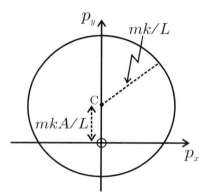

$E = 0,\ A = 1$ (parabolic trajectory in configuration space)

The origin O is situated on the hodograph.

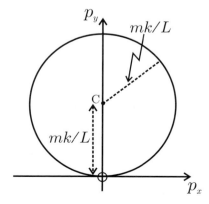

$E > 0$, $A > 1$ (hyperbolic trajectory in configuration space)

The origin O is outside the hodograph. The momentum at $r = \infty$ is $p_\infty = \sqrt{2mE} = (mk/L)\sqrt{A^2 - 1}$. This is greater[‡] than $(mk/L)(A - 1)$ if $A > 1$, and hence the hodograph is an incomplete circle. If the particle starts at $x = \infty$, $y = -\infty$ and ends at $x = \infty$, $y = \infty$ then the hodograph is traversed from a to b in the figure below.

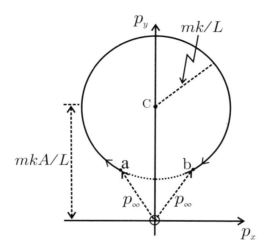

Comments

(i) If $\mathbf{L} = 0$, the hodograph is a straight line.
(ii) Starting with given initial conditions $\mathbf{r}_0$ and $\mathbf{v}_0$, and a given hodograph, one can construct the corresponding orbit in configuration space in a purely geometric ('quasi-Newtonian') manner.[3,4]
(iii) The circular nature of the hodograph was published by Hamilton in 1846.
(iv) The classical Laplace and Hamilton vectors found additional applications after the formulation of quantum mechanics in 1925. In 1926, Pauli[5] was the first to show how the new quantum mechanics could be used to obtain the energy levels of the hydrogen atom. His analysis is based on a quantum-mechanical analogue of the Laplace vector, namely the Pauli–Lenz vector operator

[‡]Because $A - 1 < \sqrt{A^2 - 1}$ if $A > 1$.

---

[3] J. Sivardière, "Comments on the dynamical invariants of the Kepler and harmonic motions," European Journal of Physics, vol. 13, pp. 64–69, 1992.
[4] A. González-Villanueva, E. Guillaumin-España, R. P. Martinez-y-Romero, H. N. Núñez-Yépez, and A. L. Salas-Brito, "From circular paths to elliptic orbits: a geometric approach to Kepler's motion," European Journal of Physics, vol. 19, pp. 431–438, 1998.
[5] W. Pauli, "Über das wasserstoffspektrum von standpunkt der neuen quantenmechanik," Zeitschrift fur Physik, vol. 36, pp. 336–363, 1926. English translation in B. L. van der Waerden (ed.), *Sources in quantum mechanics*. New York: Dover, 1968. pp. 387–415.

$$\mathbf{A} = \frac{\mathbf{r}}{r} + \frac{1}{2mk}(\mathbf{L} \times \mathbf{p} - \mathbf{p} \times \mathbf{L}). \qquad (4)$$

Here, $\mathbf{r}$ and $\mathbf{p}$ are the position and momentum operators, and the construction in brackets in (4) ensures that $\mathbf{A}$ is a Hermitian operator. It is remarkable that while the classical vector $\mathbf{A}$ played a role in the first direct solution of the Kepler problem in 1710 using the recently discovered classical mechanics,[1,2] about two centuries later the corresponding Hermitian vector operator $\mathbf{A}$ was an essential ingredient in the first solution for the hydrogen atom by the new quantum mechanics.[5]

(v) The parallel between the classical and quantum-mechanical developments extends further. A second vector operator $\mathbf{B}$ can be constructed as a quantum-mechanical analogue of the classical Hamilton vector and, together with $\mathbf{A}$, it is useful in an algebraic formulation of the Coulomb problem in an angular-momentum basis.[6] For example, the combinations $\mathbf{A} \pm i\mathbf{B}$ provide transformations (so-called shift operations) that generate all the bound-state coordinate-space and momentum-space wavefunctions of the Coulomb problem in this basis.[6,7]

Question 9.5

Starting with the equation of motion of a three-dimensional isotropic harmonic oscillator

$$\frac{dp_i}{dt} = -kr_i \qquad (i = 1, 2, 3), \qquad (1)$$

deduce the conservation equation

$$\frac{dA_{ij}}{dt} = 0, \qquad (2)$$

where

$$A_{ij} = \frac{1}{2m} p_i p_j + \tfrac{1}{2} k r_i r_j. \qquad (3)$$

(Note that we will use the notations r_1, r_2, r_3 and x, y, z interchangeably, and similarly for the components of $\mathbf{p}$.)

Solution

Multiply (1) on both sides by $p_j = m\dot{r}_j$. Then

$$p_j \dot{p}_i = -mkr_i \dot{r}_j. \qquad (4)$$

Add to (4) the same equation, but with i and j interchanged, and divide the result by $2m$. This gives

$$\frac{1}{2m}(p_i \dot{p}_j + \dot{p}_i p_j) = -\tfrac{1}{2}k(r_i \dot{r}_j + \dot{r}_i r_j), \qquad (5)$$

which is (2) with A_{ij} given by (3).

---

[6] O. L. de Lange and R. E. Raab, "Coulomb problem in an angular momentum basis: An algebraic formulation," Physical Review, vol. A37, pp. 1858–1868, 1988.

[7] O. L. de Lange and R. E. Raab, *Operator methods in quantum mechanics*. Oxford: Clarendon Press, 1991.

Comments

(i) The symmetric second-rank tensor A_{ij} is known as the Jauch–Hill–Fradkin (JHF) tensor.[8,9] It has interesting properties and applications, some of which are illustrated in Questions 9.6 to 9.9.

(ii) After studying these questions, it should be clear to the reader that the role played by A_{ij} in the theory of the isotropic harmonic oscillator is analogous to that of the Laplace vector **A** and the Hamilton vector **B** in the Coulomb problem (see Questions 9.1 to 9.4).

(iii) The reader may wonder whether Laplace- and Hamilton-type vectors can be constructed for the oscillator. The answer is that they can but it is simpler to use the tensor A_{ij}.[3,10] The Laplace- and Hamilton-type vectors are, in fact, eigenvectors of A_{ij} (see below).

Question 9.6

Prove the following algebraic properties of the JHF tensor A_{ij} defined in Question 9.5:

(a) Trace $A = E$, (1)

(b) $L_i A_{ij} = A_{ij} L_j = 0$, (2)

(c) Det $A = 0$, (3)

(d) $r_i A_{ij} r_j = r^2 E - L^2/2m$. (4)

In (2) and (4) a repeated subscript implies summation from 1 to 3.

Solution

(a) Trace $A = A_{11} + A_{22} + A_{33} = \dfrac{1}{2m}\left(p_1^2 + p_2^2 + p_3^2\right) + \tfrac{1}{2}k\left(r_1^2 + r_2^2 + r_3^2\right)$

$\qquad = E$.

(b) Recall that the components of a vector product such as $\mathbf{L} = \mathbf{r}\times\mathbf{p}$ can be written as

$$L_i = \varepsilon_{ijk} r_j p_k, \qquad (5)$$

where ε_{ijk} is the Levi-Civita tensor defined by

$$\varepsilon_{ijk} = \begin{cases} 1 & ijk = \text{any even permutation of } 1,2,3 \\ -1 & ijk = \text{any odd permutation of } 1,2,3 \\ 0 & \text{if any two subscripts are equal.} \end{cases} \qquad (6)$$

---

[8] J. M. Jauch and E. L. Hill, "On the problem of degeneracy in quantum mechanics," Physical Review, vol. 57, pp. 641–645, 1940.

[9] D. M. Fradkin, "Three-dimensional isotropic harmonic oscillator and SU_3," American Journal of Physics, vol. 33, pp. 207–211, 1965.

[10] L. H. Buch and H. H. Denman, "Conserved and piecewise-conserved Runge vectors for the isotropic harmonic oscillator," American Journal of Physics, vol. 43, pp. 1046–1048, 1975.

Then
$$L_i A_{ij} = \varepsilon_{ikl} r_k p_l \left(\frac{1}{2m} p_i p_j + \tfrac{1}{2} k r_i r_j \right) = 0 \qquad (7)$$

because $\varepsilon_{ikl} p_l p_i = \tfrac{1}{2} (\varepsilon_{ikl} p_l p_i + \varepsilon_{lki} p_i p_l) = \tfrac{1}{2} \varepsilon_{ikl} (p_l p_i - p_i p_l) = 0$, and similarly $\varepsilon_{ikl} r_k r_i = 0$. It also follows that $A_{ij} L_j = A_{ji} L_i = A_{ij} L_i = 0$, where we have used the symmetry of A_{ij} and (7).

(c) $\text{Det } A = \begin{vmatrix} A_{11} & A_{12} & A_{13} \\ A_{12} & A_{22} & A_{23} \\ A_{13} & A_{23} & A_{33} \end{vmatrix} = A_{11} \begin{vmatrix} A_{22} & A_{23} \\ A_{23} & A_{33} \end{vmatrix} - A_{12} \begin{vmatrix} A_{12} & A_{23} \\ A_{13} & A_{33} \end{vmatrix} + A_{13} \begin{vmatrix} A_{12} & A_{22} \\ A_{13} & A_{23} \end{vmatrix}$

$$= (A_{11} L_1^2 + A_{12} L_1 L_2 + A_{13} L_1 L_3) \times k/4m$$
$$= L_1 (A_{1i} L_i) \times k/4m$$
$$= 0$$

because of (2).

(d) $r_i A_{ij} r_j = \dfrac{1}{2m} r_i p_i p_j r_j + \tfrac{1}{2} k r_i r_i r_j r_j = \dfrac{1}{2m} (\mathbf{r} \cdot \mathbf{p})^2 + \tfrac{1}{2} k r^4. \qquad (8)$

Now[‡]
$$(\mathbf{r} \cdot \mathbf{p})^2 = (\mathbf{r} \cdot \mathbf{r})(\mathbf{p} \cdot \mathbf{p}) - (\mathbf{r} \times \mathbf{p})^2 = r^2 p^2 - L^2. \qquad (9)$$

Equations (8) and (9) yield (4).

Comments

(i) The oscillator possesses ten constants of the motion: the scalar E, the vector $\mathbf{L}$ (see Questions 5.3 and 6.15) and the symmetric tensor A_{ij} (see Question 9.5). Equations (1)–(3) provide five relations connecting these, and consequently there are five independent constants of the motion (which we can take to be $\mathbf{L}$ and two components of A_{ij} – see below). Thus, the oscillator, like the Coulomb problem, possesses the maximum number of constants of the motion that allows a continuous trajectory in phase space (see Question 9.2).

(ii) Equation (4) is known as the orbit equation because it contains only $\mathbf{r}$ and conserved quantities, and from it the equation of the orbit can be obtained (see Question 9.8).

Question 9.7

Consider the eigenvalue equation
$$A_{ij} u_j = \lambda u_i. \qquad (1)$$

Determine the eigenvalues λ.

[‡]Use the vector identity: $(\mathbf{a} \times \mathbf{b}) \cdot (\mathbf{c} \times \mathbf{d}) = (\mathbf{a} \cdot \mathbf{c})(\mathbf{b} \cdot \mathbf{d}) - (\mathbf{a} \cdot \mathbf{d})(\mathbf{b} \cdot \mathbf{c}).$

Solution

The condition that the secular determinant should vanish, namely

$$\begin{vmatrix} A_{11} - \lambda & A_{12} & A_{13} \\ A_{12} & A_{22} - \lambda & A_{23} \\ A_{13} & A_{23} & A_{33} - \lambda \end{vmatrix} = 0, \qquad (2)$$

yields

$$\lambda^3 - (\text{Trace } A)\lambda^2 + \left(\begin{vmatrix} A_{11} & A_{12} \\ A_{12} & A_{22} \end{vmatrix} + \begin{vmatrix} A_{22} & A_{23} \\ A_{23} & A_{33} \end{vmatrix} + \begin{vmatrix} A_{11} & A_{13} \\ A_{13} & A_{33} \end{vmatrix} \right) \lambda - (\text{Det } A) = 0. \qquad (3)$$

The three 2×2 determinants in (3) are equal to $kL_3^2/4m$, $kL_1^2/4m$ and $kL_2^2/4m$, respectively, and hence their sum is $kL^2/4m = \omega^2 L^2/4$. Also, Trace $A = E$ and Det $A = 0$ (see Question 9.6). Thus, (3) can be written

$$\lambda(\lambda^2 - E\lambda + \tfrac{1}{4}\omega^2 L^2) = 0, \qquad (4)$$

and the three eigenvalues are

$$\lambda^{(1)} = \tfrac{1}{2}\left(E + \sqrt{E^2 - \omega^2 L^2}\right), \qquad \lambda^{(2)} = \tfrac{1}{2}\left(E - \sqrt{E^2 - \omega^2 L^2}\right), \qquad \lambda^{(3)} = 0. \qquad (5)$$

Comments

(i) The two non-zero eigenvalues satisfy

$$\lambda^{(1)} + \lambda^{(2)} = E, \qquad \lambda^{(1)} \lambda^{(2)} = \tfrac{1}{4}\omega^2 L^2. \qquad (6)$$

(ii) The scalar product of **L** with both sides of (1) for the eigenvectors $\mathbf{u}^{(1)}$ and $\mathbf{u}^{(2)}$ corresponding to eigenvalues $\lambda^{(1)}$ and $\lambda^{(2)}$ yields $\mathbf{L} \cdot \mathbf{u}^{(1)} = 0$ and $\mathbf{L} \cdot \mathbf{u}^{(2)} = 0$ (because $L_i A_{ij} = 0$, see Question 9.6). Thus, $\mathbf{u}^{(1)}$ and $\mathbf{u}^{(2)}$ lie in the plane of the orbit. For the eigenvector $\mathbf{u}^{(3)}$ corresponding to eigenvalue $\lambda^{(3)} = 0$, (1) is $A_{ij} u_j^{(3)} = 0$. Thus, $\mathbf{u}^{(3)}$ is in the direction of **L**.

Question 9.8

Determine the trajectory of an isotropic harmonic oscillator by solving the orbit equation (see Question 9.6)

$$r_i A_{ij} r_j = r^2 E - L^2/2m \qquad (1)$$

in Cartesian coordinates.

Solution

Choose Cartesian axes xyz such that $\mathbf{L} = (0, 0, L)$ and A_{ij} is diagonal:

$$A_{ij} = \begin{pmatrix} p_x^2/2m + \frac{1}{2}kx^2 & 0 & 0 \\ 0 & p_y^2/2m + \frac{1}{2}ky^2 & 0 \\ 0 & 0 & 0 \end{pmatrix}. \qquad (2)$$

In this system of coordinates $A_{12} = 0$, meaning that $p_x p_y = -mkxy$. The sum of the two entries in (2) is equal to E, and their product is equal to $kL^2/4m = \omega^2 L^2/4$. Thus, (see (6) of Question 9.7)

$$A_{ij} = \begin{pmatrix} \lambda^{(1)} & 0 & 0 \\ 0 & \lambda^{(2)} & 0 \\ 0 & 0 & 0 \end{pmatrix}. \qquad (3)$$

For the above choice of axes, the trajectory lies in the xy-plane. So, $\mathbf{r} = (x, y, 0)$ and $r^2 = x^2 + y^2$. Then (1) and (3), and using also (6) of Question 9.7, gives

$$\lambda^{(1)} x^2 + \lambda^{(2)} y^2 = (x^2 + y^2)E - L^2/2m = (x^2 + y^2)(\lambda^{(1)} + \lambda^{(2)}) - 2\lambda^{(1)}\lambda^{(2)}/k.$$

Thus, we have for the trajectory in Cartesian form the ellipse

$$\frac{x^2}{a^2} + \frac{y^2}{b^2} = 1, \qquad (4)$$

where

$$a = \sqrt{2\lambda^{(1)}/k}, \qquad b = \sqrt{2\lambda^{(2)}/k}. \qquad (5)$$

Comments

(i) We see that, just as in the Coulomb problem, we are able to determine the trajectory without solving a differential equation or doing any integration, but just by using the extra constants of the motion. The trajectory in momentum space (the hodograph) can also be obtained in this manner (see Question 9.9).

(ii) In the above system of coordinates, two of the eigenvectors of A_{ij} are

$$\mathbf{u}^{(1)} = \alpha \hat{\mathbf{x}}, \qquad \mathbf{u}^{(2)} = \beta \hat{\mathbf{y}}, \qquad (6)$$

where α and β are constants (see Question 9.7). Thus, the major and minor axes of the elliptical orbit are along the eigenvectors $\mathbf{u}^{(1)}$ and $\mathbf{u}^{(2)}$ of A_{ij}, while the lengths of these axes are equal to $2a = \sqrt{8\lambda^{(1)}/k}$ and $2b = \sqrt{8\lambda^{(2)}/k}$. In this way, the tensor A_{ij} specifies the orbit of the oscillator.

(iii) In an arbitrary system of coordinates the eigenvectors are[3,10]

$$\mathbf{u}^{(1)} = \frac{\mathbf{p} \times \mathbf{L} - mka^2 \mathbf{r}}{mk\sqrt{(a^2 - b^2)(r^2 - b^2)}} \qquad (7)$$

$$\mathbf{u}^{(2)} = \frac{\mathbf{p} \times \mathbf{L} - mkb^2 \mathbf{r}}{mk\sqrt{(a^2 - b^2)(a^2 - r^2)}}. \qquad (8)$$

These eigenvectors have been normalized such that

$$\mathbf{u}^{(1)} \cdot \mathbf{u}^{(1)} = a^2 \quad \text{and} \quad \mathbf{u}^{(2)} \cdot \mathbf{u}^{(2)} = b^2.$$

In the system in which A_{ij} is diagonal, (7) and (8) reduce to (6) with $\alpha = -a\,\text{sign}\,x$, $\beta = b\,\text{sign}\,y$.

(iv) The conserved vectors (7) and (8) are Laplace vectors for the oscillator. There are two of them because the origin (the position of the centre of force) is at the centre of the ellipse, and consequently there are two aphelia and two perihelia: $\mathbf{u}^{(1)}$ is directed between the aphelia, while $\mathbf{u}^{(2)}$ is between the perihelia. Two Hamilton vectors can be constructed from $\hat{\mathbf{L}} \times \mathbf{u}^{(1)}$ and $\hat{\mathbf{L}} \times \mathbf{u}^{(2)}$. The trajectory is obtained by evaluating $\mathbf{r} \cdot \mathbf{u}^{(1)}$, and the hodograph by calculating $\mathbf{p} \cdot (\hat{\mathbf{L}} \times \mathbf{u}^{(1)})$.

Question 9.9

Determine the hodograph (the trajectory in momentum space) of an isotropic harmonic oscillator.

Solution

In the same manner as the proof of (4) in Question 9.6, we can show that

$$p_i A_{ij} p_j = p^2 E - \tfrac{1}{2}kL^2. \tag{1}$$

With the same choice of axes as in the previous question, we have from (1)

$$\lambda^{(1)} p_x^2 + \lambda^{(2)} p_y^2 = (p_x^2 + p_y^2)(\lambda^{(1)} + \lambda^{(2)}) - 2m\lambda^{(1)}\lambda^{(2)}.$$

Thus, the hodograph is the elliptical trajectory

$$\frac{p_x^2}{2m\lambda^{(1)}} + \frac{p_y^2}{2m\lambda^{(2)}} = 1. \tag{2}$$

Comments

(i) This completes our examples on extra constants of the motion for the Coulomb and oscillator problems. We conclude with some brief remarks on the connection between these constants, and symmetry and degeneracy.

(ii) The classical dynamics of a particle moving in a spherically symmetric potential $V(r)$ possesses symmetry under rotations of the coordinate system. As a result, orbits differing only in their spatial orientation are degenerate, meaning that they all have the same energy. The symmetry and the degeneracy are associated with the property that the angular momentum $\mathbf{L}$ is a constant of the motion. The three constants L_i generate the symmetry by transforming a trajectory of a given energy into another trajectory of the same energy by means of a contact transformation.[11] The existence of a constant of the motion implies a degeneracy, and vice versa.[12,13] The above degeneracy arises from an obvious geometric symmetry of all potentials $V(r)$ – it is an example of a 'geometric degeneracy'. Many, but not all, degeneracies are of this type.

(iii) Certain potentials $V(r)$ possess additional constants of the motion – for example, the Laplace vector **A** of the Coulomb potential and the JHF tensor A_{ij} for the isotropic harmonic oscillator. These constants generate additional symmetries known as 'hidden symmetries', and as a result of this extra symmetry there is additional degeneracy that is often referred to as 'accidental degeneracy'. For example, for bounded motion in the Coulomb problem the energy $E = -k/2a$ (see Question 8.9) depends only on the length a of the semi-major axis of the ellipse – it is independent of the eccentricity of the orbit. This accidental degeneracy is associated with the hidden symmetry generated by the Laplace vector **A**. If the constancy of **A** is disturbed for any reason, the hidden symmetry is broken and the accidental degeneracy is lifted.

(iv) We emphasize that **A** is constant only for non-relativistic motion in a Coulomb potential. In relativistic theory **A** is not constant: it rotates in a plane perpendicular to **L**, and consequently the elliptical orbits of the non-relativistic approximation precess in this plane (see Question 15.15). Similarly, departures from a Coulomb potential cause **A** to rotate.[14]

(v) The constants of the motion generate symmetry groups. The 'geometric' symmetry group associated with **L** is O(3), and this is enlarged to O(4) by including the hidden symmetry of **A**.[11]

(vi) All these ideas carry over to, and are even more vivid in, quantum mechanics. For example, the bound states of the non-relativistic hydrogen atom are labelled (in an angular momentum basis) by quantum numbers n, ℓ, m_ℓ. Here, ℓ and m_ℓ are quantum numbers specifying the eigenvalues of the operators $\mathbf{L}^2$ and L_z. The energy eigenvalues are independent of ℓ and m_ℓ (see (11) of Question 7.22). The degeneracy in m_ℓ (which is typical of all spherically symmetric potentials V(r)) is associated with the geometric symmetry O(3) generated by **L** (operators $L_x \pm iL_y$ change m_ℓ but not E in the eigenkets $|n\,\ell\,m_\ell\rangle$, so states with different m_ℓ have the same energy). The accidental degeneracy in ℓ is associated with the additional symmetry generated by the Pauli–Lenz vector operator **A** (see Question 9.4). This operator (which is a quantum-mechanical analogue of the classical Laplace vector) changes ℓ but not E in the eigenkets $|n\,\ell\,m_\ell\rangle$.[5,7] Relativistic effects break the hidden symmetry generated by **A** and consequently they lift the degeneracy in ℓ (see, for example, Ref. [7]).

---

[11] H. Goldstein, *Classical mechanics*. Reading: Addison-Wesley, 2nd edn, 1980.
[12] D. F. Greenberg, "Accidental degeneracy," American Journal of Physics, vol. 34, pp. 1101–1109, 1966.
[13] D. F. Greenberg, "Symmetry origin of dynamics," American Journal of Physics, vol. 35, pp. 1073–1077, 1967.
[14] K. T. McDonald, C. Farina, and A. Tort, "Right and wrong use of the Lenz vector for non-Newtonian potentials," American Journal of Physics, vol. 58, pp. 540–542, 1990.

Question 9.10

Consider motion of an isotropic harmonic oscillator in a plane with Cartesian coordinates (ξ, η) and time variable τ. The equation of motion is

$$\left(\frac{dp_\xi}{d\tau}, \frac{dp_\eta}{d\tau}\right) = -k(\xi, \eta), \tag{1}$$

where

$$p_\xi = m\frac{d\xi}{d\tau}, \qquad p_\eta = m\frac{d\eta}{d\tau}. \tag{2}$$

Let

$$\zeta = \xi + i\eta, \qquad p_\zeta = p_\xi + ip_\eta, \tag{3}$$

and consider the following transformation from coordinates (ξ, η, τ) to coordinates (x, y, t)

$$z = \zeta^2 \tag{4}$$

$$\frac{dt}{d\tau} = |\zeta|^2, \tag{5}$$

where $z = x + iy$. Show that the equation of motion (1) is transformed into

$$\left(\frac{dP_x}{dt}, \frac{dP_y}{dt}\right) = -\left(\frac{\partial V_\mathrm{D}}{\partial x}, \frac{\partial V_\mathrm{D}}{\partial y}\right), \tag{6}$$

where the momenta are

$$P_x = M\frac{dx}{dt}, \qquad P_y = M\frac{dy}{dt}, \tag{7}$$

with $M = \frac{1}{4}m$, and

$$V_\mathrm{D} = -\frac{\gamma}{r} \tag{8}$$

is a Coulomb potential with the constant γ equal to the energy of the oscillator.

Solution

According to (2) and (3)

$$p_\zeta = m\frac{d\zeta}{d\tau}, \tag{9}$$

and the equation of motion (1) is

$$\frac{dp_\zeta}{d\tau} = -k\zeta. \tag{10}$$

Let

$$P_z = P_x + iP_y = M\frac{dz}{dt}. \tag{11}$$

This can be expressed in terms of oscillator variables by using (4), (5) and (9):

$$P_z = M \frac{d\tau}{dt} \frac{d\zeta^2}{d\tau} = \frac{2M}{m} \frac{p_\zeta}{\zeta^*}, \qquad (12)$$

where $\zeta^* = \xi - i\eta$. Then

$$\frac{dP_z}{dt} = \frac{d\tau}{dt} \frac{dP_z}{d\tau}$$

$$= \frac{2M}{m} \frac{1}{|\zeta|^2} \left(\frac{1}{\zeta^*} \frac{dp_\zeta}{d\tau} - \frac{1}{(\zeta^*)^2} p_\zeta \frac{d\zeta^*}{d\tau} \right)$$

$$= \frac{2M}{m|\zeta|^2} \left(-k \frac{\zeta}{\zeta^*} - \frac{|p_\zeta|^2}{m(\zeta^*)^2} \right), \qquad (13)$$

where we have used (9) and (10) in the last step. The energy of the oscillator is

$$E_o = \frac{1}{2m} \left(p_\xi^2 + p_\eta^2 \right) + \tfrac{1}{2} k \left(\xi^2 + \eta^2 \right) \qquad (14)$$

$$= \frac{1}{2m} |p_\zeta|^2 + \tfrac{1}{2} k |\zeta|^2, \qquad (15)$$

and therefore

$$|p_\zeta|^2 = 2mE_o - mk|\zeta|^2. \qquad (16)$$

From (13) and (16) we have

$$\frac{dP_z}{dt} = -\frac{4ME_o}{m} \frac{1}{\zeta(\zeta^*)^3} = -\frac{4ME_o}{m} \frac{\zeta^2}{(\zeta\zeta^*)^3} = -\frac{4ME_o}{m} \frac{\xi^2 - \eta^2 + 2i\xi\eta}{(\xi^2 + \eta^2)^3}. \qquad (17)$$

Now, the real and imaginary parts of (4) are

$$x = \xi^2 - \eta^2, \qquad y = 2\xi\eta. \qquad (18)$$

It follows that

$$r = \sqrt{x^2 + y^2} = \xi^2 + \eta^2. \qquad (19)$$

Equations (18) and (19) enable us to express (17) in terms of x and y as

$$\frac{dP_z}{dt} = -\frac{4ME_o}{m} \frac{x + iy}{r^3}. \qquad (20)$$

If we choose $M = \tfrac{1}{4} m$ and recognize that

$$\frac{\partial}{\partial x} \frac{1}{r} = -\frac{x}{r^3} \qquad \text{and} \qquad \frac{\partial}{\partial y} \frac{1}{r} = -\frac{y}{r^3}, \qquad (21)$$

we see that the real and imaginary parts of (20) yield the desired form (6).

Comments

(i) The coordinate transformation (4)–(5) is an example of a duality transformation. Its application to the above example illustrates further the interesting connections between the oscillator and Coulomb problems.

(ii) Duality transformations have been studied by Arnold and others.[15–17] In general, these transformations relate the orbits of dual potentials. Thus, the above example shows that the Coulomb potential is dual to the isotropic oscillator potential and vice versa.

(iii) If we were to start with an anisotropic oscillator, that is with

$$\left(\frac{dp_\xi}{d\tau}, \frac{dp_\eta}{d\tau}\right) = -(k_1\xi, \, k_2\eta) \tag{22}$$

instead of (1), then the dual potential is a non-central perturbation of the Coulomb potential[17]

$$V_D(x, y) = -\frac{\gamma}{r} + \frac{1}{4}(k_1 - k_2)\frac{x}{r}, \tag{23}$$

where $\gamma = E_0$, the energy of the anisotropic oscillator. Thus, by transforming the orbits of an anisotropic oscillator one can obtain the orbits for the non-central potential (23).[17]

Question 9.11

(a) Apply the duality transformation (4)–(5) of Question 9.10 to the isotropic oscillator orbits in (ξ, η, τ) coordinates:

$$(\xi(\tau), \, \eta(\tau)) = (A\cos\omega\tau, \, B\sin\omega\tau) \tag{1}$$

where $\omega = \sqrt{k/m}$, to obtain the transformed orbits in (x, y, t) coordinates.

(b) Discuss these transformed orbits.

Solution

(a) The transformation (4) of Question 9.10 means

$$x = \xi^2 - \eta^2, \qquad y = 2\xi\eta. \tag{2}$$

From (1) and (2) we obtain

$$x(\tau) = \tfrac{1}{2}(A^2 - B^2) + \tfrac{1}{2}(A^2 + B^2)\cos 2\omega\tau \tag{3}$$

$$y(\tau) = AB\sin 2\omega\tau. \tag{4}$$

Also, by integrating (5) of Question 9.10, and requiring $t = 0$ at $\tau = 0$, we have

$$t(\tau) = \tfrac{1}{2}(A^2 + B^2)\tau + \tfrac{1}{2}(A^2 - B^2)(\sin 2\omega\tau)/2\omega. \tag{5}$$

---

[15] V. I. Arnold, *Huygens and Barrow, Newton and Hooke.* Basel: Birkhäuser, 1990.
[16] T. Needham, *Visual complex analysis.* Oxford: Oxford University Press, 1997.
[17] D. R. Stump, "A solvable non-central perturbation of the Kepler problem," European Journal of Physics, vol. 19, pp. 299–305, 1998.

(b) Equations (3)–(5) are a set of parametric equations for the Coulomb problem, provided by the duality transformation, and we now proceed to interpret them.[17] Suppose that $B > A$ (that is, B is the length of the semi-major axis of the elliptical orbit of the oscillator). Denote

$$a = \tfrac{1}{2}(A^2 + B^2), \qquad e = \frac{B^2 - A^2}{B^2 + A^2}, \qquad \psi = 2\omega\tau. \tag{6}$$

Also, note that the force constant γ in the dual potential $V_D = -\gamma/r$ (see Question 9.10) is equal to the energy of the oscillator:

$$\gamma = \tfrac{1}{2}m\omega^2(A^2 + B^2) = m\omega^2 a. \tag{7}$$

Using (6) and (7), the parametric equations (3)–(5) can be expressed as

$$x(\psi) = a(\cos\psi - e) \tag{8}$$

$$y(\psi) = a\sqrt{1 - e^2}\sin\psi \tag{9}$$

$$t(\psi) = \sqrt{Ma^3/\gamma}\,(\psi - e\sin\psi). \tag{10}$$

These are just the parametric equations for bounded motion in a Coulomb potential, expressed in terms of the eccentric anomaly ψ (see Question 8.18).

Comment

There have been many studies of transformations between the Coulomb and oscillator problems, both in classical and quantum mechanics. Some references can be found in Ref. [7]. Among these, considerable attention has been devoted to the so-called Kustaanheimo–Stiefel transformation that transforms the Coulomb (or Kepler) problem into a four-dimensional harmonic oscillator with a constraint.[18] This transformation has been known for a long time in celestial mechanics[2] and has more recently been applied also to the quantum-mechanical case.[19]

Question 9.12

Consider bounded motion in an effective potential $U(r)$ that has a minimum value $U_0 = U(r_0)$ at $r = r_0$. The turning points of the motion are $r_1(U)$ and $r_2(U)$.

(a) Show that the formal solution of Question 8.3, namely

$$\theta(r) = \theta(r_0) + \int_{r_0}^{r} \frac{L}{mr^2}\frac{dr}{\sqrt{(2/m)(E - U)}}, \tag{1}$$

---

[18] E. L. Stiefel and G. Scheifle, *Linear and regular celestial mechanics*. Berlin: Springer, 1971. See references therein.
[19] M. Kibler and T. Négadi, "Connection between the hydrogen atom and the harmonic oscillator: The zero-energy case," Physical Review, vol. A29, pp. 2891–2894 and references therein, 1984.

can be inverted to yield

$$\frac{1}{r_1(U)} - \frac{1}{r_2(U)} = \frac{1}{\pi L}\sqrt{\frac{m}{2}} \int_{U_0}^{U} \frac{\Delta\theta(E)}{\sqrt{U-E}} dE . \qquad (2)$$

Here, $\Delta\theta(E)$ is the change in θ in the complete journey $r_2 \to r_1 \to r_2$.

(b) Deduce that for closed orbits (2) becomes

$$\frac{1}{r_1(U)} - \frac{1}{r_2(U)} = \frac{2\sqrt{2m}}{\alpha L}\sqrt{U - U_0} , \qquad (3)$$

where α is a rational number $(= q/p$, where p and q are integers).

(c) By expanding both sides of (3) up to fourth order in $x = r_2(U) - r_0$, show that

$$U^{(2)} = \frac{\alpha^2 L^2}{m r_0^4} \qquad (4)$$

$$U^{(4)} = \frac{3\alpha^2 L^2}{m r_0^4}\left(5c^2 + 8\frac{c}{r_0} + \frac{8}{r_0^2}\right), \qquad (5)$$

where $U^{(n)}$ denotes the nth derivative $d^n U/dr^n$ evaluated at $r = r_0$, and

$$c = U^{(3)}/3U^{(2)} . \qquad (6)$$

(d) Use (4) and (5) to deduce that the only spherically symmetric potentials $V(r)$ that allow closed orbits for a range of initial conditions (E and L) are the Coulomb and oscillator potentials $V = -k/r$ and $V = \frac{1}{2}kr^2$.

Solution

(a) The solution given below follows that of Tikochinsky.[20] The changes in θ from $r_2 \to r_1$ and from $r_1 \to r_2$ are equal. Thus, from (1) we have

$$\Delta\theta(E) = 2\int_{r_1}^{r_2} \frac{L}{mr^2} \frac{dr}{\sqrt{(2/m)(E-U)}} . \qquad (7)$$

In (7) we change to an integration over U, taking care to integrate separately over the two branches $r_1(U)$ and $r_2(U)$ shown in the figure below:

$$\Delta\theta(E) = \sqrt{\frac{2}{m}} L \left(\int_E^{U_0} \frac{1}{r_1^2(U)} \frac{dr_1}{dU} \frac{dU}{\sqrt{E-U}} + \int_{U_0}^E \frac{1}{r_2^2(U)} \frac{dr_2}{dU} \frac{dU}{\sqrt{E-U}} \right)$$

$$= \sqrt{\frac{2}{m}} L \int_{U_0}^E \frac{dG}{dU} \frac{dU}{\sqrt{E-U}} ,$$

where

$$G(u) = \frac{1}{r_1(U)} - \frac{1}{r_2(U)} .$$

---

[20] Y. Tikochinsky, "A simplified proof of Bertrand's theorem," American Journal of Physics, vol. 56, pp. 1073–1075, 1988.

282 Solved Problems in Classical Mechanics

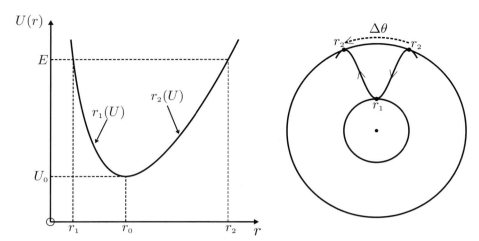

We invert this integral equation by dividing both sides by $\sqrt{\overline{U} - E}$ and integrating over E between U_0 and $\overline{U}$ to obtain

$$\int_{U_0}^{\overline{U}} \frac{\Delta\theta(E)}{\sqrt{\overline{U} - E}} dE = \sqrt{\frac{2}{m}} L \int_{U_0}^{\overline{U}} \frac{dE}{\sqrt{\overline{U} - E}} \int_{U_0}^{E} \frac{dG}{dU} \frac{dU}{\sqrt{E - U}}$$

$$= \sqrt{\frac{2}{m}} L \int_{U_0}^{\overline{U}} \frac{dG}{dU} dU \int_{U}^{\overline{U}} \frac{dE}{\sqrt{(\overline{U} - E)(E - U)}}, \qquad (8)$$

where in the last step we have used the relation

$$\int_{y=a}^{b} \int_{x=a}^{y} f(x, y) \, dxdy = \int_{x=a}^{b} \int_{y=x}^{b} f(x, y) \, dxdy \,.$$

The integral with respect to E in (8) is equal to π, and therefore the integral with respect to U is equal to $G(\overline{U}) - G(U_0) = G(\overline{U})$ because $r_1(U_0) = r_2(U_0) = r_0$. Replacing $\overline{U}$ with U, we obtain (2) from (8).

(b) For a bounded orbit to be closed we must have

$$\Delta\theta = 2\pi \frac{p}{q} \qquad (9)$$

where p and q are integers; this ensures that after the oscillatory motion completes q cycles, the angle θ will have changed by p complete revolutions. From (2) and (9) we obtain (3).

(c) In (3), U is evaluated at r_1 or r_2. Write $r_2(U) = r_0 + x$ and expand U up to fourth order in x (the reason for expanding to this order will become apparent below):

$$U - U_0 = \tfrac{1}{2} U^{(2)} x^2 + \tfrac{1}{6} U^{(3)} x^3 + \tfrac{1}{24} U^{(4)} x^4 + O(x^5), \qquad (10)$$

since $U^{(1)} = 0$. Next, we must expand the left-hand side of (3) to fourth order in x. This requires some care because $r_1(U) = r_0 - y$, where y is a function of x. To determine this function we expand

$$U - U_0 = \tfrac{1}{2}U^{(2)}y^2 - \tfrac{1}{6}U^{(3)}y^3 + \tfrac{1}{24}U^{(4)}y^4 + O(y^5). \tag{11}$$

One can easily show that consistency of (10) and (11) requires

$$y = x + cx^2 + c^2 x^3, \tag{12}$$

where c is given by (6). Then

$$\left(\frac{1}{r_0 - y} - \frac{1}{r_0 + x}\right)^2 = \frac{1}{r_0^4}\left(y + x + \frac{y^2 - x^2}{r_0} + \frac{y^3 + x^3}{r_0^2} + \cdots\right)^2$$

$$= \frac{1}{r_0^4}\left(2x + cx^2 + c^2 x^3 + \frac{2cx^3}{r_0} + \frac{2x^3}{r_0^2} + \cdots\right)^2$$

$$= \frac{1}{r_0^4}\left(4x^2 + 4cx^3 + \left\{5c^2 + \frac{8c}{r_0} + \frac{8}{r_0^2}\right\}x^4\right.$$

$$\left. + O(x^5)\right). \tag{13}$$

From (3), (10) and (13) we have

$$\frac{1}{r_0^4}\left(4x^2 + 4cx^3 + \left\{5c^2 + \frac{8c}{r_0} + \frac{8}{r_0^2}\right\}x^4\right)$$

$$= \frac{8m}{\alpha^2 L^2}\left(\tfrac{1}{2}U^{(2)}x^2 + \tfrac{1}{6}U^{(3)}x^3 + \tfrac{1}{24}U^{(4)}x^4\right).$$

Equating coefficients of the same powers of x we obtain two independent equations

$$U^{(2)} = \frac{\alpha^2 L^2}{mr_0^4} \tag{14}$$

$$U^{(4)} = \frac{3\alpha^2 L^2}{mr_0^4}\left(5c^2 + \frac{8c}{r_0} + \frac{8}{r_0^2}\right). \tag{15}$$

(Because of (6), the coefficients of x^3 yield the same relation as the coefficients of x^2, namely (14).)

(d) Equations (14) and (15) are conditions on the effective potential

$$U(r) = V(r) + \frac{L^2}{2mr^2}, \tag{16}$$

and hence on the potential $V(r)$. From (16) we have

$$U^{(2)} = V^{(2)} + \frac{3L^2}{mr_0^4}. \tag{17}$$

Thus, (14) requires

$$V^{(2)} = \frac{(\alpha^2 - 3)L^2}{mr_0^4}. \tag{18}$$

Also, $U(r)$ in (16) has a minimum at $r = r_0$, and therefore

$$V^{(1)} = \frac{L^2}{mr_0^3}. \tag{19}$$

Dividing (18) by (19) we have

$$\frac{1}{F}\frac{dF}{dr} = \frac{(\alpha^2 - 3)}{r_0}, \tag{20}$$

where $F = -dV/dr$ and dF/dr are evaluated at r_0. Here, r_0 can be regarded as a variable, dependent on the initial conditions, and (20) is a differential equation for the force with solution

$$F(r) = -\frac{dV}{dr} = -\frac{k}{r^{3-\alpha^2}}, \tag{21}$$

where k a positive constant. Thus, only a power-law force can yield closed orbits for a range of initial conditions. To determine α we use (15). From (16) and (21) we have

$$U^{(2)} = \alpha^2 V^{(1)}/r_0 \tag{22}$$

$$U^{(3)} = \alpha^2(\alpha^2 - 7)V^{(1)}/r_0^2 \tag{23}$$

$$U^{(4)} = \alpha^2(\alpha^4 - 12\alpha^2 + 47)V^{(1)}/r_0^3. \tag{24}$$

Thus, in (6)

$$c = (\alpha^2 - 7)/3r_0. \tag{25}$$

Substituting (19), (24) and (25) in (15) we obtain

$$\alpha^4 - 5\alpha^2 + 4 = 0, \tag{26}$$

and hence

$$\alpha^2 = 1 \text{ or } 4. \tag{27}$$

According to (21) and (27), the possible potentials are $V = -k/r$ (Coulomb) and $V = \frac{1}{2}kr^2$ (oscillator).

Comments

(i) The result proved above – that the only spherically symmetric potentials that possess closed orbits for a range of initial conditions are the Coulomb and oscillator potentials – is known as Bertrand's theorem.[21] The emphasis on a range of initial

---

[21] J. Bertrand, "Théorème relatif au mouvement d'un point attiré vers un centre fixe," C. R. Acad. Sci. Paris, vol. LXXVII, pp. 849–853, 1873.

conditions is essential because one can readily construct other potentials that possess closed orbits for discrete values of the angular momentum, see Question 8.16 and Ref. [22].

(ii) Bertrand's theorem is also often stated as: the only potentials $V(r)$ for which all bounded trajectories are closed are the Coulomb and oscillator potentials. The special property of these potentials that enables them to satisfy this theorem is that they possess extra constants of the motion (such as the Laplace vector **A** and the tensor A_{ij} discussed in Questions 9.1 to 9.9).[23]

(iii) In non-relativistic mechanics, the Coulomb potential has the unique property that it is the only potential $V(r)$ vanishing at infinity for which all bounded trajectories are closed. From this viewpoint one can regard the widespread observation of approximately closed orbits in astronomy as a signature of Newton's law of universal gravitation.

(iv) In an interesting paper, Pesic[24] points out that Newton was aware of some of the remarkable connections between the Kepler (Coulomb) and oscillator problems. After all, Newton showed that if the bounded motion of a body is an elliptical orbit then the body is moving in a force that is either linear and directed to the centre of the ellipse, or inverse square and directed to one of the foci of the ellipse. In this connection he was evidently aware of the concept of dual pairs of forces, and he discussed other examples of such pairs.[25] Although Newton did not refer to additional conserved quantities explicitly, he emphasizes "the quiescence of the aphelion points" (that "the aphelions are immovable; and so are the planes of the orbit"), and shows in detail that any departure from an inverse-square law results in precession.[25] (This was perhaps of more concern to him because of its application to lunar motion.) Evidently, Newton understood the consequences of what we now call the 'breaking of hidden symmetry'.

---

[22] I. Rodriguez and J. L. Brun, "Closed orbits in central forces distinct from Coulomb or harmonic oscillator type," European Journal of Physics, vol. 19, pp. 41–49, 1998.

[23] R. P. Martinez-y-Romero, H. N. Núñez-Yépez, and A. L. Salas-Brito, "Closed orbits and constants of the motion in classical mechanics," European Journal of Physics, vol. 13, pp. 26–31, 1992.

[24] P. Pesic, "Newton and hidden symmetry," European Journal of Physics, vol. 19, pp. 151–153, 1998.

[25] S. Chandrasekhar, *Newton's Principia for the common reader*, Chaps. 4, 6 and 13. Oxford: Clarendon Press, 1995.

10
Two-body problems

Two-body problems are important in their own right (for example, in astronomy and atomic and molecular physics) and also because they serve as a useful transition to the study of multi-particle systems: consideration of just two particles enables one to illustrate certain basic and pervasive features in their simplest form (see Questions 10.1, 10.2 and 10.6). Other questions in this chapter involve central, isotropic, interparticle forces; coupled oscillators; rotating oscillators and interacting charges in a magnetic field.

Question 10.1

Particles of constant mass m_1 and m_2 interact with each other and are also subject to external forces. The interparticle forces are $\mathbf{F}_{12}$ (the force that 1 exerts on 2) and $\mathbf{F}_{21}$ (the force that 2 exerts on 1). The external forces are $\mathbf{F}_1^{(e)}$ and $\mathbf{F}_2^{(e)}$. The frame of reference is assumed to be inertial.

(a) Use the equations of motion to show that

$$M \frac{d^2 \mathbf{R}}{dt^2} = \mathbf{F}^{(e)}, \tag{1}$$

where

$$M = m_1 + m_2, \qquad \mathbf{F}^{(e)} = \mathbf{F}_1^{(e)} + \mathbf{F}_2^{(e)} \tag{2}$$

are the total mass and the total external force, and

$$\mathbf{R} = \frac{(m_1 \mathbf{r}_1 + m_2 \mathbf{r}_2)}{M}. \tag{3}$$

(b) Interpret this result.

Solution

(a) The equations of motion relative to an inertial frame are

$$m_1 \frac{d^2 \mathbf{r}_1}{dt^2} = \mathbf{F}_1^{(e)} + \mathbf{F}_{21} \quad \text{and} \quad m_2 \frac{d^2 \mathbf{r}_2}{dt^2} = \mathbf{F}_2^{(e)} + \mathbf{F}_{12} \tag{4}$$

where, according to Newton's third law, $\mathbf{F}_{21} = -\mathbf{F}_{12}$. Since the masses are constant, addition of $(4)_1$ and $(4)_2$ yields (1):

$$\frac{d^2}{dt^2}(m_1\mathbf{r}_1 + m_2\mathbf{r}_2) = \mathbf{F}_1^{(e)} + \mathbf{F}_2^{(e)}. \tag{5}$$

(b) The vector $\mathbf{R}$ specifies the position (relative to a coordinate origin O) of a point in space known as the centre of mass (CM) of the two particles. According to (1) the trajectory $\mathbf{R}(t)$ of the CM relative to an inertial frame is that of a hypothetical particle of mass $M = m_1 + m_2$ acted on by the total external force $\mathbf{F}^{(e)} = \mathbf{F}_1^{(e)} + \mathbf{F}_2^{(e)}$. The interparticle forces play no role in the dynamics of the CM, which is therefore generally much simpler than the dynamics of the individual particles (see Chapter 1). Note that in $(2)_2$, $\mathbf{F}_1^{(e)}$ is to be evaluated at particle 1 and $\mathbf{F}_2^{(e)}$ at particle 2.

Comments

(i) It is clear from its definition (3) that $\mathbf{R}$ is an origin-dependent vector (it depends on the choice of coordinate origin O).
(ii) The result (1) can be extended to a system comprising an arbitrary number of particles (see Question 11.1) and it is implicit in the formulation of Newton's laws for extended objects (see Chapter 1).

Question 10.2

For the two particles in Question 10.1, prove that

$$\frac{d\mathbf{P}}{dt} = \mathbf{F}^{(e)}, \tag{1}$$

where $\mathbf{P} = m_1\dot{\mathbf{r}}_1 + m_2\dot{\mathbf{r}}_2$ is the total momentum relative to an inertial frame.

Solution

Equation (1) follows from addition of the equations of motion $\dot{\mathbf{p}}_1 = \mathbf{F}_{21} + \mathbf{F}_1^{(e)}$ and $\dot{\mathbf{p}}_2 = \mathbf{F}_{12} + \mathbf{F}_2^{(e)}$, and Newton's third law.

Comments

(i) According to (1), the rate of change of the total momentum $\mathbf{P}$ of two particles is equal to the total external force $\mathbf{F}^{(e)}$ acting on them. In particular, if $\mathbf{F}^{(e)} = 0$ then $\mathbf{P}$ is constant, which is the law of conservation of momentum for two particles relative to an inertial frame.
(ii) Equation (1) generalizes to systems comprising an arbitrary number of particles (see Question 11.1).

Question 10.3

Prove that the CM of two particles lies on the line joining the particles and between them.

Solution

The position vectors of the particles relative to the CM (labelled C in the figure below) are

$$\mathbf{r}'_1 = \mathbf{r}_1 - \mathbf{R} \quad \text{and} \quad \mathbf{r}'_2 = \mathbf{r}_2 - \mathbf{R}, \tag{1}$$

and so

$$m_1 \mathbf{r}'_1 + m_2 \mathbf{r}'_2 = m_1 \mathbf{r}_1 + m_2 \mathbf{r}_2 - (m_1 + m_2)\mathbf{R} = 0. \tag{2}$$

Therefore

$$\mathbf{r}'_2 = -(m_1/m_2)\,\mathbf{r}'_1, \tag{3}$$

meaning that $\mathbf{r}'_1$ and $\mathbf{r}'_2$ are anti-parallel as shown.

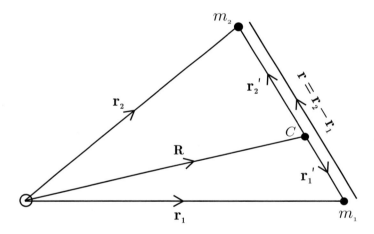

Comments

(i) For particles of constant mass, differentiation of (2) shows that

$$m_1 \dot{\mathbf{r}}'_1 + m_2 \dot{\mathbf{r}}'_2 = 0, \tag{4}$$

and therefore the total momentum relative to the CM is zero – the CM is also the centre of momentum.

(ii) It is useful to introduce the position vector

$$\mathbf{r} = \mathbf{r}_2 - \mathbf{r}_1 = \mathbf{r}'_2 - \mathbf{r}'_1 \tag{5}$$

of particle 2 relative to particle 1. In terms of this, (3) gives

$$\mathbf{r}'_1 = -\frac{m_2}{m_1 + m_2}\mathbf{r}, \qquad \mathbf{r}'_2 = \frac{m_1}{m_1 + m_2}\mathbf{r}. \qquad (6)$$

It is clear that $\mathbf{r}'_1$, $\mathbf{r}'_2$ and $\mathbf{r}$ are all origin-independent vectors (they are independent of the choice of coordinate origin O).

Question 10.4

(a) Show that the relative position vector $\mathbf{r} = \mathbf{r}_2 - \mathbf{r}_1$ for the two-body problem of Question 10.1 satisfies the equation of motion

$$\mu \frac{d^2 \mathbf{r}}{dt^2} = \mathbf{F}_{12} + \mu \left(\frac{\mathbf{F}_2^{(e)}}{m_2} - \frac{\mathbf{F}_1^{(e)}}{m_1} \right), \qquad (1)$$

where

$$\mu = \frac{m_1 m_2}{m_1 + m_2}. \qquad (2)$$

(b) Interpret (1). (Hint: Rederive (1) by considering motion of m_2 relative to a frame with origin O' at m_1, taking into account that this is a non-inertial frame.)

Solution

(a) Start with the equations of motion for each particle, as given in (4) of Question 10.1. Equation (1) follows directly from $m_1 \times (4)_2 - m_2 \times (4)_1$ and the third law.

(b) It is helpful to refer to the diagram in Question 10.3. Consider the motion of m_2 relative to a frame with origin O' at m_1 and axes parallel to the corresponding axes of the inertial frame used in (a). This is a non-inertial frame in which there is an additional force on m_2, the translational force $-m_2 \ddot{\mathbf{r}}_1$ (see Chapters 1 and 14). So

$$m_2 \frac{d^2 \mathbf{r}}{dt^2} = \mathbf{F}_2^{(e)} + \mathbf{F}_{12} - m_2 \frac{d^2 \mathbf{r}_1}{dt^2}. \qquad (3)$$

Now, $\ddot{\mathbf{r}}_1$ is given by $(4)_1$ of Question 10.1. Use of this and Newton's third law shows that (3) reduces to (1). Thus, the equation of motion (1) of the relative vector $\mathbf{r}$ is with respect to a non-inertial frame with origin located at one of the particles. This derivation is instructive in that it shows how the quantity μ enters the equation of motion via the translational force.

Comments

(i) The quantity μ defined in (2) has the unit of mass and the property $\mu < m_1$ and m_2: it is known as the reduced mass.

(ii) If the external forces $\mathbf{F}_1^{(e)}$ and $\mathbf{F}_2^{(e)}$ are zero then (1) reduces to

$$\mu \frac{d^2 \mathbf{r}}{dt^2} = \mathbf{F}_{12}, \qquad (4)$$

while the equation of motion for the CM (see (1) of Question 10.1) becomes

$$\frac{d^2\mathbf{R}}{dt^2} = 0. \tag{5}$$

If the interparticle force is central (i.e. directed along the line joining the two particles) and isotropic $(\mathbf{F}_{12} = F(r)\hat{\mathbf{r}})$ then (4) is

$$\mu \frac{d^2\mathbf{r}}{dt^2} = F(r)\hat{\mathbf{r}}. \tag{6}$$

We see that under these conditions the two-body problem separates into two one-body problems: (5) for the motion of the CM and (6) for the relative motion. This result is fundamental in analyzing this type of two-body problem (see Questions 10.7–10.9). Note that, according to (5), the CM moves with constant velocity $\dot{\mathbf{R}}$ relative to an inertial frame. Therefore, the CM can be used as the origin of an inertial frame – the so-called CM frame.

(iii) In addition to the two particles of mass m_1 and m_2 we have two fictitious particles of mass M $(= m_1 + m_2)$ located at the CM and mass μ $(= m_1 m_2/M)$ located at one of the particles. The trajectory $\mathbf{r}(t)$ of m_2 relative to m_1 can be obtained by solving (6) – see Chapter 8. Then, this is used in (6) of Question 10.3 to provide the trajectories $\mathbf{r}'_1(t)$ and $\mathbf{r}'_2(t)$ of m_1 and m_2 relative to the CM frame.

(iv) A central interparticle force conserves the total angular momentum $\mathbf{L}$ about the CM (see Question 10.6), and therefore the motion has the simple feature that $\mathbf{r}$ (and hence $\mathbf{r}'_1$ and $\mathbf{r}'_2$) is confined to a plane perpendicular to $\mathbf{L}$ and through the CM. This plane is defined by the initial values $\mathbf{r}(0)$ and $\mathbf{v}(0)$:

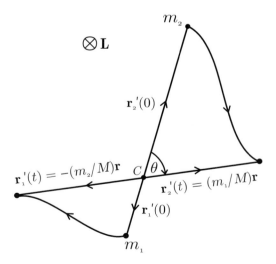

We emphasize that the relative vector $\mathbf{r}$ connects m_1 and m_2. (Sometimes, the description of this vector in the literature is misleading, being drawn from the CM or some other point.) Solutions to (6) can be obtained in the polar form $r = r(\theta)$ as described in Chapter 8. Here, r is the distance between m_1 and m_2, and the angle θ through which $\mathbf{r}$ rotates (as viewed from m_1) is also equal to the angle subtended, by either trajectory, at the CM.

(v) According to (1), equation (4) also holds for non-zero external forces that produce the same accelerations in both particles: $\mathbf{F}_2^{(e)}/m_2 = \mathbf{F}_1^{(e)}/m_1$. For example, the gravitational field due to distant sources; the gravitational field close to the surface of a planet; or charged particles with the same charge-to-mass ratio in the same electric field. See also Questions 10.16–10.19 for an interesting special case.

Question 10.5

Suppose that in the two-body problem of Question 10.1 the interparticle forces $\mathbf{F}_{12}$ and $\mathbf{F}_{21}$ are central. Prove that

$$\frac{d\mathbf{L}}{dt} = \mathbf{\Gamma}^{(e)}, \tag{1}$$

where $\mathbf{L} = \mathbf{r}_1 \times \mathbf{p}_1 + \mathbf{r}_2 \times \mathbf{p}_2$ is the total angular momentum of the particles and $\mathbf{\Gamma}^{(e)} = \mathbf{r}_1 \times \mathbf{F}_1^{(e)} + \mathbf{r}_2 \times \mathbf{F}_2^{(e)}$ is the total torque on the particles exerted by external forces.

Solution

Since $\dot{\mathbf{r}}_1 \times \mathbf{p}_1$ and $\dot{\mathbf{r}}_2 \times \mathbf{p}_2$ are both zero, we have

$$\frac{d\mathbf{L}}{dt} = \mathbf{r}_1 \times \dot{\mathbf{p}}_1 + \mathbf{r}_2 \times \dot{\mathbf{p}}_2 = (\mathbf{r}_1 - \mathbf{r}_2) \times \mathbf{F}_{21} + \mathbf{r}_1 \times \mathbf{F}_1^{(e)} + \mathbf{r}_2 \times \mathbf{F}_2^{(e)}. \tag{2}$$

In the last step we have used the second and third laws of motion. For a central force, $\mathbf{F}_{21}$ is along $\mathbf{r}_1 - \mathbf{r}_2$ and so (2) reduces to (1).

Comments

(i) According to (1), if the total torque due to external forces on two interacting particles is zero, and if the interparticle force is central, then the total angular momentum of the particles, relative to an inertial frame, is constant. This is an example of the law of conservation of angular momentum for two particles.

(ii) It is worth emphasizing the role played by Newton's third law (that there are no unbalanced interparticle forces in an inertial frame) in reaching this conclusion:[1]

"In a binary star the action exerted by one body A on the other body B is exactly balanced by the action of B on A. The mass centre of the system moves with uniform velocity, and the total angular momentum remains constant. This is only true in inertial frames of reference. In all other frames of reference the apparent forces, as measured by the apparent accelerations, will form unbalanced systems. Thus in the geocentric or Ptolemaic system of reference, with a fixed Earth, the (apparent) centripetal forces on the stars, wheeling about

---

[1]   G. Temple, in *Turning points in physics*, pp. 72–73. Amsterdam: North-Holland, 1959.

the polar axis, are not balanced by any forces of attraction acting on this purely mathematical axis.

Hence there does arise the possibility of discovering inertial frames observationally at least in the case of an isolated system such as a planetary system. This application of the third law of motion is especially valid because without it Newtonian dynamics is a fairy tale referring to some mysterious absolute space and time. But with the third law of motion we at least have the abstract possibility of discovering the inertial frame."

(iii) Equation (1) generalizes to a system containing an arbitrary number of particles interacting via central interparticle forces – see Question 11.2.

(iv) The law of conservation of angular momentum applies also to systems where the interparticle forces are not necessarily central – such as the electromagnetic interaction of charged particles, provided the angular momentum of the electromagnetic field is taken into account (see also Question 14.19).

Question 10.6

Show that for the two-body problem of Question 10.1 the total momentum $\mathbf{P}$, the total kinetic energy K and the total angular momentum $\mathbf{L}$ relative to a frame with coordinate origin at O can be expressed as

$$\mathbf{P} = M\dot{\mathbf{R}} \tag{1}$$

$$K = \tfrac{1}{2}M\dot{\mathbf{R}}^2 + \tfrac{1}{2}\mu\dot{\mathbf{r}}^2 \tag{2}$$

$$\mathbf{L} = M\mathbf{R} \times \dot{\mathbf{R}} + \mu\mathbf{r} \times \dot{\mathbf{r}}, \tag{3}$$

where $M = m_1 + m_2$, $\mu = m_1 m_2/(m_1 + m_2)$ and $\mathbf{R}$ is the position vector of the CM relative to O.

Solution

The position vectors of m_1 and m_2 relative to O are (see Question 10.3)

$$\mathbf{r}_1 = \mathbf{R} - \frac{m_2}{M}\mathbf{r} \quad \text{and} \quad \mathbf{r}_2 = \mathbf{R} + \frac{m_1}{M}\mathbf{r}. \tag{4}$$

Also

$$\mathbf{P} = m_1\dot{\mathbf{r}}_1 + m_2\dot{\mathbf{r}}_2, \quad K = \tfrac{1}{2}m_1\dot{\mathbf{r}}_1^2 + \tfrac{1}{2}m_2\dot{\mathbf{r}}_2^2, \quad \mathbf{L} = m_1\mathbf{r}_1 \times \dot{\mathbf{r}}_1 + m_2\mathbf{r}_2 \times \dot{\mathbf{r}}_2. \tag{5}$$

By substituting (4) in (5) we obtain (1)–(3) after some simplification.

Comments

(i) According to (1) the total momentum is the same as that of a particle of mass M located at the CM (the momentum of the CM).

(ii) According to (2) the total kinetic energy is that of a particle of mass M located at the CM, plus the kinetic energy of a particle of mass μ moving with the relative velocity $\dot{\mathbf{r}}$ (the kinetic energy of the relative motion). Similarly for the total angular momentum (3).

(iii) If the external forces are zero and if O is the origin of an inertial frame, then $\ddot{\mathbf{R}} = 0$ (see Question 10.4) and the angular momentum $M\mathbf{R} \times \dot{\mathbf{R}}$ of the CM is conserved. Also, if the interparticle force is central and isotropic then $\mu\ddot{\mathbf{r}} = F(r)\hat{\mathbf{r}}$ (see Question 10.4) and the angular momentum $\mu\mathbf{r} \times \dot{\mathbf{r}}$ of the relative motion is conserved.

Question 10.7

Two particles of mass m_1 and m_2 interact via an inverse-square force $\mathbf{F}_{12} = -k\hat{\mathbf{r}}/r^2$, where $\mathbf{r} = \mathbf{r}_2 - \mathbf{r}_1$ is the relative vector. There are no external forces.

(a) Use results from Questions 8.9, 10.3 and 10.4 to deduce the polar forms of the trajectories $\mathbf{r}_1(\theta)$ and $\mathbf{r}_2(\theta)$ of the particles relative to the CM frame.
(b) For an attractive force ($k > 0$) plot the trajectories relative to the CM frame for the following masses and eccentricities:
  1. $m_1 = m_2$ and $e = 0.65$;
  2. $m_1 = 2m_2$ and $e = 0.65$;
  3. $m_1 = m_2$ and $e = 2$;
  4. $m_1 = 2m_2$ and $e = 6$.
(c) For a repulsive force ($k < 0$) plot the trajectories for:
  1. $m_1 = m_2$ and $e = 4$;
  2. $m_1 = 2m_2$ and $e = 8$.

Solution

(a) According to Question 10.4 the relative vector satisfies the equation of motion $\mu\ddot{\mathbf{r}} = -k\hat{\mathbf{r}}/r^2$, and according to Question 8.9 the polar form of the solution is

$$r(\theta) = \frac{r_0}{1 - e\cos\theta}, \tag{1}$$

where

$$r_0 = \frac{L^2}{\mu k}, \qquad e = \sqrt{1 + \frac{2L^2 E}{\mu k^2}}. \tag{2}$$

Here, L is the magnitude of the angular momentum $\mathbf{L} = \mu\mathbf{r} \times \dot{\mathbf{r}}$ and E is the energy $E = \frac{1}{2}\mu\dot{r}^2 - k/r$ (both are for the CM frame). (Note the role of the reduced mass μ in these formulas.) It follows from (6) of Question 10.3 and (1) that the trajectories relative to the CM frame are

$$\mathbf{r}_1(\theta) = -\frac{m_2}{m_1 + m_2}\frac{r_0}{1 - e\cos\theta}\hat{\mathbf{r}}, \qquad \mathbf{r}_2(\theta) = \frac{m_1}{m_1 + m_2}\frac{r_0}{1 - e\cos\theta}\hat{\mathbf{r}}. \tag{3}$$

Note that here θ is the angular position of particle 2.

(b) For an attractive potential there are three possibilities for the eccentricity: $e < 1$, $e > 1$ and $e = 1$. The trajectories depicted below are obtained from (3) for the values of e shown.

1. $m_1 = m_2$ and $e = 0.65$

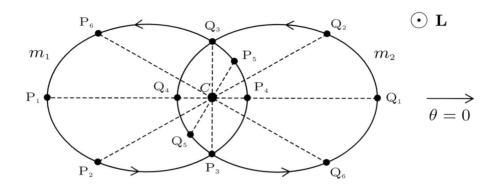

In this diagram (and those that follow) C denotes the CM, while $P_1, P_2, \cdots$ are successive positions of m_1 and $Q_1, Q_2, \cdots$ are the corresponding positions of m_2. The CM always lies on the line joining the two particles and so $P_i C Q_i$ is always a straight line (see Question 10.3). This line rotates about C (in a counter-clockwise direction in our diagrams) as the two particles orbit each other. This notation is used in all the two-body trajectories on the next few pages.

2. $m_1 = 2m_2$ and $e = 0.65$

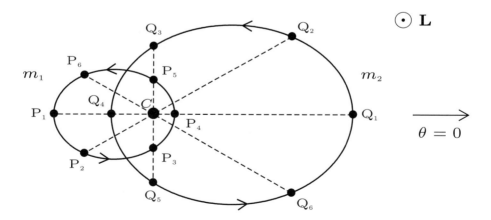

In the two diagrams above, the trajectories consist of overlapping ellipses with the CM at a common focus, and the relative vector joining m_1 and m_2 rotates about C in a counter-clockwise direction. In Question 10.9 a notebook is given that presents a dynamic output for the various possible trajectories.

3. $\quad m_1 = m_2 \quad \text{and} \quad e = 2$

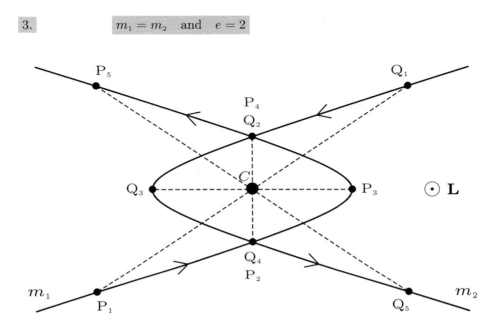

4. $\quad m_1 = 2m_2 \quad \text{and} \quad e = 6$

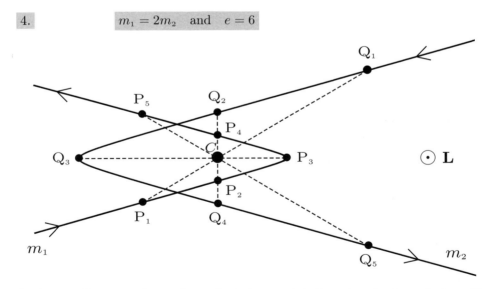

In the two diagrams above, the trajectories consist of overlapping hyperbolas with the CM at a common focus. The trajectories for $e = 1$ are similar to the preceding two figures, except that the two hyperbolas are replaced by two parabolas (see Question 10.10).

296 Solved Problems in Classical Mechanics

(c) For a repulsive potential $E > 0$ and therefore only $e > 1$ is possible.

1. $m_1 = m_2$ and $e = 4$

2. $m_1 = 2m_2$ and $e = 8$

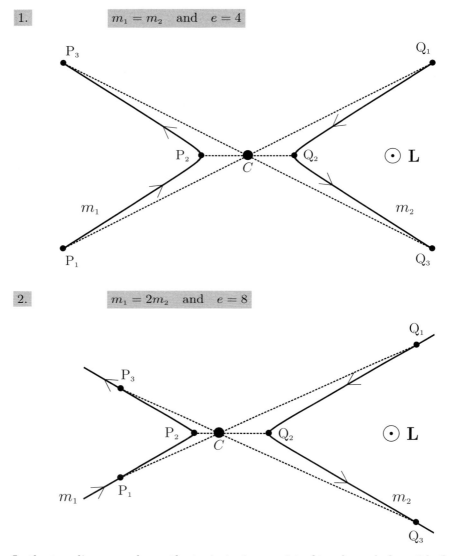

In the two diagrams above, the trajectories consist of two hyperbolas with the CM between the two focii. The line $P_i C Q_i$ rotates about C in a clockwise direction.

Comment

When $m_2 \gg m_1$ (e.g., a planet orbiting a massive star) the CM is close to m_2 and the orbit of m_2 is small compared to that of m_1. This small stellar motion provides a means for detecting planets orbiting distant stars, and for measuring their mass.

Question 10.8

Two particles of mass m_1 and m_2 interact via an attractive linear force $\mathbf{F}_{12} = -k\mathbf{r}$, where k is a positive constant and $\mathbf{r} = \mathbf{r}_2 - \mathbf{r}_1$ is the relative vector. There are no external forces.

(a) Use results from Questions 8.10, 10.3 and 10.4 to deduce the polar forms of the trajectories $\mathbf{r}_1(\theta)$ and $\mathbf{r}_2(\theta)$ of the particles relative to the CM frame.
(b) Plot the trajectories for: 1. $m_1 = m_2$; $e = 0.268$, and 2. $m_1 = 2m_2$; $e = 0.268$.

Solution

(a) According to Question 10.4 the relative vector satisfies the equation of motion $\mu\ddot{\mathbf{r}} = -k\mathbf{r}$, and according to Question 8.10 the polar form of the solution is

$$r(\theta) = \frac{r_0}{\sqrt{1 - e\cos 2\theta}}, \qquad (1)$$

where

$$r_0 = \frac{L}{\sqrt{\mu E}}, \qquad e = \sqrt{1 - \frac{kL^2}{\mu E^2}} \qquad (<1). \qquad (2)$$

Here, L is the magnitude of the angular momentum $\mathbf{L} = \mu \mathbf{r} \times \dot{\mathbf{r}}$ and $E = \frac{1}{2}\mu\dot{r}^2 + \frac{1}{2}kr^2$ is the energy. It follows from (6) of Question 10.3 and (1) that the trajectories relative to the CM frame are

$$\mathbf{r}_1(\theta) = -\frac{m_2}{m_1 + m_2}\frac{r_0}{\sqrt{1 - e\cos 2\theta}}\hat{\mathbf{r}}, \qquad \mathbf{r}_2(\theta) = \frac{m_1}{m_1 + m_2}\frac{r_0}{\sqrt{1 - e\cos 2\theta}}\hat{\mathbf{r}}, \qquad (3)$$

where θ is the angular position of particle 2.

(b) 1. $\qquad m_1 = m_2 \quad \text{and} \quad e = 0.268$

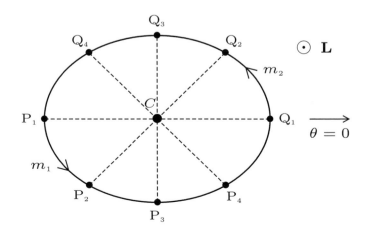

2. $\quad m_1 = 2m_2 \quad \text{and} \quad e = 0.268$

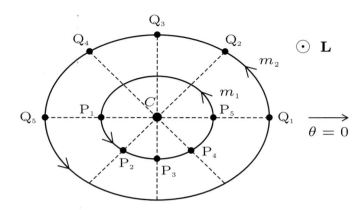

Here, $P_1, P_2, \cdots$ denote successive positions of m_1 and $Q_1, Q_2, \cdots$ are the corresponding positions of m_2. The trajectories consist of two ellipses having a common centre at the centre of mass C, and the line $P_i C Q_i$ rotates about C in a counter-clockwise direction.

Question 10.9

Two particles of mass m_1 and m_2 interact via an inverse-square force $\mathbf{F}_{21} = k\mathbf{r}/r^3$, where $\mathbf{r}$ is the relative vector $\mathbf{r} = \mathbf{r}_2 - \mathbf{r}_1$. There are no external forces and the initial conditions in the CM frame are

$$\left.\begin{array}{ll} \mathbf{r}_1(0) = (m_2 d/M,\, 0,\, 0), & \mathbf{v}_1(0) = (0,\, v_0 m_2/m_1,\, 0) \\ \mathbf{r}_2(0) = (-m_1 d/M,\, 0,\, 0), & \mathbf{v}_2(0) = (0,\, -v_0,\, 0), \end{array}\right\} \quad (1)$$

where d and v_0 are constants and $M = m_1 + m_2$.

(a) Show that the total energy E and angular momentum L are given by

$$E = \left(\frac{v_0^2}{v_e^2} - 1\right)\frac{k}{d} \quad \text{and} \quad L = m_2 v_0 d, \qquad (2)$$

where

$$v_e = \sqrt{\frac{2m_1}{m_2 M}\frac{k}{d}}. \qquad (3)$$

Use $(2)_1$ and Question 8.9 to state the conditions on v_0 for which the trajectories will be: 1. elliptical, 2. parabolic, and 3. hyperbolic.

(b) Show that the eccentricity e and semi-major axis a of an elliptical relative orbit $r(\theta)$ are

$$e = \sqrt{1 - 4\alpha^2(1 - \alpha^2)} \quad \text{and} \quad a = \frac{d}{2(1 - \alpha^2)}, \qquad (4)$$

where $\alpha = v_0/v_e$ (< 1).

(c) Use a system of units in which $d = 1$ and the period $T = 1$. Write a notebook to plot the trajectories $\mathbf{r}_1(t)$ and $\mathbf{r}_2(t)$ using (4) above and (3) of Question 10.7. Implement *Mathematica*'s Manipulate command to produce a dynamic output; select values of α in the range $0 < \alpha < 0.9$ and a mass ratio $\gamma = m_2/m_1$ in the range $1 \leq \gamma \leq 20$. (Hint: Use the eccentric anomaly and Kepler's equation to determine $\theta(t)$ – see (14) and (15) of Question 8.18 with e replaced by $-e$.)

Solution

(a) $E = K(0) + V(0)$, where $V(0) = -k/d$ and

$$K(0) = \tfrac{1}{2}m_1 \dot{y}_1^2(0) + \tfrac{1}{2}m_2 \dot{y}_2^2(0) = \tfrac{1}{2} \frac{m_2}{m_1} M v_0^2 = \frac{v_0^2}{v_e^2} \frac{k}{d}. \qquad (5)$$

L is a constant, equal to its value $m_1 v_1(0) r_1(0) + m_2 v_2(0) r_2(0)$ at $t = 0$:

$$L = m_1 v_0 \frac{m_2}{m_1} \frac{m_2 d}{M} + m_2 v_0 \frac{m_1 d}{M} = m_2 v_0 d. \qquad (6)$$

1. For elliptical orbits, $E < 0$ (see Question 8.9). According to $(2)_1$ this requires $v_0 < v_e$. **2.** For parabolic orbits $E = 0$ and so $v_0 = v_e$. **3.** For hyperbolic orbits $E > 0$ and so $v_0 > v_e$.

(b) Substituting (2) into (2) of Question 10.7 and using $r_0 = a(1 - e^2)$ gives (4).

(c) The required notebook is:

```
In[1]:= d = 1.0; T = 1.0; ω = 2π/T;

e[α_] := √(1 - 4α² (1 - α²)) ; a[α_] := d/(2(1 - α²)) ;

ψ[α_, t_] := ψ/.FindRoot[ψ + e[α] Sin[ψ] - ωt == 0, {ψ, ωt}]

θ[α_, t_] := 2 ArcTan[√((1 - e[α])/(1 + e[α])) Tan[ψ[α, t]/2]];

(* Kepler's Eqn. ψ is the eccentric anomaly.
    See Comments (iii) - (v) in Question 8.18 *)

r1[α_, γ_, θ_] := γ/(1+γ) a[α] (1 - e[α]²)/(1 - e[α] Cos[θ]) ;

r2[α_, γ_, θ_] := 1/(1+γ) -a[α] (1 - e[α]²)/(1 - e[α] Cos[θ]) ;
```

```
In[2]:= x1[α_, γ_, θ_] := r1[α, γ, θ] Cos[θ]; y1[α_, γ_, θ_] := r1[α, γ, θ] Sin[θ];
       x2[α_, γ_, θ_] := r2[α, γ, θ] Cos[θ]; y2[α_, γ_, θ_] := r2[α, γ, θ] Sin[θ];
       Min[α_] := Min[-a[α](1+e[α]), -a[α]√(1-e[α]^2)];
       Max[α_] := Max[a[α](1+e[α]), a[α]√(1-e[α]^2)];
       orbits[α_, γ_, θ_] := ParametricPlot[{{x1[α, γ, θ], y1[α, γ, θ]},
              {x2[α, γ, θ], y2[α, γ, θ]}}, {θ, 0, 2π}, PlotRange → {{Min[α],
              Max[α]}, {Min[α], Max[α]}}, PlotStyle → {Directive[Dashed,
              Thick, Red], Directive[Dashed, Thick, Blue]}, Axes → {False,
              False}];
       BinaryStar[α_, γ_, t_] := Graphics[{PointSize[0.025], Red,
              Point[{x1[α, γ, θ[α, t]], y1[α, γ, θ[α, t]]}], Blue,
              Point[{x2[α, γ, θ[α, t]], y2[α, γ, θ[α, t]]}], Purple,
              Line[{{x1[α, γ, θ[α, t]], y1[α, γ, θ[α, t]]}, {x2[α, γ, θ[α, t]],
              y2[α, γ, θ[α, t]]}}], Black, {PointSize[0.0125],
              Point[{0, 0}]}, Text[CM, {0, -0.05}]}];
       Manipulate[Show[orbits[α, γ, θ], BinaryStar[α, γ, t], Background →
              LightGray], {{α, 0.5, "velocity ratio"}, 0.1, 0.9, 0.02,
              Appearance → "Labeled"}, {{γ, 2, "mass ratio"}, 1, 20, 1,
              Appearance → "Labeled"}, {t, 0, T}]
```

Question 10.10

Consider again the two particles of Question 10.9 subject to the same initial conditions. Write a *Mathematica* notebook to obtain and plot numerical solutions to the equations of motion $m_i \ddot{\mathbf{r}}_i + (-1)^i k \mathbf{r}/r^3 = 0$, where $\mathbf{r}_i = (x_i, y_i)$ and $i = 1, 2$. Take[‡] $k = 2\pi^2$, $m_1 = m_2 = 1$, $d = 1$ and $v_0 = v_e$.

Solution

The required notebook is:

```
In[1]:= m1 = 1.0; m2 = 1.0; d = 1.0; k = 2π^2; v0 = √(2π); Tmax = 10.0;
       x10 = m2/(m1+m2) d; x20 = -m1/(m1+m2) d; y10 = 0; y20 = 0;
       vx10 = 0; vx20 = 0.0; vy10 = m2/m1 v0; vy20 = -v0;
       r1[t_] := {x1[t], y1[t]}; r2[t_] := {x2[t], y2[t]}; [t_] := r2[t] - r1[t];
```

[‡]In a system of units where $k = 2\pi^2$, $m_1 = m_2 = 1$, $d = 1$, the particles move in circles about the CM with a period $T = 1$ when $v_0 = \frac{v_e}{\sqrt{2}}$ (see (2) of Question 10.11).

```
In[2]:= EqnMotion1 = Thread[m1 r1''[t] - k r[t]/Dot[r[t].r[t]]^(3/2) == 0];

        EqnMotion2 = Thread[m2 r2''[t] + k r[t]/Dot[r[t].r[t]]^(3/2) == 0];

        InitCon1 = Join[Thread[r1[0] == {x10, y10}],
           Thread[r1'[0] == {vx10, vy10}]];

        InitCon2 = Join[Thread[r2[0] == {x20, y20}],
           Thread[r2'[0] == {vx20, vy20}]];

        EqsToSolve = Join[EqnMotion1, EqnMotion2, InitCon1, InitCon2];

        Sol = NDSolve[EqsToSolve, Join[r1[t], r2[t]], {t, 0, Tmax}];

        ParametricPlot[{{Evaluate[{x1[t], y1[t]}]/.Sol},
           {Evaluate[{x2[t], y2[t]}]/.Sol}, {Evaluate[{x1[t], -y1[t]}]
           /.Sol}, {Evaluate[{x2[t], -y2[t]}]/.Sol}},
          {t, 0, Tmax}, PlotRange → {{-5, 5}, {-5, 5}}]
```

Here, $E = 0$ and so the trajectories consist of two parabolas. In the diagram below, P_1, P_2, $\cdots$ denote successive positions of m_1 and Q_1, Q_2, $\cdots$ are the corresponding positions of m_2. The centre of mass of the system remains at rest at the origin O, and the line P_iOQ_i rotates about O in a counter-clockwise direction.

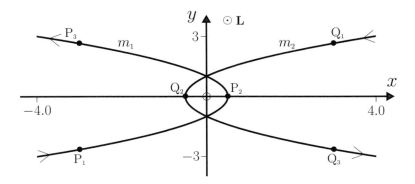

Question 10.11

Write down Kepler's laws for the gravitational two-body problem by generalizing the results in Questions 6.15, 8.9 and 8.15 for motion of a single particle in an inverse-square force-field.

Solution

I. Each particle moves in an elliptical orbit about a common focus located at the CM (see Question 8.9 and 10.7).

II. For each particle, the area per unit time swept out by its radius vector relative to the CM is a constant:

$$\frac{dA_1}{dt} = \left(\frac{m_2}{m_1+m_2}\right)^2 \frac{L}{2\mu}, \qquad \frac{dA_2}{dt} = \left(\frac{m_1}{m_1+m_2}\right)^2 \frac{L}{2\mu}. \tag{1}$$

(See Questions 6.15 and 10.7.)

III. The square of the (common) period is proportional to the cube of the semi-major axis of the ellipse traced out by the particle μ; that is, by the relative vector $\mathbf{r} = \mathbf{r}_2 - \mathbf{r}_1$ drawn from the CM to μ:

$$T^2 = (4\pi^2 \mu/k) a^3, \tag{2}$$

where $a = r_0/(1-e^2)$. (See Questions 8.15 and 10.4, and (1) of Question 10.7.)

Comment

The original statement of Kepler's laws was for the limit $m_2 \gg m_1$ (mass of the Sun $\gg$ the mass of a planet), where the CM is close to m_2 and the motion of m_2 relative to the CM was neglected.

Question 10.12

Identical springs‡ (each having force constant k) are connected to two equal masses m as shown below. The masses are constrained to move in one dimension on a frictionless horizontal surface, and the ends of the springs are attached to fixed walls at P and Q.

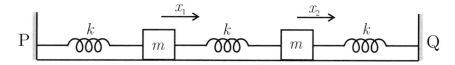

(a) Show that the general solutions for the displacements $x_1(t)$ and $x_2(t)$ of the masses from their equilibrium positions are

$$x_1(t) = A\cos(\omega_1 t + \phi_1) - B\cos(\omega_2 t + \phi_2) \tag{1}$$

$$x_2(t) = A\cos(\omega_1 t + \phi_1) + B\cos(\omega_2 t + \phi_2), \tag{2}$$

where

$$\omega_1 = \sqrt{k/m}, \qquad \omega_2 = \sqrt{3k/m}, \tag{3}$$

and A, B, ϕ_1, ϕ_2 are arbitrary constants.

‡Here, and elsewhere in this book, we assume elastic springs of negligible mass.

(b) Discuss these solutions for the initial conditions:

1. $x_1(0) = x_2(0)$, $\quad v_1(0) = v_2(0) = 0$. (4)
2. $x_1(0) = -x_2(0)$, $\quad v_1(0) = v_2(0) = 0$. (5)
3. $x_1(0) = 0$, $x_2(0) = x_0$, $\quad v_1(0) = v_2(0) = 0$. (6)

Solution

(a) The forces acting on the two masses are the external forces $F_1^{(e)} = -kx_1$ and $F_2^{(e)} = -kx_2$, and the interparticle forces $F_{21} = k(x_2 - x_1) = -F_{12}$. Consequently, the equations of motion are

$$m\frac{d^2x_1}{dt^2} = -kx_1 + k(x_2 - x_1) \tag{7}$$

and

$$m\frac{d^2x_2}{dt^2} = -kx_2 - k(x_2 - x_1). \tag{8}$$

Addition of (7) and (8) gives the equation of motion

$$\frac{d^2X}{dt^2} + \frac{k}{m}X = 0 \tag{9}$$

of the CM coordinate $X = \tfrac{1}{2}(x_1 + x_2)$, while subtraction of (7) and (8) yields the equation of motion

$$\frac{d^2x}{dt^2} + \frac{3k}{m}x = 0 \tag{10}$$

of the relative coordinate $x = x_2 - x_1$. These are the equations of a simple harmonic oscillator having angular frequencies ω_1 and ω_2, respectively. The general solutions to (9) and (10) are $X(t) = A\cos(\omega_1 t + \phi_1)$ and $x(t) = 2B\cos(\omega_2 t + \phi_2)$. Consequently, $x_1 = X - \tfrac{1}{2}x$ and $x_2 = X + \tfrac{1}{2}x$ are given by (1) and (2).

(b) 1. The initial conditions (4) require $A = x_1(0)$, $B = 0$ and $\phi_1 = 0$. Thus

$$x_1(t) = x_2(t) = x_1(0)\cos\omega_1 t. \tag{11}$$

The masses oscillate in phase and with angular frequency ω_1; this is known as the symmetric mode of oscillation.

2. The initial conditions (5) require $A = 0$, $B = -x_1(0)$ and $\phi_2 = 0$. Thus

$$x_1(t) = -x_2(t) = x_1(0)\cos\omega_2 t. \tag{12}$$

The masses oscillate out of phase and with angular frequency ω_2; this is the anti-symmetric mode of oscillation.

3. The initial conditions (6) require $A = B = \frac{1}{2}x_0$ and $\phi_1 = \phi_2 = 0$. Thus

$$x_1(t) = \tfrac{1}{2}x_0[\cos\omega_1 t - \cos\omega_2 t]$$
$$= x_0 \sin\{\tfrac{1}{2}(\omega_1 + \omega_2)t\}\sin\{\tfrac{1}{2}(\omega_2 - \omega_1)t\} \qquad (13)$$

and

$$x_2(t) = \tfrac{1}{2}x_0[\cos\omega_1 t + \cos\omega_2 t]$$
$$= x_0 \cos\{\tfrac{1}{2}(\omega_1 + \omega_2)t\}\cos\{\tfrac{1}{2}(\omega_2 - \omega_1)t\}. \qquad (14)$$

Here, we have used the identities $\cos a - \cos b = 2\sin\tfrac{1}{2}(a+b)\sin\tfrac{1}{2}(b-a)$ and $\cos a + \cos b = 2\cos\tfrac{1}{2}(a+b)\cos\tfrac{1}{2}(a-b)$.

Comments

(i) The coordinates $X(t)$ and $x(t)$ that satisfy the uncoupled (independent) equations (9) and (10) are known as normal modes. In other, more complicated, systems the normal modes differ from the CM and relative coordinates $X(t)$ and $x(t)$, and they are denoted by $\eta_1(t)$ and $\eta_2(t)$. By definition, η_1 is an oscillation at a single frequency ω_1, and η_2 is an oscillation at a single frequency ω_2. The frequencies ω_1 and ω_2 of the normal modes are known as the normal frequencies.

(ii) The result $\omega_1 < \omega_2$ (i.e. the frequency of the symmetric mode is less than that of the anti-symmetric mode) is true of coupled oscillators in general.

(iii) The symmetric mode can be excited by releasing the masses from rest at equal displacements as in (4); to excite the anti-symmetric mode the initial displacements should have equal magnitudes but opposite signs, as in (5).

(iv) The following graphs show $x_1(t)$ and $x_2(t)$ for the symmetric and anti-symmetric modes of oscillation (11) and (12). The periods are in the ratio $\sqrt{3} : 1$ – see (3).

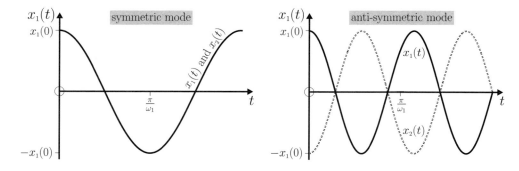

(v) The general solutions (1) and (2) are a superposition of normal modes. The graph below illustrates the superpositions (13) and (14) for the initial conditions (6) and with $\omega_2/\omega_1 = \sqrt{3}$.

(vi) This description generalizes to a system of n coupled oscillators where there are n normal modes, each with a normal frequency, and the general solution is a linear combination of the normal modes.

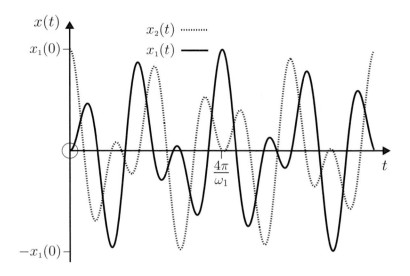

Question 10.13

A double pendulum consists of an inextensible string of negligible mass and length 2ℓ, with one end fixed and masses m attached at the midpoint and the other end.

(a) Show that for small planar oscillations the general solutions for the horizontal displacements of the masses from their equilibrium positions are

$$x_1(t) = A\cos(\omega_1 t + \phi_1) + B\cos(\omega_2 t + \phi_2) \tag{1}$$

$$x_2(t) = (\sqrt{2}+1)A\cos(\omega_1 t + \phi_1) - (\sqrt{2}-1)B\cos(\omega_2 t + \phi_2). \tag{2}$$

Here, $\omega_1 = \sqrt{2-\sqrt{2}}\,\omega_0$, $\omega_2 = \sqrt{2+\sqrt{2}}\,\omega_0$, $\omega_0 = \sqrt{g/\ell}$ (the angular frequency of a simple pendulum of length ℓ) and A, B, ϕ_1, ϕ_2 are arbitrary constants. (Hint: Determine the normal modes in terms of x_1 and x_2.)

(b) Discuss these solutions for the initial conditions:

1. $x_2(0) = (\sqrt{2}+1)\,x_1(0)$, $v_1(0) = v_2(0) = 0$. (3)
2. $x_2(0) = -(\sqrt{2}-1)\,x_1(0)$, $v_1(0) = v_2(0) = 0$. (4)

Solution

(a) Choose x- and y-coordinates in the plane of motion with origin at the fixed end of the pendulum as shown in the diagram below. The forces acting on the particles are their weight $m\mathbf{g}$ and the tensions $\mathbf{T}_1$ and $\mathbf{T}_2$ in the strings. For small oscillations ($|\theta_1|, |\theta_2| \ll 1$) the x- and y-components of the equations of motion are

$$\left.\begin{array}{r}m\ddot{x}_1 + T_1\theta_1 - T_2\theta_2 = 0 \\ m\ddot{y}_1 - mg - T_2 + T_1 = 0,\end{array}\right\} \tag{5}$$

and

$$\left.\begin{array}{l}m\ddot{x}_2 + T_2\theta_2 = 0 \\ m\ddot{y}_2 - mg + T_2 = 0.\end{array}\right\} \quad (6)$$

(Here, we have neglected terms of order θ_i^2 and higher.) For small oscillations we can neglect changes in the vertical coordinates y_1 and y_2. Also, $x_1 = \ell\theta_1$ and $x_2 = \ell(\theta_1 + \theta_2)$. Then, (5) and (6) yield $T_1 = 2mg$, $T_2 = mg$, and

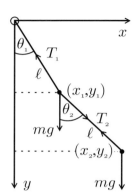

$$\ddot{x}_1 + \omega_0^2(3x_1 - x_2) = 0, \qquad \ddot{x}_2 + \omega_0^2(x_2 - x_1) = 0. \quad (7)$$

To study the solutions of these coupled equations we consider the linear combination $(7)_1 + \alpha \times (7)_2$, where α is a constant that is to be determined. That is,

$$\ddot{x}_1 + \alpha\ddot{x}_2 + \omega_0^2\left\{(3-\alpha)x_1 + (\alpha-1)x_2\right\} = 0. \quad (8)$$

Having in mind the normal modes $\eta(t)$ (see Question 10.12), we now ask: for what value(s) of α does (8) have the form of the simple harmonic equation

$$\ddot{\eta} + \omega^2\eta = 0 \quad \text{where} \quad \omega^2 = \beta\omega_0^2, \quad (9)$$

and the constant β is also to be determined? By substituting $\omega_0^2 = \omega^2/\beta$ in (8) and comparing with $(9)_1$, we see that α and β must satisfy

$$\beta = 3 - \alpha \quad \text{and} \quad \alpha\beta = \alpha - 1. \quad (10)$$

That is, $\alpha^2 - 2\alpha - 1 = 0$ and so there are two possible values of α, namely

$$\alpha_1 = 1 + \sqrt{2} \quad \text{and} \quad \alpha_2 = 1 - \sqrt{2}. \quad (11)$$

Correspondingly, $\beta_1 = 2 - \sqrt{2}$ and $\beta_2 = 2 + \sqrt{2}$, and therefore the possible values of ω in (9) are

$$\omega_1 = \sqrt{2 - \sqrt{2}}\,\omega_0, \qquad \omega_2 = \sqrt{2 + \sqrt{2}}\,\omega_0. \quad (12)$$

Thus, we have found that the linear combinations

$$\eta_i(t) = (3 - \alpha_i)x_1(t) + (\alpha_i - 1)x_2(t) \qquad (i = 1, 2), \quad (13)$$

with the α_i given by (11), satisfy $(9)_1$. We can therefore write down the general solutions

$$\eta_1(t) = (2 - \sqrt{2})x_1(t) + \sqrt{2}\,x_2(t) = 4A\cos(\omega_1 t + \phi_1) \quad (14)$$

$$\eta_2(t) = (2 + \sqrt{2})x_1(t) - \sqrt{2}\,x_2(t) = 4B\cos(\omega_2 t + \phi_2). \quad (15)$$

(Here, the factors of 4 have been inserted for convenience.) By adding and subtracting (14) and (15) we obtain the solutions (1) and (2).

(b) 1. The initial values (3) require $B = 0$, $A = x_1(0)$, $\phi_1 = \phi_2 = 0$, and so
$$x_1(t) = x_1(0) \cos\omega_1 t, \qquad x_2(t) = (\sqrt{2} + 1)x_1(0)\cos\omega_1 t. \qquad (16)$$
This is the symmetric mode where the masses oscillate in phase.

2. The initial values (4) require $A = 0$, $B = x_1(0)$, $\phi_1 = \phi_2 = 0$, and so
$$x_1(t) = x_1(0)\cos\omega_2 t, \qquad x_2(t) = -(\sqrt{2} - 1)x_1(0)\cos\omega_2 t. \qquad (17)$$
This is the anti-symmetric mode where the masses oscillate out of phase.

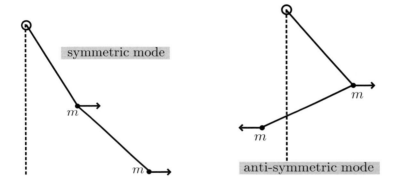

Comments

(i) The normal modes $\eta_1(t)$ and $\eta_2(t)$ are given by (14) and (15), and the respective normal frequencies are ω_1 and ω_2 given in (12). The ratio of these frequencies is $\omega_2/\omega_1 = \sqrt{2+\sqrt{2}}\big/\sqrt{2-\sqrt{2}} \approx 2.41$; again we see that the anti-symmetric mode is the faster mode.

(ii) The symmetric mode can be excited by releasing the masses from rest with $x_2(0) = (\sqrt{2}+1)x_1(0)$. The anti-symmetric mode can be excited by releasing the masses from rest with $x_2(0) = -(\sqrt{2}-1)x_1(0)$.

(iii) The general solutions (1) and (2) are linear combinations of both normal modes (14) and (15). The four arbitrary constants in (1) and (2) are fixed by the four initial conditions $x_i(0)$ and $v_i(0)$ ($i=1,2$) in the usual way (see Question 4.1).

Question 10.14

Consider a system consisting of two particles (beads), a spring and a circular wire (hoop).[2] The beads are connected by the spring and they slide without friction on the wire. The system is depicted in the figure that shows also the y- and z-axes of a Cartesian coordinate system with origin at the centre of the hoop. The yz-plane is

---

[2] F. Ochoa and J. Clavijo, "Bead, hoop and spring as a classical spontaneous symmetry breaking problem," European Journal of Physics, vol. 27, pp. 1277–1288, 2006.

horizontal and the spring is parallel to the y-axis; thus, the beads have the same z-coordinate. Each bead has mass m, the force constant of the spring is k, and the radius of the hoop is R. The equilibrium length $2r_0$ of the spring is less than the diameter of the hoop, that is $r_0 < R$. Suppose the hoop rotates about the z-axis of an inertial frame $Oxyz$ with constant angular velocity ω.

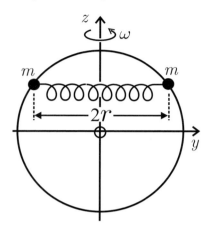

(a) Express the Lagrangian $\mathsf{L} = K - V$ in terms of cylindrical coordinates (r, ϕ, z) and show that it can be written in the one-dimensional form

$$\mathsf{L} = \tfrac{1}{2}\mu \dot z^2 - V_e(z), \qquad (1)$$

where

$$\mu = 2m(1 - z^2/R^2)^{-1} \qquad (2)$$

is a position-dependent effective mass and

$$V_e(z) = 2k(\sqrt{R^2 - z^2} - r_0)^2 - m\omega^2(R^2 - z^2) \qquad (3)$$

is a one-dimensional effective potential.

(b) Use (3) to determine the equilibrium points z_ω of the particles.
(c) Determine the stability of these equilibrium points. In this connection, show that there exists a critical angular velocity

$$\omega_c = \sqrt{\frac{2k}{m}\left(1 - \frac{r_0}{R}\right)}, \qquad (4)$$

and illustrate the significance of ω_c by plotting graphs of $V_e(z)$ versus z/R for $r_0/R = 0.5$ and $\omega = 0.4\omega_c$, ω_c and $1.4\omega_c$.

(d) Suppose the axis of rotation of the hoop is turned through an angle α about the y-axis. Determine the effect of a uniform gravitational field $\mathbf{g} = -g\,\hat{\mathbf{x}}$ on the above results.

(e) Determine the angular frequencies of small oscillations about the equilibrium points when $\alpha = 0$. Express the result in terms of ω, ω_c and $\omega_0 = \sqrt{2k/m}$, and plot its graph versus ω/ω_c for $r_0/R = 0.5$.

Solution

(a) In terms of cylindrical coordinates the Lagrangian is
$$\mathsf{L} = m(\dot r^2 + r^2\dot\phi^2 + \dot z^2) - 2k(r - r_0)^2. \tag{5}$$
(Note that the gravitational potential energy of the particles has not been included here because it is a constant – the centre of mass is always on the z-axis.) Now, $\dot\phi = \omega$, a constant. Also, the beads are constrained to move on the hoop, meaning that $r = \sqrt{R^2 - z^2}$. If we substitute these two conditions in (5) and rearrange terms, we obtain (1).

(b) At the equilibrium points, $z = $ constant and therefore $dV_\mathrm{e}/dz = 0$ (see also Question 10.15). By differentiating (3) with respect to z we have
$$\frac{dV_\mathrm{e}}{dz} = 4kr_0 z\left(\frac{1}{\sqrt{R^2 - z^2}} - \frac{1}{\xi}\right), \quad \text{where} \quad \xi(\omega) = \frac{2kr_0}{2k - m\omega^2}. \tag{6}$$

So, there are three equilibrium points:
$$z_\omega = 0, \quad \pm\sqrt{R^2 - \xi^2}. \tag{7}$$

In a non-inertial frame rotating with the hoop, the beads are at rest at these points: at $z_\omega = \pm\sqrt{R^2 - \xi^2}$ the outward centrifugal force $mr\omega^2$ (see Chapter 14) balances the inward elastic force $k(2r - 2r_0)$ and the normal reaction N of the hoop is zero; at $z_\omega = 0$ the normal reaction N balances the difference of these two forces. In the inertial frame $Oxyz$ the beads move in circles of radius ξ or R about the z-axis.

(c) It is clear from (6) and (7) that the equation $dV_\mathrm{e}/dz = 0$ possesses three real roots if $0 < \xi < R$, whereas if $\xi > R$ or $\xi < 0$ there is just one real root, namely $z_\omega = 0$. The critical condition demarcating these two results is $\xi_\mathrm{c} = R$, and this defines a critical angular velocity ω_c given by (4). If $\omega < \omega_\mathrm{c}$ then $0 < \xi < R$ and there are three real roots; if $\omega > \omega_\mathrm{c}$ there is one real root. To investigate the stability at the equilibrium points we evaluate the second derivative
$$V_\mathrm{e}'' = \frac{d^2 V_\mathrm{e}}{dz^2} = 4kr_0\left[\frac{R^2}{(R^2 - z^2)^{3/2}} - \frac{1}{\xi}\right]. \tag{8}$$

It follows that at $z_\omega = 0$, $V_\mathrm{e}'' < 0$ if $\xi < R$ (i.e. $\omega < \omega_\mathrm{c}$) and $V_\mathrm{e}'' > 0$ if $\xi > R$ (i.e. $\omega > \omega_\mathrm{c}$). It is also clear that at $z_\omega = \pm\sqrt{R^2 - \xi^2}$, $V_\mathrm{e}'' > 0$, because here $\omega < \omega_\mathrm{c}$. Thus, the equilibrium points $z_\omega = \pm\sqrt{R^2 - \xi^2}$ (which exist only if $\omega < \omega_\mathrm{c}$) are always stable, whereas $z_\omega = 0$ is unstable if $\omega < \omega_\mathrm{c}$ and stable if $\omega > \omega_\mathrm{c}$ (if $\omega = \omega_\mathrm{c}$ this point is neutral). When $\omega = 0$ the points of stable equilibrium are at $z_0 = \pm\sqrt{R^2 - r_0^2}$, corresponding to an unstretched spring. Increasing ω stretches the spring, thereby decreasing z_ω that becomes zero at $\omega = \omega_\mathrm{c}$. It is helpful to illustrate these features by plotting graphs of $V_\mathrm{e}(z)$. We write (3) in the dimensionless form
$$\frac{V_\mathrm{e}}{2kR^2} = \left(\sqrt{1 - \frac{z^2}{R^2}} - \frac{r_0}{R}\right)^2 - \left(1 - \frac{r_0}{R}\right)\left(\frac{\omega}{\omega_\mathrm{c}}\right)^2\left(1 - \frac{z^2}{R^2}\right), \tag{9}$$

which is a function of the three dimensionless quantities z/R, ω/ω_c and $\epsilon = r_0/R$. Graphs of (9) versus z/R for $\epsilon = 0.5$ and three values of ω/ω_c are shown below. We see that above ω_c there is a single minimum at $z_\omega = 0$, whereas below ω_c there are two degenerate minima, having the same energy, at $z_\omega = \pm\sqrt{R^2 - \xi^2}$ and a local maximum at $z_\omega = 0$.

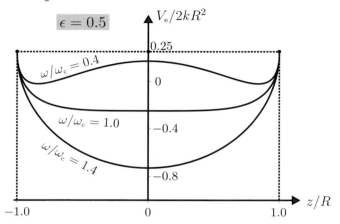

(d) If the axis of rotation is tipped up by an angle α then there is a gravitational contribution $2mgz \sin\alpha$ to the potential energy relative to the point $z = 0$. As a result, the dimensionless expression (9) changes to

$$\frac{V_e}{2kR^2} = \left(\sqrt{1 - \frac{z^2}{R^2}} - \frac{r_0}{R}\right)^2 - \left(1 - \frac{r_0}{R}\right)\left(\frac{\omega}{\omega_c}\right)^2\left(1 - \frac{z^2}{R^2}\right) + \frac{mg\sin\alpha}{kR}\frac{z}{R}. \qquad (10)$$

The gravitational term in (10) destroys the symmetry of (9) under the transformation $z \to -z$. The graphs of $V_e(z)$ are distorted as shown below: instead of two degenerate global minima there is now a global and a local minimum when $\omega < \omega_c$ and a global minimum at $z_\omega < 0$ when $\omega > \omega_c$.

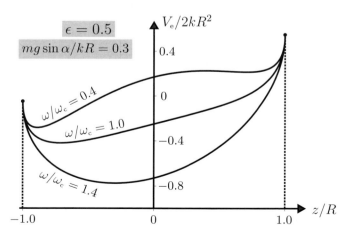

(e) The effective force constants k_e for small oscillations are equal to d^2V_e/dz^2 evaluated at the points of stable equilibrium z_ω. From (8):

$$k_e = \begin{cases} 4kr_0\left(\dfrac{R^2}{\xi^3} - \dfrac{1}{\xi}\right) & \text{at } z_\omega = \pm\sqrt{R^2 - \xi^2} \\ 4kr_0\left(\dfrac{1}{R} - \dfrac{1}{\xi}\right) & \text{at } z_\omega = 0. \end{cases} \tag{11}$$

(Recall that $(11)_1$ is for $\omega < \omega_c$, and $(11)_2$ for $\omega > \omega_c$ – see (c).) The angular frequency of small oscillations about z_ω is equal to $\sqrt{k_e/\mu}$. Here, μ is the effective mass (2): that is, $\mu = 2mR^2/\xi^2$ at $z_\omega = \pm\sqrt{R^2 - \xi^2}$ and $\mu = 2m$ at $z_\omega = 0$. Thus, we have the angular frequencies

$$\omega_< = \sqrt{\frac{2kr_0}{m}\left(\frac{1}{\xi} - \frac{\xi}{R^2}\right)} \quad \text{for } \omega < \omega_c \tag{12}$$

$$\omega_> = \sqrt{\frac{2kr_0}{m}\left(\frac{1}{R} - \frac{1}{\xi}\right)} \quad \text{for } \omega > \omega_c. \tag{13}$$

These can be expressed in terms of ω, ω_c and $\omega_0 = \sqrt{2k/m}$ by using (4) and $(6)_2$:

$$\omega_< = \sqrt{\frac{(2\omega_0^2 - \omega_c^2 - \omega^2)(\omega_c^2 - \omega^2)}{\omega_0^2 - \omega^2}} \quad \text{for } \omega < \omega_c \tag{14}$$

$$\omega_> = \sqrt{\omega^2 - \omega_c^2} \quad \text{for } \omega > \omega_c. \tag{15}$$

Here, ω_0 is the natural frequency of the two particles connected by the spring in the absence of the hoop, and ω_c is related to it by (4):

$$\omega_c = \omega_0\sqrt{1 - r_0/R}. \tag{16}$$

Thus, the ratios $\omega_</\omega_0$ and $\omega_>/\omega_0$ are functions of ω/ω_c and $\epsilon = r_0/R$:

$$\frac{\omega_<}{\omega_0} = \sqrt{1-\epsilon}\,\sqrt{\frac{\{2 - (1-\epsilon)(1 + \omega^2/\omega_c^2)\}\{1 - \omega^2/\omega_c^2\}}{1 - (1-\epsilon)\,\omega^2/\omega_c^2}} \tag{17}$$

$$\frac{\omega_>}{\omega_0} = \sqrt{1-\epsilon}\,\sqrt{\omega^2/\omega_c^2 - 1}. \tag{18}$$

Graphs of these ratios versus ω/ω_c are plotted below for $\epsilon = 0.5$. The frequency of small oscillations tends to zero as $\omega \to \omega_c$, where the equilibrium is neutral rather than stable.

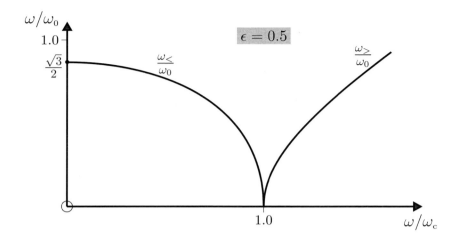

Comments

(i) The above calculations are based on the analysis given in Ref. [2]. In this article, an interesting account is presented of how this simple system provides insights into more advanced topics in physics, such as spontaneous symmetry breaking, phase transitions, order parameters, and critical exponents.

(ii) The angular velocity ω of the hoop is analogous to temperature in a thermodynamic system. The two different regions of symmetry (symmetric phase with $z_\omega = 0$ for $\omega > \omega_c$, and the broken symmetry with $z_\omega \neq 0$ for $\omega < \omega_c$) are separated by a critical point at which there is a second-order phase transition (the second derivative $d^2 V_e/d\omega^2$ has an infinite discontinuity at $\omega = \omega_c$).[2] If ω is decreased below ω_c, any external disturbance can cause the system to leave the mode $z_\omega = 0$ (which is now unstable) in favour of one of the two degenerate stable minima at $z_\omega = \pm\sqrt{R^2 - \xi^2}$.

(iii) The equilibrium positions z_ω play the role of an order parameter which 'spontaneously' acquires a non-zero value that grows as $\sqrt{\omega_c - \omega}$ just below ω_c: thus, the critical exponent for this order parameter is $\frac{1}{2}$, a value that is familiar in the Landau theory for various systems with second-order phase transitions.[2]

(iv) Another example of spontaneous symmetry breaking in a simple mechanical system is given in Question 13.14.

(v) The role of the gravitational force in causing explicit symmetry breaking illustrates the difference between explicit and spontaneous symmetry breaking.

(vi) The system is a mechanical equivalent of a thermodynamic system where the state is characterized by three variables: z_ω, ω and g, and it can be compared with a two-dimensional Ising model of a ferromagnet in the mean-field approximation.[2]

(vii) The equation of motion for z and its solutions are considered in Questions 10.15 and 13.16.

Question 10.15

For the particles, spring and rotating hoop of Question 10.14, use Lagrange's equation to obtain the equation of motion for the z-coordinate of the particles.

Solution

The z-coordinate fixes the positions of the particles in a frame rotating with the hoop. According to Question 10.14 the Lagrangian is given in terms of this coordinate by

$$L(z, \dot{z}) = m\dot{z}^2(1 - z^2/R^2)^{-1} - 2k(\sqrt{R^2 - z^2} - r_0)^2 + m\omega^2(R^2 - z^2), \qquad (1)$$

where R and ω are the radius and angular velocity of the hoop, k and $2r_0$ are the force constant and equilibrium length of the spring, and m is the mass of each particle. The corresponding Lagrange equation

$$\frac{d}{dt}\frac{\partial L}{\partial \dot{z}} - \frac{\partial L}{\partial z} = 0 \qquad (2)$$

yields the equation of motion

$$\frac{d^2 z}{dt^2} + \frac{z}{R^2 - z^2}\left(\frac{dz}{dt}\right)^2 + \frac{(\omega_0^2 - \omega^2)}{R^2}z^3 - (\omega_0^2 - \omega^2)z + \frac{\omega_0^2 r_0 z}{R^2}\sqrt{R^2 - z^2} = 0. \qquad (3)$$

Here, $\omega_0 = \sqrt{2k/m}$ is the natural frequency of the two particles connected by the spring in the absence of the hoop.

Comments

(i) In general, the non-linear equation (3) must be solved numerically (see Question 13.16). We can, however, easily identify two simple types of solution. The first is $z = $ a constant (the roots z_ω given by (7) of Question 10.14), according to which the particles move in circles of radius $\sqrt{R^2 - z_\omega^2}$ about the z-axis. The second is $z = z_\omega + A\cos\Omega t$ ($|A| \ll R$, and Ω given by (14) and (15) of Question 10.14), describing small oscillations about the points of stable equilibrium z_ω.

(ii) For numerical work it is convenient to express (3) in terms of the dimensionless coordinate $Z = z/R$ and time $\tau = t/T_0 = \omega_0 t/2\pi$, and the critical angular velocity $\omega_c = \omega_0\sqrt{1-\epsilon}$, where $\epsilon = r_0/R$ (see (4) of Question 10.14). In terms of these (3) is

$$\frac{d^2 Z}{d\tau^2} + \frac{Z}{1-Z^2}\left(\frac{dZ}{d\tau}\right)^2 + 4\pi^2\left\{1 - (1-\epsilon)\frac{\omega^2}{\omega_c^2}\right\}Z(Z^2-1) + 4\pi^2 \epsilon Z\sqrt{1-Z^2} = 0, \qquad (4)$$

where $-1 \leq Z \leq 1$.

Question 10.16

Consider two interacting charged particles, having the same mass m and charges q_1 and q_2, moving in a uniform magnetostatic field $\mathbf{B}$.[3] Neglect radiation (see Question 7.21) and gravity.

(a) Suppose the charges are identical ($q_1 = q_2 = q$). Show that the equations of motion can be separated into equations for centre-of-mass (CM) motion

$$M\ddot{\mathbf{R}} = Q\dot{\mathbf{R}} \times \mathbf{B}, \qquad (1)$$

and relative motion

$$\mu\ddot{\mathbf{r}} = \tfrac{1}{2} q\dot{\mathbf{r}} \times \mathbf{B} + \frac{kq^2}{r^2}\hat{\mathbf{r}}. \qquad (2)$$

Here, $M = 2m$ and $Q = 2q$ are the total mass and charge, $\mu = \tfrac{1}{2}m$ is the reduced mass, $\mathbf{R} = \tfrac{1}{2}(\mathbf{r}_1 + \mathbf{r}_2)$ and $\mathbf{r} = \mathbf{r}_2 - \mathbf{r}_1$ are the CM and relative position vectors, and $k = 1/4\pi\epsilon$, where ϵ is the permittivity of the medium in which the motion occurs.

(b) Suppose the charges have opposite signs ($q_1 = -q_2 = q$). Show that the equations of motion do not separate into equations for CM motion and relative motion.

Solution

(a) The equations of motion are

$$m\ddot{\mathbf{r}}_1 = -\frac{kq_1 q_2}{r^2}\hat{\mathbf{r}} + q_1 \dot{\mathbf{r}}_1 \times \mathbf{B} \qquad (3)$$

$$m\ddot{\mathbf{r}}_2 = \frac{kq_1 q_2}{r^2}\hat{\mathbf{r}} + q_2 \dot{\mathbf{r}}_2 \times \mathbf{B}. \qquad (4)$$

If we add (3) and (4) then the electrostatic forces cancel. With $q_1 = q_2 = q$ we obtain (1). Also, by subtracting (3) and (4) we obtain (2).

(b) When $q_1 = -q_2 = q$ in (3) and (4), a similar procedure gives

$$m\ddot{\mathbf{R}} = -\tfrac{1}{2} q\dot{\mathbf{r}} \times \mathbf{B} \qquad (5)$$

and

$$m\ddot{\mathbf{r}} = -2q\dot{\mathbf{R}} \times \mathbf{B} - \frac{2kq^2}{r^2}\hat{\mathbf{r}}. \qquad (6)$$

Equations (5) and (6) are coupled through the vectors $\mathbf{R}$ and $\mathbf{r}$.

Comments

(i) The separation of the equation of motion in (a) occurs even if the external forces due to $\mathbf{B}$ are non-zero.

---

[3] S. Curilef and F. Claro, "Dynamics of two interacting particles in a magnetic field in two dimensions," American Journal of Physics, vol. 65, pp. 244–250, 1997.

(ii) For the identical particles in (a), the CM motion is that of a particle of mass $2m$ and charge $2q$ in the field $\mathbf{B}$, while the relative motion is that of a mass $\tfrac{1}{2}m$ and charge $\tfrac{1}{2}q$ in the field $\mathbf{B}$ and the electrostatic field of a charge $2q$ located at the origin of the relative position vector $\mathbf{r}$.

Question 10.17

Suppose the identical interacting charged particles of Question 10.16(a) move in the xy-plane, perpendicular to the uniform magnetostatic field $\mathbf{B} = B\hat{\mathbf{z}}$, where $B > 0$.

(a) 1. What is the CM trajectory $\mathbf{R}(t)$?
 2. Express (2) of Question 10.16 in terms of plane polar coordinates (r, θ) and deduce that the quantity
$$L_\omega = \mu r^2(\dot\theta + \tfrac{1}{2}\omega) \tag{1}$$
is conserved, and that the radial equation of motion is
$$\mu\ddot r = \frac{kq^2}{r^2} + \frac{L_\omega^2}{\mu r^3} - \tfrac{1}{4}\mu r\omega^2, \tag{2}$$
where $\omega = qB/m$ is the cyclotron frequency.

(b) Use (2) to write down an expression for the effective potential $V_{\rm e}(r)$ and deduce that in the presence of a magnetic field the motion is always bounded.

Solution

(a) 1. Consider an inertial frame in which the initial conditions are $\mathbf{R}_{\rm o} = R_{\rm o}\hat{\mathbf{x}}$ and $\dot{\mathbf{R}}_{\rm o} = -\omega R_{\rm o}\hat{\mathbf{y}}$, where $\omega = QB/M = qB/m$. The solution to (1) of Question 10.16 is
$$\mathbf{R}(t) = R_{\rm o}(\cos\omega t,\ -\sin\omega t,\ 0). \tag{3}$$
The CM trajectory is a circle of radius $R_{\rm o}$ centred on the origin.

2. In plane polar coordinates the velocity and acceleration are
$$\dot{\mathbf{r}} = \dot r\hat{\mathbf{r}} + r\dot\theta\hat{\boldsymbol\theta}, \qquad \ddot{\mathbf{r}} = (\ddot r - r\dot\theta^2)\hat{\mathbf{r}} + \frac{1}{r}\frac{d}{dt}(r^2\dot\theta)\hat{\boldsymbol\theta}. \tag{4}$$
(See Question 8.1.) Also, $\hat{\mathbf{r}}\times\hat{\mathbf{z}} = -\hat{\boldsymbol\theta}$ and $\hat{\boldsymbol\theta}\times\hat{\mathbf{z}} = \hat{\mathbf{r}}$. Thus, the radial and transverse components of (2) of Question 10.16 yield
$$\mu(\ddot r - r\dot\theta^2) = \tfrac{1}{2}qBr\dot\theta + \frac{kq^2}{r^2} \tag{5}$$
and
$$\frac{d}{dt}(r^2\dot\theta) = -\frac{qB}{2\mu}r\frac{dr}{dt} = -\frac{qB}{4\mu}\frac{d}{dt}r^2. \tag{6}$$
It follows from (6) that
$$\frac{dL_\omega}{dt} = 0, \tag{7}$$

where L_ω is defined in (1). Thus, L_ω is a constant whose value is determined by the initial conditions for r and $\dot\theta$. When $B=0$ (i.e. $\omega=0$), L_o is the total angular momentum relative to the CM frame (see Question 8.2). By using (1) to eliminate $\dot\theta$ from (5), we obtain the one-dimensional radial equation (2).

(b) The radial equation can be written in terms of an effective potential as

$$\mu\ddot r = -\frac{dV_e}{dr}. \qquad (8)$$

By comparing this with (2) we see that the effective potential for the radial motion is

$$V_e(r) = \frac{kq^2}{r} + \frac{L_\omega^2}{2\mu r^2} + \tfrac{1}{8}\mu\omega^2 r^2. \qquad (9)$$

(Here, we have set an arbitrary constant of integration equal to zero.) Clearly, $V_e \to \infty$ as $1/r^2$ when $r \to 0_+$ and as r^2 when $r \to \infty$. Therefore, the radial motion is always bounded between two classical turning points that are the positive solutions to the equation $V_e(r)=$ the energy of relative motion. Despite the Coulomb repulsion, the magnetic field always binds the particles.

Comments

(i) The radial equation (2) describes motion in a rotating (non-inertial) frame: in addition to the Coulomb repulsion and the magnetic force, it contains a non-inertial force (the centrifugal force).

(ii) Relative to the CM frame the position vectors $\mathbf{r}_1$ and $\mathbf{r}_2$ are equal in magnitude and opposite in direction: $\mathbf{r}_2 = -\mathbf{r}_1$.

(iii) The simplest trajectory for the relative motion $\mathbf{r}$ is a circle. Then, r is a constant (as is $\dot\theta$ – see (1)) and (1) and (2) give

$$\mu\frac{v^2}{r} = \mu\omega v - \frac{kq^2}{r^2}. \qquad (10)$$

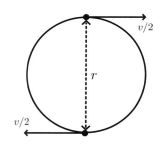

Here, $v=r|\dot\theta|$ (the speed of each particle in the CM frame is $\tfrac{1}{2}v$) and we have set $\operatorname{sign}\dot\theta = -\operatorname{sign} q$, which is the condition for the magnetic force to oppose the repulsive electrostatic interaction. By solving (10) for r we obtain for the radius of the circle in the CM frame

$$\tfrac{1}{2}r = \frac{v}{4\omega}\left[1 + \sqrt{1 + 4kq^2\omega/\mu v^3}\right]. \qquad (11)$$

(iv) It is possible to obtain solutions (in terms of elliptic integrals) for the trajectories in a plane perpendicular to **B**. However, these are complicated and the reader is referred to the literature for details.[3] It is not difficult to obtain simple solutions for closed trajectories and numerical solutions for open trajectories, and these are considered in the next two questions.

Question 10.18

Consider the two identical interacting charged particles of Question 10.17 moving in the xy-plane, perpendicular to a uniform magnetostatic field $\mathbf{B} = B\hat{\mathbf{z}}$, where $B > 0$.

(a) Show that there are closed orbits with trajectories given by

$$x_i(t) = R_0 \cos \omega t + (-1)^i A\lambda \cos \frac{\alpha}{\beta}\omega t \tag{1}$$

$$y_i(t) = -R_0 \sin \omega t + (-1)^{i-1} A\lambda \sin \frac{\alpha}{\beta}\omega t \tag{2}$$

($i = 1, 2$). Here, R_0 is a constant,

$$\omega = \frac{qB}{m}, \qquad \lambda = \left(\frac{mk}{B^2}\right)^{1/3}, \tag{3}$$

$$A = \left[4\frac{\alpha}{\beta}\left(1 - \frac{\alpha}{\beta}\right)\right]^{-1/3}, \tag{4}$$

and α and β are positive integers with $\alpha < \beta$.

(b) Express (1) and (2) in terms of the dimensionless variables $\bar{x}_i = x_i/\lambda$, $\bar{y}_i = y_i/\lambda$ and $\tau = t/T$, where $T = 2\pi/|\omega|$. Then make computer plots of the trajectories for $\bar{R}_0 = R_0/\lambda = 1$ and

$$\frac{\alpha}{\beta} = \frac{1}{2}, \frac{3}{4}, \frac{2}{3} \text{ and } \frac{1}{3}. \tag{5}$$

Solution

(a) According to the previous two questions, the trajectories of the particles are

$$\mathbf{r}_1 = \mathbf{R} - \tfrac{1}{2}\mathbf{r} \quad \text{and} \quad \mathbf{r}_2 = \mathbf{R} + \tfrac{1}{2}\mathbf{r}, \tag{6}$$

where the CM vector performs a circular motion

$$\mathbf{R} = R_0(\cos \omega t, -\sin \omega t) \tag{7}$$

and the relative vector $\mathbf{r} = (x, y)$ satisfies

$$m\ddot{\mathbf{r}} = q\dot{\mathbf{r}} \times \mathbf{B} + \frac{2kq^2}{r^3}\mathbf{r}. \tag{8}$$

The x- and y-components of (8) are

$$\ddot{x} = \omega \dot{y} + 2\omega^2\lambda^3 \frac{x}{(x^2+y^2)^{3/2}} \tag{9}$$

$$\ddot{y} = -\omega \dot{x} + 2\omega^2\lambda^3 \frac{y}{(x^2+y^2)^{3/2}}. \tag{10}$$

We are interested in solutions to (9) and (10) that will produce closed orbits for the trajectories (6). Thus, we look for solutions where the relative vector $\mathbf{r}$ performs circular motion with constant angular velocity $\dot{\theta}$:

$$x = 2A\cos\dot\theta t, \qquad y = 2A\sin\dot\theta t, \tag{11}$$

and the constants A and $\dot\theta$ are to be determined. Equations (11) are solutions to (9) and (10) provided

$$\frac{\lambda^3}{4|A|^3} = -\frac{\dot\theta}{\omega} - \frac{\dot\theta^2}{\omega^2}. \tag{12}$$

Thus, $\dot\theta$ and ω have opposite signs, meaning that $\operatorname{sign}\dot\theta = -\operatorname{sign} q$. Now, for closed orbits the ratio of the periods $2\pi/|\omega|$ and $2\pi/|\dot\theta|$ of the circular CM and relative motions must be a rational number, and so we require

$$\frac{\dot\theta}{\omega} = -\frac{\alpha}{\beta}, \tag{13}$$

where α and β are positive integers. Equations (12) and (13) yield (4). The trajectories (1) and (2) follow from (6), (7), (11) and (13).

(b) By replacing t with $2\pi\tau/|\omega|$ and x_i and y_i with $\lambda\bar{x}_i$ and $\lambda\bar{y}_i$ in (1) and (2), we have the dimensionless forms of these trajectories

$$\bar{x}_i(\tau) = \bar{R}_0 \cos 2\pi\tau + (-1)^i A\cos\left(2\pi\frac{\alpha}{\beta}\tau\right) \tag{14}$$

$$\bar{y}_i(\tau) = -\gamma\bar{R}_0 \sin 2\pi\tau + (-1)^{i-1}\gamma A\sin\left(2\pi\frac{\alpha}{\beta}\tau\right), \tag{15}$$

where $\gamma = \operatorname{sign} q$. From (14) and (15) we obtain the diagrams shown below. In these the dots labelled 1 and 2 indicate the initial positions of the particles. When β is an even integer the trajectories of the particles form a single closed path, as indicated by the solid curves in the first two figures. When β is an odd integer the trajectories of the particles are distinct: in the third and fourth figures they are indicated by a solid curve for particle 1 and a dotted curve for particle 2. The figures are for $q > 0$; for $q < 0$ the directions of the arrows are reversed.

Comments

(i) In order to obtain the closed trajectories described above, the initial conditions $\mathbf{r}_i(0)$ and $\mathbf{v}_i(0)$ must be exactly those implicit in (1) and (2).

(ii) The trajectories described by (1) and (2) are a superposition of two uniform circular motions: 'orbital' (motion of the CM about the origin with angular velocity ω) and 'spin' (rotation of the relative vector about the CM with angular velocity $\dot\theta$ that is related to ω by the closure condition (13)). The circles traced by the CM and relative vectors are called the deferent and epicycle, respectively, and the particle trajectories are known as epitrochoids. Epicycles were probably first used – albeit without a dynamical basis – in one of the early accounts of planetary orbits by Hipparchus and Ptolemy, which conformed to Aristotle's doctrine of the pre-eminence of circular motion.

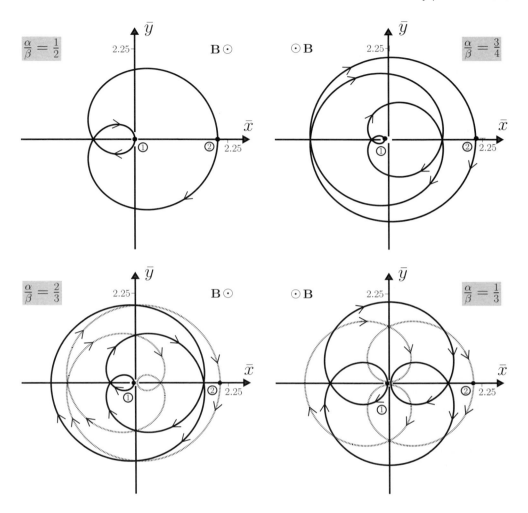

Question 10.19

Two interacting charged particles move in the xy-plane, perpendicular to a uniform magnetostatic field $\mathbf{B} = B\hat{\mathbf{z}}$. In this question we consider numerical solutions of their equations of motion (3) and (4) of Question 10.16.

(a) First, express these equations of motion in the dimensionless forms

$$\frac{d^2 \bar{\mathbf{r}}_i}{d\tau^2} = 2\pi \gamma_i \frac{d\bar{\mathbf{r}}_i}{d\tau} \times \hat{\mathbf{z}} + 4\pi^2 \gamma_1 \gamma_2 (-1)^i \frac{\bar{\mathbf{r}}}{\bar{r}^3} \qquad (i = 1, 2). \qquad (1)$$

Here, $\gamma_i = \mathrm{sign}\, q_i = \pm 1$. The quantity $\tau = t/T$ (where $T = 2\pi/\omega$ and $\omega = qB/m$ with $q = |q_1| = |q_2|$) is a dimensionless time, and $\bar{\mathbf{r}} = \mathbf{r}/\lambda$, etc., with $\lambda = (mk/B^2)^{1/3}$, are dimensionless position vectors.

(b) Use *Mathematica* to plot the particle trajectories for like charges ($\gamma_1 = \gamma_2 = 1$) and the following initial conditions at $\tau = 0$:

1. $\bar{x}_i = 0$, $\bar{y}_i = (-1)^{i+1}$, $d\bar{x}_i/d\tau = (-1)^i \pi$, $d\bar{y}_i/d\tau = 0$;
2. $\bar{x}_i = 0$, $\bar{y}_i = (-1)^{i+1}$, $d\bar{x}_i/d\tau = \pi$, $d\bar{y}_i/d\tau = 0$, where $i = 1, 2$.

(c) Use *Mathematica* to plot the particle trajectories and also the CM trajectory for unlike charges ($\gamma_1 = -\gamma_2 = 1$) and the following initial conditions at $\tau = 0$:

1. $\bar{x}_i = 0$, $\bar{y}_i = (-1)^{i+1}$, $d\bar{x}_i/d\tau = (-1)^{i+1} 2\pi$, $d\bar{y}_i/d\tau = 0$;
2. $\bar{x}_i = 0$, $\bar{y}_i = (-1)^{i+1}$, $d\bar{x}_i/d\tau = 0$, $d\bar{y}_i/d\tau = 2\pi$, where $i = 1, 2$.

Solution

(a) With $q_i = q \operatorname{sign} q_i = \gamma_i q$, we write (3) and (4) of Question 10.16 as

$$m\ddot{\mathbf{r}}_i = \gamma_i q \dot{\mathbf{r}}_i \times \mathbf{B} + k \frac{(-1)^i \gamma_1 \gamma_2 q^2}{r^3} \mathbf{r}. \tag{2}$$

Equation (1) follows if we multiply (2) by T^2/λ and set $\mathbf{B} = B\hat{\mathbf{z}}$.

(b) The *Mathematica* notebook below uses the package `VectorAnalysis` to perform the cross-product in (1). Appropriate initial conditions should be entered at the beginning of this notebook.

```
In[1]:= Needs["VectorAnalysis`"]
     γ1 = 1; γ2 = 1; τmax = 5;
     x10 = 0; y10 = 1; z10 = 0; vx10 = -π; vy10 = 0; vz10 = 0;
     x20 = 0; y20 = -1; z20 = 0; vx20 = π; vy20 = 0; vz20 = 0;
     r1[τ_] := {x1[τ], y1[τ], z1[τ]}; r2[τ_] := {x2[τ], y2[τ], z2[τ]};
     r[τ_] := r2[τ] - r1[τ];
     LorentzForce1[τ_] := 2π γ1 CrossProduct[r1'[τ], {0, 0, 1}];
     LorentzForce2[τ_] := 2π γ2 CrossProduct[r2'[τ], {0, 0, 1}];
     EqnMotion1 = Thread[r1''[τ] - LorentzForce1[τ] +
             4π² γ1 γ2  r[τ]/Dot[r[τ], r[τ]]^(3/2)  == 0];
     EqnMotion2 = Thread[r2''[τ] - LorentzForce2[τ] -
             4π² γ1 γ2  r[τ]/Dot[r[τ], r[τ]]^(3/2)  == 0];
     InitCon1 = Join[Thread[r1[0] == {x10, y10, z10}],
             Thread[r1'[0] == {vx10, vy10, vz10}]];
     InitCon2 = Join[Thread[r2[0] == {x20, y20, z20}],
             Thread[r2'[0] == {vx20, vy20, vz20}]];
```

```
In[2]:= EqsToSolve = Join[EqnMotion1, EqnMotion2, InitCon1, InitCon2];
        Sol = NDSolve[EqsToSolve, Join[r1[τ], r2[τ]], {τ, 0, τmax}];
        ParametricPlot[{Evaluate[{x1[τ], y1[τ]}]/.Sol,
            Evaluate[{x2[τ], y2[τ]}]/.Sol}, {τ, 0, τmax}, AspectRatio → 1]
        ParametricPlot[Evaluate[{x2[τ] - x1[τ], y2[τ] - y1[τ]}]/.Sol,
            {τ, 0, τmax}, AspectRatio → 1, PlotRange → {{-6, 6}, {-6, 6}}]
```

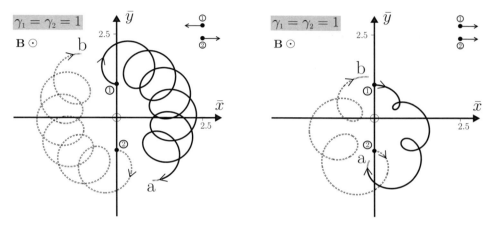

For the diagram on the left (initial conditions 1), the CM remains at rest at the origin and the relative vector oscillates between two concentric circles having an inner radius of 1 and an outer radius of about 2.32. For the diagram on the right (initial conditions 2), the CM moves in a circle of radius 0.5 and centred at $(0, -0.5)$. The relative vector again oscillates between two concentric circles having an inner radius of 1 and an outer radius of about 1.48.

(c) The following figures show the particle trajectories $(\bar{x}, \bar{y})$ and the CM trajectory $(\bar{X}, \bar{Y})$ for charges of opposite sign:

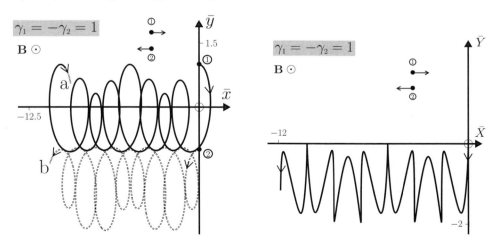

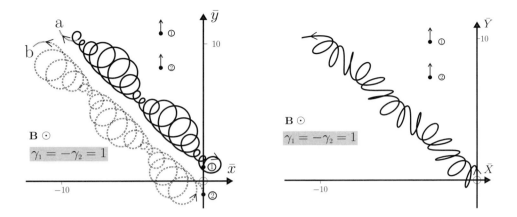

Comment

Two-dimensional motion of two interacting electrons in a plane perpendicular to an applied magnetic field is of interest in the Hall effect and other phenomena.[3]

Question 10.20

Consider the effect of a time-varying gravitational constant $G(t)$ on the Earth's orbit $\mathbf{r}(t)$ about the Sun. Neglecting the mass of the Earth in comparison with that of the Sun, the equation of motion is

$$\ddot{\mathbf{r}} + \frac{G(t)M_\text{s}}{r^3}\mathbf{r} = 0. \tag{1}$$

Use Cartesian coordinates with initial conditions

$$\mathbf{r}(0) = (r_0, 0) \quad \text{and} \quad \mathbf{v}(0) = \left(0, \sqrt{G_\text{o}M_\text{s}/r_0}\right)$$

where $r_0 = 1.49 \times 10^{11}$ m, $M_\text{s} = 2.00 \times 10^{30}$ kg, and $G_\text{o} = 6.67 \times 10^{-11}$ N m² kg⁻².

(a) **1.** Use *Mathematica* to solve (1) and plot the trajectory $y(x)$ if $G(t)$ decreases linearly with time:

$$G(t) = G_\text{o}(1 - \alpha t). \tag{2}$$

Use the value $\alpha = 3.60 \times 10^{-9}$ s⁻¹, which is large enough to have an observable effect on Earth's orbit after one year.‡ Plot the trajectory up to $t = 4$ yr.
2. Now reduce α tenfold. Plot a graph of the period T of Earth's orbit versus the number n of complete Earth orbits, up to $n = 18$.

---

‡As the Earth recedes, the time to orbit the Sun and hence the length of the year increases. In the following, one year means $365 \times 24 \times 3600$ s.

(b) Use *Mathematica* to plot the radial distance $r(t)$ for $0 \leq t \leq 64 \, \text{yr}$, and for an oscillatory dependence

$$G(t) = G_0(1 + \beta \sin 2\pi t/T_0), \tag{3}$$

where $\beta = 5.00 \times 10^{-3}$ and $T_0 = $ one year $= 3.16 \times 10^7$ s.

Solution

(a) **1.** We use the following *Mathematica* notebook to solve (1):

```
In[1]:= Ms = 2.00 × 10^30; r0 = 1.49 × 10^11; G0 = 6.67 × 10^-11;
OneYear = 3600 × 24 × 365; Tmax = 4 × OneYear; α = 3.60 × 10^-9;

x0 = r0; y0 = 0; vx0 = 0; vy0 = √((G0 Ms)/r0);

G[t_] := G0 (1 - α t); r[t_] := {x[t], y[t]};

EqnMotion = Thread[r''[t] + (G[t] Ms r[t])/(Dot[r[t], r[t]]^(3/2)) == 0];

InitCon = Join[Thread[r[0] == {x0, y0}], Thread[r'[0] == {vx0, vy0}]];

EqsToSolve = Join[EqnMotion, InitCon];

Sol = NDSolve[EqsToSolve, r[t], {t, 0, Tmax}, MaxSteps → 100000];

ParametricPlot[{Evaluate[{x[t]/r0, y[t]/r0}]/.Sol}, {t, 0, Tmax},
  PlotRange → {{-1.6, 1.6}, {-1.6, 1.6}}, AspectRatio → 1]
```

The diagram below shows Earth's orbit for $t \leq 4\,\text{yr}$. The dotted circular orbit is for G constant and equal to G_0. **2.** The graph of T/T_0 versus n is for $\alpha = 3.60 \times 10^{-10}$ s and $T_0 = 1\,\text{yr}$.

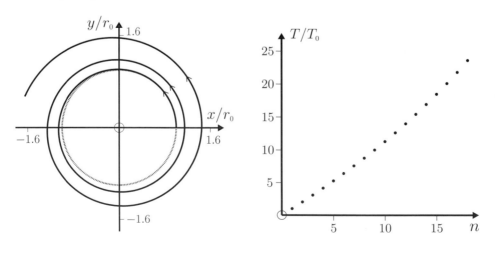

(b) We replace (2) with (3) and re-run the *Mathematica* notebook. The diagram below is for a 64-yr time span. The beat-like behaviour repeats itself and the motion is bounded; a plot of $y(x)$ shows that the trajectory fills a ring-like domain.

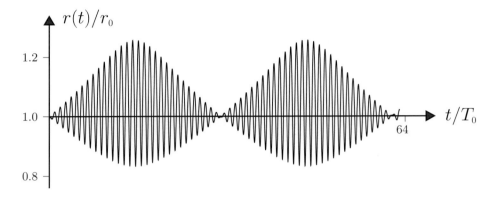

Comment

There has been considerable interest in the possibility that fundamental constants like G may vary with time.[4,5] Any actual variation in G, if it occurs, would be very small – less than about one part in 10^{11} per year. (In this question, the assumed variation of G with time has been exaggerated for illustrative purposes.)

---

[4] C. M. Will, "Experimental gravitation from Newton's *Principia* to Einstein's general relativity," in *Three hundred years of gravitation* (S. W. Hawking and W. Israel, eds.), Cambridge, Cambridge University Press, pp. 98 – 101, 1987.
[5] E. B. Norman, "Are fundamental constants really constant?," American Journal of Physics, vol. 54, pp. 317–321, 1986.

11
Multi-particle systems

The gamut of physics extends from the study of 'nothing' (the vacuum) through one-, two-, and few-body problems, to systems containing such a large number of particles that they approximate a continuum. This chapter contains examples that illustrate some aspects of classical multi-particle systems. In the study of such systems considerable licence is taken with the ubiquitous words 'particle' and 'system'. We mention three examples to illustrate this.

When Newton solved the Kepler problem he was, at first, faced by a seemingly intractable task: after all, the Sun is a complicated aggregate of particles, as is a planet. To circumvent this he first approximated the gravitational field of the Sun by that of a particle located at the centre of the Sun and with mass equal to a solar mass. He then considered, instead of the motion of each particle in a planet, the motion of a single particle having mass equal to that of the planet and located at its centre. And so Newton arrived at a tractable yet reasonable problem: the motion of two 'particles' interacting via his law of universal gravitation. This example illustrates two facets that are often encountered in many-body problems. First, there is source (or potential) theory that is used to find the external forces. Secondly, there is response theory that describes how a selected system responds to these external forces as well as to the interparticle forces within it.

In cosmology, the particles may be a dust cloud in otherwise empty space that evolves according to the laws of physics to form a star, or a solar system or even a galaxy. This can be taken to an extreme, where one considers a 'gas' comprised of galaxies – each galaxy being represented as a 'particle'. The latter is an extension of an approach used successfully in the theory of ordinary gases where, for certain properties, one can neglect the structure of molecules and treat them simply as particles.

As a third example we think of an experiment in metal physics where a crystal is subject to a tensile stress sufficient to cause plastic deformation. Here, the system of interest may be an array of parallel line defects in the crystal structure known as dislocations. These 'particles' repel each other and also move in response to external forces (the applied forces and the forces due to, for example, other crystal defects and the crystal surfaces.) Their dynamics and the dislocation structures they form are essential to understanding the mechanical properties of metals.

Question 11.1

Consider a system of N interacting particles with masses m_i, position vectors $\mathbf{r}_i$, and momenta $\mathbf{p}_i = m_i \dot{\mathbf{r}}_i$ ($i = 1, 2, \cdots, N$) relative to an inertial frame of reference. The particles are also subject to external forces $\mathbf{F}_i^{(e)}$, and the masses m_i are assumed to be constant. Prove the following:

1. That the total momentum $\mathbf{P}$ and the total external force $\mathbf{F}^{(e)}$ acting on the system, namely

$$\mathbf{P} = \sum_{i=1}^{N} m_i \dot{\mathbf{r}}_i \quad \text{and} \quad \mathbf{F}^{(e)} = \sum_{i=1}^{N} \mathbf{F}_i^{(e)}, \tag{1}$$

satisfy

$$\frac{d\mathbf{P}}{dt} = \mathbf{F}^{(e)}. \tag{2}$$

2. That the equation of motion of the centre-of-mass (CM) vector

$$\mathbf{R} = \frac{1}{M} \sum_{i=1}^{N} m_i \mathbf{r}_i \tag{3}$$

is

$$M\ddot{\mathbf{R}} = \mathbf{F}^{(e)}, \tag{4}$$

where $M = \sum_{i=1}^{N} m_i$ is the total mass of the system.

Solution

Let $\mathbf{F}_{ji}$ denote the force that particle j exerts on particle i. According to Newton's second law, the equations of motion are

$$\frac{d\mathbf{p}_i}{dt} = m_i \frac{d^2 \mathbf{r}_i}{dt^2} = \sum_{j=1}^{N} \mathbf{F}_{ji} + \mathbf{F}_i^{(e)} \quad (i = 1, 2, \cdots, N), \tag{5}$$

where in the sum the term $\mathbf{F}_{ii}$ is excluded. By adding together the N equations in (5), and using (1) and (3), we have

$$\frac{d\mathbf{P}}{dt} = M \frac{d^2 \mathbf{R}}{dt^2} = \mathbf{S} + \mathbf{F}^{(e)}, \tag{6}$$

where

$$\mathbf{S} = \sum_{i=1}^{N} \sum_{j=1}^{N} \mathbf{F}_{ji}. \tag{7}$$

In (7) terms with $j = i$ are excluded, and so $\mathbf{S}$ contains $N(N-1)$ terms. Also, for each term $\mathbf{F}_{ji}$ there is a corresponding term $\mathbf{F}_{ij}$ in (7): so $\mathbf{S}$ consists of pairs of terms $\mathbf{F}_{ji} + \mathbf{F}_{ij}$ (there are $\frac{1}{2}N(N-1)$ of them). Each pair is zero because, according to Newton's third law, $\mathbf{F}_{ji} = -\mathbf{F}_{ij}$. Therefore, $\mathbf{S} = 0$ and (6) yields (2) and (4).

Comments

(i) Equations (2) and (4) are generalizations of results obtained previously for two-body problems in Questions 10.1 and 10.2.

(ii) According to (2), the rate of change of the total momentum **P** of a system of particles is equal to the total external force $\mathbf{F}^{(e)}$ acting on them. It follows that if $\mathbf{F}^{(e)} = 0$ then **P** is constant, which is the law of conservation of momentum for a system of particles relative to an inertial frame. See also Question 14.7.

(iii) Equation (3) defines the position vector (relative to an arbitrarily chosen coordinate origin O) of the CM of the system of particles. It is clear that **R** is an origin-dependent vector – meaning that it depends on our choice of O.

(iv) If the distribution of mass is continuous (rather than discrete), with volume density $\rho = dm/dV$, then the definition (3) is replaced by

$$\mathbf{R} = \frac{1}{M} \int \mathbf{r}\, dm = \frac{1}{M} \int \mathbf{r}\rho\, dV. \tag{8}$$

(v) According to (4) the trajectory of the CM is that of a hypothetical particle of mass M acted on by the total external force $\mathbf{F}^{(e)}$. The interparticle forces play no role in the dynamics of the CM, and its trajectory is therefore generally much simpler than that of the individual particles. For example, if a balloon filled with water is thrown in a uniform gravitational field, the motions of its parts may be very complicated, but the motion of the CM is along a parabola. Note that in $(1)_2$ the external forces $\mathbf{F}_i^{(e)}$ are evaluated at the positions of the corresponding particles.

(vi) Equation (4) is the basis of Newton's (and later Maxwell's) formulation of dynamics in terms of bodies rather than particles (see Chapter 1).

(vii) The above analysis leading to (2) and (4) is so simple to perform that one can easily overlook the wonderful nature of these key results – specifically that the interparticle forces play no role in them. To emphasize this point we quote at length from an interesting discussion by Temple:[1]

"You are all familiar, of course, with the way in which Newton's laws of motion are expounded in the textbooks and the rather didactic way in which they are imposed upon us. The truth is that they are positions, hypotheses, taken up to correlate our observations, and there is nothing self-evident in them which imposes itself upon our minds. The most interesting of the laws of motion is the third. I am not a Newtonian scholar and I cannot tell you how he arrived at the third law, but there is one obvious way in which he ought to have arrived:– One of the great problems which confronts any philosopher is the crucial question 'how is it possible to know anything without knowing everything?'. In particular, if you have committed yourself to a system of mechanics in which accelerations are produced by external forces, and if you are committed to the view that these forces are interactions between particles, it would seem that before you can make any progress in the subject you would need to know not only the configuration

---

[1] G. Temple, in *Turning points in physics*, pp. 71–72. Amsterdam: North-Holland, 1959.

of the universe, but also very detailed information about the nature of the forces involved. In particular you would never be successful unless you had a complete knowledge of the interatomic forces holding the body together. How are you to solve this difficulty? It seems there is only one way. You can only make progress in the face of ignorance by assuming that all the internal interactions cancel out, and this is precisely the significance of Newton's third law. It is a bold hypothesis adopted to enable us to make progress in dynamics without elaborate knowledge of interatomic forces. Making this assumption, the only forces which we have to consider are the external forces and this at once leads to those grand theorems about the rate of change of linear and angular momentum."

Question 11.2

Suppose that the interparticle forces $\mathbf{F}_{ji}$ for the N particles in the previous question are central (meaning that $\mathbf{F}_{ji}$ is directed along the line joining particles i and j). Let $\mathbf{L}$ and $\mathbf{\Gamma}^{(e)}$ denote the total angular momentum of the particles and the total torque on the particles due to external forces:

$$\mathbf{L} = \sum_{i=1}^{N} \mathbf{r}_i \times \mathbf{p}_i \quad \text{and} \quad \mathbf{\Gamma}^{(e)} = \sum_{i=1}^{N} \mathbf{r}_i \times \mathbf{F}_i^{(e)}. \tag{1}$$

Prove that

$$\frac{d\mathbf{L}}{dt} = \mathbf{\Gamma}^{(e)}. \tag{2}$$

Solution

Differentiation of $(1)_1$ gives

$$\frac{d\mathbf{L}}{dt} = \sum_{i=1}^{N} \mathbf{r}_i \times \frac{d\mathbf{p}_i}{dt} = \sum_{i=1}^{N}\sum_{j=1}^{N} \mathbf{r}_i \times \mathbf{F}_{ji} + \sum_{i=1}^{N} \mathbf{r}_i \times \mathbf{F}_i^{(e)}, \tag{3}$$

where in the second step we have used the equations of motion (5) of Question 11.1. The double sum in (3) excludes terms with $j=i$, and consists of $\frac{1}{2}N(N-1)$ pairs of terms

$$\mathbf{r}_i \times \mathbf{F}_{ji} + \mathbf{r}_j \times \mathbf{F}_{ij} = (\mathbf{r}_i - \mathbf{r}_j) \times \mathbf{F}_{ji}. \tag{4}$$

Now, central interparticle forces $\mathbf{F}_{ji}$ are parallel to the relative position vector $\mathbf{r}_i - \mathbf{r}_j$. So, for such forces all pairs (4) are zero, and consequently (3) reduces to (2).

Comments

(i) According to (2), for a system of particles that interact via central interparticle forces, the total angular momentum is constant whenever the total torque due to external forces is zero. This is an example of the law of conservation of angular momentum of a system of particles relative to an inertial frame. See also Questions 14.18 and 14.19.

(ii) This law applies also when the interparticle forces are non-central, such as in certain electromagnetic interactions, provided the total angular momentum of the electromagnetic field is taken into account.

Question 11.3

Let $\mathbf{L}$ be the angular momentum of a system of particles relative to a frame with origin at O, and let $\mathbf{L}'$ be the angular momentum of the system relative to a frame with origin O' at the CM. Prove that

$$\mathbf{L} = \mathbf{L}' + \mathbf{R} \times \mathbf{P}, \qquad (1)$$

where $\mathbf{R}$ is the position vector of O' relative to O and $\mathbf{P} = M\dot{\mathbf{R}}$ with M the total mass of the system.

Solution

The angular momentum relative to O is

$$\mathbf{L} = \sum_{i=1}^{N} \mathbf{r}_i \times m_i \dot{\mathbf{r}}_i . \qquad (2)$$

From the figure, $\mathbf{r}_i = \mathbf{r}'_i + \mathbf{R}$ and so

$$\mathbf{L} = \sum_{i=1}^{N}(\mathbf{r}'_i + \mathbf{R}) \times m_i(\dot{\mathbf{r}}'_i + \dot{\mathbf{R}})$$

$$= \sum_{i=1}^{N} \mathbf{r}'_i \times m_i \dot{\mathbf{r}}'_i + \mathbf{R} \times \Big(\sum_{i=1}^{N} m_i\Big) \dot{\mathbf{R}} + \mathbf{R} \times \sum_{i=1}^{N} m_i \dot{\mathbf{r}}'_i + \Big(\sum_{i=1}^{N} m_i \mathbf{r}'_i\Big) \times \dot{\mathbf{R}} . \qquad (3)$$

The first term in (3) is just the angular momentum

$$\mathbf{L}' = \sum_{i=1}^{N} \mathbf{r}'_i \times m_i \dot{\mathbf{r}}'_i \qquad (4)$$

relative to O' (i.e. relative to the CM). The second term is $\mathbf{R} \times M\dot{\mathbf{R}} = \mathbf{R} \times \mathbf{P}$. The third and fourth terms are zero because

$$\sum_{i=1}^{N} m_i \mathbf{r}'_i = \sum_{i=1}^{N} m_i (\mathbf{r}_i - \mathbf{R}) = M\mathbf{R} - M\mathbf{R} = 0,$$

and similarly for the sum over $m_i \dot{\mathbf{r}}'_i$ (provided the m_i are constant). So (3) yields (1).

Comments

(i) The quantity $\mathbf{L}'$ defined in (4) is the angular momentum relative to the CM. It is independent of the choice of origin O, and is often referred to as the spin angular momentum. The quantity

$$\mathbf{L}_{\text{CM}} = \mathbf{R} \times \mathbf{P} \qquad (5)$$

is the angular momentum of a particle of mass M located at the CM – it is referred to as the orbital angular momentum. $\mathbf{L}_{\mathrm{CM}}$ depends on the choice of O, and therefore so does $\mathbf{L} = \mathbf{L}' + \mathbf{L}_{\mathrm{CM}}$.

(ii) If O is the origin of an inertial frame, then the torque equation for $\mathbf{L}_{\mathrm{CM}}$ is

$$\dot{\mathbf{L}}_{\mathrm{CM}} = \mathbf{R} \times \dot{\mathbf{P}} = \mathbf{R} \times \sum_{i=1}^{N} \mathbf{F}_i^{(e)} = \mathbf{R} \times \mathbf{F}^{(e)}, \tag{6}$$

meaning that $\mathbf{L}_{\mathrm{CM}}$ responds as if the total external force $\mathbf{F}^{(e)}$ exerts a torque $\mathbf{R} \times \mathbf{F}^{(e)}$ at $\mathbf{R}$. (In (6) we have used (2) of Question 11.1.)

(iii) If the interparticle forces are central, then the torque equation for $\mathbf{L}'$ is

$$\dot{\mathbf{L}}' = \dot{\mathbf{L}} - \dot{\mathbf{L}}_{\mathrm{CM}} = \sum_{i=1}^{N}(\mathbf{r}_i - \mathbf{R}) \times \mathbf{F}_i^{(e)} = \sum_{i=1}^{N} \mathbf{r}'_i \times \mathbf{F}_i^{(e)}, \tag{7}$$

meaning that the angular momentum relative to the CM responds to the external torque about the CM. In particular, $\mathbf{L}'$ is conserved whenever the total external torque about the CM is zero. This conclusion holds even if the frame with origin O' (the CM frame) is non-inertial.

(iv) In the definition (2) of angular momentum, the position vectors $\mathbf{r}_i$ and momentum vectors $\mathbf{p}_i = m_i \dot{\mathbf{r}}_i$ are taken with respect to the same frame with origin at O. This quantity is referred to as relative angular momentum to distinguish it from so-called absolute angular momentum, where the position and momentum vectors are with respect to different frames. Consequently, the torque equations can be different for these two types of angular momentum. The relative definition is widely used in the physics literature, whereas the absolute version is popular in engineering texts. Which one uses is essentially a matter of convenience.[2]

Question 11.4

Consider a system of N particles in a uniform gravitational field. Prove that the total gravitational torque about the CM is zero.

Solution

The force on the ith particle is $m_i \mathbf{g}$ where the gravitational acceleration $\mathbf{g}$ is a constant. The total torque about an origin O is

$$\boldsymbol{\Gamma} = \sum_{i=1}^{N} \mathbf{r}_i \times m_i \mathbf{g} = \left(\sum_{i=1}^{N} m_i \mathbf{r}_i \right) \times \mathbf{g}, \tag{1}$$

where $\mathbf{r}_i$ is the position vector of the ith particle relative to O. If O is located at the CM then $\sum_{i=1}^{N} m_i \mathbf{r}_i = 0$ and consequently $\boldsymbol{\Gamma} = 0$.

---

[2] M. Illarramendi and T. del Rio Gaztelurrutia, "Moments to be cautious of – relative versus absolute angular momentum," European Journal of Physics, vol. 16, pp. 249–255, 1995.

Comments

(i) In the above case the CM is also referred to as the centre of gravity (CG).
(ii) For a rigid body in a uniform gravitational field there is a unique point about which the total torque is zero for an arbitrary orientation of the body. For, if there are two such points separated by a vector $\mathbf{D}$ then $\mathbf{D} \times M\mathbf{g} = 0$ for any orientation of the body (and hence of $\mathbf{D}$), which is impossible.
(iii) If the field is non-uniform then the total torque $\mathbf{\Gamma} = \sum_{i=1}^{N} \mathbf{r}_i \times m_i \mathbf{g}_i$ need not be zero in a CM frame.

Question 11.5

Consider a system of N particles that interact via an attractive 'gravitational' force that is proportional to the distance between particles:

$$\mathbf{F}_{ji} = -k m_i m_j (\mathbf{r}_i - \mathbf{r}_j), \tag{1}$$

where k is a positive constant and $i, j = 1, 2, \cdots, N$. Determine the trajectories of the particles. (Hint: Choose a reference frame in which the CM is at rest at the origin.)

Solution

The equations of motion corresponding to the interparticle forces (1) are

$$m_i \ddot{\mathbf{r}}_i = -k m_i \sum_{j=1}^{N} m_j (\mathbf{r}_i - \mathbf{r}_j). \tag{2}$$

That is,

$$m_i \ddot{\mathbf{r}}_i = -kM m_i \mathbf{r}_i + kM m_i \mathbf{R}, \tag{3}$$

where $M = \sum_{j=1}^{N} m_j$ is the total mass and $\mathbf{R} = \sum_{j=1}^{N} m_j \mathbf{r}_j / M$ is the CM vector. By summing (3) over all i we find that the acceleration of the CM is zero (as it must be, because there are no external forces): $\ddot{\mathbf{R}} = 0$. So the CM moves with constant velocity relative to an inertial frame, and therefore the CM frame is also inertial. In the CM frame $\mathbf{R} = 0$ and (3) simplifies to

$$\ddot{\mathbf{r}}_i + \omega^2 \mathbf{r}_i = 0, \tag{4}$$

where $\omega = \sqrt{k/M}$ is real. We see that the trajectories $\mathbf{r}_i$ in the CM frame are described by a set of N independent, isotropic harmonic oscillator equations. The solutions for such oscillators have been studied in Questions 7.25 and 8.10. The trajectories are ellipses centred on the origin (i.e. on the CM). In general, the ellipses will differ in size, but the period $T = 2\pi/\omega = 2\pi\sqrt{M/k}$ is the same for all particles. Of course, because we are working in the CM frame, the solutions to (4) must satisfy $\sum_{i=1}^{N} m_i \mathbf{r}_i = 0$.

Comments

(i) This solution was first obtained by Newton, who used induction to show that "all bodies will describe different ellipses, with equal periodic times about their common centre of gravity, in an immovable plane."[3] Chandrasekhar feels that Newton's method of solution is superior to the modern one, in that "it exhibits the physical basis for the unfolding of the solution. And one is left marvelling at Newton's ability to explain precisely in words involved analytical arguments."[3]

(ii) For the inverse-square gravitational interaction one has, instead of (1),

$$\mathbf{F}_{ji} = -Gm_i m_j \frac{(\mathbf{r}_i - \mathbf{r}_j)}{|\mathbf{r}_i - \mathbf{r}_j|^3}, \tag{5}$$

and no analytical solution to the many-body problem exists. Only the two-body problem (see Chapter 10) can be solved exactly, and also certain special cases for more than two bodies (see below).

Question 11.6

(a) Consider the gravitational three-body problem for the special case where the three particles are always located at the corners of an equilateral triangle, the length of whose sides vary with time. Show that each particle moves on a conic section with a focus at the CM of the system. (Hint: Work in CM coordinates.)

(b) Write a *Mathematica* notebook to plot the trajectories for bounded (elliptical) motion. Use mass ratios $m_2/m_1 = 2$ and $m_3/m_1 = 10/3$, and eccentricity $e = 0.8$ and major axis $2a = 10/9$ units for particle 1 (see Question 8.9). Implement *Mathematica*'s Manipulate function to produce a dynamic output. (Hint: Use the eccentric anomaly and Kepler's equation – see (14) and (15) of Question 8.18).

Solution

(a) Let the particles have masses m_i and position vectors $\mathbf{r}_i$ ($i = 1, 2, 3$) in an inertial frame. The equation of motion of particle 1 is

$$\ddot{\mathbf{r}}_1 = -Gm_2(\mathbf{r}_1 - \mathbf{r}_2)/r_{12}^3 - Gm_3(\mathbf{r}_1 - \mathbf{r}_3)/r_{13}^3, \tag{1}$$

where $r_{ij} \equiv |\mathbf{r}_i - \mathbf{r}_j|$. Similar equations apply for the motions of particles 2 and 3. In CM coordinates, $m_1\mathbf{r}_1 + m_2\mathbf{r}_2 + m_3\mathbf{r}_3 = 0$, which means that

$$m_2(\mathbf{r}_2 - \mathbf{r}_1) + m_3(\mathbf{r}_3 - \mathbf{r}_1) = -M\mathbf{r}_1, \tag{2}$$

where $M = m_1 + m_2 + m_3$ is the total mass. If the particles always lie at the corners of an equilateral triangle, then $r_{12} = r_{13} = r_{23} = r$, say, and the square of (2) yields

---

[3] S. Chandrasekhar, *Newton's Principia for the common reader*. Oxford: Clarendon Press, 1995. Chap. 12.

$$(m_2^2 + m_3^2 + m_2 m_3) r^2 = M^2 r_1^2. \tag{3}$$

Equations (2) and (3) show that for this special case the equation of motion (1) simplifies to a single-particle equation:

$$\ddot{\mathbf{r}}_1 + GM_1 \mathbf{r}_1 / r_1^3 = 0, \quad \text{where} \quad M_1 = (m_2^2 + m_3^2 + m_2 m_3)^{3/2} M^{-2}. \tag{4}$$

Similarly, the equations of motions for particles 2 and 3 are

$$\ddot{\mathbf{r}}_2 + GM_2 \mathbf{r}_2 / r_2^3 = 0 \quad \text{and} \quad \ddot{\mathbf{r}}_3 + GM_3 \mathbf{r}_3 / r_3^3 = 0, \tag{5}$$

where $M_2 = (m_1^2 + m_3^2 + m_1 m_3)^{3/2} M^{-2}$ and $M_3 = (m_1^2 + m_2^2 + m_1 m_2)^{3/2} M^{-2}$. Equations (4) and (5) are single-particle equations for motion in attractive inverse-square fields. They have been solved in Question 8.9: each of the three particles moves on a conic section – in such a way that each is always at a corner of a rotating equilateral triangle of varying size. For bounded motion each particle describes an elliptical orbit with one focus at the common CM. For unbounded motion each particle moves on a parabolic or hyperbolic orbit with a focus at the CM.

(b) The trajectory of m_1 in plane polar coordinates is (see Question 8.9)

$$\mathbf{r}_1 = \frac{a(1 - e^2)}{1 - e \cos \theta} \hat{\mathbf{r}}(\theta), \tag{6}$$

where a and e are constants and $\hat{\mathbf{r}}(\theta)$ is a radial unit vector. Correspondingly, the trajectories of m_2 and m_3 are

$$\mathbf{r}_2 = \sqrt{\frac{m_1^2 + m_3^2 + m_1 m_3}{m_2^2 + m_3^2 + m_2 m_3}} \frac{a(1 - e^2)}{(1 - e \cos \theta)} \hat{\mathbf{r}}(\theta + \alpha_{12}), \tag{7}$$

$$\mathbf{r}_3 = \sqrt{\frac{m_1^2 + m_2^2 + m_1 m_2}{m_2^2 + m_3^2 + m_2 m_3}} \frac{a(1 - e^2)}{(1 - e \cos \theta)} \hat{\mathbf{r}}(\theta - \alpha_{13}). \tag{8}$$

Here, α_{12} is the smaller of the angles between $\mathbf{r}_1$ and $\mathbf{r}_2$. It is obtained by applying the cosine formula to a triangle with sides r, r_1, r_2:

$$\alpha_{12} = \cos^{-1} \frac{m_3^2 - m_2 m_3 - m_1 m_3 - 2 m_1 m_2}{2 \sqrt{m_1^2 + m_3^2 + m_1 m_3} \sqrt{m_2^2 + m_3^2 + m_2 m_3}}. \tag{9}$$

Similarly,

$$\alpha_{13} = \cos^{-1} \frac{m_2^2 - m_2 m_3 - m_1 m_2 - 2 m_1 m_3}{2 \sqrt{m_1^2 + m_2^2 + m_1 m_2} \sqrt{m_2^2 + m_3^2 + m_2 m_3}}. \tag{10}$$

(The square-root factors and the angles α_{12} and α_{13} in (7) and (8) ensure that the particles are at the corners of an equilateral triangle.) For the given e, a and mass ratios, (6)–(10) and the notebook below yield the figure shown on the next page. Here, the origin O is at the CM, and we have illustrated the change in orientation and size of the equilateral triangle in the xy-plane.

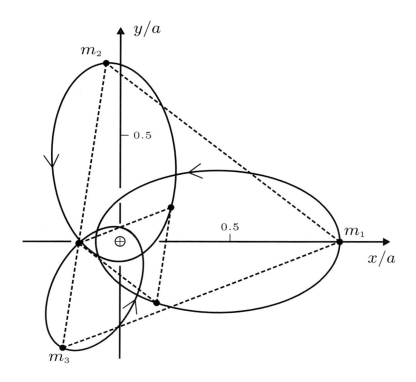

```
In[1]:= T = 1.0; ω = 2π/T;   γ2[β2_, β3_] := √((1 + β3² + β3)/(β3² + β2² + β2 β3));

γ3[β2_, β3_] := √((1 + β2² + β2)/(β3² + β2² + β2 β3));   (* Here β2 = m₂/m₁ and β3 = m₃/m₁. *)

ψ[e_, t_] := ψ/.FindRoot[ψ + e Sin[ψ] - ωt == 0, {ψ, ωt}];

(* This is Kepler's eqⁿ; ψ is the eccentric anomaly. See Question 8.9 *)

α12[β2_, β3_] := ArcCos[(β3² - β2 β3 - β3 - 2 β2)/(2√(1 + β3² + β3)√(β2² + β3² + β2 β3))];

α13[β2_, β3_] := ArcCos[(β2² - β2 β3 - β2 - 2 β3)/(2√(1 + β2² + β2)√(β2² + β3² + β2 β3))];

θ[e_, t_] := 2 ArcTan[√((1-e)/(1+e)) Tan[ψ[e, t]/2]];

r1[a_, e_, θ_] := a(1 - e²)/(1 - e Cos[θ]);

x1[a_, e_, θ_] := r1[a, e, θ] Cos[θ];

y1[a_, e_, θ_] := r1[a, e, θ] Sin[θ];
```

```
In[2]:= r2[a_, e_, θ_, β2_, β3_] := γ2[β2, β3]r1[a, e, θ];
       x2[a_, e_, θ_, β2_, β3_] := r2[a, e, θ, β2, β3] Cos[θ + α12[β2, β3]];
       y2[a_, e_, θ_, β2_, β3_] := r2[a, e, θ, β2, β3] Sin[θ + α12[β2, β3]];
       r3[a_, e_, θ_, β2_, β3_] := γ3[β2, β3]r1[a, e, θ];
       x3[a_, e_, θ_, β2_, β3_] := r3[a, e, θ, β2, β3] Cos[θ - α13[β2, β3]];
       y3[a_, e_, θ_, β2_, β3_] := r3[a, e, θ, β2, β3] Sin[θ - α13[β2, β3]];
       orbits[a_, e_, θ_, β2_, β3_] := ParametricPlot[{{x1[a, e, θ],
           y1[a, e, θ]}, {x2[a, e, θ, β2, β3], y2[a, e, θ, β2, β3]},
           {x3[a, e, θ, β2, β3], y3[a, e, θ, β2, β3]}}, {θ, 0, 2π},
           PlotRange → {{-1.2, 1.2}, {-1.2, 1.2}}, PlotStyle →
           {Directive[Dashed, Thick, Red], Directive[Dashed, Thick,
           Blue], Directive[Dashed, Thick, Green]}, Axes → {False, False}];
       planets[a_, e_, t_, β2_, β3_] := Graphics[{PointSize[0.025],
           Red, Point[{x1[a, e, θ[e, t]], y1[a, e, θ[e, t]]}], Blue,
           Point[{x2[a, e, θ[e, t], β2, β3], y2[a, e, θ[e, t], β2, β3]}],
           Green, Point[{x3[a, e, θ[e, t], β2, β3], y3[a, e, θ[e, t],
           β2, β3]}], Purple, Line[{{x1[a, e, θ[e, t]], y1[a, e,
           θ[e, t]]}, {x2[a, e, θ[e, t], β2, β3], y2[a, e, θ[e, t],
           β2, β3]}, {x3[a, e, θ[e, t], β2, β3], y3[a, e, θ[e, t],
           β2, β3]}, {x1[a, e, θ[e, t]], y1[a, e, θ[e, t]]}}], Black,
           {PointSize[0.0125], Point[{0, 0}]}, Text[CM, {0, -0.05}]}];
       Manipulate[Show[orbits[a, e, θ, β2, β3],
           planets[a, e, t, β2, β3], Background → LightGray], {{a, 0.56,
           "semi - major axis"}, 0.25, 2.25, 0.01, Appearance → "Labeled"},
           {{e, 0.8, "eccentricity"}, 0, 0.99, 0.01, Appearance → "Labeled"},
           {{β2, 2, "massratio21"}, 1, 15, 0.01, Appearance → "Labeled"},
           {{β3, 3.33, "massratio31"}, 1, 15, 0.01, Appearance → "Labeled"},
           {t, 0, T}]
```

Comments

(i) In general, the gravitational three-body problem (1) cannot be solved exactly. It is only for certain special cases that analytical solutions exist.

(ii) Equilateral triangle solutions were the first special solutions to be obtained for the general three-body problem: by Lagrange (for a rigid triangle that rotates about an axis perpendicular to its plane) and later by Euler (for a triangle that changes size).

(iii) An important question regarding these solutions is their stability with respect to small changes in initial conditions. It turns out that stability occurs only if one of the masses has more than 95% of the total mass.[4]

---

[4]   J. M. A. Danby, "Stability of the triangular points in the elliptic restricted problem of three bodies," Astronomical Journal, vol. 69, pp. 165–172, 1964.

336 Solved Problems in Classical Mechanics

(iv) An essential feature of the above analysis is that when the interparticle distances r_{ij} are all the same, the force on each particle is directed towards the common CM – see (4) and (5). In three dimensions there are only two geometric figures that provide constant r_{ij}: an equilateral triangle (three-body problem) and a regular tetrahedron (four-body problem).

(v) Dynamic plots of the orbits of the three bodies can be viewed with the notebook below. Similar plots can be made for the unbounded (parabolic and hyperbolic) trajectories.

(vi) Special two-dimensional solutions for the gravitational N-body problem also exist when the particles all have equal mass and are located at the vertices of a regular polygon. In this configuration the resultant gravitational force on a particle is directed toward the centre of the polygon and its magnitude is inversely proportional to the square of the distance to the centre. The polygon rotates non-uniformly and the lengths of its sides vary periodically as the particles move along congruent conic sections about a focus located at the centre of the polygon.[5] (This centre can either be empty or occupied by a particle.)

(vii) The three-body problem is one of the oldest topics in dynamics. Newton regarded it as the most complex problem in his *Principia*, and once remarked to a contemporary that "his head never ached but with his studies on the Moon".[6]

Question 11.7

Consider the following gravitational three-body problem. Two heavy celestial bodies (known as the primaries) move in circular orbits about their centre of mass. A third body of negligible mass moves in the same plane as the primaries under the influence of their gravitational attraction. Let m_1, m_2 and m_3 denote the masses of the bodies with $m_1 > m_2 \gg m_3$. The period of the motion of the primaries is given by $T = 2\pi\sqrt{D^3/GM}$, where D is the constant distance between them and $M = m_1 + m_2$ is the total mass.

(a) Use (1) of Question 11.6 and its cyclic permutations to express the equations of motion in the dimensionless forms

$$\frac{d^2\bar{\mathbf{r}}_i}{d\tau^2} + \frac{4\pi^2}{M}\left(\frac{m_j\bar{\mathbf{r}}_{ij}}{\bar{r}_{ij}^3} + \frac{m_k\bar{\mathbf{r}}_{ik}}{\bar{r}_{ik}^3}\right) = 0 \qquad (j,k \neq i; j \neq k), \qquad (1)$$

for $i = 1, 2, 3$, where $\bar{\mathbf{r}}_i = \mathbf{r}_i/D$ and $\tau = t/T$ are dimensionless position vectors and time.

(b) Show by a kinematical argument that the dimensionless speeds of the primaries for circular motion in the CM frame are

---

[5] E. I. Butikov, "Regular Keplerian motions in classical many-body systems," European Journal of Physics, vol. 21, pp. 465–482, 2000.
[6] R. S. Westfall, *Never at rest – A biography of Isaac Newton*, p. 543. Cambridge: Cambridge University Press, 1980.

$$\bar{v}_1 = 2\pi \frac{m_2}{M} \quad \text{and} \quad \bar{v}_2 = 2\pi \frac{m_1}{M}. \tag{2}$$

(c) Write a *Mathematica* notebook to solve (1) numerically for $m_2/M = 3/40$ and the following two sets of initial conditions:

$\bar{\mathbf{r}}_1(0) = (0, \frac{-3}{40}, 0);$ $\quad \bar{\mathbf{r}}_2(0) = (0, \frac{37}{40}, 0);$ $\quad \bar{\mathbf{r}}_3(0) = (0, \frac{45}{40}, 0);$

$\bar{\mathbf{v}}_1(0) = (\frac{-3\pi}{20}, 0, 0);$ $\quad \bar{\mathbf{v}}_2(0) = (\frac{37\pi}{20}, 0, 0);$

and 1. $\bar{\mathbf{v}}_3(0) = (\frac{62\pi}{20}, 0, 0),$ and 2. $\bar{\mathbf{v}}_3(0) = (\frac{57\pi}{20}, 0, 0).$

(d) For the initial conditions 1. , plot the trajectories of m_2 and m_3 in the 'heliocentric' frame[‡] for $0 \leq \tau \leq 1.05$.
For the initial conditions 2. , plot the trajectories of the bodies in the 'heliocentric', 'geocentric', and CM frames $0 \leq \tau \leq 3.925$.

(e) Modify the *Mathematica* notebook to animate the motion of the three bodies for the initial conditions in (c) part 2 . Then, view the motion of the bodies in the three frames referred to above.

Solution

(a) The result follows directly from (1) of Question 11.6 (and its cyclic permutations) after multiplication by T^2/D. Since the light body has negligible effect on the motions of the primaries, terms containing m_3 in the equations of motion (1) have been set to zero in the numerical calculations below.

(b) In the CM frame of the primaries, circular orbits for bodies 1 and 2 have radii $(m_2/M)D$ and $(m_1/M)D$, and therefore speeds $2\pi \frac{m_2}{M}\frac{D}{T}$ and $2\pi \frac{m_1}{M}\frac{D}{T}$, respectively. The corresponding dimensionless forms – obtained after multiplication by T/D – are (2).

(c) See the *Mathematica* notebook given in (e).

(d) The graphs are presented below, following the notebook.

(e) The notebook is for the 'heliocentric' frame. The trajectories for the 'geocentric' and CM frames are obtained by changing the appropriate variable names in the Evaluate[] commands in the Traj and Bodies[t_] sections of the notebook.

```
In[1]:= m2 = 75/1000; m1 = 1 - m2; τmax = 3.925;

       x10 = 0; y10 = -3/40; vx10 = -3π/20; vy10 = 0;

       x20 = 0; y20 = 37/40; vx20 = 37π/20; vy20 = 0;

       x30 = 0; y30 = 9/8; vx30 = 57π/20; vy30 = 0;
```

[‡]'Heliocentric' signifies the frame in which the largest mass, m_1, is at rest at the origin; 'geocentric' means that m_2 is at rest there.

```
In[2]:= r1[τ_] := {x1[τ], y1[τ]};

    r2[τ_] := {x2[τ], y2[τ]};

    r3[τ_] := {x3[τ], y3[τ]};

    r12[τ_] := r1[τ] - r2[τ]; R12[τ_] := √Dot[r12[τ], r12[τ]];

    r21[τ_] := r2[τ] - r1[τ]; R21[τ_] := √Dot[r21[τ], r21[τ]];

    r31[τ_] := r3[τ] - r1[τ]; R31[τ_] := √Dot[r31[τ], r31[τ]];

    r32[τ_] := r3[τ] - r2[τ]; R32[τ_] := √Dot[r32[τ], r32[τ]];

    EqnMot1 = Thread[r1''[τ] + 4π² m2/(m1+m2)  r12[τ]/R12[τ]³ == 0];

    EqnMot2 = Thread[r2''[τ] + 4π² m1/(m1+m2)  r21[τ]/R21[τ]³ == 0];

    EqnMot3 = Thread[r3''[τ] + 4π² m1/(m1+m2)  r31[τ]/R31[τ]³ + 4π² m2/(m1+m2)  r32[τ]/R32[τ]³ == 0];

    InCon1 = Join[Thread[r1[0] == {x10, y10}], Thread[r1'[0] == {vx10, vy10}]];

    InCon2 = Join[Thread[r2[0] == {x20, y20}], Thread[r2'[0] == {vx20, vy20}]];

    InCon3 = Join[Thread[r3[0] == {x30, y30}], Thread[r3'[0] == {vx30, vy30}]];

    EqsToSolve = Join[EqnMot1, EqnMot2, EqnMot3, InCon1, InCon2, InCon3];

    Sol = NDSolve[EqsToSolve, Join[r1[τ], r2[τ], r3[τ]], {τ, 0, τmax},
        MaxSteps → 500000, AccuracyGoal → 12, PrecisionGoal → 12];

    Traj = ParametricPlot[{{Evaluate[{x2[τ] - x1[τ], y2[τ] - y1[τ]}]/.Sol},
        {Evaluate[{x3[τ] - x1[τ], y3[τ] - y1[τ]}]/.Sol}}, {τ, 0, τmax},
        AspectRatio → 1, PlotPoints → 1000];

    Bodies[t_] := Graphics[{PointSize[0.025], Red, Point[{0, 0}],
        PointSize[0.0125], Green, Point[First[Evaluate[r21[τ]/.
            Sol/. τ → t]]], Blue, Point[First[Evaluate[r31[τ]/.
            Sol/. τ → t]]]}];

    Manipulate[Show[Traj, Bodies[t], Background → LightGray], {t, 0, τmax}]
```

Multi-particle systems 339

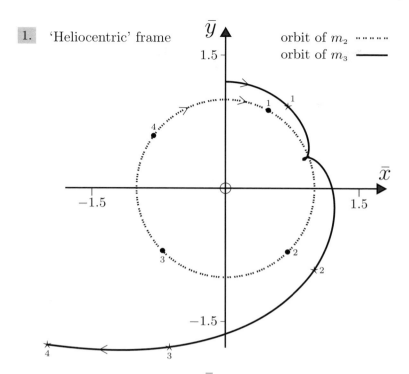

1. 'Heliocentric' frame — orbit of m_2 ······ ; orbit of m_3 ———

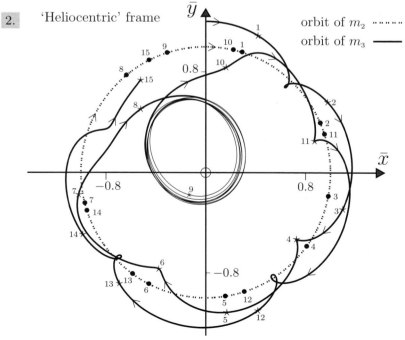

2. 'Heliocentric' frame — orbit of m_2 ······ ; orbit of m_3 ———

340 Solved Problems in Classical Mechanics

2. 'Geocentric' frame

orbit of m_1 ········
orbit of m_3 ———

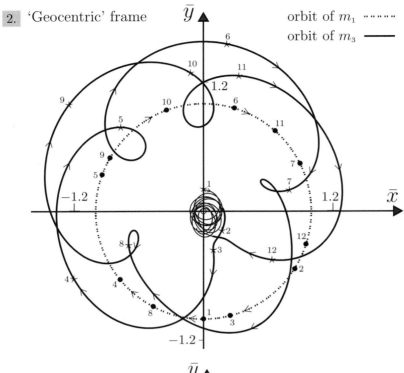

2. CM frame

orbit of m_1 ··········
orbit of m_2 ········
orbit of m_3 ———

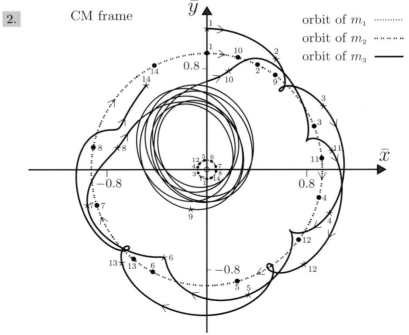

Comments

(i) The first diagram illustrates a motion where the light mass m_3 escapes from the other two masses after orbiting m_2 just once. For different initial conditions, other motions are seen, such as m_3 falling into m_1.

(ii) The second diagram depicts a trajectory for m_3 in the 'heliocentric' frame where, after orbiting the intermediate mass m_2 four times, it is pulled into an almost elliptical orbit around the largest mass m_1. After seven such orbits it is again 'captured' by m_2. This sequence is then repeated: a plot up to $\tau = 300$ contains numerous interchanges of m_3 from m_2 to m_1 and back to m_2. The third diagram is for the 'geocentric' frame and it shows clearly the orbits of m_3 around m_2. The fourth diagram shows the motion of all three bodies relative to the CM.

(iii) These plots illustrate a striking feature of the gravitational three-body problem, namely the complexity of the motion compared with the simplicity of the two-body problem, where the two trajectories follow conic sections with focii at the CM (see Chapter 10). By contrast, three-body orbits cannot be described by analytic solutions (except in special cases such as the equilateral triangle solution of the previous question).

(iv) It so happens that in our solar system the two-body Kepler laws (see Question 10.11) are a good approximation for the planets – the Sun is a single, massive star and the planets are widely separated. In a system orbiting a binary star the planetary orbits are much more complicated and "it would be an immensely difficult problem for astronomers $\cdots$ to establish the kinematical laws of planetary motion for the double-star system (Kepler), and even a much more difficult problem would be to discover that these complicated kinematical laws are generated by the simple inverse-square law of gravitational attraction to each of the stars (Newton). Our civilization has the advantage of a planet orbiting a single star."[5] (In fact, double, triple, ... star groupings are more common in the universe than are single stars.)

(v) The reader can readily check that small changes in the initial conditions have large effects on the subsequent motion of m_3 – the differences increase in the long-time limit and indicate that the motion is chaotic.[7] Because of this, it is questionable whether numerical integration can yield meaningful long-time orbits, and only periodic orbits are regarded as valid solutions at large t.[8] (See also page 499.)

(vi) The problem considered in this question is known as the circular coplanar restricted three-body (CCR3B) problem. It is part of the general, gravitational three-body problem that was first considered by Newton in connection with the Moon's motion in the fields of the Earth and Sun. Because of its importance there have been many theoretical studies of the three-body problem.[7,8] An interesting account of exact particular solutions for the gravitational many-body problem has been given by Butikov.[5]

---

[7] M. Valtonen and H. Karttunen, *The three-body problem*. Cambridge: Cambridge University Press, 2006.

[8] A. D. Bruno, *The restricted 3-body problem*. Berlin: de Gruyter, 1994.

Question 11.8

Consider again the CCR3B problem of Question 11.7. Let $\bar{S}(\bar{x}, \bar{y}, \bar{z})$ be the CM frame and $S(x, y, z)$ be a rotating (non-inertial) frame that has the same origin and z-axis as $\bar{S}$. Suppose S rotates with angular velocity $\boldsymbol{\Omega} = (0, 0, \Omega)$ relative to $\bar{S}$, where

$$\Omega = 2\pi/T = \sqrt{GM/D^3}. \tag{1}$$

Then, the primaries m_1 and m_2 are at rest in S, on the x-axis say. The light body m_3 has coordinates (x, y).

(a) Show that the equation of motion of m_3 in S has components

$$\ddot{x} - 2\Omega\dot{y} = -\partial U/\partial x \quad \text{and} \quad \ddot{y} + 2\Omega\dot{x} = -\partial U/\partial y, \tag{2}$$

where U is the effective potential (per unit mass):

$$U(x, y) = -\tfrac{1}{2}\Omega^2(x^2 + y^2) - G(m_1/r_{31} + m_2/r_{32}). \tag{3}$$

In the following it is convenient to choose the units of distance, time and mass such that $D = 1$, $T = 2\pi$ and $M = 1$.

(b) There are 5 points (the so-called Lagrange points $L_1, \cdots, L_5$) in the xy-plane at which m_3 can be at rest. Write a *Mathematica* notebook to solve for the coordinates of the Lagrange points in terms of $\mu = m_2/M$. Evaluate these coordinates when $\mu = 10^{-3}$. (Hint: Use the `VectorAnalysis` package to calculate ∇U, and then use the `NSolve` function to find its roots.)

(c) Two of the Lagrange points (labelled L_4 and L_5) are always located symmetrically off the x-axis. Show that their coordinates are

$$(a, b) = \tfrac{1}{2}(1 - 2\mu, \pm\sqrt{3}). \tag{4}$$

(d) Extend the notebook for part (b) to produce a three-dimensional graph and also a contour plot of $U(x, y)$ for $\mu = \tfrac{1}{2}$.

(e) Consider motion of m_3 in the neighbourhood of L_4 or L_5. Show that there are periodic solutions to (2), and find their frequencies. (Hint: Look for solutions

$$x = a + Ae^{\lambda t}, \quad y = b + Be^{\lambda t}, \tag{5}$$

where A, B and λ are constants. Use *Mathematica* to evaluate the derivatives of $U(x, y)$ at (a, b).)

Solution

With the CM at the origin, the coordinates of m_1 and m_2 in S are fixed at $(-\mu D, 0)$ and $((1 - \mu)D, 0)$, where $\mu = m_2/M$.

Multi-particle systems 343

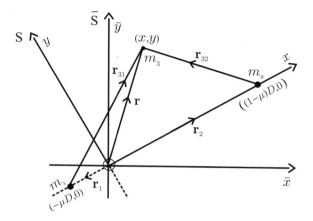

(a) The equation of motion of m_3 in the rotating frame S is (see Chapters 1 and 14)
$$m_3(\ddot{x}, \ddot{y}) = \mathbf{F} + \mathbf{F}_{\text{cf}} + \mathbf{F}_{\text{Cor}}. \tag{6}$$
Here, $\mathbf{F}$ is the gravitational force, given by
$$\mathbf{F} = \boldsymbol{\nabla}(Gm_1 m_3/r_{31} + Gm_2 m_3/r_{32}), \tag{7}$$
where $\boldsymbol{\nabla} = (\partial/\partial x, \partial/\partial y)$, and $\mathbf{F}_{\text{cf}}$ and $\mathbf{F}_{\text{Cor}}$ are the centrifugal and Coriolis forces
$$\left.\begin{array}{l} \mathbf{F}_{\text{cf}} = -m_3[\boldsymbol{\Omega} \times (\boldsymbol{\Omega} \times \mathbf{r})] = m_3\Omega^2(x, y) \\ \mathbf{F}_{\text{Cor}} = -2m_3(\boldsymbol{\Omega} \times \dot{\mathbf{r}}) = -2m_3\Omega(-\dot{y}, \dot{x}). \end{array}\right\} \tag{8}$$
Equations (6)–(8) yield (2).

(b) For a body at rest at the Lagrange points, x and y are constants in (2), and so
$$\frac{\partial U}{\partial x} = \frac{\partial U}{\partial y} = 0 \tag{9}$$
there. In terms of the above units
$$U(x, y) = -\tfrac{1}{2}(x^2 + y^2) - (1-\mu)/r_{31} - \mu/r_{32}. \tag{10}$$
The following notebook gives the solutions to (9) when $\mu = 10^{-3}$.

```
In[1]:= Needs["VectorAnalysis`"]   (* use U in the units given in (a) *)

U[x_, y_] := -1/2 (x^2 + y^2) - (1 - μ)/√((x + μ)^2 + y^2) - μ/√((x - 1 + μ)^2 + y^2);

{v1, v2, v3} = Grad[U[x, y], Cartesian[x, y, z]];   μ = 1/1000;
{x, y}/.NSolve[{v1 == 0, v2 == 0}, {x, y}, 18]
```

The solutions are:

$$L_1 = (0.93129, 0); \quad L_2 = (1.06992, 0); \quad L_3 = (-1.00042, 0); \\ L_4 = (0.49900, \tfrac{1}{2}\sqrt{3}); \quad L_5 = (0.49900, -\tfrac{1}{2}\sqrt{3}).$$ (11)

The positions of these points in the xy-plane are shown in the diagram. The first three Lagrange points are collinear with the primaries m_1 and m_2, and L_4 and L_5 are located at the corners of equilateral triangles, with the primaries at the other two vertices. This is always the case (see below).

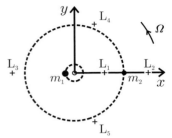

(c) According to (9) and (10), the coordinates of the Lagrange points are the solutions to the simultaneous equations

$$x - \frac{(1-\mu)(x+\mu)}{r_{31}^3} + \frac{\mu(1-\mu-x)}{r_{32}^3} = 0, \quad (12)$$

$$y\left[1 - \frac{(1-\mu)}{r_{31}^3} - \frac{\mu}{r_{32}^3}\right] = 0. \quad (13)$$

If $y \neq 0$, it is clear from (12) and (13) that $r_{31} = r_{32} = 1$. That is, L_4 and L_5 form two equilateral triangles with the primaries and so their coordinates are given by (4) – see the two diagrams above. If $y = 0$, (12) has to be solved numerically for the positions of L_1, L_2 and L_3 (see Comment (ii) below). Note that (12) and (13) express equality of the gravitational and centrifugal forces at the Lagrange points.

(d) Use the notebook for part (b) as the first input cell in the notebook below. This gives the required graph and contour plot. In these, the dots indicate the positions of the five Lagrange points.

```
In[1]:= (* CELL 1 : ENTER THE NOTEBOOK FOR PART (b) *)
In[2]:= xlist = x /.NSolve[{v1 == 0, v2 == 0}, {x, y}, 18];
        ylist = y /.NSolve[{v1 == 0, v2 == 0}, {x, y}, 18];
        zlist = MapThread[U, {xlist, ylist}];
        coords = Thread[{xlist, ylist, zlist}];
        EffectivePotential = Plot3D[U[x, y], {x, -1.5, 1.5}, {y, -1.5, 1.5},
            PlotRange → {-3, -1}, Mesh → None, Boxed → True, Axes → True];
        LagrangePoints = Graphics3D[{PointSize[0.0125], Red, Point[coords]}];
        Show[EffectivePotential, LagrangePoints]
In[3]:= contour = ContourPlot[U[x, y], {x, -1.5, 1.5}, {y, -1.5, 1.5}];
        coords = Thread[{xlist, ylist}];
        LagrangePoints = Graphics[{PointSize[0.0125], Red, Point[coords]}];
        Show[contour, LagrangePoints]
```

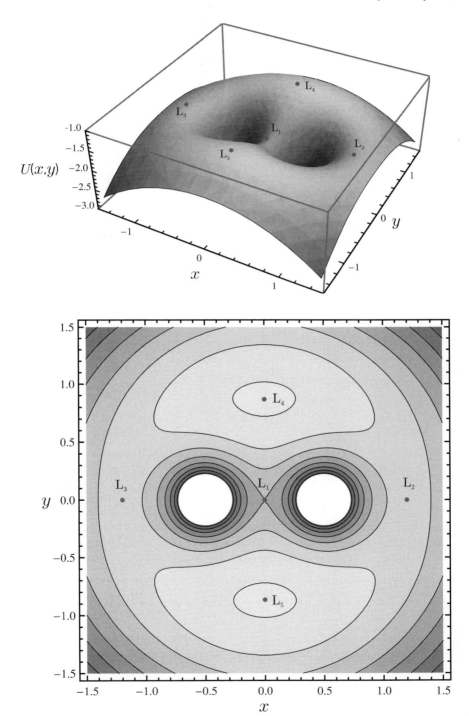

(e) For small oscillations we can expand the partial derivatives in (2) to first order in the displacements $x-a$ and $y-b$ about L_4 or L_5:

$$\left.\begin{aligned}\frac{\partial U}{\partial x} &= (x-a)\frac{\partial^2 U}{\partial x^2} + (y-b)\frac{\partial^2 U}{\partial x \partial y} \\ \frac{\partial U}{\partial y} &= (x-a)\frac{\partial^2 U}{\partial x \partial y} + (y-b)\frac{\partial^2 U}{\partial y^2}\,,\end{aligned}\right\} \quad (14)$$

where the second derivatives of U are evaluated at L_4 or L_5. Their values are (see the notebook below)

$$\frac{\partial^2 U}{\partial x^2} = -\frac{3}{4}, \quad \frac{\partial^2 U}{\partial y^2} = -\frac{9}{4}, \quad \text{and} \quad \frac{\partial^2 U}{\partial x \partial y} = -\frac{3\sqrt{3}}{4}(1-2\mu)\,. \quad (15)$$

From (2) with $\Omega = 1$, (5), (14) and (15) we find that

$$\left.\begin{aligned}(\lambda^2 - 3/4)A - (2\lambda + 3\sqrt{3}(1-2\mu)/4)B &= 0 \\ (2\lambda - 3\sqrt{3}(1-2\mu)/4)A + (\lambda^2 - 9/4)B &= 0\,.\end{aligned}\right\} \quad (16)$$

For non-trivial solutions the determinant of the coefficients of A and B in (16) must be zero. That is,

$$\lambda^4 + \lambda^2 + 27\mu(1-\mu) = 0\,, \quad (17)$$

and so

$$\lambda^2 = -\tfrac{1}{2} \pm \tfrac{1}{2}\sqrt{1 - 27\mu(1-\mu)}\,. \quad (18)$$

For periodic solutions we require that (16) has imaginary roots $\lambda_j = i\omega_j$. Therefore,

$$\Delta = 27\mu^2 - 27\mu + 1 \geq 0\,. \quad (19)$$

Thus, there is a critical value[‡]

$$\mu_c = \tfrac{1}{2}\left(1 - \sqrt{23/27}\right) \approx 0.03852 \quad (20)$$

above which there are no periodic solutions. If $\mu < \mu_c$ the two frequencies are

$$\omega_1 = \sqrt{\frac{1-\sqrt{\Delta}}{2}} \approx \sqrt{\frac{27\mu}{4}} \quad \text{and} \quad \omega_2 = \sqrt{\frac{1+\sqrt{\Delta}}{2}} \approx 1 - \frac{27\mu}{8}\,. \quad (21)$$

Small oscillations about L_4 or L_5 will, in general, be a linear combination of these slow and rapid components (see Question 11.9).

```
In[1]:= U[x_, y_] := -1/2 (x^2 + y^2) - (1-μ)/√((x+μ)^2 + y^2) - μ/√((x-1+μ)^2 + y^2);

L4 = {x → (1-2μ)/2, y → √3/2};
Uxx = Simplify[(D[U[x, y], {x, 2}, {y, 0}]/.L4)]
Uyy = Simplify[(D[U[x, y], {x, 0}, {y, 2}]/.L4)]
Uxy = Simplify[(D[U[x, y], {x, 1}, {y, 1}]/.L4)]
```

[‡]Recall that $\mu = \frac{m_2}{M} < \frac{1}{2}$ because $m_2 < m_1$.

Comments

(i) At L_1, L_2 and L_3 there are saddle points in $U(x,y)$; at L_4 and L_5 there are maxima in $U(x,y)$ – see (15) and the preceding two plots of $U(x,y)$. We have seen in part (e) that these maxima are points of stable equilibrium, and the reader may wonder whether this contradicts the familiar result that stable equilibrium occurs at potential minima. The answer is 'no': in the present case the acceleration is not equal to just the gradient of a scalar; there are also velocity-dependent contributions due to the Coriolis force in (2). Motion around the potential maxima at L_4 and L_5 is stabilized by the Coriolis force.[9]

(ii) The positions of L_1, L_2 and L_3 are determined numerically by solving $\nabla U = 0$, as in the *Mathematica* notebook for (b). They are also given as the real solutions to (12) with $y = 0$. This calculation requires that one consider each of the cases: m_3 to the left of m_1; m_3 between m_1 and m_2; and m_3 to the right of m_2. This yields for (12)

$$\frac{(1-\mu)}{(x+\mu)^2} + \alpha \frac{\mu}{(x+\mu-1)^2} + \beta x = 0, \qquad (22)$$

where $\alpha = \beta = -1$ (for L_1), $= 1$ (for L_3), and $\alpha = -\beta = 1$ (for L_2), and $\mu < \frac{1}{2}$. In each case there is a quintic with one real root that has to be found numerically – see the notebook and graphs below.

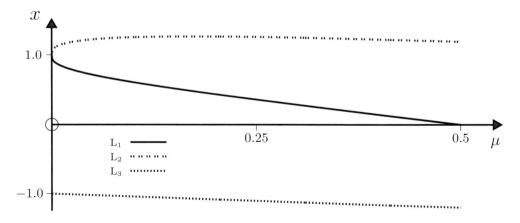

```
In[1]:= (* α = β = -1 for L1 :  α = β = 1 for L3 :  α = -β = 1 for L2 *)
        μmax = 500/1000; μmin = 0; μstep = 1/1000; α = β = -1;
        data = Table[{μ, Sol = Reduce[(1 - μ)/(x + μ)^2 + α μ/(x - 1 + μ)^2 + β x == 0, x, Reals];
                x/.ToRules[N[Sol, 18]]}, {μ, μmin, μmax, μstep}];
        graph1 = ListPlot[data];
        Show[graph1]
```

---

[9] R. Greenberg and D. R. Davis, "Stability at potential maxima: The L_4 and L_5 points of the restricted three-body problem," American Journal of Physics, vol. 10, pp. 1068–1070, 1978.

(iii) Equation $(2)_1$ times $\dot{x}$ plus $(2)_2$ times $\dot{y}$ gives

$$-\frac{1}{2}\frac{d}{dt}(\dot{x}^2+\dot{y}^2)=\frac{\partial U}{\partial x}\frac{dx}{dt}+\frac{\partial U}{\partial y}\frac{dy}{dt}=\frac{dU}{dt}, \qquad (23)$$

and so there is a conserved quantity (the Jacobi constant)

$$C=\tfrac{1}{2}v^2+U, \qquad (24)$$

where $v=\sqrt{\dot{x}^2+\dot{y}^2}$ is the speed of m_3 in S. Now, $v^2\geq 0$ and therefore $U(x,y)\leq C$. This means that the motion is confined to regions where the potential does not exceed the energy, and can therefore be described in terms of two-dimensional energy diagrams.[7,9] (The reader should note that in the literature the potential is sometimes defined as $-U$.)

(iv) The five positions $L_1,\cdots,L_5$ were discovered in the eighteenth century by Euler and Lagrange. It took more than one hundred years before astronomers observed bodies in the solar system trapped near triangular points. It is now known that there are several hundred asteroids (the Trojan asteroids) located in the vicinity of L_4 and L_5 in the Sun–Jupiter system. The other planets perturb the Trojans's orbits, causing them to drift as far as 30° or more away from L_4 and L_5.[10]

(v) No stable orbits exist near collinear Lagrange points.[7] However, a concentration of interplanetary dust at L_2 in the Sun–Earth system is responsible for the faint glow (the 'gegenschein') in the night sky in the direction opposite to the Sun.

(vi) In the 1970s it was proposed that L_4 and L_5 in the Earth–Moon system were suitable locations for large, permanent space colonies.[11]

(vii) There are certain special values of $\mu<\mu_c$ for which small-amplitude motions about L_4 and L_5 are not stable.[12]

(viii) The next question uses numerical integration to solve the equations of motion for both small and large oscillations about L_4.

Question 11.9

Consider the CCR3B problem discussed above. The following questions deal with numerical solutions for bounded motion of m_3 around the Lagrange point L_4. In these, use the units given in part (a) of Question 11.8 and take $\mu=10^{-3}$.

(a) Write a *Mathematica* notebook to solve the equations of motion (2) of Question 11.8 for the initial conditions below. Plot the trajectory $\mathbf{r}=(x,y)$ of m_3 in the rotating frame S for $0\leq t\leq t_{\max}=80\pi$ (this corresponds to 40 revolutions of the primaries). Take $\dot{\mathbf{r}}(0)=0$ and 1. $\mathbf{r}(0)=(a+0.001,b+0.002)$, and 2. $\mathbf{r}(0)=(a+0.109,b-0.073)$. Here (a,b) is the position of L_4 in S, given in (4) of Question 11.8.

---

[10] W. K. Hartmann, *Moons & planets*. United States: Thomson, 5th edn, 2005.
[11] G. K. O'Neill, "The colonization of space," Physics Today, pp. 32–40, September 1974.
[12] J. A. Blackburn, M. A. H. Nerenberg, and Y. Beaudoin, "Satellite motion in the vicinity of the triangular libration points," American Journal of Physics, vol. 45, pp. 1077 – 1081, 1977.

(b) Describe the motion of m_3 in the CM frame $\bar{S}$ for both cases in (a) above.

(c) For 1. and 2. above, create a table of values $\{t, x(t)\}$ in Mathematica for $0 \leq t \leq t_{\max} = 80\pi$, using a step size $\Delta t = t_{\max}/5000$. Use Mathematica's FindFit function to fit the function $x(t) = a_0 + a_1 \cos(\omega_1 t + \phi_1) + a_2 \cos(\omega_2 t + \phi_2)$ to the data. Here, ω_1 and ω_2 are given by (21) of Question 11.8 and a_0, a_1, a_2, ϕ_1 and ϕ_2 are adjustable parameters. Plot $x(t)$ versus t and on this graph show also a representative subset of the data table.

(d) Rerun the notebook in (a) for the following initial conditions and plot the trajectory of m_3 in the xy-plane of S for $0 \leq t \leq t_{\max} = 320\pi$:

3. $\mathbf{r}(0) = (a + 0.024, b + 0.055);$ $\dot{\mathbf{r}}(0) = (0.0778, -0.0429);$

4. $\mathbf{r}(0) = (a + 0.108, b + 0.050);$ $\dot{\mathbf{r}}(0) = (0.0868, -0.0400).$

(e) Plot the motion of m_3 in the CM frame $\bar{S}$ for case 3. above.

Solution

(a) The notebook is given below. In the diagrams obtained from it the positions of the primaries and L$_4$ are indicated by a dot •. Note that m_1 is slightly to the left of the CM, which is at the origin.

```
In[1]:= μ = 1/1000; a = (1 - 2μ)/2; b = √(3/4); Tmax = 160π;

x0 = a + 0.109;  y0 = b - 0.073;  vx0 = 0;  vy0 = 0;    (*INIT CON 2 *)

r[t_] := {x[t], y[t]}

U[x_, y_] := -1/2 (x² + y²) - (1 - μ)/√((x + μ)² + y²) - μ/√((x - 1 + μ)² + y²);

Ux[t_] := D[U[x, y], {x, 1}]/.{x → x[t], y → y[t]};

Uy[t_] := D[U[x, y], {y, 1}]/.{x → x[t], y → y[t]};

EqnMotion = Thread[{x''[t] - 2y'[t] + Ux[t], y''[t] + 2x'[t] +
            Uy[t]} == {0, 0}];

InitCon = Join[Thread[r[0] == {x0, y0}], Thread[r'[0] == {vx0, vy0}]];

EqsToSolve = Join[EqnMotion, InitCon];

Sol = NDSolve[EqsToSolve, r[t], {t, 0, Tmax}, MaxSteps → 1000000,
        AccuracyGoal → 12, PrecisionGoal → 12];

ParametricPlot[{{Evaluate[{x[t], y[t]}]/.Sol}}, {t, 0, Tmax},
        AspectRatio → 1, PlotPoints → 1000]
```

350 Solved Problems in Classical Mechanics

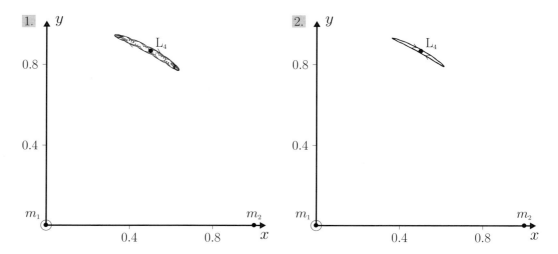

(b) Both of these initial displacements from L_4 are 'small' in the sense of Question 11.8(e). In each case, the motion of m_3 relative to $\bar{S}$ is a nearly circular orbit of radius ≈ 1 about the CM.

(c) Use the previous notebook as the the first input cell in the notebook below.

```
In[1]:= (* ENTER THE NOTEBOOK FOR (a) AS THE FIRST INPUT CELL HERE *)

In[2]:= Δ = 27μ^2 - 27μ + 1; ω1 = √((1 - √Δ)/2) ; ω2 = √((1 + √Δ)/2) ;

       f[t_] := a0 + a1 Cos[ω1 t + ϕ1] + a2 Cos[ω2 t + ϕ2];

       FourierData = Table[{t, First[x[t] /. Sol]}, {t, 0, Tmax, Tmax/100}];

       FourierCoefficients = FindFit[FourierData, f[t], {a0, a1, a2, ϕ1, ϕ2}, t];

       GraphFit = Plot[{f[t] /. FourierCoefficients}, {t, 0, Tmax},
           PlotPoints → 1000]; GraphDiffEqn = ListPlot[FourierData];

       Show[GraphFit, GraphDiffEqn]
```

The best-fit parameters are:

	a_0	a_1	a_2	ϕ_1	ϕ_2
case 1	0.4798130	−0.1442100	0.0122875	1.6941800	1.8353500
case 2	0.4850700	0.1230990	−0.0016150	0.0772048	1.0333400

The values of a_0 are close to the x-coordinate, 0.499, of L_4.

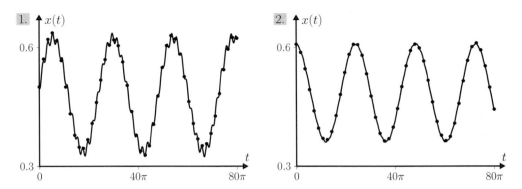

In case **1.** there is an appreciable contribution from the rapid oscillation ω_2; in case **2.** this contribution is much less and $x(t)$ is almost sinusoidal.

(d) The graphs are:

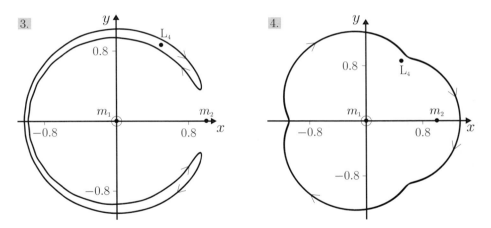

(e) The coordinates $(\bar{x}, \bar{y})$ of m_3 in the CM frame $\bar{S}$ are calculated from its coordinates (x, y) in S using the transformation: $(\bar{x}, \bar{y}) = (x\cos t - y\sin t,\ x\sin t + y\cos t)$, and the trajectory is shown below. Notice that the motion of m_3 comprises two nearly concentric orbital 'bands'. The body starts in the outer band, and between $t \approx 9.7$ and ≈ 11.7 it makes a transition to the inner band where it remains until $t \approx 105.5$; by $t \approx 107.3$ it has returned to the outer band, where it remains for the rest of the time interval.

Comments

(i) A dynamic output of the motion of m_3 in S is obtained by modifying the notebook for (a) in the usual way (see, for example, the notebook of Question 11.7).

(ii) Horseshoe orbits like that in part (d) above occur in the solar system. For example, in the motion of two inner moons about Saturn, the lighter moon (Epimetheus) performs a horseshoe orbit relative to the planet and the heavier moon (Janus). These moons are almost equidistant from Saturn and every 4 yr (approximately) they have a close encounter with each other before receding again.[13]

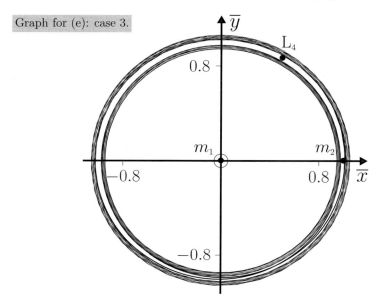

Graph for (e): case 3.

Question 11.10

Consider a system of N particles (where $N > 2$) with central, power-law interparticle forces:

$$\mathbf{F}_{ji} = -K_{ij}\frac{\mathbf{r}_i - \mathbf{r}_j}{r_{ij}^n} \qquad (j \neq i), \tag{1}$$

where $r_{ij} = |\mathbf{r}_i - \mathbf{r}_j|$. The coefficients K_{ij} and the exponent n are constants. Determine the conditions under which this N-body problem reduces to N one-body problems.[14]

Solution

For the forces (1), the equations of motion can be written

$$m_i \ddot{\mathbf{r}}_i = -\sum_{\substack{j=1 \\ j \neq i}}^{N} K_{ij} \frac{\mathbf{r}_i}{r_{ij}^n} + \sum_{\substack{j=1 \\ j \neq i,k}}^{N} K_{ij} \frac{\mathbf{r}_j}{r_{ij}^n} + K_{ik} \frac{\mathbf{r}_k}{r_{ik}^n}, \tag{2}$$

---

[13] R. S. Harrington and P. K. Seidelmann, "The dynamics of the Saturnian satellites 1980S1 and 1980S3," Icarus, vol. 47, pp. 97–99, July 1981.

[14] A. S. de Castro and C. A. Vilela, "On the regular-geometric-figure solution to the N-body problem," American Journal of Physics, vol. 22, pp. 487–490, 2001.

where $i = 1, 2, \cdots, N$ and $k \neq i$. In CM coordinates,

$$\sum_{j=1}^{N} m_j \mathbf{r}_j = 0, \quad \text{and so} \quad m_k \mathbf{r}_k = -\sum_{\substack{j=1 \\ j \neq k}}^{N} m_j \mathbf{r}_j. \tag{3}$$

Equation (3) enables us to express (2) as

$$m_i \ddot{\mathbf{r}}_i = -\sum_{\substack{j=1 \\ j \neq i}}^{N} K_{ij} \frac{\mathbf{r}_i}{r_{ij}^n} - \frac{m_i}{m_k} K_{ik} \frac{\mathbf{r}_i}{r_{ik}^n} + \sum_{\substack{j=1 \\ j \neq i, k}}^{N} \left(K_{ij} \frac{1}{r_{ij}^n} - \frac{m_j}{m_k} K_{ik} \frac{1}{r_{ik}^n} \right) \mathbf{r}_j. \tag{4}$$

In general, this is a set of coupled equations because the equation for each $\mathbf{r}_i$ contains also the position vectors of the other particles. The case $n = 0$ in (4) is special in that it allows decoupling, provided the condition

$$K_{ij}/K_{ik} = m_j/m_k \tag{5}$$

also holds. In this event (4) simplifies to a set of one-body equations

$$m_i \ddot{\mathbf{r}}_i = -M \alpha_i \mathbf{r}_i \quad (i = 1, 2, \cdots, N), \tag{6}$$

where M is the total mass and $\alpha_i = K_{ik}/m_k$. If the α_i are positive then (6) is a set of N isotropic harmonic oscillator equations with force constants $k_i = M\alpha_i$: each particle moves independently of the others and as if it were bound to the CM of the system by a spring having force constant k_i.

Comments

(i) The case $N = 2$ is special because the two-body problem with central interparticle force $F(\mathbf{r}_{12})\hat{\mathbf{r}}_{12}$ can always be reduced to two one-body problems (see Question 10.4)).

(ii) For the gravitational N-body problem, $K_{ij} = Gm_i m_j$ (and so (5) is satisfied) and $n = 3$. If the motions are such that the interparticle distances are always equal, $r_{ij} = r(t)$, then (4) has the simple form

$$\ddot{\mathbf{r}}_i = -\frac{GM}{r^3(t)} \mathbf{r}_i. \tag{7}$$

Equations (7) describe the equilateral triangle solution of the three-body problem (see Question 11.6) and the regular tetrahedron solution of the four-body problem.[15]

---

[15] H. Essén, "On the equilateral triangle solution to the three-body problem," European Journal of Physics, vol. 21, pp. 579–590, 2000.

Question 11.11

Consider a gas consisting of N non-interacting particles (molecules) that make elastic collisions with the walls of a container (taken to be a cubical box of side L). Prove that

$$PV = \tfrac{2}{3}K, \qquad (1)$$

where P and $V = L^3$ are the pressure and volume of the gas, and K is the total kinetic energy of the molecules.

Solution

Consider a face of the container that is perpendicular to the x-axis (say). For each elastic collision with this face the change in momentum of a molecule is $\Delta p = mv - (-mv) = 2mv$, where v is the x-component of its velocity. The time between collisions with the face is $\Delta t = 2L/v$ and so, for the total force exerted by all molecules on this face, we have

$$F = \sum \frac{\Delta p}{\Delta t} = \frac{1}{L} \sum mv^2, \qquad (2)$$

where the sum is over all molecules. The pressure $P = F/L^2$ is therefore

$$P = \frac{1}{L^3} \sum mv^2. \qquad (3)$$

Now the contributions of the x, y and z motions to the total kinetic energy K are equal, and so

$$\sum \tfrac{1}{2}mv^2 = \tfrac{1}{3}K. \qquad (4)$$

Equations (3) and (4) yield (1).

Comments

(i) In terms of the mean-square velocity $\overline{u^2} = (N_1 m_1 \mathbf{v}_1^2 + N_2 m_2 \mathbf{v}_2^2 + \cdots)/M$, the total kinetic energy of the gas is $K = \tfrac{1}{2} M \overline{u^2}$. Here, $M = N_1 m_1 + N_2 m_2 + \cdots$ is the mass of the gas, given in terms of its density ρ by $M = \rho V$. So (1) can be expressed as

$$P = \tfrac{1}{3} \rho \overline{u^2}. \qquad (5)$$

This means that for air at $P = 10^5$ Pa and $\rho = 1.3\,\text{kg m}^{-3}$, the root-mean-square velocity $\sqrt{\overline{u^2}}$ is about $480\,\text{m s}^{-1}$.

(ii) According to the equipartition theorem of classical statistical mechanics, the average kinetic energy of a molecule is $\tfrac{1}{2}kT$ per degree of freedom for a gas in equilibrium at temperature T, where k is Boltzmann's constant. The total kinetic

energy K is N times this average. Also, each molecule has three (translational) degrees of freedom. So,
$$K = N \times \tfrac{3}{2}kT. \tag{6}$$
Equations (1) and (6) give the equation of state of an ideal gas:
$$PV = NkT. \tag{7}$$

(iii) For an ideal, monatomic gas the total energy U (the internal energy) is entirely kinetic (no rotational, vibrational, ... energies). So, $U = K$ and (1) and (6) can be written as
$$PV = \tfrac{2}{3}U \quad \text{with} \quad U = \tfrac{3}{2}NkT. \tag{8}$$
(Alternatively, $U = \tfrac{3}{2}nRT$, where $n = N/N_A$ is the number of moles, N_A is Avogadro's number and $R = N_A k$ is the ideal gas constant.)

(iv) The study of an ideal gas has played an important role in physics. An analysis of the above sort – leading to (5) – was performed by Joule in about 1850, and soon the equation of state $PV = NkT$ and molar specific heat $n^{-1}\partial U/\partial T = \tfrac{3}{2}R$ were obtained. These classical results ignore contributions due to the finite value of Planck's constant h, and they are valid only in the limit where the average spacing between particles is large compared to the average de Broglie wavelength of the particles. Consequently, it turns out that they do not apply to the 'electron gas' in metals or to gases at low temperature and/or high pressure. Also, for some properties of a gas, such as its entropy and chemical potential, the leading-order contributions in the above limit depend on h.

Question 11.12

Consider a set of N particles that move along the x-axis. The interparticle forces are repulsive and inversely proportional to the distance between particles:
$$F_{ji} = \frac{A}{x_i - x_j}, \tag{1}$$
where A is a positive constant. A constant external force $-F$ (where $F > 0$) drives the particles in the negative x-direction. The leading particle is held fixed at $x_1 = 0$ and the remainder are in equilibrium at $x_2, \cdots, x_N$, where
$$0 = x_1 < x_2 < \cdots < x_N.$$

(a) Show that the equilibrium positions x_i of the particles are given by the N roots of the polynomial
$$P_N(u) = u + \sum_{n=2}^{N}(-1)^{n-1}\frac{(N-1)(N-2)\cdots(N-n+1)}{n!(n-1)!}u^n, \tag{2}$$
where $N \geq 2$ and $u = 2Fx/A$.

(b) Use the fact that (2) is equal (apart from a minus sign) to a Laguerre polynomial, as defined in *Mathematica*, to tabulate the equilibrium positions $u_2, u_3, \cdots, u_N$ for $N = 2, 3, \cdots, 10$. Depict the equilibrium configurations for $N = 7$ and $N = 8$.

Solution

(a) The condition for equilibrium is that the sum of the interparticle forces and the external force on each of the $N-1$ particles behind the trapped particle at $x=0$ should be zero. This yields a set of non-linear algebraic equations:

$$\sum_{\substack{j=1 \\ j \neq i}}^{N} \frac{A}{x_i - x_j} = F, \qquad \text{where } i = 2, 3, \cdots, N. \tag{3}$$

It is convenient to express (3) in the dimensionless form

$$\sum_{\substack{j=1 \\ j \neq i}}^{N} \frac{1}{u_i - u_j} = \frac{1}{2}, \qquad \text{where } i = 2, 3, \cdots, N, \tag{4}$$

and $u = 2Fx/A$. To solve these equations we introduce a polynomial of degree N whose roots are the equilibrium positions u_i ($i = 1, 2, \cdots, N$):

$$P(u) = (u - u_1)(u - u_2) \cdots (u - u_N) \tag{5}$$

with $u_1 = 0$. By differentiating (5) with respect to u we have

$$\frac{P'(u)}{P(u)} = \frac{1}{u - u_1} + \frac{1}{u - u_2} + \cdots + \frac{1}{u - u_N} \equiv S(u),$$

$$\frac{P''(u)}{P'(u)} = S(u) + \frac{S'(u)}{S(u)}, \tag{6}$$

where a prime denotes differentiation with respect to u. Now, consider the limit of (6) as $u \to u_i$ for $i = 2, 3, \cdots, N$. Use of the equilibrium condition (4) shows that in this limit $S(u)$ behaves like $\frac{1}{2} + (u - u_i)^{-1}$, and therefore

$$P''(u)/P'(u) \to 1 \quad \text{as} \quad u \to u_i. \tag{7}$$

That is, $P''(u_i) - P'(u_i) = 0$. This means that $P''(u) - P'(u)$ is a polynomial that vanishes at $u_2, u_3, \cdots, u_N$: according to (5) it must therefore be proportional to $u^{-1} P(u)$. So,

$$P''(u) - P'(u) = c u^{-1} P(u), \tag{8}$$

where c is a constant. By considering the highest term (u^N) in P, it follows by inspection of (8) that $c = -N$. Thus, $P(u)$ satisfies the differential equation

$$u P''(u) - u P'(u) + N P(u) = 0. \tag{9}$$

This possesses a polynomial solution

$$P_N(u) = \sum_{n=1}^{N} a_n u^n \tag{10}$$

with $a_1 = 1$ and

$$a_n(N) = -\frac{N-n+1}{n(n-1)} a_{n-1}(N) \qquad (n = 2, 3, \cdots, N). \tag{11}$$

That is,

$$a_n(N) = (-1)^{n-1} \frac{(N-1)(N-2)\cdots(N-n+1)}{n!(n-1)!} \qquad (n = 2, 3, \cdots, N), \tag{12}$$

and so we obtain (2).

(b) We find the roots of the Laguerre polynomials using the following notebook

```
In[1]:= Do[Print[NSolve[LaguerreL[n, -1, u] == 0, u]], {n, 1, 10, 1}]
```

and the results are tabulated below:

N	u_1	u_2	u_3	u_4	u_5	u_6	u_7	u_8	u_9	u_{10}
1	0	–	–	–	–	–	–	–	–	–
2	0	2.000	–	–	–	–	–	–	–	–
3	0	1.268	4.732	–	–	–	–	–	–	–
4	0	0.936	3.305	7.759	–	–	–	–	–	–
5	0	0.743	2.572	5.731	10.954	–	–	–	–	–
6	0	0.617	2.113	4.611	8.399	14.260	–	–	–	–
7	0	0.528	1.796	3.877	6.919	11.235	17.646	–	–	–
8	0	0.461	1.564	3.352	5.916	9.421	14.194	21.092	–	–
9	0	0.409	1.385	2.956	5.182	8.162	12.070	17.250	24.586	–
10	0	0.368	1.243	2.646	4.617	7.222	10.567	14.836	20.382	28.118

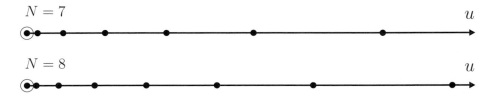

Comments

(i) This example has application to dislocation pile-ups in crystals.[16] There is a large literature dealing with the theory and application of these pile-ups to the mechanical properties of crystals.[17] The calculation also applies to the equilibrium configurations of long, parallel line charges in an electrostatic field.

---

[16] J. Eshelby, F. C. Frank, and F. R. N. Nabarro, "The equilibrium of linear arrays of dislocations," Philosophical Magazine, vol. 42, pp. 351–364, 1951.

[17] See, for example, J. P. Hirth and J. Lothe, *Theory of dislocations*. New York: Wiley, 1982.

Question 11.13

Consider again a one-dimensional array of N particles with repulsive interparticle forces $F_{ji} = A/(x_i - x_j)$, where A is a positive constant.

(a) Suppose that the outer two particles in the array are fixed at $x = -L$ and $x = L$, respectively, and the external force is zero. Show that the equilibrium positions of the $N - 2$ mobile particles are given by the roots of the polynomial

$$Q(u) = \frac{d}{du}\mathcal{L}_{N-1}(u). \qquad (1)$$

Here, $u = x/L$ and $\mathcal{L}_{N-1}(u)$ is the $(N-1)$th Legendre polynomial, which satisfies the differential equation

$$(1 - u^2)\mathcal{L}''_{N-1} - 2u\mathcal{L}'_{N-1} + N(N-1)\mathcal{L}_{N-1} = 0. \qquad (2)$$

Depict the equilibrium configurations for $N = 7$ and $N = 8$.

(b) Suppose all particles in the array are mobile and that an external force $-Dx$ (where D is a positive constant) acts on them. Show that their equilibrium positions are given by the N roots of the Hermite polynomial $\mathcal{H}_N(u)$, where $u = \sqrt{D/A}\, x$ and $\mathcal{H}_N$ satisfies the differential equation

$$\mathcal{H}''_N - 2u\mathcal{H}'_N + 2N\mathcal{H}_N = 0. \qquad (3)$$

Depict the equilibrium configurations for $N = 7$ and $N = 8$.

Solution

(a) The analysis is similar to that of the previous question. With $N \geq 3$ and $u = x/L$, the equilibrium conditions for the $N - 2$ mobile particles are

$$\sum_{\substack{j=1 \\ j \neq i}}^{N} \frac{1}{u_i - u_j} = 0, \qquad \text{where } i = 2, 3, \cdots, (N-1), \qquad (4)$$

and $u_1 = -1$ and $u_N = 1$. Let

$$P(u) = (u+1)(u - u_2)\cdots(u - u_{N-1})(u - 1). \qquad (5)$$

---

[18]   G. Szego, *Orthogonal polynomials*. New York: American Mathematical Society, 1939.

Then
$$\frac{P'(u)}{P(u)} = \frac{1}{u+1} + \frac{1}{u-u_2} + \cdots + \frac{1}{u-u_{N-1}} + \frac{1}{u-1} \equiv S(u), \qquad (6)$$
$$\frac{P''(u)}{P'(u)} = S(u) + \frac{S'(u)}{S(u)}. \qquad (7)$$

The equilibrium conditions (4) show that in the limit $u \to u_i$ the sum $S(u)$ in (6) behaves like $(u - u_i)^{-1}$, and it follows from (7) that $P''(u) \to 0$ as $u \to u_i$ (for $i = 2, 3, \cdots, N - 1$). So, $P''(u)$ is a polynomial that vanishes at $u_2, u_3, \cdots, u_{N-1}$: according to (5) it must therefore be proportional to $(u^2 - 1)^{-1} P(u)$. We write
$$P''(u) = c(u^2 - 1)^{-1} P(u), \qquad (8)$$
where c is a constant. Consider the highest term, u^N, in P. It follows from (8) that $c = N(N-1)$ and so $P(u)$ satisfies the differential equation
$$(1 - u^2) P'' + N(N-1) P = 0. \qquad (9)$$

In (9) let $P = (u^2 - 1) Q$. Then
$$(1 - u^2) Q'' - 4u Q' + \{N(N-1) - 2\} Q = 0. \qquad (10)$$

By differentiating (2) with respect to u we see that (10) is just the equation satisfied by $\mathcal{L}'_{N-1}$, and so we have the identification (1). Numerical values for the roots of $\mathcal{L}'_{N-1}$ were calculated using *Mathematica*:

```
In[1]:= Do[Print[NSolve[D[LegendreP[n-1, u], {u, 1}] == 0, u]],
          {n, 3, 10, 1}]
```

$N = 7$

$N = 8$

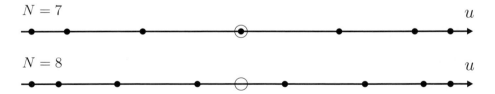

(b) In terms of the dimensionless coordinate $u = \sqrt{D/A}\, x$ the equilibrium conditions for the N particles are
$$\sum_{\substack{j=1 \\ j \neq i}}^{N} \frac{1}{u_i - u_j} = u_i, \qquad \text{where } i = 1, 2, \cdots, N. \qquad (11)$$

Let
$$P(u) = (u - u_1)(u - u_2) \cdots (u - u_N). \qquad (12)$$

Then
$$\frac{P'(u)}{P(u)} = \frac{1}{u-u_1} + \frac{1}{u-u_2} + \cdots + \frac{1}{u-u_N} \equiv S(u), \qquad (13)$$

and $P''(u)/P'(u)$ satisfies (7). The equilibrium conditions (11) show that in the limit $u \to u_i$ the sum $S(u)$ in (13) behaves like $u + (u-u_i)^{-1}$. So, $S + S'/S \to 2u$ as $u \to u_i$, and according to (7)

$$P''(u) - 2uP'(u) \to 0 \text{ as } u \to u_i \qquad (i = 1, 2, \cdots, N). \qquad (14)$$

Therefore
$$P''(u) - 2uP'(u) = cP(u), \qquad (15)$$

where the constant c is identified from the highest term, u^N, in $P(u)$ as $c = -2N$. So,

$$P'' - 2uP' + 2NP = 0, \qquad (16)$$

which is just the differential equation (3) for the Hermite polynomial $\mathcal{H}_N(u)$. Roots of $\mathcal{H}_N$ can be obtained by replacing NSolve[···] in the Mathematica notebook above with NSolve[HermiteH[n, u] == 0, u]. The equilibrium configurations for $N = 7$ and $N = 8$ are:

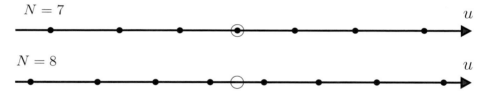

It can be shown that the largest root of $\mathcal{H}_N$ is less than $\sqrt{2N+1}$, and so the particles are located in the region

$$|x| < \sqrt{(2N+1)A/D}. \qquad (17)$$

Comments

(i) The above examples also have applications to crystal dislocations and mechanical properties of solids.[16,17]

(ii) If the number of particles N in the array is large then a continuum description in terms of a linear density $f(x)$ becomes possible. Here, $f(x)\,dx$ is the number of particles in an interval $[x, x+dx]$. The equilibrium condition for an array in $[L_1, L_2]$ subject to an external force $F(x)$ is now given by an integral equation

$$A \int_{L_1}^{L_2} \frac{f(x')}{x'-x}\,dx' = F(x), \qquad (18)$$

where the principal value of the integral is understood (in order to exclude self-interaction of the particles) and $f(x)$ is subject to the normalization

$$\int_{L_1}^{L_2} f(x)\,dx = N\,. \tag{19}$$

The integral in (18) is known as the Hilbert transform of $f(x)$, and so the determination of an equilibrium distribution $f(x)$ requires evaluation of the inverse Hilbert transform of a specified external force $F(x)$. For this purpose there are standard inversion theorems for Cauchy-type singular integral equations[19] and tables of Hilbert transforms are also available.[20] Here, we mention two examples. For an array that is driven in the negative x-direction by a constant force $-F$ (< 0) against an impenetrable barrier at $x = 0$, (18) is

$$A \int_0^L \frac{f(x')}{x' - x}\,dx' = -F\,, \tag{20}$$

where L is the equilibrium length of the array. The normalized solution

$$f(x) = \frac{F}{\pi A}\sqrt{\frac{L-x}{x}}\,, \quad \text{where} \quad L = 2NA/F\,, \tag{21}$$

is the continuum solution to Question 11.12. Equation $(21)_2$ shows how the equilibrium length depends on the number of particles, the external force and the constant A in the interparticle forces. For an array that is trapped between barriers at $x = -L$ and $x = L$ in the absence of an external force we have

$$A \int_{-L}^{L} \frac{f(x')}{x' - x}\,dx' = 0\,. \tag{22}$$

The normalized solution

$$f(x) = \frac{N}{\pi\sqrt{L^2 - x^2}} \tag{23}$$

is the continuum solution to Question 11.13(a). Graphs of the distributions (21) and (23) are shown below.

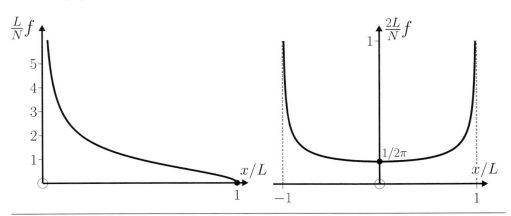

---

[19] N. I. Muskhelishvili, *Singular integral equations*. Groningen: P. Noordhoff, 1953.
[20] A. Erdélyi, ed., *Tables of integral transforms*. New York: McGraw-Hill, 1954.

Question 11.14

Consider a system comprising N particles of equal mass that move in one dimension and interact pairwise via repulsive inverse-cube interparticle forces. The equations of motion are

$$\ddot{x}_n = 2g^2 \sum_{\substack{m=1 \\ m \neq n}}^{N} (x_n - x_m)^{-3} \qquad (n = 1, 2, \cdots, N), \tag{1}$$

where g is a constant. Show that the solutions $x_n(t)$ to (1) subject to initial conditions $x_n(0)$ and $\dot{x}_n(0)$ are given by the eigenvalues of an $N \times N$ matrix $\widetilde{Q}(t)$ that has elements[21]

$$\widetilde{Q}_{nm}(t) = \delta_{nm} [x_n(0) + \dot{x}_n(0)t] + (1 - \delta_{nm}) ig [x_n(0) - x_m(0)]^{-1} t. \tag{2}$$

(Here, δ_{nm} is the Kronecker delta function: $\delta_{nm} = 1$ if $n = m$, $\delta_{nm} = 0$ if $n \neq m$.) To do this, perform the following steps:

(a) Introduce $\widetilde{Q}(t)$ by a similarity transformation

$$\widetilde{Q}(t) = U(t) Q(t) U^{-1}(t) \tag{3}$$

of a diagonal $N \times N$ matrix $Q(t)$ with elements equal to $x_n(t)$,

$$Q_{nm}(t) = \delta_{nm} x_n(t). \tag{4}$$

By differentiating (3) with respect to t, show that

$$\ddot{\widetilde{Q}} = U\left(\dot{L} - [L, M]\right) U^{-1}, \tag{5}$$

where $[L, M] = LM - ML$ and

$$M = U^{-1} \dot{U}, \qquad L = \dot{Q} - [Q, M]. \tag{6}$$

(b) Use the equations of motion (1) to show that the choices

$$L_{nm} = \delta_{nm} \dot{x}_n + (1 - \delta_{nm}) ig(x_n - x_m)^{-1} \tag{7}$$

$$M_{nm} = \delta_{nm} ig \sum_{\substack{\ell=1 \\ \ell \neq n}}^{N} (x_n - x_\ell)^{-2} - (1 - \delta_{nm}) ig(x_n - x_m)^{-2} \tag{8}$$

satisfy the key condition

$$\dot{L} = [L, M]. \tag{9}$$

(c) Hence, integrate (5) to obtain (2). (Hint: Make the convenient choice $U(0) = 1$ for the initial value of the matrix $U(t)$.)

(d) Use (2) to obtain the trajectories for $N = 2$.

---

[21] F. Calogero, *Classical many-body problems amenable to exact treatments*. Berlin: Springer, 2001.

Solution

We follow Calogero's solution.[21]

(a) Differentiation of (3) with respect to t gives

$$\begin{aligned}\dot{\tilde{Q}} &= U\dot{Q}U^{-1} + \dot{U}QU^{-1} - QU^{-1}\dot{U}U^{-1}\\ &= U(\dot{Q} + U^{-1}\dot{U}Q - QU^{-1}\dot{U})U^{-1}\\ &= U(\dot{Q} + MQ - QM)U^{-1}\\ &= ULU^{-1},\end{aligned} \qquad (10)$$

where we have used equations (6) for M and L. By the same procedure, differentiation of (10) yields

$$\ddot{\tilde{Q}} = U(\dot{L} + ML - LM)U^{-1},$$

which is the desired relation (5).

(b) We verify that the evolution equation (9) is satisfied by the choices (7) and (8). Consider first the diagonal terms

$$\dot{L}_{nm} = \sum_{\substack{\ell=1\\ \ell\neq n}}^{N}(L_{n\ell}M_{\ell n} - L_{\ell n}M_{n\ell}) \quad \text{if } m = n. \qquad (11)$$

By substituting (7) and (8) in (11) we have

$$\ddot{x}_n = -2(ig)^2 \sum_{\substack{\ell=1\\ \ell\neq n}}^{N}(x_n - x_\ell)^{-3},$$

which are just the equations of motion (1). For the off-diagonal terms, differentiation of (7) gives

$$\dot{L}_{nm} = -ig(\dot{x}_n - \dot{x}_m)(x_n - x_m)^{-2},$$

and some algebra shows (Ref. [21], pp. 24–26) that this is also the right-hand side of (9).

(c) The condition (9) reduces the evolution equation (5) to the simple form:

$$\ddot{\tilde{Q}} = 0, \qquad (12)$$

which has the general solution

$$\tilde{Q}(t) = \tilde{Q}(0) + \dot{\tilde{Q}}(0)t. \qquad (13)$$

The matrix coefficients in (13) are calculated from the similarity transformations (3) and (10), and they depend on the initial value $U(0)$. Different choices for $U(0)$ yield different matrices $\tilde{Q}(t)$, but they do not affect the eigenvalues of $\tilde{Q}(t)$.

Because we are interested in the latter, we can regard $U(0)$ as an arbitrary initial condition and make the simplifying choice $U(0) = 1$ (the unit matrix). Then, (3) and (10) yield

$$\tilde{Q}(0) = Q(0), \qquad \dot{\tilde{Q}}(0) = L(0), \tag{14}$$

and so

$$\tilde{Q}(t) = Q(0) + L(0)t. \tag{15}$$

According to (4) and (7), the elements of the matrix (15) are given – in terms of the initial conditions $x_n(0)$ and $\dot{x}_n(0)$, and the constant g in the interparticle forces – by (2). Explicit formulas for the N trajectories $x_n(t)$ can be obtained by finding the eigenvalues of (2).

(d) For $N = 2$ the eigenvalues of (2) are obtained from

$$\begin{vmatrix} x_1(0) + \dot{x}_1(0)t - x_n(t) & ig[x_1(0) - x_2(0)]^{-1}t \\ -ig[x_1(0) - x_2(0)]^{-1}t & x_2(0) + \dot{x}_2(0)t - x_n(t) \end{vmatrix} = 0 \tag{16}$$

for $n = 1, 2$. The results are

$$x_n(t) = \tfrac{1}{2}x_+(0) + \tfrac{1}{2}\dot{x}_+(0)t + (-1)^n \tfrac{1}{2}\{[x_-(0)]^2 + 2x_-(0)\dot{x}_-(0)t$$
$$+ ([\dot{x}_-(0)]^2 + 4g^2[x_-(0)]^{-2})t^2\}^{1/2}, \tag{17}$$

where $x_\pm(0) = x_1(0) \pm x_2(0)$ and similarly for $\dot{x}_\pm(0)$.

Comments

(i) The above example is a particular case of a general technique that can be used to solve certain classical many-body problems. A detailed account of this and related topics is given in Calogero's encyclopaedic study of classical many-body problems (in one, two and three dimensions) that can be solved exactly.[21] We mention that the method of solution used above is known as the Lax technique: the matrices L and M that appear in (5) to (9) are called a Lax pair, and the key evolution equation (9) is the Lax equation.

(ii) Several questions concerning the above calculations that may have occurred to the reader are discussed in Ref. [21]. These include a derivation of the forms (7) and (8) for the Lax matrices, and the non-uniqueness of Lax pairs corresponding to a given set of equations of motion.

(iii) The system considered above experiences only repulsive interparticle forces, and therefore the motions are unbounded: the particles will eventually disperse and move freely, cf. (17). The system can be bounded by including also some attractive interaction – such as a harmonic interaction – and this interesting problem is analyzed next.

Question 11.15

Suppose Question 11.14 is modified to include an external harmonic interaction that attracts each particle towards the origin.[21] The equations of motion are

$$\ddot{x}_n = -\omega^2 x_n + 2g^2 \sum_{\substack{m=1 \\ m \neq n}}^{N} (x_n - x_m)^{-3} \qquad (n = 1, 2, \cdots, N), \tag{1}$$

where ω and g are constants. Extend the analysis of the previous question to the equations of motion (1). In particular, show that the solutions $x_n(t)$ to (1) with initial conditions $x_n(0)$ and $\dot{x}_n(0)$ are given by the eigenvalues of an $N \times N$ matrix $\widetilde{Q}(t)$ that has elements

$$\begin{aligned}\widetilde{Q}_{nm}(t) = {} & \delta_{nm}\left[x_n(0)\cos\omega t + \dot{x}_n(0)\omega^{-1}\sin\omega t\right] \\ & + (1-\delta_{nm})ig\left[x_n(0) - x_m(0)\right]^{-1}\omega^{-1}\sin\omega t\,. \end{aligned} \tag{2}$$

(Hint: Equations (3)–(8) of Question 11.14 are unaltered, but the Lax equation (9) must be modified to read

$$\dot{L} = [L, M] - \omega^2 Q, \tag{3}$$

with $Q_{nm} = \delta_{nm} x_n$, as in (4) of Question 11.14. Show that (3) is consistent with (1), then obtain the evolution equation for $\widetilde{Q}(t)$ and show that it leads to (2).)

Solution

To demonstrate that (3) is consistent with (1), we need consider only the diagonal terms of (3), because Q is a diagonal matrix:

$$\dot{L}_{nm} = \sum_{\substack{\ell=1 \\ \ell \neq n}}^{N} (L_{n\ell}M_{\ell n} - L_{\ell n}M_{n\ell}) - \omega^2 x_n \qquad (m = n), \tag{4}$$

which, for the Lax pair (7) and (8), reduce to (1). The term $-\omega^2 Q$ in (3) is the essential modification that incorporates the effect of the harmonic interactions. It leads – via (5) and (3) of Question 11.14 – to an evolution equation

$$\ddot{\widetilde{Q}} + \omega^2 \widetilde{Q} = 0, \tag{5}$$

instead of $\ddot{\widetilde{Q}} = 0$. The general solution to (5) is

$$\widetilde{Q} = \widetilde{Q}(0)\cos\omega t + \dot{\widetilde{Q}}(0)\omega^{-1}\sin\omega t = Q(0)\cos\omega t + L(0)\omega^{-1}\sin\omega t, \tag{6}$$

where the calculation in the last step is the same as that leading to (15) in the previous question. Use of (4) and (7) of Question 11.14 in (6) yields (2).

Comments

(i) It is evident from (2) that when $\omega \neq 0$ the motions are bounded and periodic, with period $T = 2\pi/\omega$. In the limit $\omega = 0$ all the above results reduce to those obtained in the previous question, where the motions are unbounded. The binding is provided by the harmonic forces that attract each particle to the origin.

(ii) One can also consider a system of particles where the harmonic interactions are pairwise between the particles, rather than an external harmonic attraction toward the origin. Then, (1) is replaced by

$$\ddot{x}_n = -\bar{\omega}^2 \sum_{m=1}^{N} (x_n - x_m) + 2g^2 \sum_{\substack{m=1 \\ m \neq n}}^{N} (x_n - x_m)^{-3} \quad (n = 1, 2, \cdots, N). \quad (7)$$

There is a simple relation between these two systems: the substitution $\bar{\omega}^2 = \omega^2/N$ and a shift in coordinates $x_n \to x_n + X$, where $X = \sum_{m=1}^{N} x_m/N$ is the CM coordinate, reduces (7) to (1).

(iii) The system possesses a unique set of equilibrium points given (according to (1) with $x_n = $ constant) by solutions to the non-linear equations

$$x_n = 2\left(\frac{g}{\omega}\right)^2 \sum_{\substack{m=1 \\ m \neq n}}^{N} (x_n - x_m)^{-3} \quad (n = 1, 2, \cdots, N). \quad (8)$$

It is not difficult to see that these x_n are also solutions for the equilibrium of a system with inverse first-power, instead of inverse-cube, repulsive forces:[22]

$$x_n = 2\left(\frac{\bar{g}}{\omega}\right)^2 \sum_{\substack{m=1 \\ m \neq n}}^{N} (x_n - x_m)^{-1} \quad (n = 1, 2, \cdots, N). \quad (9)$$

The latter problem has been solved in Question 11.13(b): the equilibrium positions are the roots of a Hermite polynomial

$$\mathcal{H}_N(\sqrt{\omega/g}\, x_n) = 0. \quad (10)$$

(iv) The system (7) with pairwise harmonic and repulsive inverse-cube interactions is known as the Calogero model. It was first analyzed in its quantum-mechanical form,[23] and the classical problem was solved later.[24] Much research continues to be devoted to understanding this model and its variations and applications.[25]

---

[22] F. Calogero, "Equilibrium configuration of the one-dimensional n-body problem with quadratic and inversely quadratic pair potentials," Lettere al Nuovo Cimento, vol. 20, pp. 251–253, 1977.

[23] F. Calogero, "Solution of the one-dimensional N-body problems with quadratic and/or inversely quadratic pair potentials," Journal of Mathematical Physics, vol. 12, pp. 419–436, 1971.

[24] M. Olshanetsky and A. Perelomov, "Explicit solution of the Calogero model in the classical case and geodesic flows on symmetric spaces of zero curvature," Lettere al Nuovo Cimento, vol. 16, pp. 333–339, 1976.

[25] See Journal of Nonlinear Mathematical Physics, vol. 12, supplement 1, 2005.

Question 11.16

Consider a particle of mass m with position vector $\mathbf{r}$ in the gravitational field of a stationary system of N particles. The gravitational potential $\Phi(\mathbf{r})$ of the system is defined as the potential energy per unit mass: $\Phi(\mathbf{r}) = V(\mathbf{r})/m$. Suppose that the distribution of mass in the system is continuous, with mass density $\rho(\mathbf{r})$.

(a) Show that $\Phi(\mathbf{r})$ satisfies Poisson's equation

$$\nabla^2 \Phi(\mathbf{r}) = 4\pi G \rho(\mathbf{r}). \tag{1}$$

(Hint: Start with the expression for the potential energy of two particles and then use the result

$$\nabla^2 (1/r) = -4\pi \delta(\mathbf{r}), \tag{2}$$

where $\delta(\mathbf{r})$ is the Dirac delta function that has the properties

$$\delta(\mathbf{r}) = 0 \text{ if } \mathbf{r} \neq 0, \text{ and } \int \delta(\mathbf{r}) \, d^3\mathbf{r} = 1 \tag{3}$$

if the region of integration includes the origin.)

(b) The gravitational field of the system is the force per unit mass: $\mathbf{g}(\mathbf{r}) = \mathbf{F}(\mathbf{r})/m$, where $\mathbf{F}$ is the total gravitational force exerted on m. Prove that

$$\nabla \times \mathbf{g}(\mathbf{r}) = 0 \tag{4}$$

and

$$\nabla \cdot \mathbf{g}(\mathbf{r}) = -4\pi G \rho(\mathbf{r}). \tag{5}$$

Solution

(a) The potential energy of a mass m located at $\mathbf{r}$ due to a system of particles with masses m_i and position vectors $\mathbf{r}_i$ ($i = 1, 2 \cdots, N$) is a sum of N two-particle potentials:

$$V(\mathbf{r}) = -Gm \sum_{i=1}^{N} \frac{m_i}{|\mathbf{r} - \mathbf{r}_i|}. \tag{6}$$

The corresponding gravitational potential is

$$\Phi(\mathbf{r}) = -G \sum_{i=1}^{N} \frac{m_i}{|\mathbf{r} - \mathbf{r}_i|}. \tag{7}$$

For a continuous distribution of mass the sum in (7) can be replaced by a volume integral:

$$\Phi(\mathbf{r}) = -G \int \frac{\rho(\mathbf{r}')}{|\mathbf{r} - \mathbf{r}'|} \, d^3\mathbf{r}'. \tag{8}$$

Then

$$\nabla^2 \Phi(\mathbf{r}) = -G \int \rho(\mathbf{r}') \left\{ \nabla^2 |\mathbf{r} - \mathbf{r}'|^{-1} \right\} d^3\mathbf{r}'$$

$$= 4\pi G \int \rho(\mathbf{r}') \delta(\mathbf{r} - \mathbf{r}') d^3\mathbf{r}'$$

$$= 4\pi G \rho(\mathbf{r}), \qquad (9)$$

where we have used (2) and (3).

(b) The gravitational force on m is conservative and related to the potential energy (6) by $\mathbf{F}(\mathbf{r}) = -\boldsymbol{\nabla} V(\mathbf{r})$ (see Chapter 5). Equation (4) follows directly from this (see Question 5.7). It also follows that $\boldsymbol{\nabla} \cdot \mathbf{F} = -\boldsymbol{\nabla} \cdot (\boldsymbol{\nabla} V) = -\nabla^2 V$. This, combined with (1), yields (5).

Comments

(i) The interpretation and derivation of the key result (2), as well as the two steps prior to (9), require careful treatment.[26]

(ii) The integral and differential forms (8) and (9) are equivalent. Either may be used to determine the potential $\Phi(\mathbf{r})$ of a given mass distribution $\rho(\mathbf{r})$, and then the field

$$\mathbf{g}(\mathbf{r}) = -\boldsymbol{\nabla} \Phi(\mathbf{r}). \qquad (10)$$

Which one chooses is essentially a matter of convenience: in practice it can be simpler to solve the differential equation (9) subject to appropriate boundary conditions rather than to perform the integrations (8) over all space. Some examples are given below.

(iii) By making the replacements $\mathbf{g} \to \mathbf{E}$ and $G \to -1/4\pi\epsilon_0$ in (4) and (5) we obtain corresponding equations for the electrostatic field of a stationary charge distribution:

$$\boldsymbol{\nabla} \times \mathbf{E}(\mathbf{r}) = 0, \qquad \boldsymbol{\nabla} \cdot \mathbf{E}(\mathbf{r}) = \rho(\mathbf{r})/\epsilon_0. \qquad (11)$$

Question 11.17

A sphere of mass M and radius R has uniform density $\rho = M \div \frac{4}{3}\pi R^3$. Use Poisson's equation to calculate the gravitational potential Φ and hence the gravitational field $\mathbf{g}$ of this sphere at all points in space.

Solution

We use spherical polar coordinates (r, θ, ϕ). Because of the spherical symmetry of the system, Φ cannot depend on θ and ϕ. With $\Phi = \Phi(r)$, Poisson's equation (see (1) of the previous question) is

$$\frac{d}{dr}\left(r^2 \frac{d\Phi}{dr}\right) = \begin{cases} kr^2 & \text{if } r < R \\ 0 & \text{if } r > R \end{cases}, \qquad (1)$$

---

[26] V. Hnizdo, "On the Laplacian of $1/r$," European Journal of Physics, vol. 21, pp. L1–L3, 2000.

where k is the constant
$$k = 4\pi G\rho = 3GM/R^3. \tag{2}$$

1. $r < R$

By integrating $(1)_1$ between $r = 0$ and r we have $d\Phi/dr = \tfrac{1}{3}kr$, and so
$$\Phi(r) = \Phi(0) + \tfrac{1}{6}kr^2 \qquad (r < R). \tag{3}$$

2. $r > R$

We use $(1)_1$ in the integration from $r = 0$ to R, and $(1)_2$ in the integration from $r = R$ to r. Thus, $d\Phi/dr = kR^3/3r^2$ for $r > R$. This, together with $d\Phi/dr = \tfrac{1}{3}kr$ for $r < R$, means
$$\Phi(r) = \Phi(0) + \tfrac{1}{2}kR^2 - \tfrac{1}{3}kR^3/r \qquad (r > R). \tag{4}$$

One often chooses $\Phi(\infty) = 0$: according to (4) this requires $\Phi(0) = -\tfrac{1}{2}kR^2$. Then, from (2), (3) and (4) the solution for the gravitational potential is
$$\Phi(r) = -\tfrac{4}{3}\pi R^2 \rho G\left(\frac{3}{2} - \frac{r^2}{2R^2}\right) \tag{5}$$
$$= -\frac{GM}{R}\left(\frac{3}{2} - \frac{r^2}{2R^2}\right) \tag{6}$$

if $r < R$, and
$$\Phi(r) = -\tfrac{4}{3}\pi R^3 \rho G/r = -GM/r \tag{7}$$

if $r > R$. The resulting gravitational field $\mathbf{g}(\mathbf{r}) = -\boldsymbol{\nabla}\Phi(r) = -\hat{\mathbf{r}}\, d\Phi/dr$ is
$$\mathbf{g}(\mathbf{r}) = \begin{cases} -\dfrac{GM}{R^2}\dfrac{r}{R}\hat{\mathbf{r}} & \text{if } r < R \\ -\dfrac{GM}{r^2}\hat{\mathbf{r}} & \text{if } r > R. \end{cases} \tag{8}$$

Comments

(i) According to (3) and (4), the gravitational potential and its derivative are continuous at the surface $r = R$. This continuity appears naturally in the calculation and is not imposed as an added condition.

(ii) Equations (7) and $(8)_2$ show that the potential and field outside the sphere are those of a particle of mass M located at the centre of the sphere. Actually, this result is valid for any spherically symmetric density $\rho(r)$, as is apparent from Poisson's equation (1) and the relation
$$\int_0^R 4\pi r^2 \rho(r)\, dr = M.$$

(iii) Inside the sphere the field decreases linearly to zero at the origin:

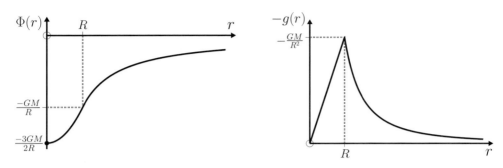

Question 11.18

Determine the gravitational potential Φ and the gravitational field **g** at all points in space due to a spherical shell with inner and outer radii R_1 and R_2, and uniform density ρ. (Hint: Use the results of the previous question and superposition.)

Solution

By superposition, the gravitational potential of the shell is equal to the potential of a uniform sphere of radius R_2 minus the potential of a concentric uniform sphere of radius R_1. Therefore, by using the potentials (5) and (7) of the previous question we have:

1. $r < R_1$

$$\Phi(r) = -\tfrac{4}{3}\pi R_2^2 \rho G\bigl(3/2 - r^2/2R_2^2\bigr) + \tfrac{4}{3}\pi R_1^2 \rho G\bigl(3/2 - r^2/2R_1^2\bigr)$$
$$= -2\pi\bigl(R_2^2 - R_1^2\bigr)\rho G. \tag{1}$$

2. $R_1 < r < R_2$

$$\Phi(r) = -\tfrac{4}{3}\pi R_2^2 \rho G\bigl(3/2 - r^2/2R_2^2\bigr) + \tfrac{4}{3}\pi R_1^3 \rho G/r$$
$$= -4\pi\bigl(\tfrac{1}{2}R_2^2 - \tfrac{1}{6}r^2 - \tfrac{1}{3}R_1^3/r\bigr)\rho G. \tag{2}$$

3. $r > R_2$

$$\Phi(r) = -\tfrac{4}{3}\pi\bigl(R_2^3 - R_1^3\bigr)\rho G/r. \tag{3}$$

The resulting gravitational field $\mathbf{g}(\mathbf{r}) = -\boldsymbol{\nabla}\Phi(r) = -\hat{\mathbf{r}}\,d\Phi/dr$ is

$$\mathbf{g}(\mathbf{r}) = \begin{cases} 0 & \text{if } r < R_1 \\ -\tfrac{4}{3}\pi\bigl(r - R_1^3/r^2\bigr)\rho G\,\hat{\mathbf{r}} & \text{if } R_1 < r < R_2 \\ -\tfrac{4}{3}\pi\bigl(R_2^3 - R_1^3\bigr)\rho G/r^2\,\hat{\mathbf{r}} & \text{if } r > R_2. \end{cases} \tag{4}$$

Comments

(i) Equations (1)–(4) yield the graphs:

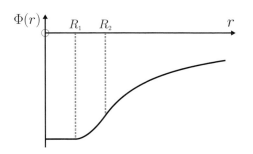
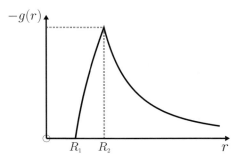

(ii) The result that the gravitational field inside a uniform spherical shell is zero clearly applies also if the density is spherically symmetric: $\rho = \rho(r)$.
(iii) This is a consequence of the inverse-square property of the gravitational field, and therefore it applies also to the electric field of a shell containing a spherically symmetric distribution of static electric charge.

Question 11.19

Calculate the gravitational potential and field on the axis of:

(a) A thin uniform disc of mass M and radius R.
(b) A thin uniform annulus of mass M having inner and outer radii R_1 and R_2 respectively.

Solution

(a) Choose the z-axis along the symmetry axis of the disc with coordinate origin at the disc. In terms of the constant mass per unit area, $\sigma = M/\pi R^2$, (8) of Question 11.16 gives

$$\Phi(z) = -G\sigma \int \frac{d^2\mathbf{r}'}{|z\hat{\mathbf{z}} - \mathbf{r}'|}. \tag{1}$$

We integrate over annular rings of radius r' and width dr', from $r' = 0$ to $r' = R$. Then, $d^2\mathbf{r}' = 2\pi r' dr'$ and $|z\hat{\mathbf{z}} - \mathbf{r}'| = \sqrt{z^2 + r'^2}$, and so

$$\Phi(z) = -2\pi G\sigma \int_0^R \frac{r' \, dr'}{\sqrt{z^2 + r'^2}} = -2\pi G\sigma \left[\sqrt{R^2 + z^2} - |z|\right]. \tag{2}$$

The corresponding gravitational field $\mathbf{g}(z) = -\boldsymbol{\nabla}\Phi(z) = -\hat{\mathbf{z}} \, d\Phi/dz$ is

$$\mathbf{g}(z) = \mp 2\pi G\sigma \left[1 - \frac{|z|}{\sqrt{R^2 + z^2}}\right]\hat{\mathbf{z}}, \tag{3}$$

where the upper (lower) sign is for $z > 0$ ($z < 0$).

(b) By changing the lower and upper limits in (2) to R_1 and R_2 we obtain

$$\Phi(z) = -2\pi G\sigma\left[\sqrt{R_2^2 + z^2} - \sqrt{R_1^2 + z^2}\right], \tag{4}$$

and then

$$\mathbf{g}(z) = -2\pi G\sigma\left[\frac{z}{\sqrt{R_1^2 + z^2}} - \frac{z}{\sqrt{R_2^2 + z^2}}\right]\hat{\mathbf{z}}. \tag{5}$$

Comments

(i) In the limit $|z| \gg R$, (3) reduces to $\mp GM\hat{\mathbf{z}}/z^2$. The opposite limit ($|z| \ll R$) gives a constant $\mathbf{g} = \mp 2GM\hat{\mathbf{z}}/R^2$ near the surface of the disc. (Again, in these expressions the upper (lower) sign is for $z > 0$ ($z < 0$).)

(ii) Equations (4) and (5) can also be obtained from (2) and (3) by superposition.

Question 11.20

A spherical galaxy consists of a core surrounded by a 'halo'. The core has mass M_C, radius R_C and constant density. The halo has mass M_H, radius R_H and density $\rho_H(r) \sim r^{n-2}$, where r is the radial distance to the centre of the galaxy and n ($\neq -1$) is a constant. Let $v(r)$ be the speed of an object of mass m in a circular orbit of radius r about the galactic centre.

(a) Calculate $v(r)$ at all points in space. Express the results in terms of the dimensionless ratios r/R_C, R_H/R_C, M_H/M_C, and the velocity at the edge of the core,

$$v_C = \sqrt{\frac{GM_C}{R_C}}. \tag{1}$$

(b) Use the answer to (a) to deduce what is special about the case $n = 0$ (an inverse-square dependence of the halo density on radial distance).

(c) Take $R_H = 5R_C$, $M_H = 4M_C$ and make a graphical comparison of the results for $n = 0$ and $n = 2$ (a constant halo density). Also indicate on the graphs the curves when the halo is absent ($M_H = 0$).

Solution

(a) According to Question 11.17, the gravitational force on a mass m at a distance r from the centre of a spherically symmetric mass distribution is

$$F = \frac{GmM(r)}{r^2}, \tag{2}$$

where

$$M(r) = \int_0^r \rho(r)4\pi r^2\,dr \tag{3}$$

is the mass inside a sphere of radius r. (The mass outside this sphere exerts no force on m – see Question 11.18.) By equating (2) to the centripetal force $F = mv^2(r)/r$ for circular motion we have

$$v(r) = \sqrt{\frac{GM(r)}{r}}. \tag{4}$$

(Note that $v(R_\text{C}) = v_\text{C}$ in (1).) Thus, the solution to this problem requires an evaluation of the mass $M(r)$ given by (3) at all points in space. Inside the core, $M(r) \sim r^3$: therefore $M(r) = (r/R_\text{C})^3 M_\text{C}$, and (4) and (1) yield

$$\frac{v(r)}{v_\text{C}} = \frac{r}{R_\text{C}} \qquad (r \leq R_\text{C}). \tag{5}$$

At all points outside the halo, $M(r) = M_\text{C} + M_\text{H}$ and (4) and (1) give

$$\frac{v(r)}{v_\text{C}} = \sqrt{\left(1 + \frac{M_\text{H}}{M_\text{C}}\right) \frac{R_\text{C}}{r}} \qquad (r \geq R_\text{H}). \tag{6}$$

(Equation (6) is independent of the form of the halo density $\rho_\text{H}(r)$.) Inside the halo, and for a density $\rho_\text{H} \sim r^{n-2}$, (3) shows that

$$M(r) = M_\text{C} + \frac{r^{n+1} - R_\text{C}^{n+1}}{R_\text{H}^{n+1} - R_\text{C}^{n+1}} M_\text{H} \qquad (R_\text{C} \leq r \leq R_\text{H}), \tag{7}$$

where $n \neq -1$. Equations (4), (7) and (1) give

$$\frac{v(r)}{v_\text{C}} = \sqrt{\frac{\frac{M_\text{H}}{M_\text{C}} \left(\frac{r}{R_\text{C}}\right)^n}{\left(\frac{R_\text{H}}{R_\text{C}}\right)^{n+1} - 1} + \left(1 - \frac{\frac{M_\text{H}}{M_\text{C}}}{\left(\frac{R_\text{H}}{R_\text{C}}\right)^{n+1} - 1}\right) \frac{R_\text{C}}{r}} \qquad (R_\text{C} \leq r \leq R_\text{H}). \tag{8}$$

Equations (5), (6) and (8) are the solutions to part (a) of the question.

(b) According to (8), the case $n = 0$ is special in that it allows the possibility for $v(r)$ to be independent of r: by inspection, this occurs when

$$M_\text{H}/M_\text{C} = (R_\text{H}/R_\text{C}) - 1. \tag{9}$$

(c) From (5), (6) and (8) with $n = 0$ or 2, and for $R_\text{H} = 5R_\text{C}$ and $M_\text{H} = 4M_\text{C}$, we obtain the following graphs:

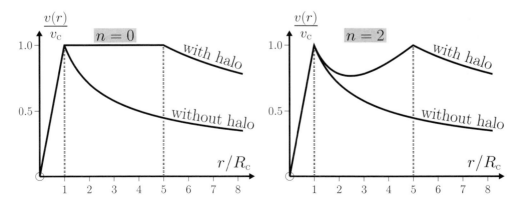

In each case, the velocity in the absence of a halo is given by (5) and (6) with $M_H = 0$ and $R_H = R_C$. The first graph illustrates the constancy of the velocity for orbits within the halo when $n = 0$ and (9) is satisfied; the second graph shows the effect of changing n.

Comments

(i) For galaxies, plots of measured stellar velocities $v(r)$ versus radial distance r are known as rotation curves. Often, the observed rotation curves have a 'flat' appearance, with an approximate plateau extending out to several times the visible galactic radius. This feature is currently thought to indicate the presence of a halo of 'invisible matter' around the galactic core, a connection that is apparent in the above rough model. Such matter is generally referred to as 'dark', meaning matter whose electromagnetic interaction is weak, or non-existent, and whose presence is inferred by its gravitational interaction on visible matter or light. The ratio M_H/M_C of dark matter to visible matter can be estimated from the width of the plateau in the rotation curve by using (9), and values up to about 1000 have been obtained, although some galaxies seem to contain little dark matter.

(ii) Other techniques exist for detecting the presence of dark matter on larger scales. It is thought that as much as 95% of the matter/energy in the universe may be dark. The nature of this 'missing' matter is a major unanswered question in physics.[27]

Question 11.21

(a) Suppose a particle of mass m is dropped into a tunnel drilled along a diameter through a planet of uniform density (see the first diagram below). Show that the resulting motion is simple harmonic and determine the period. (Assume the motion is frictionless.)

(b) Also determine the period for a straight-line tunnel connecting any two points on the surface of the planet.

Solution

(a) Let M and R be the mass and radius of the planet. The gravitational force on the particle when it is a distance r from the centre O of the planet is (see Question 11.17) $F = -kr$, where $k = GMm/R^3$ is a positive constant; that is, a Hooke's-law-type restoring force. The resulting equation of motion $m\ddot{r} = -kr$ is that of a simple harmonic oscillator with angular frequency $\omega = \sqrt{k/m}$ and period $T = 2\pi/\omega$. So

$$T = 2\pi\sqrt{R^3/GM}. \tag{1}$$

---

[27] See, for example, V. Trimble, "Dark matter in the universe: where, what and why?," Contemporary Physics, vol. 29, pp. 373–392, 1988.

(b) Consider motion along the straight-line tunnel $AO'B$ (shown in the second diagram below). The component of the gravitational force along $AO'B$ is $-kr\cos\theta = -kx$. Consequently, for frictionless motion along $AO'B$ we have $m\ddot{x} = -kx$ and the period is again given by (1).

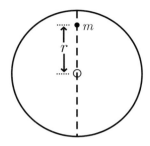
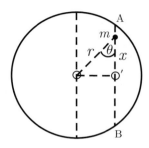

Comments

(i) Numerical values of T, obtained from (1) with $G = 6.67 \times 10^{-11}\,\text{N}\,\text{m}^2\,\text{kg}^{-2}$, are tabulated alongside. Note that the transit time for a 'one-way trip' is $\tfrac{1}{2}T$; so for Earth it is about $42\,\text{min}$.

	M (kg)	R (m)	T (min)
Earth	6.00×10^{24}	6.37×10^6	84
Moon	7.35×10^{22}	1.73×10^6	108
Mars	6.42×10^{23}	3.39×10^6	100
Jupiter	1.90×10^{27}	7.13×10^7	177
Deimos‡	2.24×10^{15}	6.3×10^3	135

(ii) The formula (1) is the same as the period of a satellite orbiting the planet at low altitude.

(iii) It is interesting to determine the equation of a tunnel that minimizes the transit time between two points on the surface. It can be shown that the curve is a hypocycloid and that for two points A and B which subtend an angle α at the centre, the minimum transit time is

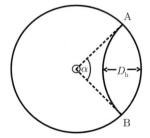

$$\tau_{\min} = \tau_s \sqrt{\frac{2\alpha}{\pi}\left(1 - \frac{\alpha}{2\pi}\right)}, \qquad (2)$$

where τ_s is the transit time along the chord connecting A and B (that is, $\tau_s = \tfrac{1}{2}T$ with T given by (1)).[28–30] The maximum depth reached is $D_h = \alpha R/\pi$, which can be compared with the maximum depth $D_s = (1 - \cos\tfrac{1}{2}\alpha)R$ reached

‡Deimos, the smaller and outermost of the two moons of Mars, is rather ellipsoidal with dimensions $\sim 15\,\text{km} \times 12\,\text{km} \times 10\,\text{km}$. We have approximated this by a sphere of radius $6.3\,\text{km}$.

---

[28] G. Venezian, "Terrestrial brachistochrone," American Journal of Physics, vol. 34, p. 701, 1966.
[29] R. L. Mallett, "Comments on 'Through the Earth in forty minutes'," American Journal of Physics, vol. 34, p. 702, 1966.
[30] L. J. Laslett, "Trajectory for minimum transit time through the Earth," American Journal of Physics, vol. 34, p. 702, 1966.

on the straight-line path from A to B. When $\alpha \ll 1$, $\tau_{\min}/\tau_s \approx \sqrt{2\alpha/\pi}$ and $D_h/D_s \approx 8/\pi\alpha$: for nearby points A and B, a substantial reduction in the transit time τ_s requires going much deeper below the surface. The graph of (2) is:

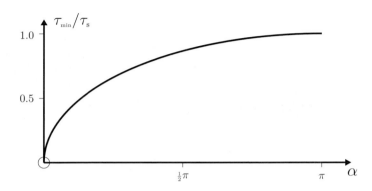

Question 11.22

(a) Show that the escape velocity of a particle of mass m from a uniform sphere of mass M and radius R that is fixed in space is given by

$$v_e = \sqrt{2GM/R}. \tag{1}$$

(b) Determine v_e if both objects are free to move. (Hint: Refer to Question 10.6.)

Solution

(a) For a non-relativistic particle of mass m with velocity v at the surface of the sphere, and velocity v_∞ at $r = \infty$, conservation of energy requires that

$$\frac{1}{2}mv^2 - \frac{GMm}{R} = \frac{1}{2}mv_\infty^2. \tag{2}$$

The escape velocity v_e is, by definition, the initial velocity v for which $v_\infty = 0$ in (2), and this yields (1).

(b) If both masses move, then the kinetic energy of the relative motion is $\frac{1}{2}\mu v^2$, where $\mu = Mm/(M+m)$ is the reduced mass and $v = |\dot{\mathbf{r}}|$ is the relative speed (see Question 10.6). Then (2) is replaced by

$$\frac{1}{2}\frac{Mm}{M+m}v^2 - \frac{GMm}{R} = \frac{1}{2}\frac{Mm}{M+m}v_\infty^2, \tag{3}$$

and so

$$v_e = \sqrt{2G(M+m)/R}. \tag{4}$$

Clearly, (1) is a good approximation when $M \gg m$.

Comments

(i) Equations (1) and (4) apply also when the density of the sphere is spherically symmetric: $\rho = \rho(r)$ for $r \leq R$.

(ii) Values of (1) for some astronomical bodies are:

	M (kg)	R (m)	v_e (kms^{-1})
Earth	6.00×10^{24}	6.37×10^{6}	11.2
Moon	7.35×10^{22}	1.73×10^{6}	2.38
Mars	6.42×10^{23}	3.39×10^{6}	5.03
Jupiter	1.90×10^{27}	7.13×10^{7}	59.6
Deimos	2.24×10^{15}	6.3×10^{3}	0.0069

(iii) Light gases (such as hydrogen and helium) in a planetary atmosphere have a higher average speed (in thermal equilibrium) than heavier elements such as nitrogen and oxygen. Consequently, in time a larger proportion of the lighter species will escape from the gravitational field. Small bodies like the Moon (which has a relatively low v_e) have no atmosphere at all, whereas the heavier Earth retains its nitrogen, oxygen, etc. but has lost most of its initial hydrogen.

(iv) In 1798, Laplace considered the possibility that the theoretical expression for the escape velocity from a star could exceed c (the speed of light in vacuum). That is,

$$\sqrt{2GM/R} > c. \tag{5}$$

This result suggests that even light cannot escape from such a star, which would therefore be invisible to a distant observer. One may question the applicability of the non-relativistic relation (2) to light quanta (photons), which are massless particles ($m = 0$). However, it turns out that the result obtained from general relativity is identical to (5). Objects for which (5) applies are known as black holes. For a black hole with mass equal to that of the Sun ($M = 1.99 \times 10^{30}$ kg), (5) shows that its radius R should be less than 295 m.

Question 11.23

Two particles are released from rest at a distance R apart. They have inertial masses m^{I} and M^{I}, and gravitational masses m^{G} and M^{G} – see Question 2.4. Determine the relative acceleration a in terms of these masses, R and the gravitational constant G.

Solution

The equation of motion for the relative position vector $\mathbf{r}$ is

$$\mu \ddot{\mathbf{r}} = -(GM^{\mathrm{G}} m^{\mathrm{G}}/R^2)\hat{\mathbf{r}}, \tag{1}$$

where $\mu = m^{\mathrm{I}} M^{\mathrm{I}}/(M^{\mathrm{I}} + m^{\mathrm{I}})$, see Question 10.4. Therefore

$$a = \left(1 + \frac{m^{\mathrm{I}}}{M^{\mathrm{I}}}\right) \frac{m^{\mathrm{G}}}{m^{\mathrm{I}}} \frac{GM^{\mathrm{G}}}{R^2}. \tag{2}$$

Comments

(i) Equation (2) shows that a depends on m^{I} through the factor $(1 + m^{\text{I}}/M^{\text{I}})$. This factor is a correction to Galileo's famous result that all objects fall with the same gravitational acceleration. (See (3) of Question 2.4, which is based on the assumption that the position of the mass M is fixed.)

(ii) Of course, for objects falling near the Earth's surface this correction is unmeasurable. But for cannonballs falling on a small, spherical asteroid it could be significant.

Question 11.24

Consider the CM motion of an object whose mass $m(t)$ is either (a) increasing (e.g. a developing hailstone) or (b) decreasing (e.g. a rocket). The CM velocity of the object at time t relative to an inertial frame S is $\mathbf{v}(t)$, and an external force $\mathbf{F}$ acts on the object. Let $\mathbf{u}(t)$ be the velocity relative to S of the mass that is gained or lost. Show that in both instances the equation of motion is

$$m \frac{d\mathbf{v}}{dt} = \mathbf{F} + (\mathbf{u} - \mathbf{v}) \frac{dm}{dt}. \tag{1}$$

Solution

Let $\mathbf{P}$ be the total momentum of the particles comprising the object. Then, relative to S the equation of motion of the CM is (see Question 11.1)

$$\frac{d\mathbf{P}}{dt} = \mathbf{F}. \tag{2}$$

(a) Suppose a mass δm moving with velocity $\mathbf{u}$ relative to S is accreted in a small time interval from t to $t + \delta t$. The change in the total momentum is

$$\delta \mathbf{P} = \mathbf{P}(t + \delta t) - \mathbf{P}(t)$$
$$= (m + \delta m)(\mathbf{v} + \delta \mathbf{v}) - (m\mathbf{v} + \delta m\, \mathbf{u})$$
$$= m\, \delta \mathbf{v} + (\mathbf{v} - \mathbf{u})\, \delta m + \delta m\, \delta \mathbf{v}, \tag{3}$$

leading to an instantaneous rate of change

$$\frac{d\mathbf{P}}{dt} = \lim_{\delta t \to 0} \frac{\delta \mathbf{P}}{\delta t} = m \frac{d\mathbf{v}}{dt} + (\mathbf{v} - \mathbf{u}) \frac{dm}{dt}, \tag{4}$$

provided $\delta m \to 0$ as $\delta t \to 0$. Equations (2) and (4) yield (1).

(b) Let $(\delta m < 0)$ be the mass lost in a time interval δt. Then,

$$\delta \mathbf{P} = (m + \delta m)(\mathbf{v} + \delta \mathbf{v}) + (-\delta m)(\mathbf{u} + \delta \mathbf{u}) - m\mathbf{v}$$
$$= m\, \delta \mathbf{v} + (\mathbf{v} - \mathbf{u})\, \delta m + \delta m(\delta \mathbf{v} - \delta \mathbf{u}), \tag{5}$$

and so $d\mathbf{P}/dt$ is again given by (4).

Comments

(i) Equation (1) describes the motion of a particle of varying mass $m(t)$ located at the CM of the object and acted on by the external force $\mathbf{F}$.

(ii) The quantity $\mathbf{u} - \mathbf{v}$ is the velocity of the accreted or emitted material relative to the object: for rockets and jet engines it is the velocity of the exhaust gases, and is referred to as the exhaust velocity $\mathbf{v}_e = \mathbf{u} - \mathbf{v}$. In terms of $\mathbf{v}_e$, (1) is

$$m\frac{d\mathbf{v}}{dt} = \mathbf{F} + \mathbf{v}_e \frac{dm}{dt}. \tag{6}$$

(iii) The term $\mathbf{v}_e\, dm/dt$ in (6) is known as the thrust of a rocket or jet, and (6) indicates how this thrust alters the velocity $\mathbf{v}$. Here, $dm/dt < 0$, and so a rocket or jet is accelerated if $\mathbf{v}$ and $\mathbf{v}_e$ are anti-parallel, and decelerated if $\mathbf{v}$ and $\mathbf{v}_e$ are parallel.

(iv) Although it is customary to treat variable-mass systems in terms of Newtonian theory (as we have done here), the analysis and its applications can also be presented in the Lagrangian formulation.[31]

Question 11.25

A spherical raindrop falls from rest through a uniform, stationary mist. Its radius r increases from an initial value r_0 by accretion of all the mist that the drop encounters. Prove that in the absence of air drag, the acceleration of the drop varies with r according to

$$\frac{dv}{dt} = \frac{g}{7}\left\{1 + 6\left(\frac{r_0}{r}\right)^7\right\}. \tag{1}$$

Solution

The instantaneous mass $m(t)$ of the raindrop and its rate of increase are given by

$$m = \rho_w \tfrac{4}{3}\pi r^3 \quad \text{and} \quad dm/dt = \rho_m \pi r^2 v, \tag{2}$$

where ρ_w and ρ_m are the densities of water and mist. It follows from (2) that the rate of increase of radius is proportional to the velocity:

$$dr/dt = Av, \quad \text{where} \quad A = \rho_m/4\rho_w. \tag{3}$$

Consider now the equation of motion. Because the mist is stationary and the motion is one-dimensional, we have from (1) of Question 11.24

$$\frac{d}{dt}(mv) = mg. \tag{4}$$

Since $m \propto r^3$, we can perform the following manipulations on (4):

---

[31] C. Leubner and P. Krumm, "Lagrangians for simple systems with variable mass," European Journal of Physics, vol. 11, pp. 31–34, 1990.

$$\frac{d}{dt}(r^3 v) = r^3 g = \frac{g}{7r^3}\frac{dr^7}{dr} = \frac{g}{7r^3}\frac{dr^7}{dt}\Big/\frac{dr}{dt} = \frac{g}{7Ar^3 v}\frac{dr^7}{dt}, \qquad (5)$$

where in the last step we have used $(3)_1$. Equation (5) is a differential equation

$$\frac{1}{2}\frac{d}{dt}(r^3 v)^2 = \frac{g}{7A}\frac{dr^7}{dt}, \qquad (6)$$

which can be integrated to yield $v(r)$. The solution that satisfies the initial condition $v = 0$ when $r = r_0$ is

$$v^2 = \frac{2g}{7A}\frac{(r^7 - r_0^7)}{r^6}. \qquad (7)$$

Differentiation of (7) with respect to t and use of $(3)_1$ to eliminate dr/dt gives (1).

Comments

(i) The mechanical energy of the raindrop is not conserved due to the inelastic nature of the accretion.[32]

(ii) According to (1), for a drag-free drop whose radius r has grown to several times its initial value r_0, the acceleration is approximately $g/7$.

(iii) When air drag is included, the equation of motion (4) becomes

$$\frac{d}{dt}(mv) = mg - \tfrac{1}{2}C_\mathrm{d}\pi\rho_\mathrm{a} r^2 v^2, \qquad (8)$$

where C_d is the drag coefficient and ρ_a is the density of air (see Question 3.8). Solutions to this equation show that for raindrops encountered in practice, "the inertia of the air dominates over that of the mist" and consequently the asymptotic acceleration is only of order $g/1000$ rather than the drag-free value $g/7$.[33]

Question 11.26

A rocket of mass m_0, which is moving in a straight line with constant speed v_0 in free space, is accelerated by igniting its engines at time $t = 0$. The exhaust speed v_e is constant. Calculate the speed $v(t)$ of the rocket in terms of v_0, v_e, m_0 and $m(t)$.

Solution

In free space the external force $\mathbf{F} = 0$. Also the motion is one-dimensional. So, the equation of motion is $m\dot{v} = -v_\mathrm{e}\dot{m}$, where $v_\mathrm{e} > 0$ (see (6) of Question 11.24). Integrating this with respect to t gives

---

[32] K. Krane, "The falling raindrop: variations on a theme of Newton," American Journal of Physics, vol. 49, pp. 113–117, 1981.

[33] B. F. Edwards, J. W. Wilder, and E. E. Scime, "Dynamics of falling raindrops," European Journal of Physics, vol. 22, pp. 113–118, 2001.

$$\int_{v_0}^{v(t)} dv = -v_e \int_{m_0}^{m(t)} \frac{dm}{m}, \tag{1}$$

and so

$$v(t) = v_0 + v_e \ln\left(\frac{m_0}{m(t)}\right). \tag{2}$$

Comments

(i) According to (2), the speed $v(t)$ attained depends on the ratio of the initial and final masses of the rocket, and is independent of the rate at which fuel is consumed.

(ii) Also, the increase in speed will exceed the exhaust velocity if the mass of fuel consumed $m_f = m_0 - m(t)$ exceeds $(1 - e^{-1})m_0 \approx 0.63 m_0$.

Question 11.27

Suppose the rocket in the previous question takes off vertically from the surface of a planet by igniting its engines at time $t = 0$. The burn rate $-\dot{m} = k$ is constant and the fuel is consumed in a time t_b while the rocket is still close to the surface of the planet. Find the speed $v(t)$ of the rocket and the height $y(t)$ it reaches, for $t \leq t_b$, in terms of v_e, m_0, $m(t)$, k and the gravitational acceleration g. (Neglect atmospheric drag and rotation of the planet.)

Solution

The motion is one-dimensional and the equation of motion is $m\dot{v} = -mg - v_e \dot{m}$ (see (6) of Question 11.24). Here, v_e is constant and for motion close to the surface of the planet g is nearly constant. So, integration with respect to t gives

$$v(t) = -gt + v_e \ln\{m_0/m(t)\} \tag{1}$$

$$= -(g/k)\{m_0 - m(t)\} + v_e \ln\{m_0/m(t)\}, \tag{2}$$

because $m(t) = m_0 - kt$ if the burn rate is constant. By integrating (1) with respect to t we obtain the vertical height

$$y(t) = v_e t - \tfrac{1}{2}gt^2 - \frac{v_e(m_0 - kt)}{k}\ln\left(\frac{m_0}{m_0 - kt}\right) \tag{3}$$

$$= \frac{v_e}{k}\{m_0 - m(t)\} - \frac{g}{2k^2}\{m_0 - m(t)\}^2 - \frac{v_e}{k}m(t)\ln\left(\frac{m_0}{m(t)}\right). \tag{4}$$

Note that k, v_e, and g are all positive.

Comments

(i) For the rocket to lift off, the thrust $-v_e \dot{m}$ must exceed the initial weight $m_0 g$. That is, $k > m_0 g/v_e$. This condition ensures that $v(t) > 0$ in (2).

(ii) The second term in (2) is the increase in speed for a rocket in gravity-free space. The first term represents the retarding effect of gravity. For given initial and final masses m_0 and $m(t)$, this retardation is minimized by maximizing k, and this accounts for the large burn rate in practice.

(iii) Equations (1)–(4) apply for $t \leq t_\mathrm{b}$, the burnout time. The subsequent fate of the rocket depends on whether $v(t_\mathrm{b})$ is less than the escape velocity of the planet. If it is, the rocket continues to some maximum distance H_max and then it falls back to the surface. Otherwise, it escapes from the planet.

Question 11.28

A rocket is fired vertically (in the y-direction) from the Earth's surface. The initial mass of the rocket is 50 000 kg and 90% of this is fuel, which is consumed at a constant rate. The exhaust velocity is 5000 ms^{-1}. The rocket lifts off when the engines ignite at time $t = 0$, and burnout occurs at $t_\mathrm{b} = 300$ s.

(a) Use (6) of Question 11.24 to write down an equation of motion for each of the following cases:

1. Neglect air resistance and assume the gravitational acceleration is constant, equal to its value g_0 at the surface.

2. Neglect air resistance and use an altitude-dependent gravitational acceleration $g_0 R_\mathrm{e}^2/(R_\mathrm{e} + y)^2$, where R_e is Earth's radius.

3. Suppose that $g(y)$ varies with altitude as in case **2.** and that the atmospheric drag force exerted on the rocket is $F_\mathrm{d} = -\beta \dot{y}^2$, where β is a constant. (See also Question 3.8.)

4. Repeat **3.** but replace β with $\beta e^{-y/Y}$ (where Y is a constant) to include the effect of decreasing air density with altitude. (See also Questions 3.14 and 7.9.)

(b) Use *Mathematica* to calculate the displacement $y(t)$, the velocity $\dot{y}(t)$ and the acceleration $\ddot{y}(t)$ for $0 \leq t \leq t_\mathrm{b}$ for cases **2.** – **4.** above. For case **1.** use the results of the previous question. Take $g_0 = 9.8$ ms^{-2}, $R_\mathrm{e} = 6371$ km, $\beta = 2$ kg m^{-1}, and $Y = 7460$ m.

(c) On the same axes, plot graphs of $y(t)$ for each of these four cases. Show also $y(t)$ for the same rocket accelerating in free space. Repeat these for $\dot{y}(t)$ and $\ddot{y}(t)$.

Solution

(a) The equations of motion are:

1. $$\frac{d^2 y}{dt^2} + g_0 - \frac{k v_\mathrm{e}}{m(t)} = 0 \,. \qquad (1)$$

2. $$\frac{d^2 y}{dt^2} + \frac{g_0 R_\mathrm{e}^2}{(R_\mathrm{e} + y)^2} - \frac{k v_\mathrm{e}}{m(t)} = 0 \,. \qquad (2)$$

3. $$\frac{d^2 y}{dt^2} + \frac{g_0 R_\mathrm{e}^2}{(R_\mathrm{e} + y)^2} + \frac{\beta \dot{y}^2 - k v_\mathrm{e}}{m(t)} = 0 \,. \qquad (3)$$

4. As in (3) but with $\beta \to \beta e^{-y/Y}$. $\qquad (4)$

(b) For case 1. we use (4) of the previous question. For cases 2. – 4. we obtain numerical solutions using the notebook listed below.

(c) The graphs are:

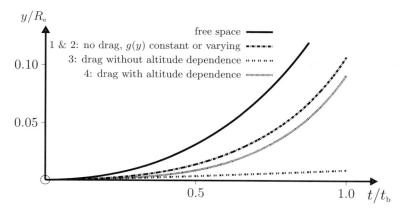

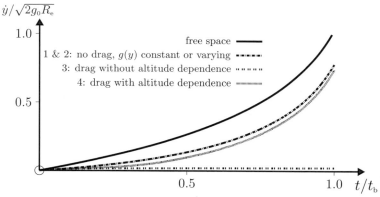

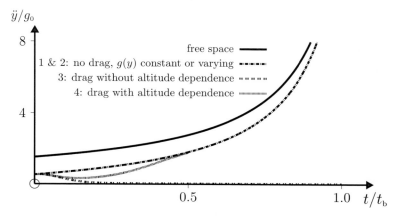

```
In[1]:= g0 = 9.8; Re = 6371000; m0 = 1.0×10^4; mfuel = 9/10 m0; tb = 300.0;

       Ve = 5000.0; β = 2.0; Y = 7462.1; T0 = tb;

       m[t] := m0 - mfuel t/tb;

       Sol = NDSolve[{y''[t] + (g0 Re^2)/(Re + y[t])^2 + (β e^(-Y[t]/Y) y'[t]^2 - Ve mfuel/tb)/m[t] == 0,

       y'[0] == 0, y[0] == 0}, {y[t], y'[t], y''[t]}, {t, 0, T0}];

       Plot[Evaluate[y[t]/Re/.Sol], {t, 0, T0}]

       Plot[Evaluate[y'[t]/√(2 g0 Re)/.Sol], {t, 0, T0}]

       Plot[Evaluate[y''[t]/g0/.Sol], {t, 0, T0}]
```

Comments

(i) In the above example, burnout occurs at an altitude $y \ll R_e$ and therefore the rocket moves in an almost uniform gravitational field. Consequently, the graphs for cases 1. and 2. are essentially coincident, except for a small deviation (not shown) for $t \approx t_b$.

(ii) The graphs for cases 3. and 4. show the importance of including the decrease of drag force with altitude.

(iii) Dynamic pressure (a quantity used in aerospace engineering) is defined as $Q = \frac{1}{2} \times$ air density $\times$ speed^2. Missiles, rockets and other vehicles are all designed to withstand only a certain maximum dynamic pressure before they suffer structural damage. The phrase 'passing through max Q' can sometimes be heard during the broadcasting of rocket launches, where Q increases rapidly from zero at $t = 0$ to a maximum, and then it decreases as the air density decreases. The graph below shows $Q(t)$ for the rocket in case 4. of the above question and for an initial air density of $1.29\,\mathrm{kg\,m^{-3}}$.

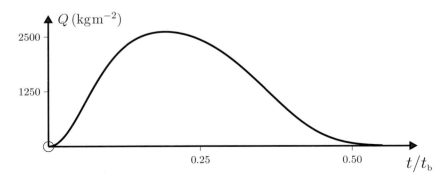

Question 11.29

An object suspended from a spring having force constant k oscillates vertically along the y-axis and is subject to a linear drag $-\alpha \dot{y}$, where $\alpha > 0$. The mass $m(t)$ of the object decreases with time due to a loss of material.

(a) Suppose that relative to the object the velocity v_e of the exiting mass is sufficiently small that its thrust $-v_e \dot{m}$ is negligible compared to the other forces acting. Use (6) of Question 11.24 to obtain the equation of motion

$$m(t)\ddot{y} + \alpha \dot{y} + ky = 0, \tag{1}$$

where $y = 0$ corresponds to the equilibrium position when $\dot{m} = 0$.

(b) Suppose the mass decreases at a constant rate $\dot{m} = -c$ and has initial value m_0. Show that in the limit of weak damping the solution to (1) is

$$y(t) = A_0 (1 - ct/m_0)^{\gamma} \cos\{h(t) + \phi\}, \tag{2}$$

where A_0 and ϕ are arbitrary constants, and

$$\gamma = \frac{\alpha}{2c} + \frac{1}{4}, \tag{3}$$

$$h(t) = \frac{2\sqrt{k}}{c}\left(\sqrt{m_0} - \sqrt{m_0 - ct}\right) - \frac{(\alpha + \frac{1}{2}c)(\alpha + \frac{3}{2}c)}{4c\sqrt{k}}\left(\frac{1}{\sqrt{m_0 - ct}} - \frac{1}{\sqrt{m_0}}\right). \tag{4}$$

The answer should stipulate the condition for 'weak damping'. (Hint: Look for a solution to (1) of the form

$$y(t) = A(t)\cos\{h(t) + \phi\}, \quad \text{where} \quad A(t) = A_0 f(t), \tag{5}$$

and show that the dimensionless functions $f(t)$ and $h(t)$ must satisfy the coupled equations

$$m\ddot{f} + \alpha \dot{f} + (k - m\dot{h}^2)f = 0 \tag{6}$$

$$2m\dot{h}\dot{f} + (\alpha \dot{h} + m\ddot{h})f = 0. \tag{7}$$

Then, obtain approximate solutions to (6) and (7) in the limit of weak damping.)

(c) Determine how the shape of the envelope in (2) depends on γ.

Solution

(a) With $\mathbf{v}_e = 0$ and $\mathbf{F} = -(\alpha \dot{y} + ky)\hat{\mathbf{y}}$, (6) of Question 11.24 yields (1).

(b) For the ansatz (5), equation (1) gives $C_1 \cos\{h(t) + \phi\} + C_2 \sin\{h(t) + \phi\} = 0$ for all t, where C_1 and C_2 denote the left-hand sides of (6) and (7). Thus, (6) and (7) must hold for all t if (5) is to be a solution to (1). These two coupled, non-linear equations for f and h are intractable in general, but they can be solved in the limit of weak damping and for a linear decrease of mass with time:

$$m(t) = m_\mathrm{o} - ct. \tag{8}$$

We proceed as follows. For weak damping we can, as a first approximation, neglect the terms in $\dot{f}$ and $\ddot{f}$ in (6). Then

$$\dot{h} = \sqrt{k/m(t)}. \tag{9}$$

From (8) and (9) we have $m\ddot{h} = \tfrac{1}{2} c\dot{h}$, and so (7) becomes

$$(m_\mathrm{o} - ct)\dot{f} + (\tfrac{1}{2}\alpha + \tfrac{1}{4}c)f = 0. \tag{10}$$

The solution to (10) that satisfies the initial condition $f(0) = 1$ is

$$f(t) = (1 - ct/m_\mathrm{o})^{\gamma}, \tag{11}$$

where γ is given by (3). Next, we improve the approximation (9) for $\dot{h}$ by using (11) in (6). This gives

$$\dot{h}^2 = \frac{k}{m} - \frac{(\alpha + \tfrac{1}{2}c)(\alpha + \tfrac{3}{2}c)}{4m^2}. \tag{12}$$

We now assume that

$$\frac{(\alpha + \tfrac{1}{2}c)(\alpha + \tfrac{3}{2}c)}{4m} \ll k, \tag{13}$$

a condition that defines the weak-damping limit. Then, (12) can be approximated as

$$\dot{h} = \sqrt{\frac{k}{m}} - \frac{(\alpha + \tfrac{1}{2}c)(\alpha + \tfrac{3}{2}c)}{8\sqrt{km^3}}. \tag{14}$$

With m given by (8) it is elementary to integrate (14) between $t = 0$ and t, and the result is (4). This completes the proof of (2).

(c) The envelope of the damped oscillations (2) is determined by the time-dependent amplitude

$$A(t) = A_\mathrm{o}(1 - ct/m_\mathrm{o})^{\gamma}. \tag{15}$$

Therefore the sign of $\ddot{A}$ establishes the concavity of the envelope of the damped oscillations. From (15) we have

$$\ddot{A} = \frac{A_\mathrm{o} c^2}{m_\mathrm{o}^2} \gamma(\gamma - 1)(1 - ct/m_\mathrm{o})^{\gamma-2}. \tag{16}$$

It follows that for $\gamma > 1$, $\ddot{A} > 0$ and the envelope is concave up. For $\gamma = 1$, $\ddot{A} = 0$ and the envelope decreases linearly with time. For $\gamma < 1$, $\ddot{A} < 0$ and the envelope is concave down.

Comments

(i) According to the equation of motion (1), the mechanical energy $E = \tfrac{1}{2}m\dot{y}^2 + \tfrac{1}{2}ky^2$ of the oscillator decreases at a rate

$$\dot{E} = -(\alpha + \tfrac{1}{2}c)\dot{y}^2. \tag{17}$$

So, both the drag and the mass loss contribute to the dissipation.

(ii) In the limit $c \to 0$ we have $\gamma \approx \alpha/2c$, and (2) becomes

$$y(t) = A_o e^{-t/\tau} \cos\left\{\sqrt{\frac{k}{m_o}}\left(1 - \frac{m_o/k}{2\tau^2}\right)t + \phi\right\}, \tag{18}$$

where $\tau = 2m_o/\alpha$. This is just the familiar solution for a weakly damped oscillator of constant mass m_o (see Question 4.5).

(iii) Flores et al.[34] have described a simple variable-mass oscillator constructed from an inverted, sand-filled soda bottle hung from a spring. A constant flow rate c was achieved at low acceleration, $\ddot{y} \ll g$. A sensor interfaced to a computer was used to measure $y(t)$ for different values of c, and good agreement was found with theory. This indicates that the assumption $v_e \approx 0$ is reasonable in their experiment.

(iv) The accuracy of the solution (2) is discussed in the next question.

Question 11.30

Consider the variable-mass oscillator of Question 11.29 with $k = 10.0\,\mathrm{N\,m^{-1}}$, $\alpha = 0.050\,\mathrm{kg\,s^{-1}}$, final mass $m_\mathrm{f} = 0.1\,\mathrm{kg}$, initial conditions $y_0 = 0.1\,\mathrm{m}$, $\dot{y}_0 = 0$, and the following initial mass m_o and flow rate $c\,(= -\dot{m})$:

1. $m_o = 1.0\,\mathrm{kg},\ c = 1.0 \times 10^{-3}\,\mathrm{kg\,s^{-1}}$; 2. $m_o = 1.0\,\mathrm{kg},\ c = 33\tfrac{1}{3} \times 10^{-3}\,\mathrm{kg\,s^{-1}}$;
3. $m_o = 20.0\,\mathrm{kg},\ c = 0.20\,\mathrm{kg\,s^{-1}}$.

(a) Determine whether the oscillators satisfy the condition for weak damping – that is, (13) of Question 11.29.
(b) Calculate the value of the exponent $\gamma = \alpha/2c + \tfrac{1}{4}$ in the envelope of the damped oscillations.
(c) Use *Mathematica* to plot graphs of $y(t)$ versus t obtained from equations (2)–(4) of Question 11.29. Include the envelope and also a set of points $y(t)$ obtained from a numerical solution of the equation of motion (1).
(d) From these graphs determine the periods T_n of the nth cycle of the oscillations up to $n = 20$. Do this as follows: Use *Mathematica*'s drawing tool to locate the approximate times at which $y(t)$ has maxima. Then refine these values using the FindRoot function to calculate the roots of $\dot{y}(t) = 0$. From these obtain values of the T_n. Also plot graphs of T_n^2 versus $m(t)$, the mass at the midpoint of each cycle. Give a simple, approximate formula for these graphs.

---

[34] J. Flores, G. Solovey, and S. Gil, "Variable mass oscillator," American Journal of Physics, vol. 71, pp. 721–725, 2003.

Solution

(a) The condition for weak damping is $N \ll k$, where $N = (\alpha + \tfrac{1}{2}c)(\alpha + \tfrac{3}{2}c)/4m(t)$. For the above oscillators the initial and final values of N/k are

1. $N_i/k = 6.5 \times 10^{-5}$, $\quad N_f/k = 6.5 \times 10^{-4}$;
2. $N_i/k = 3.1 \times 10^{-4}$, $\quad N_f/k = 3.1 \times 10^{-3}$;
3. $N_i/k = 6.5 \times 10^{-5}$, $\quad N_f/k = 1.3 \times 10^{-2}$,

showing that the condition for weak damping is well satisfied (though less so near the end of the motion in case **3**).

(b) **1.** $\gamma = 25.25$ **2.** $\gamma = 1$ **3.** $\gamma = 0.375$.

(c) The following graphs were obtained using the first cell of the *Mathematica* notebook below. In these graphs, the envelope of the oscillations is calculated using (15) of the previous question and is indicated by a dotted curve. For case **1.** we plot $y(t)$ up to $t = 100\,\text{s}$ (the mass stops decreasing at $t = 1000\,\text{s}$). For cases **2.** and **3.** the arrow $\downarrow$ on the graph indicates the time at which the mass reaches m_f.

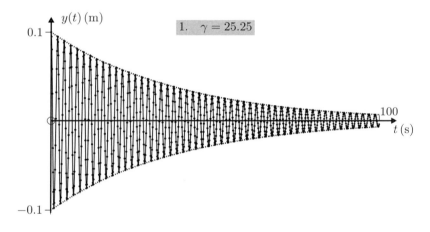

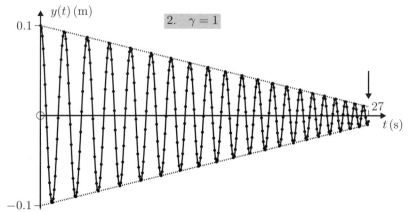

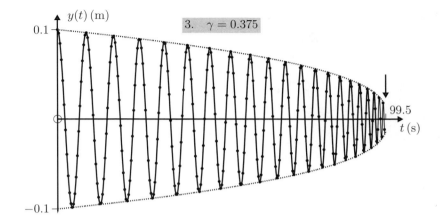

(d) Including a second cell (see below) in the notebook gives the graph of T_n^2 versus $m(t)$ for case 3. The straight line is a plot of $T^2 = (4\pi^2/k)m(t)$, which follows from the approximation $\omega \approx \sqrt{k/m(t)}$. Graphs for the other two cases are similar.

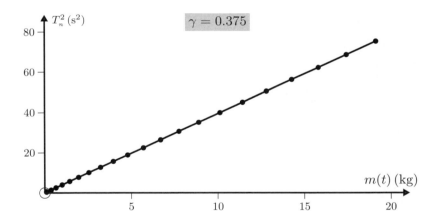

```
In[1]:= y0 = 0.1; k = 10.0; α = 0.05; m_f = 0.1;
        m_0 = 20.0; c = 0.2;      (* change these values for cases 1 and 2 *)

        ε = (m_0 - m_f)/m_f;  τ = (m_0 - m_f)/c      (* leak time *);   T_0 = τ;  (* calcⁿ time *)

        m[t_] := Piecewise[{{m_f (1 + ε (τ-t)/τ), t ≤ τ}, {m_f, t > τ}}]

        Sol = NDSolve[{ m[t] y''[t] + α y'[t] + k y[t] == 0,
                y[0] == y0, y'[0] == 0}, {y[t], y'[t]}, {t, 0, T_0}, MaxSteps → 100000];

        Plot[Evaluate[y[t]/.Sol], {t, 0, T_0}, PlotPoints → 100]
```

```
In[2]:= roots = t /. FindRoot[y'[t] == 0 /. Sol, {t, {8.7, 17, 24.9, 32.4,
                   39.5, 46.2, 52.5, 58.5, 64, 69.1, 73.9, 78.2, 82.2, 85.7,
                   88.9, 91.6, 94., 96, 97.6, 98.8}}];

        PeriodSq[t_] := (4π²/k) m[t];  data = {};

        Do[If[n == 1, T = roots[[1]]; time = T ÷ 2; data = Append[data, {m[time], T²}],
           T = roots[[n]] - roots[[n - 1]]; time = roots[[n - 1]] + T ÷ 2;
           data = Append[data, {m[time], T²}]], {n, 1, 20, 1}]

        gr1 = ListPlot[data, PlotRange → {{0, 20}, {0, 80}}];
        gr2 = ParametricPlot[{m[t], PeriodSq[t]}, {t, 0, 100}];
        Show[gr1, gr2]
```

Comment

The analytical values of $y(t)$ for weak damping are in good agreement with the numerical values. Also, the graphs illustrate the dependence of the shape of the envelope on the exponent $\gamma = \alpha/2c + \frac{1}{4}$ as discussed in (c) of Question 11.29.

Question 11.31

A uniform flexible rope of length ℓ and mass M is stretched on a table with a segment of length x_0 hanging over the edge. The rope is released from rest.

(a) Suppose the motion is frictionless. By solving the equation of motion, show that the time taken for the rope to slide off the table is

$$t_s = \tau \cosh^{-1}(\ell/x_0), \qquad \text{where } \tau = \sqrt{\ell/g}. \tag{1}$$

(A frictionless vertical barrier close to the end of the table prevents the rope from overshooting the edge.)

(b) Solve the same problem by applying conservation of energy, and compare the result with (1).

(c) Suppose the motion is subject to friction, with a coefficient of sliding friction μ between the rope and table. Determine the modification to (1) that this requires.

(d) Calculate the total loss in mechanical energy of the rope at time t_s in terms of μ, x_0/ℓ, and $Mg\ell$.

Solution

(a) The diagram below shows the rope at time t, when the length that hangs over the edge of the table is $x(t)$. The equation of motion of the rope,

$$M\frac{d^2x}{dt^2} = \frac{x}{\ell}Mg, \tag{2}$$

can be expressed as
$$\frac{d^2x}{dt^2} - \frac{x}{\tau^2} = 0, \qquad (3)$$

where $\tau = \sqrt{\ell/g}$ is a characteristic time for the motion. The general solution to (3) is

$$x(t) = A\cosh t/\tau + B\sinh t/\tau, \qquad (4)$$

where A and B are arbitrary constants. The initial conditions $x = x_0$ and $\dot{x} = 0$ at $t = 0$ require $A = x_0$ and $B = 0$, and so

$$x(t) = x_0 \cosh t/\tau. \qquad (5)$$

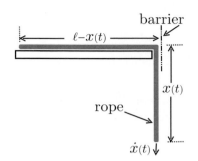

The total time t_s for which the rope slides on the table is obtained by putting $x(t_s) = \ell$ in (5), and this yields (1).

(b) The initial and subsequent energies of a uniform rope are

$$E_i = -\tfrac{1}{2}Mg\frac{x_0^2}{\ell} + C \quad \text{and} \quad E_f = -\tfrac{1}{2}Mg\frac{x^2}{\ell} + \tfrac{1}{2}M\left(\frac{dx}{dt}\right)^2 + C, \qquad (6)$$

where C is an arbitrary constant. Therefore, conservation of energy requires

$$\frac{dx}{dt} = \frac{\sqrt{x^2 - x_0^2}}{\tau},$$

and so

$$t_s = \tau \int_{x_0}^{\ell} \frac{dx}{\sqrt{x^2 - x_0^2}} = \tau \ln\left(\ell/x_0 + \sqrt{\ell^2/x_0^2 - 1}\right). \qquad (7)$$

The identity $\ln(u + \sqrt{u^2 - 1}) = \cosh^{-1} u$ shows that (7) is the same as (1).

(c) The segment of length $\ell - x$ that is in contact with the table has weight equal to $\{(\ell - x)/\ell\}Mg$, and it experiences a frictional force $\mu\{(\ell - x)/\ell\}Mg$. So, the equation of motion (2) is modified to read

$$M\frac{d^2x}{dt^2} = \frac{x}{\ell}Mg - \mu\frac{\ell - x}{\ell}Mg. \qquad (8)$$

That is,

$$\frac{d^2x}{dt^2} - \frac{x}{\tau^2} = -\mu g, \qquad \text{where } \tau = \sqrt{\frac{\ell}{(1+\mu)g}}. \qquad (9)$$

By inspection, the general solution to $(9)_1$ is equal to (4) $+ \mu g\tau^2$. The initial conditions require $A = x_0 - \mu g\tau^2$ and $B = 0$, so that

$$x(t) = (x_0 - \mu g\tau^2)\cosh t/\tau + \mu g\tau^2. \qquad (10)$$

It follows that the sliding time, defined by $x(t_s) = \ell$, is now

$$t_\text{s} = \tau \cosh^{-1} \frac{\ell - \mu g \tau^2}{x_0 - \mu g \tau^2} = \tau \cosh^{-1}\left\{(1+\mu)\frac{x_0}{\ell} - \mu\right\}^{-1}. \tag{11}$$

In terms of the length of rope $D = \ell - x_0$ on the table at $t = 0$ and a characteristic length

$$\lambda = g\tau^2 = \ell/(1+\mu), \tag{12}$$

$(11)_1$ becomes

$$t_\text{s}/\tau = \cosh^{-1}(1 - D/\lambda)^{-1}, \tag{13}$$

showing that the dimensionless sliding time t_s/τ is a universal function of the dimensionless ratio D/λ.[35] Note that $t_\text{s}/\tau \to \infty$ as $D/\lambda \to 1$, as one expects (the critical condition for slipping is $\mu(\ell - x_0) = x_0$).

(d) From (6) we see that the total loss in mechanical energy of the rope is

$$E_\text{i} - E_\text{f} = \frac{Mg}{2\ell}(\ell^2 - x_0^2) - \frac{M}{2}\{\dot{x}(t_\text{s})\}^2. \tag{14}$$

Here, $\dot{x}(t_\text{s})$ is the velocity of the rope at the instant when it loses contact with the table, and according to (10) it is given by

$$\dot{x}(t_\text{s}) = (x_0/\tau - \mu g\tau)\sinh t_\text{s}/\tau. \tag{15}$$

This can be expressed in terms of the dimensionless ratio D/λ introduced above by using the relations $\sinh\theta = \sqrt{\cosh^2\theta - 1}$, $\cosh t_\text{s}/\tau = (1 - D/\lambda)^{-1}$, and (12). A little algebra leads to

$$\dot{x}(t_\text{s}) = \frac{\lambda}{\tau}\sqrt{\frac{D}{\lambda}\left(2 - \frac{D}{\lambda}\right)}, \tag{16}$$

showing that $\dot{x}(t_\text{s})$ expressed in units of λ/τ is also a universal function of D/λ. The first term in (14) can be expressed in terms of D/λ by making the factorization $\ell^2 - x_0^2 = (\ell - x_0)(\ell + x_0) = D(2\ell - D)$. By using also $(9)_2$, (12) and (16) in (14), a short calculation shows that

$$E_\text{i} - E_\text{f} = \frac{\mu}{1+\mu}\frac{M}{2}\frac{D^2}{\tau^2} = \tfrac{1}{2}\mu(1 - x_0/\ell)^2 Mg\ell. \tag{17}$$

Comments

(i) For the rope to move, the initial value of the force in (8) must be positive, and this requires

$$x_0 > \mu\ell/(1+\mu). \tag{18}$$

(This condition is also apparent in (11).) Consequently, $D < \dfrac{\ell}{(1+\mu)}$ and $\dfrac{D}{\lambda} < 1$.

---

[35] F. Behroozi, "The sliding chain problem with and without friction: a universal solution," European Journal of Physics, vol. 18, pp. 15–17, 1997.

(ii) Equations (11) and (13) yield the following plots of the dimensionless sliding time t_s/τ versus x_0/ℓ and D/λ. For the first graph, the values of μ range from $\mu = 0$ (curve (1)) to $\mu = 1.0$ (curve (6)) in steps of 0.2.

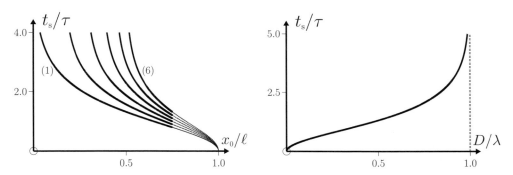

(iii) According to (17) and (18), the total loss in mechanical energy is less than

$$\frac{\mu}{(1+\mu)^2} \tfrac{1}{2} Mg\ell. \tag{19}$$

This upper bound is approached in the limit $x_0 \to \mu\ell/(1+\mu)$ and, for given M, g and ℓ, it has a maximum value $\tfrac{1}{8} Mg\ell$ when $\mu = 1$.

(iv) The above problem can be solved with other initial conditions, such as $x = 0$ and $\dot{x} = v_0$ at $t = 0$. Then, the solution to (9) is

$$x(t) = \mu g \tau^2 (1 - \cosh t/\tau) + v_0 \tau \sinh t/\tau. \tag{20}$$

The sliding time is determined by setting $x(t_s) = \ell$, and some calculation shows that

$$t_s = \tau \cosh^{-1} \frac{\mu(1+\mu) + \theta\sqrt{\theta^2 + 1 + 2\mu}}{\theta^2 - \mu^2}, \qquad \text{where } \theta = v_0\tau/\lambda > \mu. \tag{21}$$

Question 11.32

A uniform flexible chain of length ℓ and mass M is initially suspended with its two ends close together and at the same elevation, and then one end is released. Consider a one-dimensional approximation to this two-dimensional problem, in which the chain is represented by two vertical segments connected by a horizontal cross-piece which is sufficiently short that its contribution to the kinetic and potential energies may be neglected.

(a) Show that the Lagrangian of the chain is

$$\mathsf{L} = \tfrac{1}{4}\mu(\ell - x)\dot{x}^2 + \tfrac{1}{4}\mu g(\ell^2 + 2\ell x - x^2), \tag{1}$$

where $x(t)$ is the vertical distance that the falling end has travelled in a time t, and the constant $\mu = M/\ell$ is the linear mass density of the chain.

(b) Use Lagrange's equation to prove that the energy $E = 2K - L$ of the chain is conserved, and to obtain the equation of motion

$$\frac{dp_1}{dt} = M_1 g - \tfrac{1}{4}\mu \dot{x}^2 \qquad (2)$$

of the falling segment. Here, $M_1(t)$ is the mass of this segment at time t, and $p_1 = M_1 \dot{x}$ is its momentum.

(c) Use energy conservation to show that the speed $v = \dot{x}$ at position x is given by

$$v(x) = \sqrt{2gx}\,\sqrt{\frac{1 - x/2\ell}{1 - x/\ell}}. \qquad (3)$$

(d) Show that the inverse form of the trajectory for the free end is

$$t = \frac{1}{2}\sqrt{\frac{2\ell}{g}} \int_0^{x(t)/\ell} \sqrt{\frac{1 - u}{u(1 - \tfrac{1}{2}u)}}\, du. \qquad (4)$$

Use *Mathematica* to evaluate the total time τ for which the free end falls (in units of $\sqrt{2\ell/g}$) and plot a graph of $x(t)/\ell$ versus $(t/\tau)^2$.

(e) Show that the tension at the fixed end is given by

$$T = \frac{Mg}{4}\left\{\frac{2 + 2x/\ell - 3x^2/\ell^2}{1 - x/\ell}\right\}. \qquad (5)$$

Solution

(a) The figure depicts the one-dimensional model of the chain a time t after release, when the free end has fallen a distance $x(t)$. In this model we neglect the small horizontal segment. So, at time t the falling segment has length $\ell_1 = \tfrac{1}{2}(\ell - x)$, mass $M_1 = \mu \ell_1$, and velocity $\dot{x}(t)$; the segment that is fixed at O has length $\ell_2 = \tfrac{1}{2}(\ell + x)$ and mass $M_2 = \mu \ell_2$. The total kinetic energy is that of M_1:

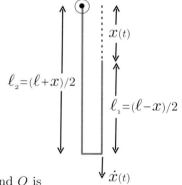

$$K = \tfrac{1}{2} M_1 \dot{x}^2 = \tfrac{1}{4}\mu(\ell - x)\dot{x}^2, \qquad (6)$$

and the potential energy relative to the fixed end O is

$$V = -M_1 g(x + \tfrac{1}{2}\ell_1) - M_2 g \tfrac{1}{2}\ell_2$$
$$= -\tfrac{1}{4}\mu g(\ell^2 + 2\ell x - x^2). \qquad (7)$$

The Lagrangian $L = K - V$ obtained from (6) and (7) is (1).

(b) With $E = K + V = 2K - \mathsf{L}$ and $\mathsf{L} = \mathsf{L}(x, v)$, where $v = \dot{x}$, we have

$$\frac{dE}{dt} = \frac{d}{dt}(2K) - \frac{\partial \mathsf{L}}{\partial x}\frac{dx}{dt} - \frac{\partial \mathsf{L}}{\partial v}\frac{dv}{dt}.$$

According to Lagrange's equation

$$\frac{\partial \mathsf{L}}{\partial x} = \frac{d}{dt}\frac{\partial \mathsf{L}}{\partial v}, \tag{8}$$

and so

$$\frac{dE}{dt} = \frac{d}{dt}(2K) - \left(\frac{d}{dt}\frac{\partial \mathsf{L}}{\partial v}\right)v - \frac{\partial \mathsf{L}}{\partial v}\frac{dv}{dt} = \frac{d}{dt}\left(2K - v\frac{\partial \mathsf{L}}{\partial v}\right) = 0 \tag{9}$$

because $v\,\partial\mathsf{L}/\partial v = 2K$ (see (1) and (6)). The equation of motion (2) follows directly from the Lagrangian (1) and Lagrange's equation (8).

(c) Because the energy is conserved, the sum of K and V given by (6) and (7) can be equated to the initial energy (when $x = 0$ and $\dot{x} = 0$):

$$\tfrac{1}{4}\mu(\ell - x)\dot{x}^2 - \tfrac{1}{4}\mu g\left(\ell^2 + 2\ell x - x^2\right) = -\tfrac{1}{4}\mu g\,\ell^2. \tag{10}$$

By solving (10) for the velocity $v = \dot{x}$ we obtain (3).

(d) By setting $v = dx/dt$ in (3) and integrating with respect to t, we find that the inverse equation $t = t(x)$ for the trajectory of the free end is given by (4). The total time τ for which the free end falls is obtained by setting $x(\tau) = \ell$ in (4), and a numerical integration yields

$$\tau \approx 0.847213\sqrt{2\ell/g}. \tag{11}$$

The trajectory $x(t)$ of the free end obtained from (4), and the *Mathematica* notebook used to calculate it, are given below.

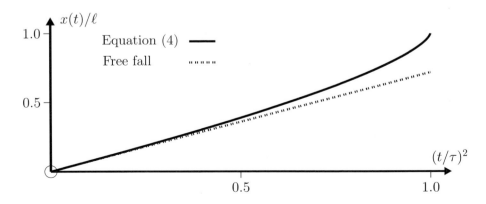

```
In[1]:= DimTime[X_] := NIntegrate[Sqrt[(1-u)/(u(1-1/2 u))], {u, 0, X}] / NIntegrate[Sqrt[(1-u)/(u(1-1/2 u))], {u, 0, 1}];

ListPlot[Table[{DimTime[X]^2, X}, {X, 0, 1, 0.001}]]
```

(e) The equation of motion (2) can be written

$$M_1 \frac{dv}{dt} = M_1 g - \tfrac{1}{4}\mu \dot{x}^2 - \dot{x}\frac{dM_1}{dt}. \tag{12}$$

Here, $M_1 = \mu \tfrac{1}{2}(\ell - x)$ and therefore $dM_1/dt = -\tfrac{1}{2}\mu\dot{x}$. So, (12) becomes

$$M_1 \frac{dv}{dt} = M_1 g + \tfrac{1}{4}\mu v^2, \tag{13}$$

showing that in addition to the gravitational force $M_1 g$ there is a downward tension $\tfrac{1}{4}\mu v^2$ in the falling segment. Because energy is conserved, the same downward tension must act in the fixed segment. Therefore, the force that the chain exerts on the fixed end O is

$$T = M_2 g + \tfrac{1}{4}\mu v^2. \tag{14}$$

With $M_2 = \mu \tfrac{1}{2}(\ell + x)$ and v given by (3), equation (14) yields (5).

Comments

(i) There is a rather extensive literature on the subject of falling chains, and a review has been given in Ref. [36]. These authors emphasize how useful the Lagrange formulation is in obtaining correct solutions: "Lagrange's equation of motion contains a unique description of what happens when masses are transferred between the two parts of a falling chain, a description that actually enforces energy conservation in the falling chain."

(ii) Some of the early solutions to the problem considered above (the fall of a tightly folded chain) are incorrect in that they assume the released end falls freely and is brought to rest by inelastic impacts with the fixed segment, thereby not conserving mechanical energy.[36] In fact, energy conservation (see (9)) requires that the force on the falling segment should exceed its weight (see (13)), and the resulting velocity and acceleration are greater than in free fall – see (3), where it is clear that $v(x) > \sqrt{2gx}$, and from which it follows that the acceleration

$$a(x) = \dot{v} = \left\{1 + \frac{1}{(1 - x/\ell)^2}\right\}\frac{g}{2} \tag{15}$$

increases monotonically from an initial value g. Experiments confirm the energy-conserving model.[36]

---

[36] C. W. Wong and K. Yasui, "Falling chains," American Journal of Physics, vol. 74, pp. 490–496, 2006.

(iii) In the one-dimensional model the velocity (3), the acceleration (15), and the tension (5) become infinite at the end of the fall (i.e. as $x \to \ell$). This is an artefact of the model; the actual values are, of course, finite and there are more realistic treatments of the closing stages of the motion.[36,37] Measured values of the maximum tension exceed (by more than an order of magnitude) the value $2Mg$ for a freely falling chain.[37]

Question 11.33

Consider a system of particles for which the Lagrangian may be an explicit function of time: $\mathsf{L}(q_i, \dot{q}_i, t)$.

(a) Starting with the differential $d\mathsf{L}$, and using Lagrange's equations (see Chapter 1), show that
$$d\left(\mathsf{L} - \sum p_i \dot{q}_i\right) = \sum \dot{p}_i \, dq_i - \sum \dot{q}_i \, dp_i + \frac{\partial \mathsf{L}}{\partial t} \, dt, \qquad (1)$$
where the summations are over all i and
$$p_i = \partial \mathsf{L}/\partial \dot{q}_i. \qquad (2)$$

(b) What properties can be deduced from (1) for the function
$$\mathsf{H} = \sum p_i \dot{q}_i - \mathsf{L} \ ? \qquad (3)$$

Solution

(a)
$$d\mathsf{L} = \sum \left(\frac{\partial \mathsf{L}}{\partial q_i} \, dq_i + \frac{\partial \mathsf{L}}{\partial \dot{q}_i} \, d\dot{q}_i\right) + \frac{\partial \mathsf{L}}{\partial t} \, dt$$
$$= \sum \left(\dot{p}_i \, dq_i + p_i \, d\dot{q}_i\right) + \frac{\partial \mathsf{L}}{\partial t} \, dt, \qquad (4)$$

where in the last step we have used Lagrange's equations ($\dot{p}_i = \partial \mathsf{L}/\partial q_i$) and (2). Now use the relation
$$d\left(\sum p_i \dot{q}_i\right) = \sum \left(p_i \, d\dot{q}_i + \dot{q}_i \, dp_i\right) \qquad (5)$$
to eliminate the second sum from (4), and then rearrange terms to obtain (1).

(b) It follows from (1) that:
- ☞ H is a function of the q_i, p_i and t:
$$\mathsf{H} = \mathsf{H}(q_i, p_i, t). \qquad (6)$$
- ☞ $\dot{q}_i = \partial \mathsf{H}/\partial p_i, \qquad \dot{p}_i = -\partial \mathsf{H}/\partial q_i. \qquad (7)$
- ☞ $\partial \mathsf{H}/\partial t = -\partial \mathsf{L}/\partial t. \qquad (8)$
- ☞ $d\mathsf{H}/dt = -\partial \mathsf{L}/\partial t = \partial \mathsf{H}/\partial t. \qquad (9)$

Equation (9) means that if L (and consequently H) does not depend explicitly on t then H is a constant (a conserved quantity).

---

[37] J.-C. Géminard and L. Vanel, "The motion of a freely falling chain tip: force measurements," American Journal of Physics, vol. 76, pp. 541–545, 2008.

Comments

(i) Equation (3) is a transformation that changes a function L of q_i, $\dot{q}_i$, and t into a function H of q_i, p_i and t. It is an example of a Legendre transformation.

(ii) The function H defined in (3) is known as the Hamiltonian of the system, and it plays an important role in mechanics. Equations (7) are Hamilton's equations of motion and they are the basis for Hamiltonian dynamics.

(iii) For a system with a time-independent Lagrangian L = K − V, where the kinetic energy K is a homogeneous quadratic function of the generalized velocities $\dot{q}_i$, the Hamiltonian (3) is equal to the energy of the system,

$$\mathsf{H} = E = K + V, \qquad (10)$$

and the property H = a constant (see (9)) is just a statement of conservation of energy. For a single particle moving in a potential V the Hamiltonian in Cartesian coordinates is

$$\mathsf{H} = \frac{1}{2m}\left(p_x^2 + p_y^2 + p_z^2\right) + V(x, y, z). \qquad (11)$$

(iv) In an inertial frame time is homogeneous and the Lagrangian of a closed (isolated) system is time independent. Consequently, in an inertial frame the energy of a closed system is conserved.

(v) For systems where the Lagrangian depends explicitly on t the question of the connection between H and E requires some care. Such Lagrangians can represent more than one physical system and the relationship between H and E has to be determined for each system. An example is given in Question 4.16.

(vi) In addition to its role in classical mechanics, the Hamiltonian is indispensable in quantum mechanics. For example, from the Hamiltonian one constructs (according to a prescription and usually working in Cartesian coordinates) a Hamiltonian operator H_{op} that enters the famous Schrödinger equation $\mathsf{H}_{\text{op}}\Psi = i\hbar\dot{\Psi}$.

12
Rigid bodies

A rigid body is an idealized multi-particle system in which the distance between each pair of particles is a constant: $|\mathbf{r}_i(t) - \mathbf{r}_j(t)| = c_{ij}$, where the c_{ij} are independent of t. So, in a rigid body the relative positions of all particles are fixed. All real objects undergo deformations, however small, when subjected to stresses. Nevertheless, the assumption of rigidity is a useful approximation in many cases of interest in physics and engineering.

The simplest motion of a rigid body is rotation about an axis that is fixed in space: the trajectories of the particles are circles in planes perpendicular to the axis, and the velocity of a particle is proportional to its distance from the axis. Also simple is a translation of the body: here, all displacements $\delta\mathbf{r}_i$ in a time δt are equal. Every particle has the same velocity and the trajectories differ by constant displacements $\mathbf{c}_{ij}$ and therefore have the same shape.

Next in simplicity is two-dimensional motion (also known as planar or laminar motion). Here, each particle moves parallel to a fixed plane; that is, each particle performs a translation parallel to the plane and a rotation about an instantaneous axis perpendicular to the plane. The most general motion of a rigid body consists of translation of a point that is fixed in (or relative to) the body, together with rotation about an instantaneous axis passing through that point. (Often this is referred to as 'rotation about the point' – the equivalence of the two statements is Euler's theorem.)

Even though the theory of rigid bodies is a considerable simplification of the general many-body problem, it is still one of the most intricate parts of classical mechanics. The theory is highly developed and there have been extensive studies of even apparently elementary systems such as gyroscopes and tops, and there are also widespread applications in atomic, molecular and nuclear physics, and astronomy and engineering.

The questions in this chapter deal first with some general properties of rigid bodies. Then, there are applications to both planar motion and questions in which the axis of rotation changes with time.

Question 12.1

How many degrees of freedom f does a rigid body have (that is, how many independent coordinates are required to uniquely determine its position in space at a given instant):

(a) In rotation about an axis that is fixed in space?
(b) In translation?
(c) In two-dimensional (planar or laminar) motion?
(d) In general?

Solution

(a) Just one coordinate is required, the angle of rotation: $f = 1$.
(b) Since there is no rotation, it is sufficient to specify the trajectory of just one point in (or relative to) the body: $f = 3$.
(c) Planar motion consists of translations parallel to a fixed plane and rotation about an axis perpendicular to that plane. Two coordinates are required for the translations and one for the rotation: $f = 3$.
(d) In general, the position of a rigid body is specified if the coordinates of three non-collinear points fixed in (or relative to) it are known. This requires $3 \times 3 = 9$ coordinates. However, not all these coordinates are independent. The rigidity requirement provides three relations between them, and so $f = 6$. (This can also be deduced by first specifying the position of one point – for example, the centre of mass C. This requires 3 coordinates. The position of a second point A requires 2 coordinates because AC is fixed, and the position of a third non-collinear point B requires just one coordinate because BC and AB are fixed. That is, $f = 3 + 2 + 1 = 6$.)

Question 12.2

According to (2) of Question 11.1 and 11.2, the rates of change of the momentum **P** and the angular momentum **L** of a rigid body relative to an inertial frame satisfy‡

$$\dot{\mathbf{P}} = \mathbf{F} \qquad \text{and} \qquad \dot{\mathbf{L}} = \mathbf{\Gamma}. \tag{1}$$

Here, **F** and **Γ** are the total force and the total torque due to external forces acting on the body. Discuss the role of (1) in rigid-body dynamics.

Solution

☞ First, note that the interparticle forces play no role in (1). This is a consequence of Newton's third law and, in the case of $(1)_2$, the assumption that the interparticle forces are central – see Questions 11.2 and 14.19.

☞ Equation $(1)_1$ is the equation of motion

$$M\ddot{\mathbf{R}} = \mathbf{F} \tag{2}$$

for the position vector **R** of the centre of mass relative to some arbitrarily chosen origin O of an inertial frame. So, $(1)_1$ describes the translational motion of one

‡For convenience we omit the superscript e on **F** and **Γ**.

point associated with the body – its solution provides 3 of the 6 parameters needed to specify the trajectory for general motion of a rigid body.

☞ The remaining three parameters are associated with the rotational motion of the body, and they are determined by $(1)_2$.

☞ The two independent equations (1), which comprise six component equations, are the equations of motion of a rigid body: they provide a complete description of its general motion. By comparison, the general N-body problem involves $3N$ component equations of motion: the reduction from $3N$ to 6 equations is a consequence of the rigidity condition.

☞ Often, the rotational motion is analyzed in terms of rotation about the centre of mass C, rather than the coordinate origin O to which $\mathbf{R}$ is referred. Then, $(1)_2$ is written

$$\dot{\mathbf{L}}_C = \mathbf{\Gamma}_C, \qquad (3)$$

where $\mathbf{L}_C$ and $\mathbf{\Gamma}_C$ are the angular momentum and torque about the CM. (Equation (3) holds even if C is accelerating relative to an inertial frame – see (7) of Question 11.3 and (5) below.) So, (2) describes the motion of the CM and (3) describes rotation about the CM.

☞ In certain cases – for example, if one point in the body is fixed – it is preferable to use $(1)_2$ instead of (3), with $\mathbf{L}$ and $\mathbf{\Gamma}$ being the angular momentum and torque about that point.

Comment

In the rotational equation $\dot{\mathbf{L}} = \mathbf{\Gamma}$, the angular momentum and torque are to be evaluated with respect to either the CM or a point that is fixed in an inertial frame, but is otherwise arbitrary. Relative to an arbitrary moving point (Q, say) the angular momentum and torque are[†]

$$\mathbf{L}_Q = \sum_i m_i (\mathbf{r}_i - \mathbf{r}_Q) \times (\dot{\mathbf{r}}_i - \dot{\mathbf{r}}_Q), \qquad \mathbf{\Gamma}_Q = \sum_i (\mathbf{r}_i - \mathbf{r}_Q) \times \mathbf{F}_i^{(e)}, \qquad (4)$$

where the $\mathbf{r}_i$ are position vectors in an inertial frame. By differentiating $(4)_1$ with respect to t and using the equation of motion $m_i \ddot{\mathbf{r}}_i = \mathbf{F}_i$, and assuming central interparticle forces, we obtain the rotational equation of motion relative to Q:

$$\dot{\mathbf{L}}_Q = \mathbf{\Gamma}_Q - M(\mathbf{R} - \mathbf{r}_Q) \times \ddot{\mathbf{r}}_Q, \qquad (5)$$

where M is the total mass and $\mathbf{R}$ is the position vector of the CM relative to the inertial frame. So, in general, there is an additional term associated with the acceleration $\ddot{\mathbf{r}}_Q$ of Q. By inspection, this term is zero if (i) $\ddot{\mathbf{r}}_Q$ is zero, or (ii) if Q is the CM, regardless of its acceleration (both of which we know already), or (iii) if $\ddot{\mathbf{r}}_Q$ is parallel or antiparallel to the vector that joins Q to the CM. In calculations where Q does not satisfy any of these three conditions it is essential to use the correct form (5) of the rotational equation of motion, and not $\dot{\mathbf{L}} = \mathbf{\Gamma}$ (see Question 12.14).

[†]Here, and in what follows, the summations over i are from 1 to N.

Question 12.3

(a) Define the angular velocity $\boldsymbol{\omega}$ of a rigid body rotating about some axis.
(b) Then derive the relation
$$\mathbf{v} = \boldsymbol{\omega} \times \mathbf{r} \tag{1}$$
for the velocity of a point in the body with position vector $\mathbf{r}$ relative to an origin on this axis. (Hint: First, sketch a diagram showing the axis of rotation, the vector $\mathbf{r}$, and its trajectory.)

Solution

(a) The diagram shows the axis of rotation and the position vector $\mathbf{r}$ relative to a point A on the axis. Also shown are the components of $\mathbf{r}$ that are parallel and perpendicular to the axis. The endpoint of $\mathbf{r}$ traces out a circle centred on the axis. In a time dt the endpoint rotates through an angle $d\theta$. The unit vectors $\hat{\mathbf{r}}_\perp$ and $\hat{\boldsymbol{\theta}}$ are in the directions of increasing $r_\perp$ and θ. By definition, the angular speed is the rate of change $\omega = d\theta/dt$ and the angular velocity is $\boldsymbol{\omega} = (d\theta/dt)\mathbf{n}$, where the unit vector $\mathbf{n}$ satisfies a right-hand rule: $\mathbf{n} \times \hat{\mathbf{r}}_\perp = \hat{\boldsymbol{\theta}}$.

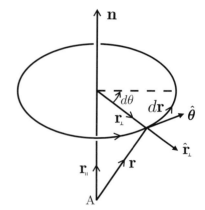

(b) Let $d\mathbf{r}$ be the change in $\mathbf{r}$ in a time dt. From the diagram and the definition of $\boldsymbol{\omega}$:

$$d\mathbf{r} = r_\perp d\theta\, \hat{\boldsymbol{\theta}} = \boldsymbol{\omega} \times \mathbf{r}_\perp dt\,. \tag{2}$$

Now, $\mathbf{r}_\perp = \mathbf{r} - \mathbf{r}_\parallel$ and $\boldsymbol{\omega} \times \mathbf{r}_\parallel = 0$. So (2) yields (1).

Comments

(i) The angle $d\theta$ turned through in a time dt is the same for all points in the body, and therefore so is $\boldsymbol{\omega}$. Also, the relation (1) is independent of where the point A is located on the axis of rotation (cf. the step from (2) to (1)).

(ii) From (1), if B and C are any two points fixed in the body then
$$\mathbf{v}_\text{B} = \mathbf{v}_\text{C} + \boldsymbol{\omega} \times (\mathbf{r}_\text{B} - \mathbf{r}_\text{C})\,. \tag{3}$$

(iii) Equation (3) can be applied to a body that is translating as well as rotating. Often, C is taken to be the CM and (3) is written (we omit the subscript B on $\mathbf{v}$ and set $\mathbf{v}_\text{C} = \dot{\mathbf{R}}$)
$$\mathbf{v} = \dot{\mathbf{R}} + \boldsymbol{\omega} \times \mathbf{r}\,. \tag{4}$$

Equation (4) is the extension of (1) to include translation: it gives the velocity $\mathbf{v}$ of a point in the body in terms of the velocity $\dot{\mathbf{R}}$ of the CM (relative to some frame) and the position vector **r** of that point relative to the CM. So, the velocity of any point in the body is known once the CM velocity and the angular velocity are known (i.e. six degrees of freedom – see Question 12.1).

(iv) The translational equation of motion $M\ddot{\mathbf{R}} = \mathbf{F}$ provides a first-order differential equation for $\dot{\mathbf{R}}$, and we will see below that the rotational equation of motion $\dot{\mathbf{L}} = \boldsymbol{\Gamma}$ provides a first-order differential equation for $\boldsymbol{\omega}$.

Question 12.4

Consider a rigid body that is rotating about a fixed axis (the z-axis, say) with angular velocity $\boldsymbol{\omega} = \omega\hat{\mathbf{z}}$.

(a) Show that the z-component of the angular momentum of the body about the axis of rotation is

$$L_z = I_z\omega, \qquad \text{where} \qquad I_z = \sum_i m_i(x_i^2 + y_i^2). \tag{1}$$

(b) Deduce that the z-component of the rotational equation of motion is

$$I_z\dot{\omega} = \Gamma_z, \tag{2}$$

where Γ_z is the total torque about the z-axis due to external forces.

(c) Show that the kinetic energy of rotation is

$$K = \tfrac{1}{2}I_z\omega^2. \tag{3}$$

Solution

(a) The velocity of the ith particle is (see Question 12.3)

$$\mathbf{v}_i = \boldsymbol{\omega} \times \mathbf{r}_i = (0, 0, \omega) \times (x_i, y_i, z_i) = (-\omega y_i, \omega x_i, 0), \tag{4}$$

and so

$$L_z = \sum_i m_i(\mathbf{r}_i \times \mathbf{v}_i)_z = \sum_i m_i(x_i^2 + y_i^2)\omega, \tag{5}$$

which is (1) because ω is the same for all particles (see Question 12.3).

(b) The z-component of the rotational equation of motion $\dot{L}_z = \Gamma_z$ and $(1)_1$ give (2) because for a rigid body I_z is independent of t.

(c) $K = \tfrac{1}{2}\sum_i m_i \mathbf{v}_i^2 = \tfrac{1}{2}\sum_i m_i(x_i^2 + y_i^2)\omega^2 = \tfrac{1}{2}I_z\omega^2.$

Comments

(i) The quantity I_z defined in $(1)_2$ is called the moment of inertia about the z-axis. The derivative $\dot{\omega} = \ddot{\theta}$ is the angular acceleration.

(ii) Equations $(1)_1$, (2) and (3) are rotational analogues of the equations $p = mv$, $m\dot{v} = F$ and $K = \tfrac{1}{2}mv^2$ for one-dimensional translational motion. It is apparent from this comparison that the moment of inertia is analogous to mass: for example, (2) shows that the moment of inertia about an axis is a measure of the rotational inertia relative to that axis.

(iii) The moment of inertia is therefore an important dynamical property of a rigid body. In general, the moment of inertia about an axis is defined by

$$I = \sum_i m_i r_i^2, \qquad (6)$$

where r_i is the perpendicular distance of the ith particle from the axis. If the distribution of mass is continuous then (6) is replaced by

$$I = \int r^2 \, dm, \qquad (7)$$

where the integration is taken over the entire body. Some calculations of moments of inertia are given in Questions 12.8 and 12.10.

(iv) The kinetic energy (3) changes at a rate $\dot{K} = I_z \dot{\omega}\omega = \Gamma_z \dot{\theta}$, and therefore

$$dK = \Gamma_z \, d\theta. \qquad (8)$$

Now, $\Gamma_z \, d\theta = dW$ is the work done by the external forces during the infinitesimal rotation $d\theta$, and (8) is the work–energy theorem for rotational motion. If the external forces are conservative then there exists a potential-energy function $V(\theta)$ such that $dV = -dK$ (see Chapter 5). Thus, the mechanical energy

$$E = \tfrac{1}{2} I_z \omega^2 + V(\theta), \qquad \text{with} \qquad V(\theta) = -\int \Gamma_z \, d\theta, \qquad (9)$$

is conserved during the motion.

Question 12.5

Extend the results of Question 12.4 for rotation about a fixed axis to planar motion of a rigid body.

Solution

In planar motion of a rigid body, each particle moves parallel to a fixed plane. The motion can be specified by two-dimensional motion of the CM and rotation about the CM (meaning rotation about an axis through the CM and perpendicular to the fixed plane). Motion of the CM is governed by the translational equation $M\ddot{\mathbf{R}} = \mathbf{F}$. For the rotational motion we consider the angular momentum about the CM:

$$\mathbf{L}_C = \sum_i m_i \mathbf{r}_i \times \mathbf{v}_i, \tag{1}$$

where $\mathbf{v}_i = \boldsymbol{\omega} \times \mathbf{r}_i$ and the $\mathbf{r}_i$ are position vectors relative to the CM. Let the fixed plane be parallel to the xy-plane. Then, $\boldsymbol{\omega} = (0, 0, \omega)$ and the evaluation of (1) proceeds as in Question 12.4. The result for the z-component of $\mathbf{L}_C$ is

$$L_C = I_C \omega \quad \text{and} \quad I_C = \sum_i m_i (x_i^2 + y_i^2). \tag{2}$$

I_C is the moment of inertia about the CM (that is, about the axis Cz). For a rigid body, I_C is independent of t and therefore the z-component of the rotational equation of motion $\dot{\mathbf{L}}_C = \boldsymbol{\Gamma}_C$ is

$$I_C \dot{\omega} = \Gamma_C. \tag{3}$$

Here, Γ_C is the torque about the axis Cz. The total kinetic energy consists of translational and rotational contributions:

$$K = \tfrac{1}{2} M \dot{R}^2 + \tfrac{1}{2} I_C \omega^2, \tag{4}$$

and for conservative external forces, the mechanical energy relative to an inertial frame is

$$E = \tfrac{1}{2} M \dot{R}^2 + \tfrac{1}{2} I_C \omega^2 + V(\theta), \tag{5}$$

where

$$V(\theta) = -\int \Gamma_C \, d\theta. \tag{6}$$

Comment

The separation of the total kinetic energy into translational and rotational parts always occurs in rigid-body motion. In general,

$$K = \tfrac{1}{2} \sum_i m_i v_i^2, \tag{7}$$

where $\mathbf{v}_i = \dot{\mathbf{R}} + \boldsymbol{\omega} \times \mathbf{r}_i$ and $\mathbf{r}_i$ is the position vector of the ith particle relative to the CM (see Question 12.3). So

$$K = \tfrac{1}{2} \sum_i m_i \mathbf{v}_i \cdot (\dot{\mathbf{R}} + \boldsymbol{\omega} \times \mathbf{r}_i)$$

$$= \tfrac{1}{2} M \dot{R}^2 + \tfrac{1}{2} \boldsymbol{\omega} \cdot \sum_i m_i \mathbf{r}_i \times \mathbf{v}_i$$

$$= \tfrac{1}{2} M \dot{R}^2 + \tfrac{1}{2} \boldsymbol{\omega} \cdot \mathbf{L}_C. \tag{8}$$

(Here, we have used the identity $\mathbf{a} \cdot (\mathbf{b} \times \mathbf{c}) = \mathbf{b} \cdot (\mathbf{c} \times \mathbf{a})$ and the relation $\sum_i m_i \mathbf{v}_i = M \dot{\mathbf{R}}$, that holds because $\sum_i m_i \mathbf{r}_i = 0$.) Equation (8) is valid for any rigid-body motion. In planar motion, $\mathbf{L}_C = I_C \boldsymbol{\omega}$ and (8) reduces to (4).

Question 12.6

Suppose one point O of a rigid body is fixed in inertial space. Let $Oxyz$ be a coordinate system fixed within the body and let $\boldsymbol{\omega}$ be the angular velocity of the body relative to an inertial frame with origin at O. The angular momentum $\mathbf{L}$ about O, expressed in terms of body coordinates, is

$$\mathbf{L} = \sum_i m_i \mathbf{r}_i \times \mathbf{v}_i, \qquad \text{where} \qquad \mathbf{v}_i = \boldsymbol{\omega} \times \mathbf{r}_i. \tag{1}$$

(a) Show that $\mathbf{L}$ is given in terms of the components of $\boldsymbol{\omega}$ by

$$\mathbf{L} = (I_{xx}\omega_x + I_{xy}\omega_y + I_{xz}\omega_z)\hat{\mathbf{x}} + (I_{yx}\omega_x + I_{yy}\omega_y + I_{yz}\omega_z)\hat{\mathbf{y}}$$
$$+ (I_{zx}\omega_x + I_{zy}\omega_y + I_{zz}\omega_z)\hat{\mathbf{z}}, \tag{2}$$

where

$$I_{xx} = \sum_i m_i(y_i^2 + z_i^2), \quad I_{yy} = \sum_i m_i(x_i^2 + z_i^2), \quad I_{zz} = \sum_i m_i(x_i^2 + y_i^2), \tag{3}$$

$$\left. \begin{array}{c} I_{xy} = I_{yx} = -\sum_i m_i x_i y_i, \qquad I_{yz} = I_{zy} = -\sum_i m_i y_i z_i, \\ \\ I_{xz} = I_{zx} = -\sum_i m_i x_i z_i. \end{array} \right\} \tag{4}$$

(b) Hence, express the equations of motion in the inertial frame as first-order equations in the time derivatives of the components of $\boldsymbol{\omega}$.
(c) What is the form of (2) and the equations of motion for rotation about a fixed axis (the z-axis, say)?
(d) What is the form of (2) and the equations of motion in a body coordinate system in which the off-diagonal components I_{xy}, I_{yz} and I_{xz} are zero (that is, in principal axes)?

Solution

(a) We start with the vector product $\mathbf{r}_i \times (\boldsymbol{\omega} \times \mathbf{r}_i)$, where $\mathbf{r}_i$ and $\boldsymbol{\omega}$ are expressed in the body system as

$$\mathbf{r}_i = x_i \hat{\mathbf{x}} + y_i \hat{\mathbf{y}} + z_i \hat{\mathbf{z}}, \qquad \boldsymbol{\omega} = \omega_x \hat{\mathbf{x}} + \omega_y \hat{\mathbf{y}} + \omega_z \hat{\mathbf{z}}. \tag{5}$$

A direct calculation yields

$$\mathbf{r}_i \times (\boldsymbol{\omega} \times \mathbf{r}_i) = \{\omega_x(y_i^2 + z_i^2) - \omega_y x_i y_i - \omega_z x_i z_i\}\hat{\mathbf{x}}$$
$$+ \{-\omega_x x_i y_i + \omega_y(x_i^2 + z_i^2) - \omega_z y_i z_i\}\hat{\mathbf{y}}$$
$$+ \{-\omega_x x_i z_i - \omega_y y_i z_i + \omega_z(x_i^2 + y_i^2)\}\hat{\mathbf{z}}. \tag{6}$$

By substituting (6) in (1) we obtain (2)–(4).

(b) The equation of motion in the inertial frame is $\dot{\mathbf{L}} = \boldsymbol{\Gamma}$. To express this in terms of the components of $\boldsymbol{\omega}$, we must differentiate (2) with respect to t, taking into account that in this frame the time-dependent quantities are $\boldsymbol{\omega}$ and the unit vectors $\hat{\mathbf{x}}$, $\hat{\mathbf{y}}$, $\hat{\mathbf{z}}$ of the body frame. (The six components $I_{xx}, \cdots, I_{zz}$ in (4) are constants because they are evaluated in the body frame, where x_i, y_i and z_i are constant.) The unit vectors are fixed in the body, and so they rotate with angular velocity $\boldsymbol{\omega}$ with respect to the inertial frame; that is (see Question 12.3)

$$d\hat{\mathbf{x}}/dt = \boldsymbol{\omega} \times \hat{\mathbf{x}}, \qquad d\hat{\mathbf{y}}/dt = \boldsymbol{\omega} \times \hat{\mathbf{y}}, \qquad d\hat{\mathbf{z}}/dt = \boldsymbol{\omega} \times \hat{\mathbf{z}}. \tag{7}$$

By differentiating (2) and using (7), we find that the x-component of the equation of motion is

$$I_{xx}\dot{\omega}_x + I_{xy}\dot{\omega}_y + I_{xz}\dot{\omega}_z + (I_{zz} - I_{yy})\omega_y\omega_z$$
$$+ I_{zx}\omega_x\omega_y - I_{yx}\omega_x\omega_z + I_{zy}\omega_y^2 - I_{yz}\omega_z^2 = \Gamma_x. \tag{8}$$

The y- and z-components of the equation of motion are obtained from (8) by the cyclic permutations $x \to y \to z \to x$.

(c) If the rotation is about the z-axis then $\boldsymbol{\omega} = (0, 0, \omega)$ and (2) simplifies to

$$\mathbf{L} = I_{xz}\omega\,\hat{\mathbf{x}} + I_{yz}\omega\,\hat{\mathbf{y}} + I_{zz}\omega\,\hat{\mathbf{z}}. \tag{9}$$

Similarly, the equation of motion (8), and the y- and z-components obtained from (8) by cyclic permutation, simplify to

$$\left.\begin{array}{l} I_{xz}\dot{\omega} - I_{yz}\omega^2 = \Gamma_x \\ I_{yz}\dot{\omega} + I_{xz}\omega^2 = \Gamma_y \\ I_{zz}\dot{\omega} = \Gamma_z. \end{array}\right\} \tag{10}$$

Note that here the fixed point O can be any point on the axis of rotation.

(d) The principal axes are a special set of body axes $OXYZ$ in which the off-diagonal elements I_{XY}, I_{YZ} and I_{XZ} are zero. The diagonal elements are denoted $I_1 = I_{XX}$, $I_2 = I_{YY}$ and $I_3 = I_{ZZ}$, and the unit vectors of the principal axes are $\mathbf{e}_1$, $\mathbf{e}_2$, $\mathbf{e}_3$. In principal axes the angular momentum (2) has the simple form

$$\mathbf{L} = I_1\omega_1\,\mathbf{e}_1 + I_2\omega_2\,\mathbf{e}_2 + I_3\omega_3\,\mathbf{e}_3. \tag{11}$$

Similarly, the equation of motion (8) and its permutations become

$$\left.\begin{array}{l} I_1\dot{\omega}_1 + (I_3 - I_2)\,\omega_2\omega_3 = \Gamma_1 \\ I_2\dot{\omega}_2 + (I_1 - I_3)\,\omega_1\omega_3 = \Gamma_2 \\ I_3\dot{\omega}_3 + (I_2 - I_1)\,\omega_1\omega_2 = \Gamma_3. \end{array}\right\} \tag{12}$$

Equations (12) are known as Euler's equations for a rigid body. We remark that the rotational kinetic energy $K = \tfrac{1}{2}\boldsymbol{\omega} \cdot \mathbf{L}$ also has a simple form in principal axes, namely

$$K = \tfrac{1}{2}\left(I_1\omega_1^2 + I_2\omega_2^2 + I_3\omega_3^2\right). \tag{13}$$

Comments

(i) In general, the angular momentum $\mathbf{L}$ is not parallel to the axis of rotation – see (2). Even in the simple case of rotation about a fixed axis, $\mathbf{L}$ can have components perpendicular to the axis of rotation – see (9).

(ii) The quantities I_{xx}, I_{yy}, I_{zz} defined in (3) are referred to as the moments of inertia about the x-, y- and z-axes, respectively. The quantities I_{xy}, I_{yz}, I_{xz} defined in (4) are known as the products of inertia with respect to the same axes.

(iii) For a continuous distribution of mass of density $\rho(\mathbf{r})$ in a volume V, the sums in (3) and (4) can be replaced by integrals:

$$I_{xx} = \int (y^2 + z^2)\, dm, \qquad I_{xy} = -\int xy\, dm, \qquad \text{etc.} \qquad (14)$$

(iv) Suppose a rigid body possesses an axis of symmetry, meaning that the masses dm at $\mathbf{r}_\perp$ and $-\mathbf{r}_\perp$ are the same.‡ Then, some of the products of inertia are zero. For example, if Oz is an axis of symmetry then I_{xz} and I_{yz} are zero;† consequently the angular momentum (9) has a component only along the axis of rotation:

$$\mathbf{L} = (0,\, 0,\, I_{zz}\omega), \qquad (15)$$

and the equations of motion (10) become

$$\Gamma_x = \Gamma_y = 0 \qquad \text{and} \qquad I_{zz}\dot{\omega} = \Gamma_z. \qquad (16)$$

(v) Equation (2) for $\mathbf{L}$ and the equations of motion (8), etc. are cumbersome, and often they are written in a concise tensor form. For this purpose the components of $\mathbf{r}$ and $\boldsymbol{\omega}$, and the moments and products of inertia are denoted by x_α, ω_α, $I_{\alpha\beta}$, where α, $\beta = 1, 2$ or 3. Also, the unit vectors of the body frame are denoted by $\mathbf{e}_\alpha$. Then (2)–(4) can be abbreviated as

$$\mathbf{L} = I_{\alpha\beta}\omega_\beta \mathbf{e}_\alpha. \qquad (17)$$

Here, $I_{\alpha\beta}$ is the inertia tensor: for a continuous body it has components

$$I_{\alpha\beta} = \int_V (\delta_{\alpha\beta} x_\gamma x_\gamma - x_\alpha x_\beta)\rho(\mathbf{r})\, dV, \qquad (18)$$

where $\delta_{\alpha\beta}$ is the Kronecker delta function. In equations such as (17) it is always understood that a repeated index implies summation from 1 to 3 (so (17) contains a double sum over α and β). The equations of motion (8), etc. abbreviate to

$$I_{\alpha\beta}\dot{\omega}_\beta + \varepsilon_{\alpha\beta\gamma}\omega_\beta I_{\gamma\delta}\omega_\delta = \Gamma_\alpha, \qquad (19)$$

where $\varepsilon_{\alpha\beta\gamma}$ is the Levi-Civita tensor (see Question 9.6).

‡ So, an axis of symmetry must pass through the CM.
† So, a symmetry axis is a principal axis.

(vi) There is an important special case where the point O (mentioned in the statement of this question) need not be fixed in inertial space, namely when O is the CM of the rigid body (see Question 12.2). All the results obtained above apply also when O is the CM, and they describe the general rotation of a rigid body about its CM. When taken together with the translational equation of motion $M\ddot{\mathbf{R}} = \mathbf{F}$, they provide a complete description of the dynamics of the body.

Question 12.7

Let I_O be the moment of inertia of a rigid body about an axis through a point O, and let I_C be the moment of inertia about a parallel axis through the centre of mass C. Prove that
$$I_O = I_C + MD^2, \tag{1}$$
where M is the mass of the body and D is the distance between the axes.

Solution

Consider, for example, the moment of inertia about the Oz-axis:
$$I_z = \sum_i m_i(x_i^2 + y_i^2). \tag{2}$$

The coordinates of the particles relative to the CM are $\bar{x}_i = x_i - x_C$, $\bar{y}_i = y_i - y_C$, $\bar{z}_i = z_i - z_C$, where (x_C, y_C, z_C) is the position vector of C relative to $Oxyz$. So (2) can be expressed as
$$I_z = \sum_i m_i \left\{ (\bar{x}_i + x_C)^2 + (\bar{y}_i + y_C)^2 \right\}$$
$$= \sum_i m_i (\bar{x}_i^2 + \bar{y}_i^2) + 2x_C \sum_i m_i \bar{x}_i + 2y_C \sum_i m_i \bar{y}_i + \sum_i m_i (x_C^2 + y_C^2). \tag{3}$$

The first sum in (2) is the moment of inertia about an axis through C and parallel to the z-axis. The second and third sums are zero because they are CM coordinates in the CM frame. The fourth sum is equal to MD^2. This proves (1) for the moment of inertia about the z-axis, and by extension for any axis through O.

Comments

(i) Equation (1) is known as the parallel-axis theorem. It reduces the task of calculating the moment of inertia about any axis to the computation of the moment of inertia about a parallel axis through the CM.
(ii) A similar result holds for the products of inertia. For example, from (4) of Question 12.6 we have
$$I_{Oxy} = I_{Cxy} - Mx_C y_C, \tag{4}$$
and similarly for I_{Oyz} and I_{Oxz}.

(iii) Therefore, for the components $I_{\alpha\beta}$ of the inertia tensor, the parallel-axis theorem states: the moments and products of inertia of a body of mass M with respect to Cartesian axes $Oxyz$ are equal to those with respect to a set of parallel axes through the CM plus those of a particle of mass M, situated at the CM, with respect to $Oxyz$. See also Question 14.3.

Question 12.8

A truncated sphere is formed by removing a cap of height h from a uniform sphere of radius a. (a) Calculate the position of the centre of mass. (b) Calculate the moment of inertia about the axis of symmetry. (Express the results in terms of a and $\epsilon = h/a$.)

Solution

(a) From (8) of Question 11.1, the Cartesian coordinates of the CM are given by

$$X = \frac{1}{M}\int x\,dm\,, \qquad Y = \frac{1}{M}\int y\,dm\,, \qquad Z = \frac{1}{M}\int z\,dm\,, \qquad (1)$$

where M is the total mass and the integrals are taken over all elements dm comprising the truncated sphere. To evaluate (1) we choose the coordinates shown, with x-axis along the symmetry axis and origin O at the centre of the sphere prior to truncation. First consider Y. From the symmetry of the problem it is clear that for an element dm at y there is always an equal element dm at $-y$, and therefore $Y = 0$. Similarly, $Z = 0$. For the elements of mass dm in $(1)_1$ for X, choose slices perpendicular to the x-axis and of thickness dx as shown. Because the truncated sphere is uniform,

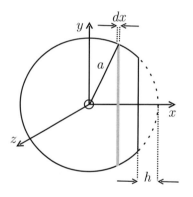

$$dm = M\pi(a^2 - x^2)\,dx \div \int_{-a}^{a-h} \pi(a^2 - x^2)\,dx\,. \qquad (2)$$

From $(1)_1$ and (2) we have

$$X = \int_{-a}^{a-h} x(a^2 - x^2)\,dx \div \int_{-a}^{a-h} (a^2 - x^2)\,dx$$

$$= -\frac{3}{4}\frac{\epsilon^2(2-\epsilon)^2}{(4 - 3\epsilon^2 + \epsilon^3)}\,a\,, \qquad \text{where } \epsilon = h/a. \qquad (3)$$

(b) The axis of symmetry is the x-axis, and the moment of inertia – of the infinitesimal shaded disc shown in the figure – about this axis is

$$dI_x = \tfrac{1}{2}(a^2 - x^2)\,dm\,. \qquad (4)$$

From (2) and (4) we have the total moment of inertia about the x-axis:

$$I_x = \tfrac{1}{2}M \int_{-a}^{a-h} (a^2 - x^2)^2 \, dx \; \div \; \int_{-a}^{a-h} (a^2 - x^2) \, dx \,. \tag{5}$$

The integrals in (5) are elementary and a short calculation yields

$$I_x = \frac{4 + 4\epsilon + 3\epsilon^2 - 3\epsilon^3}{10(1 + \epsilon)} Ma^2 \,. \tag{6}$$

Comments

(i) When $\epsilon = 0$ (a sphere), (3) gives $X = 0$ as required. When $\epsilon = 1$ (a hemisphere), (3) gives $X = -3a/8$. In the limit $\epsilon \to 2$ (a 'speck' at $x = -a$), (3) is indeterminate, and L'Hôpital's rule yields $X = -a$, as expected. When $\epsilon = 0$ or 1, (6) yields the familiar expression $I_x = \tfrac{2}{5}Ma^2$.

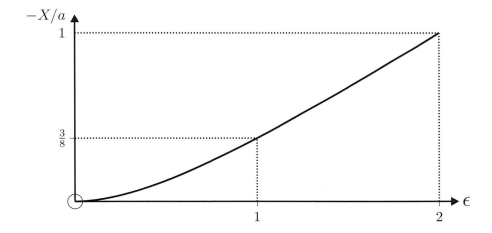

(ii) Equation (3) shows that truncation of a uniform sphere displaces the centre of mass (and hence the centre of gravity) away from the centre of the sphere, along the axis of symmetry. Consequently, a truncated sphere in a uniform gravitational field experiences a torque that tends to orient the flat face perpendicular to the field. This property makes it possible to construct a simple gyroscope from a truncated sphere that is spinning about its symmetry axis (see Question 12.24).

Question 12.9

Consider a homogeneous solid of revolution generated by rotating a function $y = f(x)$, for $x \in [x_1, x_2]$, about the x-axis. Show that the moments of inertia I_x and I_y about the x- and y-axes can be expressed in terms of single integrals:

$$I_x = \tfrac{1}{2}\pi\rho \int_{x_1}^{x_2} f^4(x)\,dx \tag{1}$$

$$I_y = \tfrac{1}{2}I_x + \pi\rho \int_{x_1}^{x_2} x^2 f^2(x)\,dx, \tag{2}$$

where $\rho = M/V$ is the density of the body.

Solution

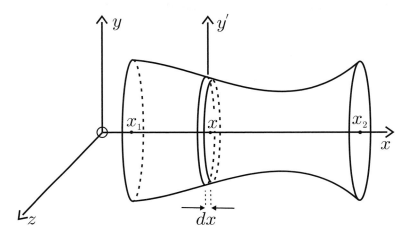

Consider first a thin disc perpendicular to the x-axis, located at x and of thickness dx and mass dm. Its moment of inertia about the x-axis is $dI_x = \tfrac{1}{2}f^2(x)\,dm$, where $dm = \rho\,dV = \rho\pi f^2(x)\,dx$. So, the total moment of inertia about the x-axis is given by (1). To calculate I_y we start with the moment of inertia of the thin disc about the y' axis. Clearly,[‡] $dI_{y'} + dI_{z'} = dI_x$ and $dI_{z'} = dI_{y'}$. Therefore, $dI_{y'} = \tfrac{1}{2}dI_x$. According to the parallel-axis theorem $dI_y = dI_{y'} + x^2\,dm$, and so we obtain (2).

Comments

(i) Usually, the evaluation of the moment of inertia of a rigid body requires multiple integration (often a triple integration must be performed). However, for a homogeneous solid of revolution, (1) and (2) show that just one integration is required.[1] Some examples are given in the next question.

(ii) This approach, where moments of inertia are expressed in terms of a generating function $f(x)$, allows one to study the minimization of certain moments of inertia by using the calculus of variations.[1]

[‡]This is an example of the perpendicular-axis theorem.

---

[1] R. Diaz, W. J. Herrera, and R. Martinez, "Moments of inertia for solids of revolution and variational methods," European Journal of Physics, vol. 27, pp. 183–192, 2006.

(iii) For solids of revolution such as shells and pipes, which are generated by two functions $f_1(x)$ and $f_2(x)$, we have instead of (1) and (2),

$$I_x = \tfrac{1}{2}\pi\rho \int_{x_1}^{x_2} [f_1^4(x) - f_2^4(x)]\,dx \tag{3}$$

$$I_y = \tfrac{1}{2}I_x + \pi\rho \int_{x_1}^{x_2} x^2 [f_1^2(x) - f_2^2(x)]\,dx . \tag{4}$$

(iv) The moments of inertia (1)–(4) are expressed in terms of the density ρ of the body. They can also be written in terms of the mass M because

$$\rho = \frac{M}{V} = M \div \pi \int_{x_1}^{x_2} f^2(x)\,dx , \tag{5}$$

when there is one generating function, and

$$\rho = M \div \pi \int_{x_1}^{x_2} [f_1^2(x) - f_2^2(x)]\,dx , \tag{6}$$

when there are two.

Question 12.10

Use the method of Question 12.9 to obtain the following moments of inertia for homogeneous solids of revolution:

	Solid	Axis	Moment of inertia
1.	sphere (radius = a)	any diameter	$\tfrac{2}{5}Ma^2$
2.	spherical shell (outer radius = a, inner radius = b)	any diameter	$\tfrac{2}{5}M\left(\dfrac{a^5 - b^5}{a^3 - b^3}\right)$
3.	thin spherical shell (radius = a)	any diameter	$\tfrac{2}{3}Ma^2$
4.	circular cylinder (radius = a, height = h)	longitudinal axis	$\tfrac{1}{2}Ma^2$
		perpendicular axis through the CM	$\tfrac{1}{4}Ma^2 + \tfrac{1}{12}Mh^2$
5.	circular cylindrical pipe (outer radius = a, inner radius = b, height = h)	longitudinal axis	$\tfrac{1}{2}M(a^2 + b^2)$
		perpendicular axis through the CM	$\tfrac{1}{4}M(a^2 + b^2) + \tfrac{1}{12}Mh^2$

	Solid	Axis	Moment of inertia
6.	thin cylindrical pipe (radius = a, height = h)	longitudinal axis	Ma^2
		perpendicular axis through the CM	$\frac{1}{2}Ma^2 + \frac{1}{12}Mh^2$
7.	ellipsoid of revolution (semi-axes = a, b, b)	x-axis	$\frac{2}{5}Mb^2$
		y-axis	$\frac{1}{5}M(a^2 + b^2)$
8.	circular cone (radius = a, height = h)	axis of symmetry	$\frac{3}{10}Ma^2$
		perpendicular axis through the vertex	$\frac{3}{20}Ma^2 + \frac{3}{5}Mh^2$
9.	torus formed by rotating the circle $x^2 + (y-a)^2 = b^2$ ($a > b$) about the x-axis	x-axis	$Ma^2 + \frac{3}{4}Mb^2$

Solution

In each case we identify the generating function(s) $y = f(x)$ and apply the appropriate equations from (1) to (6) of Question 12.9:

1. $f(x) = \sqrt{a^2 - x^2}$ and $\rho = 3M/4\pi a^3$. So

$$I_x = \frac{3M}{8a^3} \int_{-a}^{a} (a^2 - x^2)^2 \, dx = \frac{2}{5}Ma^2. \tag{1}$$

2. $f_1(x) = \sqrt{a^2 - x^2}$, $f_2(x) = \sqrt{b^2 - x^2}$ and $\rho = 3M/4\pi(a^3 - b^3)$. So

$$I_x = \frac{3M}{8(a^3 - b^3)} \left\{ \int_{-a}^{a} (a^2 - x^2)^2 \, dx - \int_{-b}^{b} (b^2 - x^2)^2 \, dx \right\}$$

$$= \frac{2}{5}M \left(\frac{a^5 - b^5}{a^3 - b^3} \right). \tag{2}$$

3. In (2) set $b = a - \epsilon$. The limit $\epsilon \to 0$ gives

$$I_x = \frac{2}{3}Ma^2. \tag{3}$$

4. $f(x) = a$ and $\rho = M/\pi a^2 h$. So

$$I_x = \frac{M}{2a^2h} \int_0^h a^4 \, dx = \tfrac{1}{2}Ma^2. \tag{4}$$

$$I_y = \tfrac{1}{2}I_x + \frac{M}{a^2h} \int_{-h/2}^{h/2} x^2 a^2 \, dx = \tfrac{1}{4}Ma^2 + \tfrac{1}{12}Mh^2. \tag{5}$$

5. $f_1(x) = a$, $f_2(x) = b$ and $\rho = M/\pi(a^2 - b^2)h$. So

$$I_x = \frac{M}{2(a^2 - b^2)h} \int_0^h (a^4 - b^4) \, dx = \tfrac{1}{2}M(a^2 + b^2). \tag{6}$$

$$I_y = \tfrac{1}{2}I_x + \frac{M}{(a^2 - b^2)h} \int_{-h/2}^{h/2} x^2(a^2 - b^2) \, dx$$

$$= \tfrac{1}{4}M(a^2 + b^2) + \tfrac{1}{12}Mh^2. \tag{7}$$

6. In (6) and (7) set $b = a$.

7. $f(x) = (b/a)\sqrt{a^2 - x^2}$ and $\rho = 3M/4\pi ab^2$. So

$$I_x = \frac{3M}{8ab^2} \int_{-a}^{a} (b/a)^4 (a^2 - x^2)^2 \, dx = \tfrac{2}{5}Mb^2. \tag{8}$$

$$I_y = \tfrac{1}{2}I_x + \frac{3M}{4a^3} \int_{-a}^{a} x^2(a^2 - x^2) \, dx = \tfrac{1}{5}M(a^2 + b^2). \tag{9}$$

8. $f(x) = ax/h$ and $\rho = 3M/\pi a^2 h$. So

$$I_x = \frac{3M}{2a^2h} \int_0^h (ax/h)^4 \, dx = \tfrac{3}{10}Ma^2. \tag{10}$$

$$I_y = \tfrac{1}{2}I_x + \frac{3M}{a^2h} \int_0^h x^2 (ax/h)^2 \, dx = \tfrac{3}{20}Ma^2 + \tfrac{3}{5}Mh^2. \tag{11}$$

9. $f_1(x) = a + \sqrt{b^2 - x^2}$, $f_2(x) = a - \sqrt{b^2 - x^2}$.

$$\rho = M \div \pi \int_{-b}^{b} 4a\sqrt{b^2 - x^2} \, dx$$

$$= \frac{M}{2\pi^2 ab^2}.$$

$$I_x = \frac{M}{4\pi ab^2} \int_{-b}^{b} 8a\sqrt{b^2 - x^2} \left\{ a^2 + b^2 - x^2 \right\} dx$$

$$= Ma^2 + \tfrac{3}{4}Mb^2. \tag{12}$$

Question 12.11

Deduce the conditions for a rigid body to be in equilibrium in an inertial frame.

Solution

A rigid body is in equilibrium under the action of a set of external forces $\mathbf{F}_i$ in some frame if all particles are at rest for all t; that is, $\dot{\mathbf{r}}_i = 0$ (or, equivalently, $\dot{\mathbf{R}} = 0$ and $\boldsymbol{\omega} = 0$). Consequently, the total momentum and angular momentum are zero: $\mathbf{P} = 0$ and $\mathbf{L} = 0$. If the frame is inertial then $\dot{\mathbf{P}} = \mathbf{F}$ and $\dot{\mathbf{L}} = \boldsymbol{\Gamma}$, and it follows that

$$\mathbf{F} = 0 \quad \text{and} \quad \boldsymbol{\Gamma} = 0 \tag{1}$$

for all t. That is, a rigid body is in equilibrium in an inertial frame if the total external force is zero and if the total torque due to external forces is zero.

Comments

(i) In an inertial frame, $\dot{\mathbf{r}}_i = 0 \Rightarrow \mathbf{F} = 0$ and $\boldsymbol{\Gamma} = 0$. The converse of this result is: $\mathbf{F} = 0$ and $\boldsymbol{\Gamma} = 0$, together with the initial conditions $\dot{\mathbf{R}} = 0$ and $\boldsymbol{\omega} = 0$ at $t = 0$, $\Rightarrow \dot{\mathbf{R}} = 0$ and $\boldsymbol{\omega} = 0$ (and hence $\dot{\mathbf{r}}_i = 0$) for all t.

(ii) The origin O to which $\boldsymbol{\Gamma}$ is referred can be any point in the inertial frame, or it could be the CM. This arbitrariness is often exploited in equilibrium calculations by making a convenient choice of O.

(iii) The equilibrium conditions (1) consist, in general, of a set of six equations. There are instances in which only some of these are satisfied: for a car accelerating along a straight, horizontal road, five of the conditions hold. In such a case, where the total force is not zero, it is the torque about the CM ($\boldsymbol{\Gamma}_C$) that is zero – the torque about an arbitrary point need not vanish.

Question 12.12

A system of coplanar forces $\mathbf{F}_i$ acting in the xy-plane has a non-zero resultant $\sum_i \mathbf{F}_i = (F_x, F_y)$ and the total torque with respect to the origin O is $\boldsymbol{\Gamma} = \sum_i \mathbf{r}_i \times \mathbf{F}_i = \Gamma \hat{\mathbf{z}}$. Equilibrium can be established by applying an appropriate force at a (non-unique) point B. (a) What force must be applied at B? (b) Show that the locus of the points B is the straight line

$$y = (F_y/F_x)x - (\Gamma/F_x). \tag{1}$$

Solution

For equilibrium the total force and the total torque must be zero. So, (a) a force $-(F_x, F_y)$ must be applied, and (b) the torque due to $-(F_x, F_y)$ applied at $\mathbf{r}_B = (x, y)$ must cancel the existing torque $\boldsymbol{\Gamma}$. That is, $-(x, y) \times (F_x, F_y) = -\Gamma \hat{\mathbf{z}}$, which is (1).

Comments

(i) The non-uniqueness of B is a consequence of the fact that $\mathbf{r}_B \times \mathbf{F} = (\mathbf{r}_B + \mathbf{r}_\|) \times \mathbf{F}$, where $\mathbf{r}_\|$ is any vector parallel to $\mathbf{F}$.

(ii) The locus is $x = \Gamma/F_y$ if $F_x = 0$; $y = -\Gamma/F_x$ if $F_y = 0$; and $y = xF_y/F_x$ if $\Gamma = 0$.

Question 12.13

A uniform ladder of length 2ℓ and weight w leans in a vertical plane against a wall at an angle θ to the horizontal. The floor is rough with coefficient of static friction μ. A person of weight W stands on the ladder at a distance D from its base. The reaction forces are $\mathbf{P}$ at the wall and $\mathbf{Q}$ at the floor.

(a) If the wall is frictionless, determine $\mathbf{P}$ and $\mathbf{Q}$ in terms of W, w, ℓ, D and θ. Hence, deduce that the person can climb a maximum distance

$$D_{\max} = \{2\mu(1 + w/W)\tan\theta - w/W\}\ell \tag{1}$$

up the ladder before it slips.

(b) If the wall is rough, explain why the reaction forces $\mathbf{P}$ and $\mathbf{Q}$ cannot be determined by the conditions for static equilibrium alone.

Solution

(a) The external forces acting on the ladder are the weights W and w and the reactions $\mathbf{P}$ and $\mathbf{Q}$. Because the wall is frictionless, $\mathbf{P}$ has only a horizontal component. The conditions for equilibrium are first that the total external force must be zero:

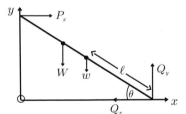

$$Q_x = P_x, \qquad Q_y = W + w, \tag{2}$$

and second that the total torque about any point must be zero. It is convenient to take the torque about the base of the ladder. Then, for equilibrium

$$P_x 2\ell \sin\theta - WD\cos\theta - w\ell\cos\theta = 0, \tag{3}$$

and so

$$P_x = \tfrac{1}{2}(w + WD/\ell)\cot\theta. \tag{4}$$

Equations (2) and (4) are the desired expressions for the components of $\mathbf{P}$ and $\mathbf{Q}$. When $D = D_{\max}$ the static frictional force Q_x reaches its maximum value μQ_y. It then follows from (2) and (4) that

$$\tfrac{1}{2}(w + WD_{\max}/\ell)\cot\theta = \mu(W + w), \tag{5}$$

and therefore $D_{\max}$ is given by (1).

(b) If the wall is rough then $P_y \neq 0$ and the equilibrium conditions (2) and (4) are modified to read
$$Q_x = P_x, \qquad Q_y + P_y = W + w, \tag{6}$$
$$P_x + P_y \cot\theta = \tfrac{1}{2}(w + WD/\ell)\cot\theta. \tag{7}$$

These three equations cannot be solved for the four unknowns P_x, P_y, Q_x and Q_y. Furthermore, no additional independent equations can be obtained by evaluating the torque about any other point in space.

Comments

(i) The solution (1) for a frictionless wall is a monotonic increasing function of θ that varies between $D_{\max} = 0$ at
$$\theta_1 = \tan^{-1}\left[\frac{w}{2\mu(w+W)}\right], \tag{8}$$
and $D_{\max} = 2\ell$ at
$$\theta_2 = \tan^{-1}\left[\frac{w+2W}{2\mu(w+W)}\right]. \tag{9}$$

Note that the condition for an unloaded ladder ($W = 0$) not to slip is $\theta \geq \theta_3 = \tan^{-1}(1/2\mu)$, and according to (1) a ladder inclined at θ_3 can be climbed to $D_{\max} = \ell$ (i.e. the midpoint). Also, $\theta_1 < \theta_3$ and when $\theta_1 < \theta < \theta_3$ in (1) the ladder is on the verge of slipping when the load W is at $D_{\max} < \ell$. To achieve this, the ladder should be held in position until the person is in place.

(ii) The extension of the above analysis to a rough wall turns out to be unexpectedly challenging. It is sometimes assumed that the critical conditions for the onset of slipping are
$$P_y = \mu' P_x, \qquad Q_x = \mu Q_y, \tag{10}$$
where μ' is the coefficient of static friction of the wall. The five equations (6), (7) and (10) can be solved for the limiting reactions and the critical distance
$$D_{\max} = \left(\frac{2\mu(\mu' + \tan\theta)(1 + w/W)}{1 + \mu\mu'} - \frac{w}{W}\right)\ell. \tag{11}$$

However, there is no evident reason why the maximum values (10) should be reached together, and, in fact, the correct description is more subtle than this.

(iii) Systems in which reaction forces cannot be determined solely from the conditions for equilibrium are referred to as 'statically indeterminate'. Such systems are important in engineering, and useful references are given in Ref. [2]. It is known that the reaction forces depend on both the elasticity of the system and its history (the manner in which it is set up). For the ladder against a rough wall a

---

[2] K. S. Mendelson, "Statics of a ladder leaning against a rough wall," American Journal of Physics, vol. 63, pp. 148–150, 1995.

detailed analysis has been performed by Gonzaléz and Gratton.[3] They find that, in general, it is necessary to take account of both compression and flexion, and "that the reactions can never be determined using only statics, because there is no way to ascertain which kind of deformation dominates, and so which limiting condition will be attained, if any."

Question 12.14

A uniform, rigid rod of mass M and length 2ℓ is released from rest with its lower end in contact with a horizontal frictionless surface. At time t the rod makes an angle $\theta(t)$ with the vertical, and the initial value at $t = 0$ is θ_0.

(a) Show that at time t the point of contact with the surface has moved a distance

$$d(t) = \ell(\sin\theta - \sin\theta_0). \tag{1}$$

(b) Show that the rotational equation of motion yields the following differential equation for $\theta(t)$:

$$\frac{\ell}{g}\left(\frac{1}{3} + \sin^2\theta\right)\ddot\theta + \frac{\ell}{g}\dot\theta^2\cos\theta\sin\theta = \sin\theta. \tag{2}$$

Do this for each of the following choices of reference point for calculating torque:
1. the CM, 2. the point of contact between the rod and the surface.

(c) Deduce from (2) that

$$\dot\theta^2 = \frac{6g}{\ell}\frac{\cos\theta_0 - \cos\theta}{1 + 3\sin^2\theta} \tag{3}$$

and show that the force that the rod exerts on the surface is

$$N(\theta) = \frac{4 + 3\cos^2\theta - 6\cos\theta_0\cos\theta}{(4 - 3\cos^2\theta)^2}Mg. \tag{4}$$

(d) Use (3) to show that the time taken for the CM to reach the plane is given by

$$\tau_f = \frac{1}{\sqrt{6}}\int_{\theta_0}^{\pi/2}\sqrt{\frac{1 + 3\sin^2\theta}{\cos\theta_0 - \cos\theta}}\,d\theta, \tag{5}$$

where $\tau_f = \sqrt{\dfrac{g}{\ell}}\,t_f$ is a dimensionless time.

---

[3] A. G. González and J. Gratton, "Reaction forces on a ladder leaning on a rough wall," American Journal of Physics, vol. 64, pp. 1001–1005, 1996.

Solution

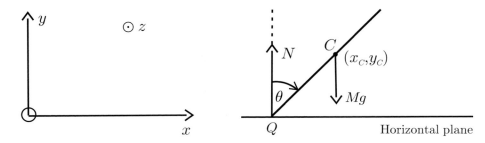

(a) In the diagram, C is the CM and Q is the (moving) point of contact with the surface. The only external forces acting on the rod are its weight Mg and the reaction N from the plane. $Oxyz$ is an inertial frame at rest with respect to the plane. The motion is parallel to the xy-plane and therefore planar. Because there is no horizontal force and the rod is released from rest, the x-coordinate of the CM is constant: $x_C = \text{constant}$. Also, $x_Q(t) = x_C - \ell\sin\theta(t)$. Therefore, Q moves a distance $x_Q(0) - x_Q(t) = \ell\sin\theta - \ell\sin\theta_0$ in a time t.

(b) 1. The y-component of the translational equation of motion of the CM, $M\ddot{\mathbf{R}} = \mathbf{F}$, is
$$M\ddot{y}_C = N - Mg, \qquad \text{where} \quad y_C = \ell\cos\theta. \tag{6}$$
That is,
$$-M\ell\ddot{\theta}\sin\theta - M\ell\dot\theta^2\cos\theta = N - Mg. \tag{7}$$
Also, the rotational equation of motion $I_C\ddot\theta = \Gamma_C$ for planar motion about the CM (see Question 12.5) gives
$$I_C\ddot\theta = N\ell\sin\theta, \tag{8}$$
where $I_C = \tfrac{1}{3}M\ell^2$ is the moment of inertia of the rod about a perpendicular axis (Cz) through the CM. Eliminating N between (7) and (8) gives (2).

2. The point of contact Q accelerates relative to the inertial frame $Oxyz$ and so the rotational equation of motion is given by (5) of Question 12.2:
$$\dot{\mathbf{L}}_Q = \mathbf{\Gamma}_Q - M(\mathbf{R} - \mathbf{r}_Q)\times \ddot{\mathbf{r}}_Q. \tag{9}$$
Now
$$\dot{\mathbf{L}}_Q = -I_Q\ddot\theta\hat{\mathbf{z}} \qquad \text{and} \qquad \mathbf{\Gamma}_Q = -Mg\ell\sin\theta\hat{\mathbf{z}}, \tag{10}$$
where $I_Q = I_C + M\ell^2 = \tfrac{4}{3}M\ell^2$ (by the parallel-axis theorem of Question 12.7). Also, $\mathbf{R} = x_C\hat{\mathbf{x}} + y_C\hat{\mathbf{y}}$ and $\mathbf{r}_Q = (x_C - \ell\sin\theta)\hat{\mathbf{x}}$, where x_C is a constant. So
$$\ddot{\mathbf{r}}_Q = \ell(\dot\theta^2\sin\theta - \ddot\theta\cos\theta)\hat{\mathbf{x}} \tag{11}$$
and
$$(\mathbf{R}-\mathbf{r}_Q)\times\ddot{\mathbf{r}}_Q = \ell^2(\ddot\theta\cos^2\theta - \dot\theta^2\cos\theta\sin\theta)\hat{\mathbf{z}}. \tag{12}$$
Equations (9), (10) and (12) also yield (2).

(c) The identity

$$\frac{d}{d\theta}\left\{(1+3\sin^2\theta)\left(\frac{d\theta}{dt}\right)^2\right\} = 2(1+3\sin^2\theta)\ddot{\theta} + 6\dot{\theta}^2\cos\theta\sin\theta \qquad (13)$$

enables us to write (2) as

$$\frac{d}{d\theta}\left\{(1+3\sin^2\theta)\left(\frac{d\theta}{dt}\right)^2\right\} = \frac{6g}{\ell}\sin\theta. \qquad (14)$$

By integrating both sides of (14) with respect to θ between θ_0 and $\theta(t)$, and setting $\dot{\theta} = 0$ at $t = 0$, we obtain (3).

The point of contact Q is in vertical equilibrium, and therefore the force that the rod exerts on the surface is equal in magnitude to the reaction $N(\theta)$. The latter can be calculated by first using (8) to eliminate $\ddot{\theta}$ from (7):

$$-3N\sin^2\theta - M\ell\dot{\theta}^2\cos\theta = N - mg, \qquad (15)$$

and then substituting (3) in (15), and solving for N. The result is (4).

(d) Solve (3) for $\dot{\theta}$ and then integrate with respect to t. This yields (5).

Comments

(i) In the calculations leading to (1), both of the frames used – the CM frame and the frame with origin at Q – are non-inertial. However, there is a striking difference between the two calculations: the CM frame is special in that there is no contribution due to non-inertiality (the term involving $\ddot{\mathbf{r}}_Q$ in (9) vanishes when $\mathbf{r}_Q = \mathbf{R}$) whereas for the point of contact Q there is such a contribution – cf. (12).

(ii) The relation (3), which was deduced by integrating the equation of motion (2), can also be obtained directly from conservation of energy:

$$\tfrac{1}{2}M\dot{y}_C^2 + \tfrac{1}{2}I_C\dot{\theta}^2 + Mgy_C = Mg\ell\cos\theta_0, \qquad (16)$$

with $y_C = \ell\cos\theta$ and $I_C = \tfrac{1}{3}M\ell^2$.

Question 12.15

Use *Mathematica* to obtain numerical solutions of $\tau_f(\theta_0)$ and $\theta(\tau)$ for the sliding rod of Question 12.14, and to plot graphs of the following:

☞ $\tau_f(\theta_0)$ for $\pi/120 \leq \theta_0 \leq \pi/2$.

☞ $\theta(\tau)$, $d(\tau)/\ell$, $N(\tau)/Mg$ versus τ for $0 \leq \tau \leq \tau_f$ and $\theta_0 = \pi/3$, $\pi/4$ and $\pi/6$. (For τ_f use (5); for $\theta(\tau)$ use (2); for $d(\tau)$ use (1); and for $N(\tau)$ use (4) of Question 12.14.)

Solution

We obtain the following graphs using the *Mathematica* notebook below:

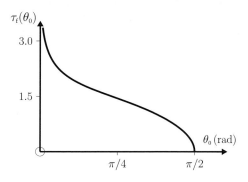
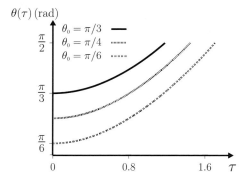
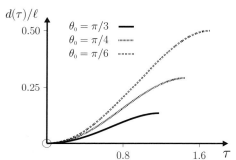
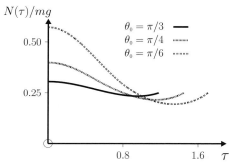

```
In[1]:= θmin = π/120;  θmax = π/2;

       τmax[θ0_] := (1/√6) Chop[NIntegrate[√((1 + 3 Sin[θ]^2)/(Cos[θ0] - Cos[θ])), {θ, θ0, π/2}], 10^-6];

       Plot[τmax[θ0], {θ0, θmin, θmax}]

       θ0 = π/6;  θ0dot = 0;

       Sol = NDSolve[{((1/3) + Sin[θ[τ]]^2) θ''[τ] + (θ'[τ])^2 Cos[θ[τ]] Sin[θ[τ]]
              - Sin[θ[τ]] == 0, θ[0] == θ0, θ'[0] == θ0dot}, {θ[τ]},
              {τ, 0, τmax[θ0]}, MaxSteps -> 10000];

       Plot[Evaluate[θ[τ]/.Sol], {τ, 0, τmax[θ0]}]

       Plot[Evaluate[(Sin[θ[τ]] - Sin[θ0])/.Sol], {τ, 0, τmax[θ0]}]

       Plot[Abs[Evaluate[(4 + 3 Cos[θ[τ]]^2 - 6 Cos[θ0] Cos[θ[τ]])/((4 - 3 Cos[θ[τ]]^2)^2)/.Sol]],
              {τ, 0, τmax[θ0]}]
```

Question 12.16

A rigid, axially symmetric wheel has mass M, radius a and moment of inertia I about its axis (i.e. the CM). The wheel is spun about this axis with constant angular speed ω_0, and is then released in an upright position on a horizontal plane. It slips for a time τ and then rolls without slipping. The coefficient of kinetic friction between the wheel and the plane is μ. Show that

$$\tau = \frac{\omega_0 a}{\mu g(1 + Ma^2/I)}, \tag{1}$$

and determine the CM speed of the rolling wheel.

Solution

While the wheel is slipping there is a frictional force μMg acting as shown below. This causes a horizontal acceleration μg of the CM that therefore acquires a horizontal speed

$$v = \mu g t \quad (t \leq \tau). \tag{2}$$

The frictional force also produces a torque μMga about the CM and therefore an angular deceleration $\dot{\omega} = -\mu Mga/I$. So

$$\omega(t) = \omega_0 - \mu Mgat/I \quad (t \leq \tau). \tag{3}$$

Therefore, ω decreases while the wheel is slipping until it reaches a value $\omega = v/a$, which is the condition for rolling without slipping. Equating (2) $\div a$ to (3) and setting $t = \tau$, gives (1). For $t > \tau$ the point of contact between the wheel and the surface is instantaneously at rest. So, the wheel rolls without slipping: the horizontal force is zero and $v(t)$ is a constant $= v(\tau)$. From (1) and (2) we have

$$v(t) = v(\tau) = \frac{\omega_0 a}{(1 + Ma^2/I)} \quad (t > \tau). \tag{4}$$

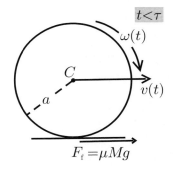

Comments

(i) The graphs of $F(t)$, $v(t)$ and $\omega(t)/a$ versus t are:

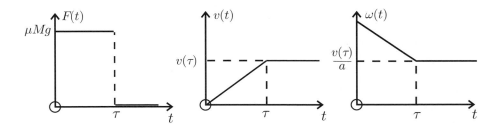

(ii) The speed (4) of the rolling wheel is independent of the coefficient of kinetic friction μ. In the above calculation this arises because v is proportional to μ and t – see (2) – while $t = \tau$ is inversely proportional to μ – see (1).

(iii) Equation (4) can also be obtained from conservation of angular momentum, using the initial point of contact O with the plane as origin. The initial angular momentum about O is $I\omega_0$ (being entirely spin angular momentum about the CM) and the final angular momentum is the sum of spin angular momentum $I\omega(t)$ and orbital angular momentum $Mv(t)a$ about O (see Question 11.3). The frictional force passes through O and therefore the angular momentum about O is conserved:

$$I\omega(t) + Mv(t)a = I\omega_0 . \tag{5}$$

For $t \geq \tau$ the wheel rolls without slipping: then $\omega = v/a$ and (5) yields (4).

(iv) A survey has been published of the response of both physics professors and students to solving the above question, and the results make for interesting reading.[4]

Question 12.17

Suppose that the wheel in Question 12.16 has an initial horizontal CM speed v_0 in addition to an initial topspin ω_0. Determine the instant τ at which the wheel rolls without slipping and the CM speed $v(\tau)$ of the rolling wheel.

Solution

Here, it is important to take into account that the direction of the frictional force during slipping depends on which of $\omega_0 a$ and v_0 is larger, as indicated in the figure. Consequently, (2) and (3) of Question 12.16 are replaced by

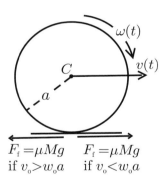

$$v(t) = v_0 \mp \mu g t \quad \text{and} \quad \omega(t) = \omega_0 \pm \frac{\mu M g a t}{I} \quad (t \leq \tau), \tag{1}$$

where the upper sign applies if $v_0 > \omega_0 a$ and the lower sign if $v_0 < \omega_0 a$. At the instant $t = \tau$ when pure rolling commences, we have $v(\tau) = a\omega(\tau)$, and (1) gives

$$\tau = \frac{|\omega_0 a - v_0|}{\mu g (1 + Ma^2/I)} . \tag{2}$$

Equations (1) and (2) yield the constant speed for $t \geq \tau$:

$$v(t) = v(\tau) = \frac{\omega_0 a + v_0 Ma^2/I}{1 + Ma^2/I} \quad (t \geq \tau). \tag{3}$$

---

[4] C. Singh, "When physical intuition fails," American Journal of Physics, vol. 70, pp. 1103–1109, 2002.

Comments

(i) The speed (3) is independent of μ and, just as in Question 12.16, this is associated with conservation of angular momentum about the initial point of contact with the plane:
$$I\omega(t) + Mv(t)a = I\omega_0 + Mv_0 a. \qquad (4)$$
When pure rolling starts, $\omega = v/a$ and (4) yields (3).

(ii) The initial and final kinetic energies are
$$K_i = \tfrac{1}{2}Mv_0^2 + \tfrac{1}{2}I\omega_0^2 \quad \text{and} \quad K_f = \tfrac{1}{2}Mv^2 + \tfrac{1}{2}I\omega^2, \qquad (5)$$
with $\omega = v/a$. From (3) and (5) we find a loss in kinetic energy
$$K_i - K_f = \tfrac{1}{2}M\frac{(\omega_0 a - v_0)^2}{1 + Ma^2/I}, \qquad (6)$$
which is equal to the work done against friction during slipping.

Question 12.18

Suppose that the wheel in Question 12.16 has an initial horizontal CM speed v_0 and an initial backspin $-\omega_0$. Determine the possible motions of the wheel. Illustrate your answer by sketching graphs of the CM speed $v(t)$ and $a\omega(t)$ versus t.

Solution

The wheel slips for some time τ and then rolls without slipping. For $t < \tau$ the translational and rotational equations of motion are $M\dot{v} = -\mu Mg$ and $I\dot{\omega} = \mu Mga$. These, together with the initial conditions $v(0) = v_0$ and $\omega(0) = -\omega_0$, give

$$v(t) = v_0 - \mu g t, \quad \omega(t) = -\omega_0 + \frac{\mu M g a t}{I} \quad (t \leq \tau). \qquad (1)$$

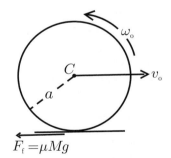

The value of τ is determined from (1) by setting $v(\tau) = a\omega(\tau)$:

$$\tau = \frac{(1 + \omega_0 a/v_0)}{(1 + Ma^2/I)} \frac{v_0}{\mu g}. \qquad (2)$$

For $t > \tau$, the frictional force is zero: consequently, $v(t)$ and $\omega(t)$ are constant and equal to $v(\tau)$ and $\omega(\tau)$:

$$v(t) = a\omega(t) = v(\tau) = \frac{(v_0 Ma^2/I) - \omega_0 a}{1 + Ma^2/I} \quad (t \geq \tau). \qquad (3)$$

It is apparent that there are three cases to consider, according to whether $v(\tau)$ is positive, zero or negative:

1. $v(\tau) > 0$ (i.e. $v_\circ > I\omega_\circ/Ma$) The wheel always moves to the right. The spin changes sign at time $I\omega_\circ/\mu Mga < \tau$.

2. $v(\tau) = 0$ (i.e. $v_\circ = I\omega_\circ/Ma$) The wheel comes to rest ($v = 0$ and $\omega = 0$) at time $\tau = v_\circ/\mu g$, and remains stationary. The distance travelled is $x(\tau) = v_\circ^2/2\mu g$.

3. $v(\tau) < 0$ (i.e. $v_\circ < I\omega_\circ/Ma$) The spin does not change sign. Instead, the wheel reverses direction (starts moving to the left) at time $v_\circ/\mu g < \tau$, when $x = v_\circ^2/2\mu g$. It passes its starting point after a time $2v_\circ/\mu g$ and continues moving to the left.

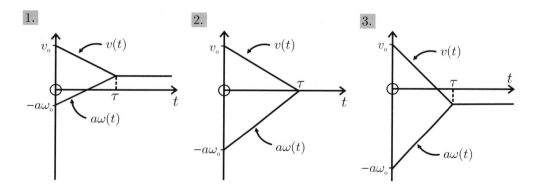

Comments

(i) These results apply to any axially symmetric wheel or spherically symmetric sphere or shell. One should simply use the appropriate moment of inertia I.

(ii) For the wheel to be rolling when it returns to its starting point (case 3) we require $\tau < 2v_\circ/\mu g$, meaning that $a\omega_\circ < (1 + 2Ma^2/I)v_\circ$.

Question 12.19

A thin uniform rod AB of mass M and length 2ℓ is placed with the end A against the base of a wall and inclined at an angle $\theta_\circ$ to the vertical. A point mass m is attached to the end B and the rod is released from rest. Let $\tau(m)$ denote the time taken for the rod to reach the ground, and $\tau(0)$ its value when the mass m is removed. Show that

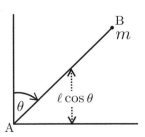

$$\tau(m) = \sqrt{\frac{M + 3m}{M + 2m}}\, \tau(0). \qquad (1)$$

Solution

This question can be solved using conservation of energy:

$$Mg\ell \cos\theta + mg2\ell \cos\theta + \tfrac{1}{2}I_A\dot\theta^2 + \tfrac{1}{2}m(2\ell\dot\theta)^2 = Mg\ell \cos\theta_0 + mg2\ell \cos\theta_0, \qquad (2)$$

where I_A is the moment of inertia of the rod about A. (The first two terms in (2) are the potential energies of the rod and mass m, and the next two terms are their kinetic energies.) From the parallel-axis theorem (see Question 12.7),

$$I_A = I_C + M\ell^2 = \tfrac{4}{3}M\ell^2 \qquad (3)$$

because $I_C = \tfrac{1}{3}M\ell^2$. After substituting (3) in (2), we solve for the angular speed $d\theta/dt$ of the system:

$$\frac{d\theta}{dt} = \sqrt{\frac{3g}{2\ell}\frac{M+2m}{M+3m}\frac{1}{(\cos\theta_0 - \cos\theta)}}. \qquad (4)$$

If $dt(m)$ denotes the time for the rod plus mass m to fall through an angle $d\theta$, and $dt(0)$ the corresponding time when $m=0$, then according to (4)

$$dt(m) = \sqrt{\frac{M+3m}{M+2m}}\, dt(0), \qquad (5)$$

and (1) follows by integration.

Comments

(i) The effect (1) is easily observed, for example by using two metre sticks, one of which is weighted. It is noticeable that the unweighted stick reaches the ground first. This effect is largest for $m \gg M$, when it amounts to about 23%.

(ii) We can apply (2) to a non-uniform rod or lamina. Here, it is helpful to write $I_A = Mk^2$, where k is the so-called radius of gyration ($=\sqrt{\tfrac{4}{3}}\,\ell$ for a uniform rod). Then, instead of (1), equation (2) yields

$$\tau(m) = \sqrt{\frac{M + 4m\ell^2/k^2}{M+2m}}\,\tau(0). \qquad (6)$$

It is apparent that the difference in falling times can be increased by making k small – that is, by concentrating the mass M near A. We also see that $\tau(m) > \tau(0)$ provided $k < \sqrt{2}\,\ell$, and that the maximum effect is now $\approx \sqrt{2}\,\ell/k$ when $m \gg M$.

Question 12.20

A round, rigid object (such as a cylinder or sphere whose densities are axially or spherically symmetric, respectively) has mass M, radius a and moment of inertia I (about its axis or centre). The object is released from rest on a plane that is inclined at an angle α to the horizontal and for which the coefficient of static friction is μ_s.

(a) Suppose the object rolls without slipping. Determine the acceleration of the CM and the required frictional force in terms of M, g, a, α and I. Deduce that there is a critical angle of inclination

$$\alpha_c = \tan^{-1}(1 + Ma^2/I)\mu_s, \qquad (1)$$

below which there is pure rolling and above which there is slipping.

(b) Suppose $\alpha > \alpha_c$, so that slipping occurs. Determine the acceleration of the CM and the angular acceleration, and then deduce a relation between the CM velocity and the angular velocity.

(c) Show that the mechanical energy of the object is constant during rolling but decreases during slipping.

Solution

The motion is planar and can be described by two coordinates: the distance x that the centre of mass C moves and the angle θ through which the object rotates about C. The forces acting are the weight Mg and the reaction from the plane, which consists of a normal component N and the friction F_f. There is no motion perpendicular to the plane and therefore

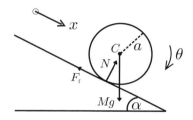

$$N = Mg\cos\alpha. \qquad (2)$$

The equations of motion for translation of C and rotation about C are

$$M\ddot{x} = Mg\sin\alpha - F_f \qquad (3)$$

$$I\dot\omega = F_f a \qquad (\omega = \dot\theta). \qquad (4)$$

(a) For rolling without slipping we must have $\dot x = a\omega$. Then, (3) and (4) can be solved for the two unknowns $\ddot x$ and F_f:

$$\ddot x = \frac{g\sin\alpha}{1 + I/Ma^2} \qquad (5)$$

$$F_f = \frac{Mg\sin\alpha}{1 + Ma^2/I}. \qquad (6)$$

Equation (6) gives the frictional force that is required to prevent the object from slipping. Now, F_f cannot exceed $\mu_s N$, where μ_s is the coefficient of static friction (static because the instantaneous point of contact with the surface is at rest). So, it follows from (2) and (6) that there is a critical value of α, given by

$$\tan\alpha_c = (1 + Ma^2/I)\mu_s, \qquad (7)$$

below which there is pure rolling and above which there is slipping.

(b) When slipping occurs the frictional force is
$$F_\text{f} = \mu N = \mu Mg\cos\alpha, \tag{8}$$
where μ is the coefficient of kinetic friction. Then, (3) and (4) can be solved for the two unknown accelerations $\ddot{x}$ and $\dot{\omega}$:
$$\ddot{x} = g(\sin\alpha - \mu\cos\alpha) \tag{9}$$
$$\dot{\omega} = (\mu Mga/I)\cos\alpha. \tag{10}$$

The object starts from rest and so the linear and angular velocities $\dot{x}$ and ω are equal to t times the constant accelerations (9) and (10), respectively. Therefore,
$$\dot{x} = \beta a\omega, \quad \text{where } \beta = (I/\mu Ma^2)(\tan\alpha - \mu). \tag{11}$$

For slipping, $\dot{x} > a\omega$ and so $\beta > 1$. The limit $\dot{x} \to a\omega$ marks the onset of pure rolling, and (11) with $\mu = \mu_\text{s}$ yields (1).

(c) The mechanical energy of the object,
$$E = \tfrac{1}{2}M\dot{x}^2 + \tfrac{1}{2}I\omega^2 - Mgx\sin\alpha, \tag{12}$$
has a rate of change
$$\dot{E} = M\dot{x}\ddot{x} + I\omega\dot{\omega} - Mg\dot{x}\sin\alpha. \tag{13}$$

During rolling, $\dot{x} = a\omega$ and also (5) holds. Consequently, $\dot{E} = 0$: even though there is friction, the point of contact with the plane is always at rest and therefore the frictional force F_f exerted by the plane does no work. During slipping, (9)–(11) and (13) show that
$$\dot{E} = \{(\mu^2 M^2 g^2 a^2/I)(1-\beta)\cos^2\alpha\}t, \tag{14}$$
which is negative because $\beta > 1$. The energy decreases quadratically with t.

Comments

(i) According to (5), for two objects that have the same mass M and radius a, but different moments of inertia I, the object with the smaller I has the larger acceleration during rolling. This is the basis for a well-known demonstration involving a race down an inclined plane between two metal cylinders having identical external dimensions and mass. The first is a uniform aluminium cylinder. The second is a hollow aluminium cylinder of the same dimensions, with a steel liner attached to its inner surface to achieve the required mass. The first cylinder always wins the race, which is just what one would expect from (5) because $I_1 < I_2$.

(ii) A race in which both cylinders slip should result in a tie – see (9), which is independent of I.

(iii) Slipping starts at a lower critical angle for the object with larger I (see (1)).

(iv) If we make $I \ll Ma^2$ (by concentrating high-density material near the axis) then $\alpha_\text{c} \to \tfrac{1}{2}\pi$ and the acceleration (5) of a rolling object is increased, and approaches the value $g\sin\alpha$ attained during sliding on a frictionless plane.

Question 12.21

A spool is constructed from two identical cylinders of radius a connected by a cylindrical axle of radius b ($< a$). The spool is placed at rest on a horizontal plane and a constant external force T, making an angle α with the horizontal, is applied by pulling on an inextensible string wound around the axle. Determine the possible responses of the spool that involve pure rolling for $0 \leq \alpha \leq \frac{1}{2}\pi$. Include a calculation of the maximum CM acceleration in pure rolling. (Neglect the mass of the string.)

Solution

The motion is planar and described by two coordinates: the horizontal distance x moved by the centre of mass C and the angle θ through which the spool rotates. The equations of motion for horizontal motion of C and rotation about C are

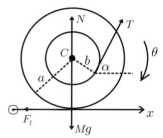

$$M\ddot{x} = T\cos\alpha - F_{\mathrm{f}} \quad (1)$$
$$I\dot{\omega} = F_{\mathrm{f}}a - Tb, \quad (2)$$

where $\omega = \dot{\theta}$ and I is the moment of inertia of the spool about its axis. For pure rolling, F_{f} is the force of static friction. Also, $\dot{x} = a\omega$, and (1) and (2) can be solved for $\ddot{x}$ and F_{f}:

$$\ddot{x} = \frac{T/M}{1 + I/Ma^2}\left(\cos\alpha - b/a\right) \quad (3)$$

$$F_{\mathrm{f}} = \frac{T}{1 + I/Ma^2}\left(\frac{I}{Ma^2}\cos\alpha + b/a\right). \quad (4)$$

(Note that F_{f} is positive if $0 \leq \alpha \leq \frac{1}{2}\pi$, meaning that friction acts to the left as drawn in the diagram.) It follows from (3) that there exists a critical angle of inclination

$$\alpha_{\mathrm{c}} = \cos^{-1} b/a, \quad (5)$$

which distinguishes two types of response. If $\alpha < \alpha_{\mathrm{c}}$ (that is, $\cos\alpha > b/a$) then $\ddot{x} > 0$: the spool moves to the right. But if $\alpha > \alpha_{\mathrm{c}}$ (that is, $\cos\alpha < b/a$) then $\ddot{x} < 0$ and the spool moves to the left. (In this case the frictional force is in the direction of motion of the CM.) When $\alpha = \alpha_{\mathrm{c}}$, both $\ddot{x}$ and $\dot{\omega}$ are zero and the spool remains at rest – a force applied at this angle cannot cause the spool to roll without slipping. Equation (4) gives the frictional force required to prevent slipping. Because $F_{\mathrm{f}} \leq \mu_{\mathrm{s}} N = \mu_{\mathrm{s}}(Mg - T\sin\alpha)$, it follows from (4) that pure rolling occurs up to a maximum tension

$$T_{\max} = \frac{\mu_{\mathrm{s}}(1 + I/Ma^2)Mg}{(I/Ma^2)\cos\alpha + \mu_{\mathrm{s}}(1 + I/Ma^2)\sin\alpha + b/a}. \quad (6)$$

For $T > T_{\max}$ slipping will occur. From (3) and (6) the maximum CM acceleration of a spool in pure rolling is

$$\ddot{x}_{\max} = \frac{\mu_{\mathrm{s}}(\cos\alpha - b/a)g}{(I/Ma^2)\cos\alpha + \mu_{\mathrm{s}}(1 + I/Ma^2)\sin\alpha + b/a}. \quad (7)$$

Comments

(i) There is a simple geometric interpretation of α_c. When $\alpha = \alpha_c$ the line of action of T passes through the point of contact P between the spool and the plane: T therefore exerts no torque about P. When $\alpha < \alpha_c$ this line passes to the left of P, and T exerts a clockwise torque about P; while if $\alpha > \alpha_c$ the line passes to the right of P and the torque is counter-clockwise.

(ii) Below α_c the horizontal component $T\cos\alpha$ of the applied force is greater than the frictional force F_f given by (4), and the spool is accelerated to the right. The opposite is true above α_c.

(iii) The friction (4) exerted during rolling reduces to zero at an angle α_0 for which

$$\cos\alpha_0 = -Mab/I. \qquad (8)$$

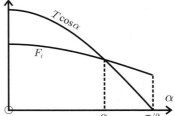

This angle exists (and lies between $\tfrac{1}{2}\pi$ and π) if $I \geq Mab$: that is, if the radius of gyration $k \geq \sqrt{ab}$. When $\alpha = \alpha_0$ and $T < T_{\max}$ there is frictionless pure rolling along the negative x-axis ($\ddot{x}$ is negative, see (1)). If $\alpha = \alpha_0$ and $T = T_{\max}$ then $\mu_s N = F_f = 0$. That is, $N = 0$: the weight Mg is supported by the vertical component $T_{\max}\sin\alpha_0$. For $\alpha > \alpha_0$, F_f is negative, meaning that friction acts to the right in the first diagram above.

(iv) If the tension T is created by passing the string over a frictionless pulley and connecting the free end to a device that exerts a constant force, then the tension will always be directed to a fixed point (the pulley). It is not difficult to show that as a result the spool will perform an anharmonic oscillatory motion (which becomes harmonic for small oscillations).[5]

Question 12.22

Consider rotational motion of an axially symmetric, homogeneous rigid body in the absence of external forces and relative to an inertial frame with origin at the CM. The total angular momentum $\mathbf{L}$ and the angular velocity $\boldsymbol{\omega}$ are not necessarily parallel.

(a) Let $\mathbf{n}$ be a unit vector along the axis of symmetry. Prove that if $\boldsymbol{\omega}$ is not along $\mathbf{n}$ then $\boldsymbol{\omega}$, $\mathbf{L}$ and $\mathbf{n}$ are coplanar. (Hint: Start with $\mathbf{L}$ and $\boldsymbol{\omega}$ expressed in terms of principal axes, and recall that a symmetry axis is a principal axis – see Question 12.6.)

(b) Hence, give a geometric description of the motion relative to the inertial frame.[6]

---

[5] C. Carnero, P. Carpena, and J. Aguiar, "The rolling body paradox: an oscillatory motion approach," European Journal of Physics, vol. 18, pp. 409–416, 1997.

[6] E. Butikov, "Inertial rotation of a rigid body," European Journal of Physics, vol. 27, pp. 913–922, 2006.

Solution

(a) In principal axes having unit vectors $\mathbf{e}_1$, $\mathbf{e}_2$, $\mathbf{e}_3$ the total angular momentum is

$$\mathbf{L} = I_1\omega_1\mathbf{e}_1 + I_2\omega_2\mathbf{e}_2 + I_3\omega_3\mathbf{e}_3 , \qquad (1)$$

and $\boldsymbol{\omega} = \omega_1\mathbf{e}_1 + \omega_2\mathbf{e}_2 + \omega_3\mathbf{e}_3$. For a homogeneous, axially symmetric body the axis of symmetry $\mathbf{n} = n_1\mathbf{e}_1 + n_2\mathbf{e}_2 + n_3\mathbf{e}_3$ coincides with a principal axis (i.e. two of the n_i are zero). This means that

$$\mathbf{n} \cdot (\boldsymbol{\omega} \times \mathbf{L}) = n_1\omega_2\omega_3(I_3 - I_2) + n_2\omega_1\omega_3(I_1 - I_3) + n_3\omega_1\omega_2(I_2 - I_1) \qquad (2)$$

is always zero: for example, if $n_3 = 1$ then $n_1 = n_2 = 0$ and $I_2 = I_1$. Therefore, $\boldsymbol{\omega}$, $\mathbf{L}$ and $\mathbf{n}$ are coplanar (unless $\boldsymbol{\omega}$ is along $\mathbf{n}$, in which case they are collinear).

(b) In the absence of external forces $\mathbf{L}$ is conserved.

1. Consider first the simplest possibility – that $\boldsymbol{\omega}$ is along the symmetry axis $\mathbf{n}$ (or, in general, along any of the principal axes). Then, $\boldsymbol{\omega}$ and $\mathbf{L}$ are proportional, and $\boldsymbol{\omega}$ is also conserved (in both magnitude and direction): relative to an inertial frame with origin at the CM the body spins around a fixed axis with constant angular speed ω. A principal axis is also an axis of free rotation. This is true even if the body is not axially symmetric.

2. If $\boldsymbol{\omega}$ is not along the symmetry axis then $\mathbf{L}$ deviates from $\boldsymbol{\omega}$ and the motion is more complicated. To analyze it we use the result in (a). First note that there is an ordering of the vectors $\boldsymbol{\omega}$, $\mathbf{L}$ and $\mathbf{n}$ in their common plane. To see this, let $I_{\parallel}$ be the moment of inertia about the symmetry axis and $I_{\perp}$ be the moment of inertia about a transverse axis. If $I_{\perp} > I_{\parallel}$ (an elongated or prolate body) then $\mathbf{L}$ given by (1) deviates from $\mathbf{n}$ by a larger amount than does $\boldsymbol{\omega}$; and vice versa if $I_{\perp} < I_{\parallel}$ (an oblate body).[‡] This is illustrated in the following two diagrams, where $Oxyz$ is an inertial frame with O at the CM. We emphasize that $\mathbf{L}$ is a conserved (fixed) vector and for convenience we have drawn it vertically. (The body has been drawn schematically.)

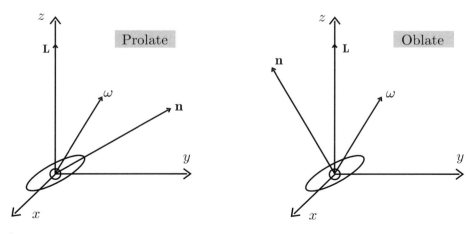

[‡] For example, if $\mathbf{n} = \mathbf{e}_3$ then $\boldsymbol{\omega} = \omega_1\mathbf{e}_1 + \omega_2\mathbf{e}_2 + \omega_3\mathbf{n}$ and $\mathbf{L} = I_{\perp}(\omega_1\mathbf{e}_1 + \omega_2\mathbf{e}_2) + I_{\parallel}\omega_3\mathbf{n}$.

We now resolve $\boldsymbol{\omega}$ into components $\boldsymbol{\omega}_s$ and $\boldsymbol{\omega}_p$ along $\mathbf{n}$ and $\mathbf{L}$, respectively, and also into components $\boldsymbol{\omega}_\parallel$ and $\boldsymbol{\omega}_\perp$ along and perpendicular to $\mathbf{n}$. This is illustrated for a prolate body in the first of the two diagrams below. The resolution $\boldsymbol{\omega} = \boldsymbol{\omega}_s + \boldsymbol{\omega}_p$ and the fact that $\boldsymbol{\omega}$, $\mathbf{L}$ and $\mathbf{n}$ are always coplanar shows that the motion consists of two superimposed rotations: spin with angular velocity $\boldsymbol{\omega}_s$ about the symmetry axis $\mathbf{n}$ and rotation of $\mathbf{n}$ about the fixed vector $\mathbf{L}$ with angular velocity $\boldsymbol{\omega}_p$. The unit vector $\mathbf{n}$ traces out a cone centred on $\mathbf{L}$ and with vertex at O – we say that it precesses about $\mathbf{L}$. The resolution $\boldsymbol{\omega} = \boldsymbol{\omega}_\parallel + \boldsymbol{\omega}_\perp$ enables us to calculate the precessional angular velocity $\boldsymbol{\omega}_p$ by simple geometry (similar triangles in the diagram): $\mathbf{L}/L_\perp = \boldsymbol{\omega}_p/\boldsymbol{\omega}_\perp$. Now, $L_\perp = I_\perp \omega_\perp$ and so

$$\boldsymbol{\omega}_p = \mathbf{L}/I_\perp. \tag{3}$$

Thus, $\boldsymbol{\omega}_p$ is a constant vector. Therefore, the component $\omega_\perp$ is also constant and, by conservation of $\mathbf{L}$, so is $\omega_\parallel$. That is, the magnitude ω is constant, as is ω_s. The vector $\boldsymbol{\omega}$ rotates about $\mathbf{L}$ with constant angular speed ω_p (to remain coplanar with $\mathbf{L}$ and $\mathbf{n}$), tracing out a cone with axis along $\mathbf{L}$ and vertex at O. Thus, both $\mathbf{n}$ and $\boldsymbol{\omega}$ precess uniformly about $\mathbf{L}$. The corresponding diagram representing the decomposition $\boldsymbol{\omega} = \boldsymbol{\omega}_s + \boldsymbol{\omega}_p$ for an oblate body shows that $\boldsymbol{\omega}_s$ points in the opposite direction to $\mathbf{n}$. Consequently, when $\boldsymbol{\omega}$ is almost along $\mathbf{n}$ the precession and spin are in opposite senses. For both types of body and for small deviations of $\boldsymbol{\omega}$ (and hence $\mathbf{L}$) from $\mathbf{n}$ we have $\omega_\perp \ll \omega$ and $L \approx I_\parallel \omega$. Then, (3) gives $\omega_p \approx (I_\parallel/I_\perp)\omega$. So, for a prolate body $\omega_p < \omega$ and for an oblate body $\omega_p > \omega$. In the limit of a thin disc $I_\parallel \approx 2I_\perp$ and therefore $\omega_p \approx 2\omega$.

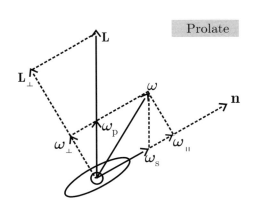
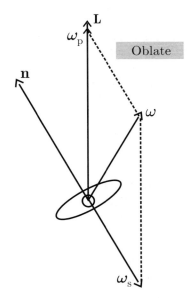

Comments

(i) The motion of a free rigid body in an inertial frame is more intricate than that of a free particle. In addition to uniform rectilinear motion of the CM, we have seen that an axially symmetric free body can undergo uniform rotation about an axis that is either fixed in space or precessing uniformly about a fixed axis. This regular precession occurs if the angular velocity is not aligned with a principal axis of inertia. A computer simulation program is available to illustrate the dynamics.[6]

(ii) The customary approach to this problem is algebraic, rather than geometric, and involves solving the Euler equations of motion. These equations describe the dynamics in a non-inertial frame that rotates with the body (the coordinate axes are the principal axes – see Question 12.6), and the results must be transformed to obtain solutions in the CM inertial frame.[7]

(iii) Free motion of a rigid body has applications to freely falling bodies, orbiting satellites, the precession of the Earth's axis of rotation (Chandler wobble), etc.

(iv) When a torque is applied to a body it can undergo a forced precession in addition to the torque-free precession found above (as, for example, in the top and the gyroscope – see Questions 12.23 and 12.24). In this event, the unforced precession is referred to as a nutation.

Question 12.23

A top is a homogeneous, axially symmetric body with a sharp point at one end of its axis, where it is pivoted. Consider a top of mass M that is rotating about its axis with angular velocity $\boldsymbol{\omega}_s$ in a uniform gravitational field. The axis is inclined to the vertical and the pivot is frictionless and located at a fixed point O in an inertial frame.

(a) Show that for a rapidly spinning top a possible motion is a steady precession about a vertical axis through O with angular velocity

$$\boldsymbol{\omega}_p = -(M\ell/I\omega_s)\,\mathbf{g}, \tag{1}$$

where ℓ is the distance from the pivot to the CM, I is the moment of inertia about the symmetry axis and $\mathbf{g}$ is the acceleration due to gravity.

(b) For what initial conditions is this steady precession attained?

(c) Give a geometric description, involving precession and nutation, of the motion of an inclined, spinning top released with its axis at rest.[8] (Hint: For the nutation use the results of Question 12.22).

---

[7] See, for example, A. P. Arya, *Introduction to classical mechanics*. Boston: Allyn and Bacon, 1990.

[8] E. Butikov, "Precession and nutation of a gyroscope," *European Journal of Physics*, vol. 27, pp. 1071–1081, 2006.

Solution

(a)

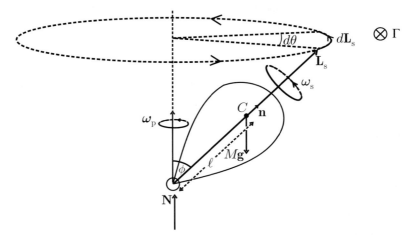

The CM is at C. The forces acting are the weight $M\mathbf{g}$ and the reaction $\mathbf{N}$ at the pivot. The total angular momentum $\mathbf{L}$ of the top changes at a rate

$$\dot{\mathbf{L}} = \mathbf{\Gamma}, \tag{2}$$

where $\mathbf{\Gamma}$ is the total torque due to external forces. Relative to O we have

$$\mathbf{\Gamma} = \ell\mathbf{n} \times M\mathbf{g}, \tag{3}$$

where $\mathbf{n}$ is a unit vector along the symmetry axis. For a rapidly spinning top $\mathbf{L} \approx \mathbf{L}_s$ (the spin angular momentum) and (2) can be approximated as

$$\dot{\mathbf{L}}_s = \mathbf{\Gamma}. \tag{4}$$

Now, $\mathbf{L}_s$ is perpendicular to $\mathbf{\Gamma}$ and therefore the magnitude $L_s = \sqrt{\mathbf{L}_s \cdot \mathbf{L}_s}$ is constant: $dL_s^2/dt = 2\mathbf{L}_s \cdot \dot{\mathbf{L}}_s = 2\mathbf{L}_s \cdot \mathbf{\Gamma} = 0$. Also, $d\mathbf{L}_s$ is along $\mathbf{\Gamma}$, which is perpendicular to the plane containing $\mathbf{n}$ and $\mathbf{g}$ at each instant. Therefore, the tip of the vector $\mathbf{L}_s$ moves in a circle centred on the vertical axis through O: the vector $\mathbf{L}_s$ traces out a cone with vertex at O. The angular velocity of this precession follows from (3) and (4):

$$\dot{\mathbf{L}}_s = \ell\mathbf{n} \times M\mathbf{g} = (I\omega_s\mathbf{n}) \times (M\ell\mathbf{g}/I\omega_s) = \boldsymbol{\omega}_p \times \mathbf{L}_s, \tag{5}$$

where $\mathbf{L}_s = I\omega_s\mathbf{n}$ and $\boldsymbol{\omega}_p$ is given by (1). The meaning of $\boldsymbol{\omega}_p$ in (5) is easily found. From (5), $|d\mathbf{L}_s| = (\omega_p L_s \sin\phi)dt$, while from the diagram $|d\mathbf{L}_s| = (L_s \sin\phi)d\theta$. Therefore, $\omega_p = d\theta/dt$ is the angular speed of the precession. The direction of $\boldsymbol{\omega}_p$ is along the axis of precession through O (parallel or anti-parallel to $\mathbf{g}$: if the initial spin $\boldsymbol{\omega}_s$ is 'up' – directed from the pivot to the upper end of the axis – then $\boldsymbol{\omega}_p$ is also 'up', as depicted in the diagram; if $\boldsymbol{\omega}_s$ is reversed then $\boldsymbol{\omega}_p$ also reverses). Therefore, $\boldsymbol{\omega}_p$ is the angular velocity of precession. It is constant (because of our assumption that the pivot is frictionless) and so the precession is termed steady (or uniform). Note that this steady precession is, in fact, a possible exact solution to (2) because $\mathbf{L} = \mathbf{L}_s + \mathbf{L}_p$ and $\mathbf{L}_p$ (the angular momentum of precession) is a constant. Both $\mathbf{L}_s$ and $\mathbf{L}$ can perform a steady precession with angular velocity $\boldsymbol{\omega}_p$. Other motions of the top are possible, such as that described in (c) below.

(b) For the top to perform a steady precession, specific initial conditions are required: the top should be given a spin $\boldsymbol{\omega}_s$ about its axis (which is inclined to the vertical) and also an angular velocity $\boldsymbol{\omega}_p$ (determined by (1)) about the vertical axis through O. The latter amounts to giving the free end of the axis (located a distance D from O) an initial velocity

$$\omega_p D \sin\phi = (Mg\ell D/I\omega_s)\sin\phi \tag{6}$$

perpendicular to the plane of $\mathbf{n}$ and $\mathbf{g}$.

(c) For other initial conditions the motion is more complicated than a steady precession. Suppose, for example, that the upper end of the axis of a tilted, spinning top is released from rest instead of at the initial velocity (6). The total angular momentum $\mathbf{L}$ – which is initially directed along the axis $\mathbf{n}$ – will start to precess with the angular velocity $\boldsymbol{\omega}_p$ given in (1). So, $\mathbf{L}$ consists of a superposition of the spin angular momentum and a precessional component:

$$\mathbf{L} = \mathbf{L}_s + \mathbf{L}_p, \tag{7}$$

as illustrated below. (Note that it is $\mathbf{L}$ that precesses for $t \geq 0$; the axis $\mathbf{n}$ and the centre of mass C are at rest at $t = 0$.) Because $\mathbf{L}_s$ deviates from the symmetry axis $\mathbf{n}$, we conclude from Question 12.22 that $\mathbf{n}$ will perform an additional precession, namely a rotation about $\mathbf{L}_s$ with angular velocity

$$\boldsymbol{\omega}_n = \mathbf{L}_s/I_\perp, \tag{8}$$

where $I_\perp$ is the moment of inertia about an axis through C and perpendicular to $\mathbf{n}$. The subscript n in (8) indicates that this precession is referred to as a nutation (a nodding). The nutation is illustrated in the second diagram below.

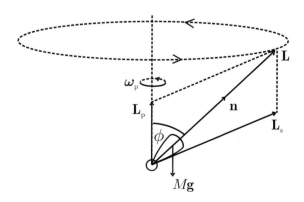
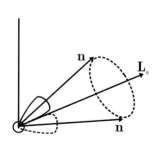

For a rapidly spinning top $L_s \approx I\omega_s$ and so

$$\omega_n \approx (I/I_\perp)\omega_s. \tag{9}$$

It follows that if I and $I_\perp$ are comparable then so are the angular speeds of nutation and spin. Consequently, $\omega_n \gg \omega_p$ (see (1)) – meaning that the nutation is rapid compared to the precession. Also, the amplitude of the nutation is small because $\mathbf{L}_s$ for a rapidly spinning top lies almost along $\mathbf{L}$ (see also (12) below). We conclude that if the axis $\mathbf{n}$ of a rapidly spinning, inclined top is released from rest then its subsequent motion is the superposition of two precessions. First, a steady, slow precession along a vertical cone: this is a forced precession due to the gravitational torque. Secondly, a rapid nutation of small amplitude about the (precessing) spin angular momentum vector $\mathbf{L}_s$. The respective angular frequencies ω_p and ω_n are independent of the angle of inclination ϕ. Further discussion of the motion, and also computer simulations, can be found in Ref. [8].

Comments

(i) There is a critical value of the spin angular frequency

$$\omega_{sc} = \sqrt{(4Mg\ell I_\perp/I^2)\cos\phi} \tag{10}$$

below which steady precession is not possible.[9]

(ii) The reader may have noticed that there is a second possibility for steady precession. For example, when the axis of a rapidly spinning top is almost vertical, the gravitational torque is small and the top can precess with the rapid frequency $\omega_p \approx L/I_\perp \approx (I/I_\perp)\omega_s$ of a free top (see Question 12.22).

(iii) Usually the dynamics of a top in a gravitational field is analyzed by using Lagrange's equations, or by solving Euler's equations for motion relative to a non-inertial frame.[9]

(iv) For the slow precession of a rapidly spinning top the angle of inclination varies according to[9]

$$\phi(t) = \phi_0 + \Delta\phi(1 - \cos\omega_n t), \tag{11}$$

where

$$\Delta\phi = |(\omega_p - \omega_0)/\omega_n|\sin\phi_0 \tag{12}$$

is the half-angle of the nutation cone depicted in the last figure above. In (11), ω_0 and ϕ_0 are initial conditions: ω_0 is the initial angular frequency of precession about the vertical, and ϕ_0 is the initial inclination of $\mathbf{n}$. For the initial condition $\omega_0 = \omega_p$ we have $\Delta\phi = 0$ and therefore no nutation, in agreement with the discussion in (b) above. The initial condition $\omega_0 = 0$ discussed in (c) produces $\Delta\phi = (\omega_p/\omega_n)\sin\phi_0$, which is small for a rapidly spinning top, where $\omega_p \ll \omega_n$.

---

[9] See, for example, V. D. Barger and M. G. Olsson, *Classical mechanics*. New York: McGraw-Hill, 1995.

438 Solved Problems in Classical Mechanics

(v) It is interesting to also consider precession of a top in a non-inertial frame that undergoes a uniform vertical acceleration **a** relative to the inertial frame used above. The effect is to replace **g** in the above analysis with an effective gravitational acceleration $\mathbf{g}-\mathbf{a}$. So, an upward acceleration increases ω_p and a downward acceleration decreases it (see (1)). In particular, in free fall ($\mathbf{a}=\mathbf{g}$) there is no forced precession ($\boldsymbol{\omega}_p = 0$ and the motion is that described in Question 12.22).

(vi) The above analysis is for a frictionless pivot. Friction has two effects on the motion: it damps out the nutation and it causes the spin ω_s to decrease with time. The next three questions deal with precession of a gyroscope in which the effect of friction is reduced to such an extent that one can detect even the influence of the non-inertial nature of a laboratory frame due to the Earth's rotation.

Question 12.24

In an air-suspension gyroscope the rotor consists of a precision steel ball on which a flat face has been ground. The rotor is levitated on an air 'cushion' in a hemispherical cup and spins with a constant frequency f about its axis, which is horizontal.[‡] Consider the motion of the rotor relative to an inertial frame in which there is a uniform gravitational field. Explain why the rotor precesses and show that the period of precession of a rapidly spinning rotor is

$$T = kf, \qquad \text{where } k = 4\pi^2 I/MgR. \tag{1}$$

Here, g is the acceleration due to gravity, M and I are the mass of the rotor and its moment of inertia about the symmetry axis, and R is the distance from the CM of the rotor to the centre O of the steel ball from which it was formed.

Solution

The diagram shows a side view of a rotor whose spin angular velocity is directed out of the flat face. The origin O is a fixed point located at the centre of the steel ball. The coordinate axes $Oxyz$ are those of the inertial frame – for clarity they have been drawn displaced from O. Because of the flat face, the CM is located away from O as indicated, and there is a gravitational torque $\boldsymbol{\Gamma}_g$ that is always horizontal and perpendicular to the spin angular momentum $\mathbf{L}_s = I\boldsymbol{\omega}_s$ (at the instant depicted in the diagram, $\boldsymbol{\Gamma}_g = -MgR\hat{\mathbf{y}}$). This torque would tip a stationary rotor until its flat face is upward. However, a spinning rotor

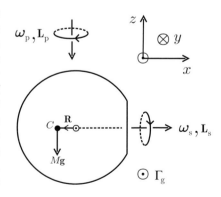

[‡]In practice, the magnitude of the spin angular frequency is maintained against air friction, and the axis is kept nearly horizontal, by means of a magnetic drive – see Comment (i) below.

possesses angular momentum and so, instead of tipping, it precesses. This precession is simple for sufficiently large $\boldsymbol{\omega}_{\rm s}$ (see Comment (ii) below). Then, we can neglect the angular momentum of precession $\mathbf{L}_{\rm p}$ (and also any angular momentum perpendicular to $\mathbf{L}_{\rm s}$ and $\mathbf{L}_{\rm p}$) and approximate the rotational equation $\dot{\mathbf{L}} = \boldsymbol{\Gamma}_{\rm g}$, where $\mathbf{L}$ is the total angular momentum, as $\dot{\mathbf{L}}_{\rm s} = \boldsymbol{\Gamma}_{\rm g}$. It follows that $\mathbf{L}_{\rm s} \cdot \dot{\mathbf{L}}_{\rm s} = 0$ and therefore $dL_{\rm s}^2/dt = 0$. That is, the magnitude $L_{\rm s}$ is a constant. Also, because $\mathbf{L}_{\rm s}$ is constrained to be horizontal we can express the rotational equation in terms of plane polar coordinates:

$$d\mathbf{L}_{\rm s}/dt = -\eta \Gamma_{\rm g} \hat{\boldsymbol{\theta}}. \qquad (2)$$

Here, $\Gamma_{\rm g} = MgR$ and η specifies the spin direction: $\eta = 1$ if $\boldsymbol{\omega}_{\rm s}$ points out of the flat face and $\eta = -1$ if $\boldsymbol{\omega}_{\rm s}$ points into the flat face. The interpretation of (2) is facilitated by the following two diagrams, which are views looking down on the rotor. In each, $\mathbf{L}_{\rm s}$ traces a circle in the xy-plane – clockwise if $\eta = 1$ and anti-clockwise if $\eta = -1$:

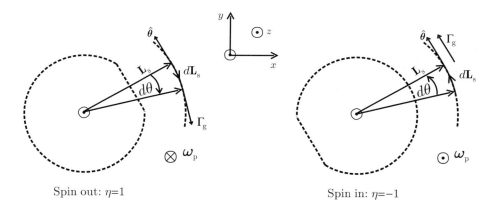

Spin out: $\eta=1$ Spin in: $\eta=-1$

It is clear that in both cases $|d\mathbf{L}_{\rm s}| = L_{\rm s} d\theta$. From (2), $|d\mathbf{L}_{\rm s}| = \Gamma_{\rm g} dt$ and therefore the angular frequency $\omega_{\rm p} = \dot{\theta}$ of the precession is given by

$$\omega_{\rm p} = \Gamma_{\rm g}/L_{\rm s} = MgR/I\omega_{\rm s}. \qquad (3)$$

Now, $\omega_{\rm p} = 2\pi/T$ and $\omega_{\rm s} = 2\pi f$, and so we obtain (1). Note that $\omega_{\rm p}$ is independent of the direction of spin η, and therefore the periods of clockwise and anti-clockwise precession are the same in an inertial frame:

$$T_{\rm c} = T_{\rm a} = T. \qquad (4)$$

Comments

(i) In practice, the spin $\omega_{\rm s}$ of the rotor is maintained by a magnetic drive. To this end, the rotor is magnetized in such a way that it has a magnetic dipole moment parallel to its flat face. This magnetized rotor is then driven as a synchronous motor by an encircling field coil excited by a suitable oscillator/power amplifier combination. In the above analysis we have ignored the torque $\boldsymbol{\Gamma}_d$ due to the drive. If this is included then (3) is modified to read

$$\omega_p = (\Gamma_g + \Gamma_d)/L_s = (\Gamma_g + \Gamma_d)/I\omega_s. \tag{5}$$

In practice, $\Gamma_d \ll \Gamma_g$ and so k in (1) is replaced by

$$k = \frac{4\pi^2 I}{MgR}(1 - \Gamma_d/\Gamma_g). \tag{6}$$

The period (1) is extracted by measuring T as a function of the rms current in the magnetic drive and extrapolating to zero current (i.e. $\Gamma_d = 0$).[10]

(ii) Equation (3) shows that ω_p is inversely proportional to ω_s, and therefore at large spin the precession is slow and $L_p \ll L_s$, as assumed above. At low spin the rotor precesses more rapidly and experiment shows that the motion becomes increasingly 'wobbly'. This nutation (nodding) is neglected in the simple analysis given above, which describes only the (almost) steady precession achieved at large spin, where the rotor axis deviates from the horizontal by less than about $0.5°$.[10]

(iii) Careful measurements of T show departures from (1); in particular, the result $T_c = T_a$ is not supported by experiment.[10] The reason for this breaking of the invariance with respect to the spin direction is discussed in the next question.

Question 12.25

(a) Let S' be a frame rotating with angular velocity $\boldsymbol{\Omega}$ with respect to an inertial frame S that shares a common origin with S'. Let $\mathbf{L}$ be the angular momentum of a rigid body in S'. Show that the rotational equation in S' is

$$\dot{\mathbf{L}} = \boldsymbol{\Gamma} + \mathbf{L} \times \boldsymbol{\Omega}. \tag{1}$$

(Hint: Use (7) of Question 14.20.)

(b) Hence, extend the results of Question 12.24 to an air-suspension gyroscope in a laboratory frame on Earth. In particular, show that for a rapidly spinning gyroscope the periods for clockwise and anti-clockwise precession (when viewed from above) are

$$T_c = kf + \tfrac{1}{2}k'f^2 \quad \text{and} \quad T_a = kf - \tfrac{1}{2}k'f^2. \tag{2}$$

Here, k is given by (1) of Question 12.24 and

$$k' = -(\gamma\Omega k^2/\pi)\sin\lambda, \tag{3}$$

where $\Omega = |\boldsymbol{\Omega}|$ is the magnitude of the Earth's angular velocity, λ is the latitude of the laboratory, and

$$\left.\begin{array}{rl}\gamma = & 1 \text{ in the northern hemisphere} \\ = & -1 \text{ in the southern hemisphere.}\end{array}\right\} \tag{4}$$

---

[10] O. L. de Lange and J. Pierrus, "Measurement of inertial and non-inertial properties of an air suspension gyroscope," American Journal of Physics, vol. 61, pp. 974–981, 1993.

Solution

(a) According to (7) of Question 14.20, for any vector $\mathbf{L}$:

The rate of change of $\mathbf{L}$ in S′ = The rate of change of $\mathbf{L}$ in S + $(\mathbf{L} \times \boldsymbol{\Omega})$. (5)

If $\mathbf{L}$ denotes angular momentum of a rigid body, then $\dot{\mathbf{L}}$ in S $= \boldsymbol{\Gamma}$ and (5) yields (1).

(b) We proceed as in Question 12.24. So, for a rapidly spinning gyroscope we again neglect the angular momentum of precession $\mathbf{L}_\mathrm{p}$ in comparison with the spin angular momentum $\mathbf{L}_\mathrm{s}$, and write (1) as

$$d\mathbf{L}_\mathrm{s}/dt = \boldsymbol{\Gamma}_\mathrm{g} + \mathbf{L}_\mathrm{s} \times \boldsymbol{\Omega}. \qquad (6)$$

It follows that $\mathbf{L}_\mathrm{s} \cdot \dot{\mathbf{L}}_\mathrm{s} = 0$ and so the magnitude $\mathbf{L}_\mathrm{s}$ is constant here as well. Also, recall that $\mathbf{L}_\mathrm{s}$ is constrained to be horizontal: in cylindrical coordinates $\mathbf{L}_\mathrm{s} = L_\mathrm{s}\hat{\boldsymbol{\rho}}$, and we see from (6) that (2) of Question 12.24 is modified to read

$$d\mathbf{L}_\mathrm{s}/dt = -(\eta\Gamma_\mathrm{g} + L_\mathrm{s}\Omega_z)\hat{\boldsymbol{\theta}}. \qquad (7)$$

The new feature here is the term in Ω_z (the vertical component of Earth's angular velocity); this term represents a perturbation because in practice $|L_\mathrm{s}\Omega_z/\Gamma_\mathrm{g} \ll 1|$. The analysis of (7) proceeds as in Question 12.24 and we conclude that the angular frequencies of clockwise and anti-clockwise precession (corresponding to $\eta = 1$ and $\eta = -1$, respectively) are

$$\omega_\mathrm{pc} = (\Gamma_\mathrm{g} + L_\mathrm{s}\Omega_z)/L_\mathrm{s}, \qquad \omega_\mathrm{pa} = (\Gamma_\mathrm{g} - L_\mathrm{s}\Omega_z)/L_\mathrm{s}. \qquad (8)$$

Thus, the period $T_\mathrm{c} = 2\pi/\omega_\mathrm{pc}$ of the clockwise precession is

$$T_\mathrm{c} = (2\pi L_\mathrm{s}/\Gamma_\mathrm{g})(1 + L_\mathrm{s}\Omega_z/\Gamma_\mathrm{g})^{-1} \approx (2\pi L_\mathrm{s}/\Gamma_\mathrm{g})(1 - L_\mathrm{s}\Omega_z/\Gamma_\mathrm{g}). \qquad (9)$$

Now, $L_\mathrm{s} = I\omega_\mathrm{s} = I2\pi f$ and $\Gamma_\mathrm{g} = MgR$. So

$$T_\mathrm{c} = kf - k^2 f^2 \Omega_z/2\pi. \qquad (10)$$

In the northern hemisphere the angular velocity $\boldsymbol{\Omega}$ of the Earth points out of the ground, whereas in the southern hemisphere it points into the ground. Therefore, $\Omega_z = \gamma\Omega\sin\lambda$, where γ is defined in (4), and (10) yields (2)$_1$. In a similar way we find that the period $T_\mathrm{a} = 2\pi/\omega_\mathrm{pa}$ of anti-clockwise precession is given by (2)$_2$.

Comments

(i) The torque $\mathbf{L} \times \boldsymbol{\Omega}$ in (1) is due to the well-known Coriolis force that acts in a rotating frame (see Chapters 1 and 14). Thus, the above analysis shows that the Coriolis force breaks the invariance of the precessional motion with respect to the direction of spin $\boldsymbol{\omega}_\mathrm{s}$ relative to the rotor; if $\Omega_z \neq 0$ in (8) then $T_\mathrm{c} \neq T_\mathrm{a}$. The effect is largest at the poles ($\lambda = 90°$) and vanishes at the equator ($\lambda = 0°$).

(ii) This property of the gyroscope – where its precession is affected by the rotation of the Earth – is reminiscent of the famous Foucalt pendulum, see Question 14.29.
(iii) The steady precessional motion of the gyroscope is accompanied by some nutation, whose amplitude is small for rapid spin ($\omega_s \gg \omega_p, \Omega$).[10]
(iv) Equations (2) apply in the limit where the torque $\mathbf{\Gamma}_d$ due to the magnetic drive is zero (see also Question 12.24). In practice, the effect of this torque can be comparable to that of the Coriolis force, and the values of T_c and T_a at $\mathbf{\Gamma}_d = 0$ are obtained by extrapolation.[10] The following question involves a comparison of the theoretical results (2) with experimental results obtained by this method.

Question 12.26

Measurements of the precessional periods T_c and T_a as functions of the spin frequency f have been reported for the air-suspension gyroscope discussed in Questions 12.24 and 12.25.[10] The rotor had radius $a = 25.438$ mm and the removed cap had thickness $h = 2.027$ mm. The laboratory was located at latitude $\lambda = 29.62°$ in the southern hemisphere and the gravitational acceleration was $g = 9.794\,\mathrm{m\,s^{-2}}$.

(a) Calculate the constants k and k' in (2) of Question 12.25. To do this, make use of the formulae for the CM and the moment of inertia given in (3) and (6) of Question 12.8. Also, take $\Omega = 7.292 \times 10^{-5}\,\mathrm{rad\,s^{-1}}$ for the Earth.
(b) Use the following measurements for this gyroscope to obtain experimental values of k and k', and compare them with the theoretical values in (a).

f (Hz)	T_c (s)	T_a (s)	f (Hz)	T_c (s)	T_a (s)	f (Hz)	T_c (s)	T_a (s)
10.00	93.51	93.42	35.00	327.73	326.57	60.00	562.53	559.07
15.00	140.28	140.12	40.00	374.56	372.96	65.00	609.78	609.39
20.00	187.21	186.77	45.00	421.47	419.62	70.00	656.64	651.88
25.00	234.08	233.52	50.00	468.70	466.17	—	—	—
30.00	280.89	279.99	55.00	515.54	512.54	—	—	—

Solution

(a) From (1) of Question 12.24 and (3) and (6) of Question 12.8 we have

$$k = \frac{4\pi^2 I}{MgR} = \frac{8\pi^2 a}{15g}\left(\frac{4 + 4\epsilon + 3\epsilon^2 - 3\epsilon^3}{\epsilon^2}\right), \tag{1}$$

where $\epsilon = h/a = 2.027/25.438 = 7.9684 \times 10^{-2}$. This, together with the given values of a and g, yields

$$k_{\mathrm{th}} = 9.337\,\mathrm{s}^2. \tag{2}$$

From (3) of Question 12.25, and using (2) and the given Ω and λ, we have

$$k'_{\mathrm{th}} = (\Omega k_{\mathrm{th}}^2/\pi)\sin\lambda = 1.000 \times 10^{-3}\,\mathrm{s}^3. \tag{3}$$

The estimated uncertainty in k_{th} is $\pm 0.017\,\mathrm{s}^2$, and in k'_{th} it is $\pm 0.004 \times 10^{-3}\,\mathrm{s}^3$.[10]

(b) From (2) of Question 12.25 we have

$$\tfrac{1}{2}(T_c + T_a) = kf \quad \text{and} \quad T_c - T_a = k'f^2. \tag{4}$$

In the following graphs the straight lines are the theoretical expressions (4) with k and k' given by (2) and (3), and the data points are from the experimental values of T_c and T_a in the above table. A regression analysis of these data points gives

$$k_{\text{exp}} = 9.346 \pm 0.001 \, \text{s}^2, \qquad k'_{\text{exp}} = (0.997 \pm 0.017) \times 10^{-3} \, \text{s}^3, \tag{5}$$

which are in good agreement with the theoretical values (2) and (3).

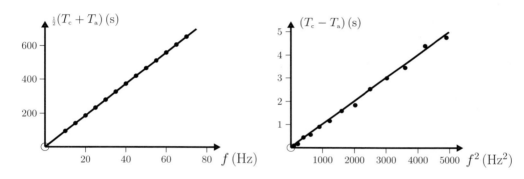

Comment

The result $(4)_1$ is almost entirely an inertial property due to the gravitational field of the Earth (it contains a small non-inertial effect due to the contribution of the centrifugal force to the gravitational acceleration g, and hence to the value of k in (1)). By contrast, $(4)_2$ is a non-inertial effect due to the Coriolis force on the gyroscope.

Question 12.27

Consider a gyrocompass that is fashioned from the air-suspension gyroscope of Question 12.24 by machining a second, parallel and identical flat face on the rotor. The rotor spins with constant angular frequency ω_s about its axis, which is constrained to be horizontal by a torque $\boldsymbol{\Gamma}$. The rotor is free to rotate about the vertical.

(a) Show that in a laboratory frame on Earth a rapidly spinning rotor performs non-linear oscillations about true North.
(b) Determine the period of small oscillations in terms of ω_s and the angular frequency Ω of the Earth, the latitude λ, and moments of inertia of the gyrocompass.

444 Solved Problems in Classical Mechanics

Solution

(a) The diagram shows a view of the rotor from above. The CM is at O, which is a fixed point. $Oxyz$ is a laboratory frame: the xy-plane is horizontal with the x- and y-axes oriented due East and due North. The only torque due to external forces is $\boldsymbol{\Gamma} = (\Gamma_x, \Gamma_y, 0)$. Now, $Oxyz$ is a non-inertial frame and the z-component of the rotational equation is (see Question 12.25)

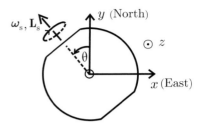

$$I_z \dot{\omega}_z = (\mathbf{L} \times \boldsymbol{\Omega})_z, \qquad (1)$$

where $\boldsymbol{\Omega}$ is the angular velocity of the Earth and I_z is the moment of inertia of the rotor about the z-axis. With the above choice of axes, $\boldsymbol{\Omega} = (0, \Omega_y, \Omega_z)$ and the horizontal part of $\mathbf{L}$ is $\mathbf{L}_s$. So

$$(\mathbf{L} \times \boldsymbol{\Omega})_z = -L_s \Omega_y \sin\theta = -I\omega_s \Omega \cos\lambda \sin\theta, \qquad (2)$$

where I is the moment of inertia of the rotor about its axis and λ is the latitude. Equations (1) and (2) with $\omega_z = \dot\theta$ yield

$$\ddot\theta + \{(I/I_z)\omega_s \Omega \cos\lambda\}\sin\theta = 0, \qquad (3)$$

and so the rotor oscillates in a horizontal plane about true North ($\theta = 0$).

(b) For small oscillations ($|\theta| \ll 1$), the non-linear equation (3) can be approximated as

$$\ddot\theta + \{(I/I_z)\omega_s \Omega \cos\lambda\}\theta = 0, \qquad (4)$$

showing that the oscillations become harmonic with period

$$T = 2\pi\{(I/I_z)\omega_s \Omega \cos\lambda\}^{-1/2}. \qquad (5)$$

Comments

(i) For the Earth, $\Omega \approx 7.3 \times 10^{-5}\,\mathrm{rad\,s^{-1}}$. So, if I/I_z in (5) is of order unity, a gyrocompass near the equator will have a period $T \approx 10\,\mathrm{s}$ if $\omega_s \approx 5500\,\mathrm{rad\,s^{-1}}$.

(ii) The oscillatory behaviour of the gyrocompass is due to the horizontal component Ω_y of the Earth's angular velocity $\boldsymbol{\Omega}$ – see (2). By contrast, it is the vertical component of $\boldsymbol{\Omega}$ that affects the precession of a gyroscope (see Question 12.25).

(iii) The rotational equation for the spin angular momentum $\mathbf{L}_s$ of a rapidly spinning gyroscope is

$$\dot{\mathbf{L}}_s = \boldsymbol{\Gamma} + \mathbf{L}_s \times \boldsymbol{\Omega} = (\Gamma + L_s \Omega_z)\hat{\boldsymbol\theta}. \qquad (6)$$

That is, $|d\mathbf{L}_s| = (\Gamma + L_s \Omega_z)dt$. But $|d\mathbf{L}_s| = L_s d\theta$. Therefore, $\dot\theta = \Omega_z + \Gamma/L_s$. This equation determines the constraining torque Γ because $\dot\theta$ can be calculated from (3) and the relevant initial conditions. In practice, the rotor of a gyrocompass is constrained mechanically (by gimbals).[11]

---

[11] See, for example, R. F. Deimel, *Mechanics of the gyroscope*. New York: Dover, 1950.

Question 12.28

Two identical wheels, each of radius a, are connected by an axle of length b, about which they can turn freely. This system rolls without slipping on a plane inclined at an angle α to the horizontal. Choose Cartesian coordinates in the plane, with x-axis horizontal and y-axis directed up the plane. The initial conditions are that the system is placed on the plane, with spin angular velocity $\boldsymbol{\omega}$ about an axis through the CM and perpendicular to the plane, at an instant when the axle is horizontal (i.e. directed along the x-axis).

(a) Show that the trajectory of the CM is the cycloid[12]
$$x(t) = (g_0/4\omega^2)(2\omega t - \sin 2\omega t), \qquad y(t) = (g_0/4\omega^2)\cos 2\omega t. \qquad (1)$$
Here
$$g_0 = \frac{g\sin\alpha}{(1 + 2I_A/Ma^2)}, \qquad (2)$$
M is the total mass of the system, I_A is the moment of inertia of a wheel about the axle, and g is the gravitational acceleration. (Hint: This requires the use of four equations of motion – for translation of the CM, rotation about the CM, and rotation of each wheel about the axle – and two constraint equations associated with the condition for no slipping.)

(b) Express (1) in dimensionless form $\bar{x}(\tau)$, $\bar{y}(\tau)$ and plot $\bar{y}(\bar{x})$.

Solution

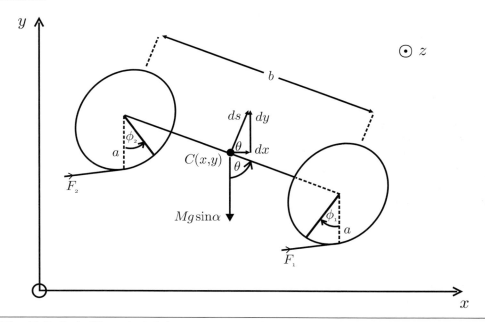

---

[12] E. D. Peck, "Cart wheels," American Journal of Physics, vol. 46, pp. 509–512, 1978.

The diagram shows the wheels and axle relative to a Cartesian coordinate system Oxy on the inclined plane. C is the CM, located at the midpoint of the axle. Three angles are shown: θ between the axle and the y-axis, and the corresponding angles ϕ_1 and ϕ_2 through which a point on the rim of each wheel rotates. Also shown is an infinitesimal displacement ds of the CM. Because there is no slipping, ds is always directed perpendicular to the axle, and this implies the constraint

$$ds = \tfrac{1}{2}a(d\phi_1 - d\phi_2). \tag{3}$$

A second constraint is[‡]

$$d\theta = \frac{a}{b}(d\phi_1 + d\phi_2). \tag{4}$$

(a) The forces acting are gravity and the contact forces between the wheels and the plane. For the latter we are concerned with the tangential components F_1 and F_2 in the plane of each wheel. The total force in the direction of ds is $F_1 + F_2 - Mg\sin\alpha\sin\theta$, and so the acceleration of the CM along the trajectory is given by

$$M\ddot{s} = F_1 + F_2 - Mg\sin\alpha\sin\theta. \tag{5}$$

Also, for rotation about a perpendicular axis (Cz) through the CM we have

$$I_z\ddot{\theta} = \tfrac{1}{2}b(F_1 - F_2), \tag{6}$$

where I_z is the total moment of inertia about this axis. There is another pair of equations of motion, namely those for rotation of each wheel about the axle:

$$I_A\ddot{\phi}_1 = -aF_1, \qquad I_A\ddot{\phi}_2 = aF_2. \tag{7}$$

Equations (7) enable us to eliminate the unknown forces F_i from (5) and (6):

$$M\ddot{s} = -(I_A/a)(\ddot{\phi}_1 - \ddot{\phi}_2) - Mg\sin\alpha\sin\theta \tag{8}$$

$$I_z\ddot{\theta} = -(bI_A/2a)(\ddot{\phi}_1 + \ddot{\phi}_2). \tag{9}$$

Finally, the angular accelerations $\ddot{\phi}_i$ in (8) and (9) can be eliminated by using the constraint equations (3) and (4). Consequently, (8) and (9) become

$$\ddot{s} = -g_0\sin\theta \tag{10}$$

$$[I_z + (b^2/2a^2)I_A]\ddot{\theta} = 0, \tag{11}$$

where g_0 is the constant (2). The quantity in square brackets in (11) is non-zero. Therefore, $\ddot{\theta} = 0$, and so the angular speed $\dot{\theta}$ is a constant of the motion:

$$\dot{\theta} = \omega \quad\text{and}\quad \theta = \omega t + \theta_0. \tag{12}$$

From (10) and (12) we have $\ddot{s} = -g_0\sin(\omega t + \theta_0)$. Therefore

$$\dot{s} = (g_0/\omega)\cos(\omega t + \theta_0) + v_0, \tag{13}$$

[‡]Because $d\theta = d\theta' + d\theta''$, where $bd\theta' = ad\phi_1$ and $bd\theta'' = ad\phi_2$.

where v_o is the value of $\dot{s}$ when $\theta = \frac{1}{2}\pi$ (i.e. when the axle is horizontal). We can obtain the trajectory (x, y) of the CM from (13) by noting that $dx = ds \cos\theta$ and $dy = ds \sin\theta$ (see the diagram). Then, $\dot{x} = \dot{s}\cos\theta$ and $\dot{y} = \dot{s}\sin\theta$, and so

$$\dot{x} = (g_o/2\omega)[1 + \cos 2(\omega t + \theta_o)] + v_o \cos(\omega t + \theta_o) \tag{14}$$

$$\dot{y} = (g_o/2\omega)\sin 2(\omega t + \theta_o)] + v_o \sin(\omega t + \theta_o). \tag{15}$$

(Here we have used the relations $\cos^2\theta = \frac{1}{2}(1 + \cos 2\theta)$ and $\cos\theta\sin\theta = \frac{1}{2}\sin 2\theta$.) Integration of (14) and (15) gives the parametric equations of the CM trajectory:

$$x(t) = x_o + (g_o/2\omega)t + (g_o/4\omega^2)[\sin 2(\omega t + \theta_o) - \sin 2\theta_o]$$
$$+ (v_o/\omega)[\sin(\omega t + \theta_o) - \sin\theta_o] \tag{16}$$

$$y(t) = y_o - (g_o/4\omega^2)[\cos 2(\omega t + \theta_o) - \cos 2\theta_o]$$
$$- (v_o/\omega)[\cos(\omega t + \theta_o) - \cos\theta_o]. \tag{17}$$

Now choose initial conditions such that $x_o = 0$ and $y_o = g_o/4\omega^2$, and the axle is along the x-axis and has no linear velocity at $t = 0$ (i.e. $\theta_o = \frac{1}{2}\pi$ and $v_o = 0$). Then (16) and (17) yield (1).

(b) The dimensionless form of (1) is

$$\bar{x}(\tau) = 4\pi\tau - \sin 4\pi\tau, \qquad \bar{y}(\tau) = \cos 4\pi\tau, \tag{18}$$

where $\bar{x} = 4\omega^2 x/g_o$, $\bar{y} = 4\omega^2 y/g_o$ and $\tau = \omega t/2\pi$. The graph of $\bar{y}(\bar{x})$ is:

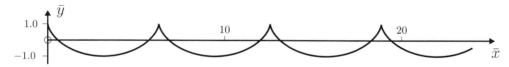

Comments

(i) Of the four coordinates ϕ_1, ϕ_2, s and θ associated with the system, just two are independent, namely ϕ_1 and ϕ_2 or s and θ – see (3) and (4) – and they suffice to specify the motion of the system.

(ii) The motion can also be analyzed using Lagrange's equations.[12]

(iii) On a horizontal plane, $g_o = 0$ and so the CM is either at rest (if $v_o = 0$) or it moves in a circle of radius v_o/ω – see (16) and (17).

(iv) According to (1), the motion of the CM along an inclined plane consists of uniform circular motion with frequency 2ω about a point that moves horizontally along the positive x-axis at constant speed $g_o/2\omega$. This horizontal motion is perpendicular to the component $Mg\sin\alpha$ of the weight down the plane. Also, the radius of the circular motion is equal to $g_o/4\omega^2$, which is small for rapid spin. In this regard, Peck has remarked that the motion "is reminiscent of the precession of a gyroscope in a horizontal plane. However, it will appear that a superposed oscillatory motion, like the nutation of the gyroscope, is not optional but intrinsic to the motion of the cart wheels."[12]

Question 12.29

A sphere of mass M and diameter D moves with CM velocity $\mathbf{v} = \dot{\mathbf{r}}$ and spin $\boldsymbol{\omega}$ through air of density ρ. It experiences drag and so-called lift (Magnus) forces:

$$\mathbf{F}_{\mathrm{d}} = -\tfrac{1}{8}C_{\mathrm{d}}\pi D^2 \rho v \mathbf{v} \quad \text{and} \quad \mathbf{F}_{\ell} = \tfrac{1}{8}C_{\ell}\pi D^2 \rho v (\hat{\boldsymbol{\omega}} \times \mathbf{v}), \quad (1)$$

where the dimensionless quantities C_{d} and C_{ℓ} are the drag and lift coefficients (see Comment (ii) below), $\hat{\boldsymbol{\omega}} = \boldsymbol{\omega}/\omega$ and $v = |\mathbf{v}|$. The diagram is for a sphere with backspin, moving in a vertical plane.

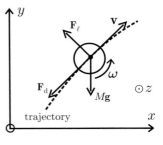

(a) Express the equation of motion for the CM vector $\mathbf{r}(t)$ in the form

$$\ddot{\mathbf{r}} - \mathbf{g} + \left(\frac{C_{\mathrm{d}}\pi D^2 \rho}{8M}\right)v\mathbf{v} - \left(\frac{C_{\ell}\pi D^2 \rho}{8M}\right)v(\hat{\boldsymbol{\omega}} \times \mathbf{v}) = 0, \quad (2)$$

where $\mathbf{g}$ is the acceleration due to gravity. Suppose coordinates are chosen so that $\mathbf{g} = (0, -g, 0)$ and the initial conditions are $\mathbf{r}_{\mathrm{o}} = 0$ and $\mathbf{v}_{\mathrm{o}} = v_{\mathrm{o}}(\cos\theta_{\mathrm{o}}, \sin\theta_{\mathrm{o}}, 0)$. What can be stated regarding the plane in which the CM moves if the sphere has either backspin or topspin, i.e. if $\hat{\boldsymbol{\omega}} = (0, 0, \pm 1)$?

(b) Consider a golf ball struck with backspin: $\hat{\boldsymbol{\omega}} = (0, 0, 1)$. Write a *Mathematica* notebook to solve (2) for $\mathbf{r}(t)$, taking $M = 0.050\,\mathrm{kg}$, $D = 0.042\,\mathrm{m}$, $\rho = 1.3\,\mathrm{kg\,m^{-3}}$, $v_{\mathrm{o}} = 60\,\mathrm{m\,s^{-1}}$, $g = 9.8\,\mathrm{m\,s^{-2}}$, $\theta_{\mathrm{o}} = 5°$, $C_{\mathrm{d}} = 0.28$ and $C_{\ell} = 0.17$ (the latter two coefficients are typical for golf balls spinning at 3500 rpm).[13] Repeat for $\theta_{\mathrm{o}} = 10°$, $15°$, $20°$, $25°$ and $30°$, and plot the trajectories $y(x)$ on the same axes.

(c) Plot the trajectories for $v_{\mathrm{o}} = 60\,\mathrm{m\,s^{-1}}$, $\theta_{\mathrm{o}} = 10°$ and for the following drag and lift coefficients:

1. $C_{\mathrm{d}} = 0$; $C_{\ell} = 0$, **2.** $C_{\mathrm{d}} = 0.28$; $C_{\ell} = 0$, and **3.** $C_{\mathrm{d}} = 0.28$; $C_{\ell} = 0.17$.

(d) Suppose that the golf ball is now either hooked or sliced, so that it is spinning about a vertical axis: $\hat{\boldsymbol{\omega}} = (0, \pm 1, 0)$. Adapt the notebook to calculate $\mathbf{r}(t)$ for both directions of spin, and for $\theta_{\mathrm{o}} = 45°$, $v_{\mathrm{o}} = 60\,\mathrm{m\,s^{-1}}$, $C_{\mathrm{d}} = 0.28$ and $C_{\ell} = 0.17, 0.11, 0.06$ and 0. Plot the projections of these trajectories on the horizontal (xz-) plane.

Solution

(a) Equation (2) follows directly from Newton's second law for the CM motion, $M\ddot{\mathbf{r}} = M\mathbf{g} + \mathbf{F}_{\mathrm{d}} + \mathbf{F}_{\ell}$. The motion is restricted to the plane defined by $\mathbf{v}_{\mathrm{o}}$ and $\mathbf{g}$ because the CM starts moving in this plane and there are no forces perpendicular to it when the sphere has just backspin or topspin. With the above choice of axes the plane of motion is the xy-plane, with y-axis vertical.

(b) We use the following *Mathematica* notebook to obtain the trajectories below:

---

[13] W. M. MacDonald and S. Hanzley, "The physics of the drive in golf," *American Journal of Physics*, vol. 59, pp. 213–218, 1991.

Rigid bodies

```
In[1]:= (* USE THIS CELL FOR THE GOLF BALL PROBLEM OF QUESTION 12.29 *)

        M = 0.05;  d = 0.042;  ρ = 1.3;
        C_D[t_] := 0.28;
        C_L[t_] := 0.17;

In[2]:= (* USE THIS CELL FOR THE TENNIS BALL PROBLEM OF QUESTION 12.30 *)
```

$$(*\; M = 0.057;\; d = 0.067;\; r0 = \frac{d}{2};\; \omega 0 = \frac{3500.0}{60};\; \rho = 1.3;$$

$$C_D[t\_] := 0.508 + \frac{1}{\left(22.5 + 4.196 \left(\frac{\text{speed}[t]}{r0\,\omega 0}\right)^{2.5}\right)^{0.4}};$$

$$C_L[t\_] := \frac{1}{2.022 + 0.981 \frac{\text{speed}[t]}{r0\,\omega 0}};\; *)$$

```
In[3]:= Needs["VectorAnalysis`"]
        M = 0.05;  d = 0.042;  ρ = 1.3;
        xmax = 1; ymax = 1;    (* Req^d for plot. Program refines these guesses *)
        tmax = 2;    (* time that ball is airborne. Program refines this value *)
        x0 = 0; y0 = 0; z0 = 0; v0 = 60.0; g0 = 9.8; g = {0, -g0, 0}; ω_hat = {0, 0, 1};
        r[t_] := {x[t], y[t], z[t]};
        v[t_] := r'[t];  speed[t_] := √Dot[v[t], v[t]];
```

$$\text{Drag}[t\_] := -C_D[t]\,\frac{\pi d^2 \rho}{8M}\,\text{speed}[t]\,v[t];$$

$$\text{Lift}[t\_] := C_L[t]\,\frac{\pi d^2 \rho}{8M}\,\text{speed}[t]\,\text{CrossProduct}[\omega_{\text{hat}}, v[t]];$$

```
        plotarray = Table[
            InitCon = Join[Thread[r[0] == {x0, y0, z0}],
                Thread[r'[0] == {v0 Cos[θ0], v0 Sin[θ0], 0}]];
            EqnMotion = Thread[r''[t] - g - Drag[t] - Lift[t] == 0.];
            EqnsToSolve = Join[EqnMotion, InitCon];
            Sol = NDSolve[EqnsToSolve, Join[r[t], v[t], r''[t]], {t, 0, 10tmax}];
            tmax = t /. FindRoot[y[t] == 0 /. Sol, {t, 6}];
            xmax = Max[xmax, First[x[t] /. Sol /. t → tmax]];
```

$$\text{ymax} = \text{Max}\left[\text{ymax}, \text{Max}\left[\text{Table}\left[\{\text{First}[y[t] /. \text{Sol}]\}, \left\{t, 0, \text{tmax}, \frac{\text{tmax}}{100}\right\}\right]\right]\right];$$

```
            ParametricPlot[{{Evaluate[{x[t], y[t]}] /. Sol}}, {t, 0, 10tmax},
                AspectRatio → 1, PlotRange → {{0, xmax}, {0, ymax}}],
```

$$\left\{\theta 0,\; 30\frac{\pi}{180},\; 5\frac{\pi}{180},\; -5\frac{\pi}{180}\right\}\Bigr];$$

```
        Show[plotarray]
```

Maximum range occurs for an angle of projection $\theta_0 \approx 27°$.

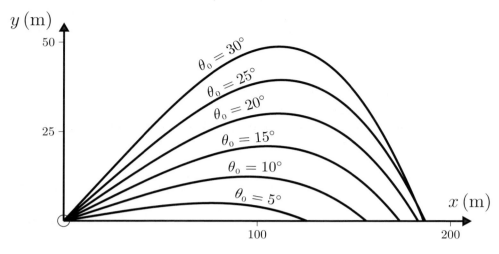

(c) The trajectories are:

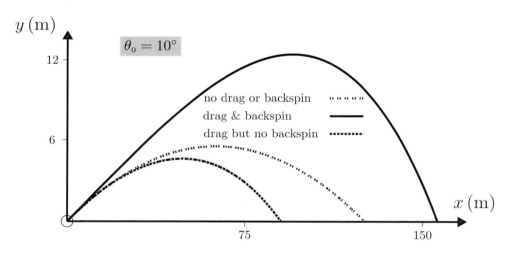

(d) For slice or hook shots, the 'lift' force has lateral components but no vertical component – the ball is therefore deflected instead of lifted. The diagram below illustrates that for a sliced shot by a right-handed golfer ($\boldsymbol{\omega}$ points into the ground) the deflection is to the right. For a hooked shot ($\boldsymbol{\omega}$ points out of the ground) the deflection is to the left.

Comments

(i) The phenomenon whereby a spinning sphere moving through a fluid experiences a force perpendicular to both its spin and velocity is known as the Magnus effect (*ca.* 1853). It contributes to the non-parabolic motion of spinning balls in sports such as golf, tennis and baseball. Newton noticed that the trajectories of tennis balls with spin could deviate and wondered whether rays of light "could swerve the same way – if they 'should possibly be globular bodies' spinning against the ether."[14]

(ii) In general, the drag and lift coefficients depend on $\mathbf{v}$ and $\boldsymbol{\omega}$, and they are usually specified by empirical relations (see Question 12.30). Here, we have taken them to be constants, equal to their initial values.

(iii) The lift due to backspin enables the ball to remain airborne for longer and to travel further than in the absence of spin. For example, for the trajectories in part (c) the time of flight is more than double and the range is about 70% greater when the ball has backspin.

(iv) If C_ℓ is increased in (b) above (for example, by increasing the backspin) then the initial vertical component of the total force on the ball can become upward. This produces trajectories $y(x)$ with an initial upward curvature.

(v) The effects of spin damping on the motion of a golf ball have been studied and found to be small.[13]

(vi) The effect of topspin on the motion of a tennis ball is considered next.

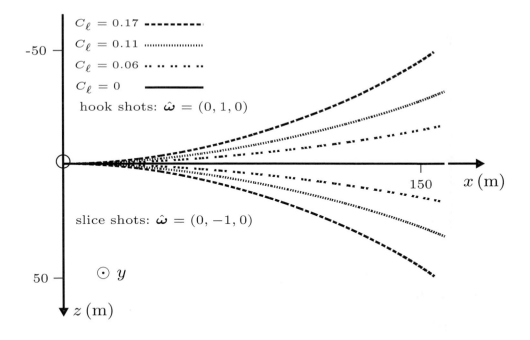

[14]   J. Gleick, *Isaac Newton*, p. 81. London: Harper Fourth Estate, 2003.

Question 12.30

Suppose the sphere in Question 12.29 is a tennis ball with topspin: $\hat{\boldsymbol{\omega}} = (0,\,0,\,-1)$. For the calculations below, take $M = 0.057\,\text{kg}$, $D = 0.067\,\text{m}$, $\rho = 1.3\,\text{kg m}^{-3}$ and, unless stated otherwise, $\omega = 3500\,\text{rpm}$. Use the following empirical drag and lift coefficients that are typical for tennis balls:[15]

$$\left.\begin{array}{l} C_\text{d} = 0.508 + \dfrac{1}{[22.50 + 4.196(2v/\omega D)^{5/2}]^{2/5}} \\[1em] C_\ell = \dfrac{1}{2.202 + 0.981(2v/\omega D)} \end{array}\right\} \quad (1)$$

(a) Modify the notebook of Question 12.29 to use the C_d and C_ℓ given by (1).
(b) Hence, solve the equation of motion for $\mathbf{r}(t)$. Use the initial conditions $\mathbf{r}(0) = (0,\,0.8,\,0)\,\text{m}$, $v_0 = 20.0\,\text{m s}^{-1}$ and $\theta_0 = 15°$. Repeat for $\theta_0 = 20°$, $25°$, $30°$, $35°$ and $40°$, and plot the trajectories $y(x)$ on the same axes.
(c) Plot the trajectories $y(x)$ for the following cases: **1.** without drag or topspin, **2.** with drag but no topspin, and **3.** with both drag and topspin. Take $v_0 = 20.0\,\text{m s}^{-1}$ and $\theta_0 = 20°$.
(d) Plot the trajectories $y(x)$ for a tennis ball with **1.** no spin: $\omega = 0$, **2.** topspin: $\omega = 3500\,\text{rpm}$, and **3.** topspin: $\omega \to \infty$. Take $v_0 = 20.0\,\text{m s}^{-1}$ and $\theta_0 = 20°$. Also, plot the trajectory with **4.** backspin: $\omega = 3500\,\text{rpm}$.

Solution

(a) Use cell 2 of the preceding notebook instead of cell 1.
(b) This yields the following trajectories:

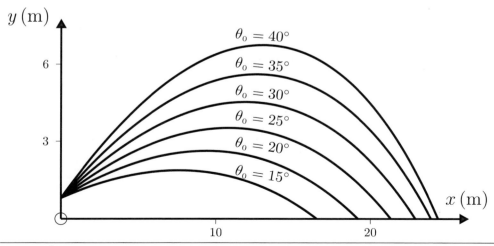

[15] A. Štěpánek, "The aerodynamics of tennis balls – the topspin lob," American Journal of Physics, vol. 56, pp. 138–142, 1988.

(c) In the diagram: 'no drag' means $C_d = 0$ and 'no topspin' means $C_\ell = 0$.

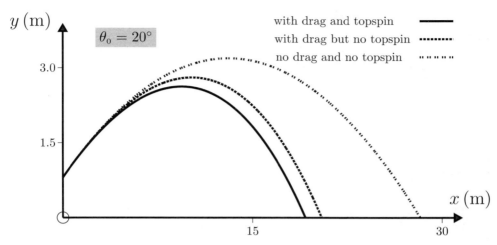

(d) The trajectories are:

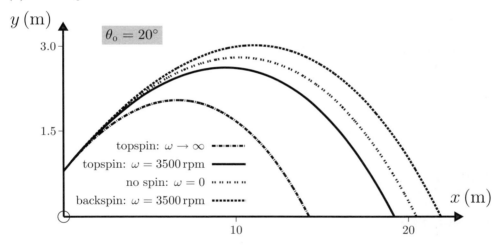

Comment

Topspin reduces the maximum height and horizontal range of the ball. In this respect, it has the opposite effect to backspin.

13

Non-linear oscillations

If one were to list some characteristic features of non-linear oscillators then the following would come to mind:

- ☞ Most systems are described by non-linear equations of motion.
- ☞ Some non-linear oscillators possess a linear (harmonic) limit, but many do not.
- ☞ In most instances non-linear equations of motion are intractable – analytic solutions cannot be found, even in simple cases, so the use of computers and numerical techniques is essential in this field.
- ☞ The superposition principle of a linear theory does not apply (but approximate solutions in the form of a truncated Fourier series are sometimes reasonable).
- ☞ Non-linear systems can possess extreme sensitivity to initial conditions, indicative of chaotic behaviour.
- ☞ In general, non-linear systems exhibit a far richer range of behaviour than do linear systems, and even simple non-linear systems can have surprising and complex properties. Among these are chaos, amplitude-dependent frequencies, multi-valued amplitudes, amplitude jumps, hysteresis, phase locking, and inverted oscillations.
- ☞ Many properties of non-linear equations are universal.

The following questions illustrate some of these features.

Question 13.1

A particle of mass m moves in a one-dimensional potential given by the series expansion

$$V(x) = \tfrac{1}{2}k_1 x^2 + \tfrac{1}{3}k_2 x^3 + \tfrac{1}{4}k_3 x^4 \cdots, \qquad (1)$$

where the k_i are constants.

(a) Show that the equation of motion is

$$\ddot{x} + \alpha x + \beta x^2 + \gamma x^3 + \cdots = 0, \qquad (2)$$

where α, β, γ, $\cdots$ are constants.

(b) Sketch the possible energy diagrams when the only non-zero coefficients are: 1. $\alpha (> 0)$ and β, 2. $\alpha (> 0)$ and γ, 3. $\gamma (> 0)$. In each case comment on the nature of the possible motions (whether bounded or unbounded.)

Solution

(a) The force $F = -dV/dx$ corresponding to (1) is
$$F(x) = -k_1 x - k_2 x^2 - k_3 x^3 - \cdots, \qquad (3)$$
and therefore the equation of motion $m\ddot{x} = F$ is (2) with $\alpha = k_1/m$, $\beta = k_2/m$, $\gamma = k_3/m$, etc.

(b)   1. $V(x) = m(\tfrac{1}{2}\alpha x^2 + \tfrac{1}{3}\beta x^3)$.

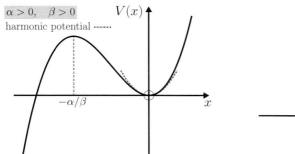
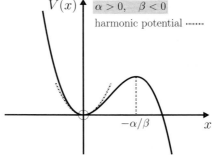

The motion can be either unbounded or bounded (periodic). The harmonic approximation is reasonable for $|x| \ll \alpha/|\beta|$.

   2. $V(x) = m(\tfrac{1}{2}\alpha x^2 + \tfrac{1}{4}\gamma x^4)$.

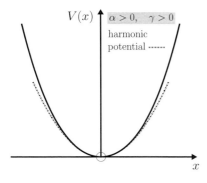
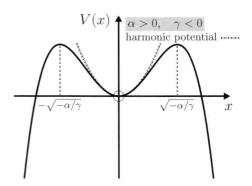

The motion is always bounded if $\gamma > 0$ (the system is referred to as 'hard'); but if $\gamma < 0$ (a 'soft' system) it can be either bounded or unbounded. The harmonic approximation is reasonable if $|x| \ll \sqrt{\alpha/|\gamma|}$.

   3. $V(x) = \tfrac{1}{4}m\gamma x^4$.

The graph of $V(x)$ is similar to the above for $\gamma > 0$, and the motion is always bounded.

Comments

(i) In general, if $\alpha = k_1/m > 0$ then the potential (1) has a minimum at $x = 0$, and for small oscillations (2) describes a perturbed (anharmonic) oscillator. A linear (harmonic) approximation is reasonable in the limit when energy $\to 0_+$.

(ii) In general, if $x = 0$ is to be a point of stable equilibrium then $F(x)$ must change sign there: the leading non-zero term in (3) must involve an odd power of x – and correspondingly an even power in $V(x)$.

(iii) If the leading term in (1) is even and a quartic or higher ($x^4, x^6, \ldots$), then the oscillations have no linear (harmonic) approximation and they are referred to as 'intrinsically non-linear'.[1]

(iv) Intrinsically non-linear oscillations can also occur when $V(x)$ is not differentiable at $x = 0$. For example, the function $V(x) = V_0 |x|^n$, where V_0 is a positive constant and n is an odd integer. The corresponding oscillations are sometimes called 'dynamically shifted'.[2]

(v) The period T of the oscillatory motion can be calculated (either analytically or numerically) from the formula (1) of Question 5.17.

Question 13.2

A particle of mass m moves in the one-dimensional potential $V(x) = F_0|x|$, where F_0 is a positive constant.

(a) Determine the solution $x(t)$ for the initial conditions $x(0) = x_0 (> 0)$ and $\dot{x}(0) = 0$. Deduce that the angular frequency of the oscillations is

$$\omega = \tfrac{1}{2}\pi\sqrt{F_0/2mx_0}. \tag{1}$$

(b) Plot a graph of $x(t)$ versus t up to $t = 7\,\text{s}$, for $x_0 = 1$, $m = 1$ and $F_0 = 1$ (in some units). On the same axes, plot the harmonic oscillation $x_0 \cos \omega t$.

(c) Include a frictional force $-\alpha \dot{x}$ (where α is a positive constant), and use *Mathematica* to solve the equation of motion

$$\ddot{x} + \frac{2}{\tau}\dot{x} + \frac{F_0}{m}\operatorname{sign} x = 0, \tag{2}$$

where $\tau = 2m/\alpha$ and $\operatorname{sign} x = 1$ if $x > 0$, and -1 if $x < 0$. Use the same initial conditions and parameters as in (a) and (b), and take $\tau = 20\,\text{s}$. Plot $x(t)$ versus t up to $t = 20\,\text{s}$. On the same axes plot also the damped harmonic oscillation $x_0 e^{-2t/\tau} \cos \omega t$.

---

[1] P. Mohazzabi, "Theory and examples of intrinsically nonlinear oscillators," American Journal of Physics, vol. 72, pp. 492–498, 2004.

[2] W. M. Hartmann, "The dynamically shifted oscillator," American Journal of Physics, vol. 54, pp. 28–32, 1986.

Solution

(a) $F = -dV/dx = -F_o$ if $x > 0$ and $F = F_o$ if $x < 0$. For $x > 0$ the equation of motion $\ddot{x} + F_o/m = 0$ and the initial conditions give

$$x(t) = x_0 - F_o t^2/2m \tag{3}$$

for $0 \leq t \leq \frac{1}{4}T$, where

$$T = 4\sqrt{2mx_0/F_o}. \tag{4}$$

For $t > \frac{1}{4}T$, x becomes negative. The solution to $\ddot{x} - F_o/m = 0$ that matches (3) at $t = \frac{1}{4}T$ is

$$x(t) = -\{x_0 - (F_o/2m)(t - \tfrac{1}{2}T)^2\} \tag{5}$$

for $\frac{1}{4}T \leq t \leq \frac{3}{4}T$. In general,

$$x(t) = (-1)^n\{x_0 - (F_o/2m)(t - \tfrac{1}{2}nT)^2\} \tag{6}$$

for $\frac{1}{4}(2n-1)T \leq t \leq \frac{1}{4}(2n+1)T$ and $n = 0, 1, 2, \ldots$. The period of the motion is T, and $\omega = 2\pi/T$ is given by (1).

(b) Cell 1 of the *Mathematica* notebook given below yields the graph:

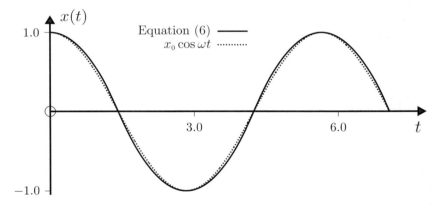

The solution (6) has a Fourier series[3]

$$x(t) = \frac{32x_0}{\pi^3}\left[\cos\omega t - \frac{1}{3^3}\cos 3\omega t + \frac{1}{5^3}\cos 5\omega t - \cdots\right], \tag{7}$$

which is reasonably approximated by the harmonic oscillation $x_0 \cos \omega t$, as is evident in the figure.

(c) In (2) we use the *Mathematica* function Sign[x] that includes the value $\text{sign}\, 0 = 0$. (This does not affect the numerical calculations.) The relevant code is given in cell 2 of the notebook below.

---

[3] I. R. Gatland, "Theory of a nonharmonic oscillator," American Journal of Physics, vol. 59, pp. 155–158, 1991.

458 Solved Problems in Classical Mechanics

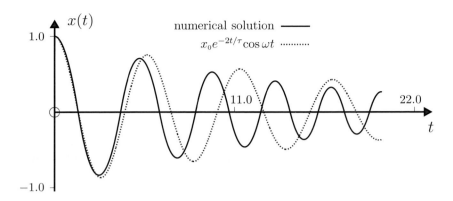

Comments

(i) The dependence of ω on the amplitude x_0 is a feature of non-linear oscillations, and contrasts with harmonic oscillations, where $\omega = \sqrt{k/m}$ is independent of x_0.

(ii) Approximate analytical solutions to (2) have been obtained.[3]

(iii) The potential $V = F_o|x|$ has been used in the study of various non-linear systems including an elastic ball bouncing vertically.[1] Here, $V(x) = mgx$ for $x \geq 0$ and $V(x) = \infty$ if $x < 0$. The oscillations are intrinsically non-linear. Instead of (4) the period is $T = 2\sqrt{2x_0/g}$.

```
In[1]:= x0 = 1; F0 = 1; m = 1; T = 4 Sqrt[(2 m x0)/F0]; ω0 = (2π)/T; x[t_] := x0 - (F0 t^2)/(2 m);

f[t_] := Piecewise[{{x[t], t ≤ T/4}, {-x[t - T/2], T/4 < t ≤ 3T/4},
     {x[t - T], 3T/4 < t ≤ 5T/4}}];

g1 = Plot[{f[t]}, {t, 0, 5T/4}, PlotStyle → {Blue, DotDashed, Thin}];

g2 = Plot[{x0 Cos[ω0 t]}, {t, 0, 5T/4}, PlotStyle → {Orange, Dashed, Thin}];

Show[g1, g2]

In[2]:= Clear[x]; τ = 20; tmax = 4 T;
Sol = NDSolve[{x''[t] + (2 x'[t])/τ + (F0 Sign[x[t]])/m == 0, x[0] == x0, x'[0] == 0},
     {x[t], x'[t], x''[t]}, {t, 0, tmax}, MaxSteps → 100000];

g1 = Plot[{x[t]/.Sol}, {t, 0, tmax}, PlotRange → All, PlotStyle → {Black}];
g2 = Plot[{x0 Exp[-(2t)/τ] Cos[ω0 t]}, {t, 0, tmax}, PlotRange → All,
     PlotStyle → {Orange, Dashed, Thin}];
Show[g1, g2]
```

Question 13.3

Consider a particle of mass m oscillating in the one-dimensional potential

$$V(x) = \begin{cases} \frac{1}{2}m\omega_0^2(x+X)^2 & \text{if } x \geq 0 \\ \frac{1}{2}m\omega_0^2(x-X)^2 & \text{if } x \leq 0, \end{cases} \qquad (1)$$

where ω_0 and X are constants

(a) Plot $V(x)$ for $X > 0$ and $X < 0$. Show also the corresponding graphs of $F(x)$.
(b) Use (1) of Question 5.17 to show that the angular frequency of oscillations of amplitude x_0 is given by

$$\omega(x_0) = \omega_0 \left[1 - \frac{2}{\pi} \sin^{-1}\left(\frac{X}{x_0+X}\right) \right]^{-1}, \qquad (2)$$

where $x_0 > -2X$ if $X < 0$. Plot the graphs of ω/ω_0 versus $x_0/|X|$.

Solution

(a) The first pair of diagrams below shows the two parabolas in (1) and the potential $V(x)$ (solid curves). If $X > 0$ the minimum at the origin is approximately V-shaped and becomes more so as X increases, whereas for $X < 0$ the potential is bimodal with minima at $\pm X$ and a Λ-shaped local maximum at the origin. The second pair of diagrams shows the restoring force versus displacement.

(b) The angular frequency is $\omega = 2\pi/T$, where

$$T = \int_{x_1}^{x_2} \sqrt{\frac{2m}{\{V(x_2) - V(x)\}}}\, dx. \qquad (3)$$

Here, $V(x)$ is given by (1), and x_1 and x_2 are the classical turning points. Let $x_2 = x_0$. Then, $V(x_2) = \frac{1}{2}m\omega_0^2(x_0+X)^2$, and $x_1 = -x_0$ (provided $x_0 > -2X$ when $X < 0$). By evaluating the integral in (3) we obtain (2). If $x_0 < -2X$ when $X < 0$, the particle oscillates harmonically with angular frequency ω_0 in either one of the two wells: ω has a discontinuity equal to $\frac{1}{2}\omega_0$ at $x_0 = -2X$ as shown in the graph of ω/ω_0 versus $x_0/|X|$.

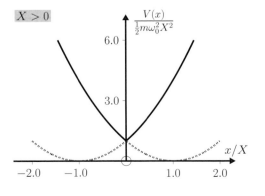
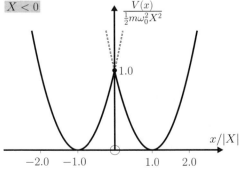

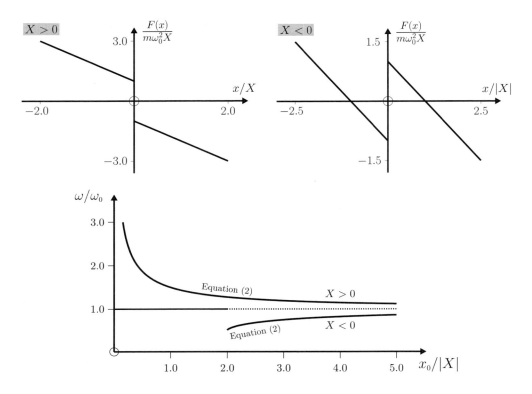

Comment

The above theory has been applied to the vibrations of a two-point librator and of a spring doorstop.[2]

Question 13.4

Consider the arrangement shown below, where a mass m is attached to two light identical springs that are in turn attached to two rigid walls a distance 2ℓ apart. The springs have force constant k and equilibrium length ℓ_0. Neglect the force of gravity and assume that the system is frictionless.

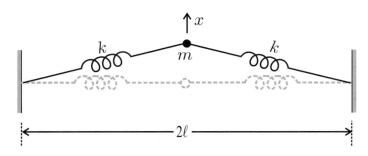

(a) Suppose $\ell = \ell_0$. Show that for small transverse displacements ($|x| \ll \ell_0$) the equation of motion is
$$\ddot{x} + \gamma x^3 = 0, \qquad (1)$$
where $\gamma = k/m\ell_0^2$ is a positive constant.

(b) Suppose $\ell > \ell_0$. Show that the equation of motion (1) changes to
$$\ddot{x} + \alpha x + \gamma x^3 = 0, \qquad (2)$$
where $\alpha = 2k(\ell - \ell_0)/m\ell$ and $\gamma = k/m\ell^2$ are positive constants.

Solution

(a) When m is displaced by an amount x, the extension of each spring is
$$\sqrt{\ell_0^2 + x^2} - \ell_0 = \ell_0\sqrt{1 + x^2/\ell_0^2} - \ell_0 \approx x^2/2\ell_0.$$
This produces a tension $T = kx^2/2\ell_0$ in each spring and thus a net restoring force on m equal to
$$F = -2Tx/\ell_0 = -kx^3/\ell_0^2. \qquad (3)$$
The corresponding equation of motion is given by (1).

(b) When $\ell > \ell_0$ the extension is $\sqrt{\ell^2 + x^2} - \ell_0 \approx \ell - \ell_0 + x^2/2\ell$. The tension T in each spring, which is k times this extension, results in a net restoring force equal to $-2Tx/\ell$. So, the equation of motion is
$$m\ddot{x} + \{2k(\ell - \ell_0)/\ell\}x + (k/\ell^2)x^3 = 0. \qquad (4)$$

Comments

(i) The effect of giving the springs an initial tension is to change the equation of motion from (1) to (2).
(ii) Equation (2) has a linear (harmonic) approximation in the limit of small oscillations ($|x| \ll \sqrt{\alpha/\gamma}$, see Question 13.1), whereas the oscillations corresponding to (1) are intrinsically non-linear.
(iii) Approximate solutions to the non-linear equations (1) and (2) are of general interest, and they are studied in the next four questions.

Question 13.5

Consider the equation of motion
$$\ddot{x} + \gamma x^3 = 0 \qquad (1)$$
(γ is a positive constant) with initial conditions
$$x = x_0 \quad \text{and} \quad \dot{x} = 0 \quad \text{at } t = 0. \qquad (2)$$

(a) Show that the period of the oscillations is
$$T = \sqrt{\frac{2}{\gamma}} B(\tfrac{1}{4}, \tfrac{1}{2}) \frac{1}{x_0} \approx \frac{7.4163}{\sqrt{\gamma}} \frac{1}{x_0}. \qquad (3)$$

Here, B denotes the beta function
$$B(\tfrac{1}{4}, \tfrac{1}{2}) = \int_0^1 u^{-3/4}(1-u)^{-1/2} du \approx 5.2441. \qquad (4)$$

(b) Show that the Fourier expansion for $x(t)$ can be expressed as
$$x(t) = \sum_{n=0}^{\infty} a_{2n+1} \cos(2n+1)\omega t \qquad (\omega = 2\pi/T). \qquad (5)$$

(c) Use an integration by parts (twice) to express the Fourier coefficients
$$a_n = \frac{\omega}{\pi} \int_0^{2\pi/\omega} x(t) \cos n\omega t\, dt \qquad (6)$$

as
$$a_n = \frac{\gamma}{\pi n^2 \omega} \int_0^{2\pi/\omega} x^3(t) \cos n\omega t\, dt. \qquad (7)$$

(d) Use the ansatz
$$x(t) = a_1 \cos \omega t \qquad (8)$$

in (7) to obtain the estimates
$$a_1 = 2\omega/\sqrt{3\gamma}, \quad a_3 = a_1/27 \approx 0.03704 a_1. \qquad (9)$$

Impose the initial condition $(2)_1$ on the corresponding approximate solution $x = a_1 \cos \omega t + a_3 \cos 3\omega t$ and deduce that this yields
$$T_1 \approx 1.014\, T \qquad (10)$$

for the period, where T is the actual value (3). (Hint: Use the identity
$$\cos^3 \omega t = \tfrac{3}{4} \cos \omega t + \tfrac{1}{4} \cos 3\omega t\,.) \qquad (11)$$

(e) Repeat (d), starting with the more accurate approximation
$$x(t) = a_1 \cos \omega t + a_3 \cos 3\omega t, \qquad (12)$$

instead of (8). Show that this yields
$$a_1 \approx 0.9782(2\omega/\sqrt{3\gamma}), \quad a_3 \approx 0.04501 a_1, \quad a_5 \approx 0.001723 a_1, \qquad (13)$$

and an improved estimate for the period
$$T_2 \approx 1.002\, T. \qquad (14)$$

(Hint: Use also the identity
$$\cos^2 \omega t \cos 3\omega t = \tfrac{1}{4} \cos \omega t + \tfrac{1}{2} \cos 3\omega t + \tfrac{1}{4} \cos 5\omega t\,.) \qquad (15)$$

Solution

(a) According to Question 5.17,

$$T = \int_{-x_0}^{x_0} \sqrt{\frac{2m}{V(x_0) - V(x)}}\, dx. \qquad (16)$$

Here, $V(x) = \tfrac{1}{4} m\gamma x^4$. With the substitution $u = (x/x_0)^4$, (16) yields (3):

$$T = \sqrt{\frac{2}{\gamma}\frac{1}{x_0}} \int_0^1 u^{-3/4}(1-u)^{-1/2}\, du. \qquad (17)$$

(b) Start with the standard Fourier expansion

$$x(t) = \tfrac{1}{2} a_0 + \sum_{n=1}^{\infty}(a_n \cos n\omega t + b_n \sin n\omega t), \qquad (18)$$

where

$$a_n = (\omega/\pi)\int_0^{2\pi/\omega} x(t)\cos n\omega t\, dt \qquad (n=0,1,\ldots), \qquad (19)$$

$$b_n = (\omega/\pi)\int_0^{2\pi/\omega} x(t)\sin n\omega t\, dt \qquad (n=1,2,\ldots). \qquad (20)$$

Now, time-reversal invariance of a non-dissipative system and the initial condition $\dot{x} = 0$ at $t = 0$ require that the motion from $t = 0$ to $T \,(= 2\pi/\omega)$ and the reversed motion from $t = T$ to 0 be the same. Therefore, in the interval $[0, T]$, $x(t)$ is an even function about $t = \pi/\omega$ and an odd function about $t = \pi/2\omega$ and $3\pi/2\omega$:

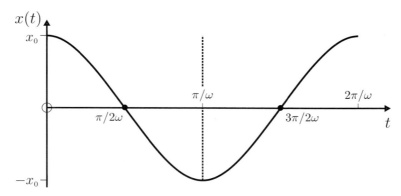

It is helpful to shift the time origin to $t = \pi/\omega$ by letting $t = t' + \pi/\omega$. Then

$$a_n = (-1)^n(\omega/\pi)\int_{-\pi/\omega}^{\pi/\omega} x(t' + \pi/\omega)\cos n\omega t'\, dt' \qquad (21)$$

$$b_n = (-1)^n(\omega/\pi)\int_{-\pi/\omega}^{\pi/\omega} x(t' + \pi/\omega)\sin n\omega t'\, dt'. \qquad (22)$$

Consider first (22): x is an even function of t', while $\sin n\omega t'$ is an odd function – therefore, $b_n = 0$ for all n. In (21), if n is even then the integrals from $t' = -\pi/\omega$ to $-\pi/2\omega$ and 0 to $\pi/2\omega$ cancel, as do those from $-\pi/2\omega$ to 0 and $\pi/2\omega$ to π/ω. If n is odd no such cancellation occurs. So the Fourier series for $x(t)$ is a cosine series containing only odd harmonics, as in (5).

(c) We integrate (6) by parts in the following manner:

$$a_n = \frac{1}{\pi n}\int_0^{2\pi/\omega} x(t)\left(\frac{d}{dt}\sin n\omega t\right)dt = -\frac{1}{\pi n}\int_0^{2\pi/\omega} \frac{dx}{dt}\sin n\omega t\, dt.$$

A repetition of this process, and use of the initial condition $(2)_2$, yields

$$a_n = -\frac{1}{\pi n^2 \omega}\int_0^{2\pi/\omega}\frac{d^2 x}{dt^2}\cos n\omega t\, dt, \qquad (23)$$

which, in view of the equation of motion (1), is (7).

(d) With the approximation (8) and the identity (11), equation (7) with $n=1$ gives

$$a_1 = \frac{\gamma}{\pi\omega}a_1^3\int_0^{2\pi/\omega}\left(\tfrac{3}{4}\cos\omega t + \tfrac{1}{4}\cos 3\omega t\right)\cos\omega t\, dt = \frac{3\gamma a_1^3}{4\omega^2}, \qquad (24)$$

because of the orthonormality condition

$$\frac{\omega}{\pi}\int_0^{2\pi/\omega}\cos m\omega t\cos n\omega t\, dt = \delta_{mn}. \qquad (25)$$

Thus, $a_1 = 2\omega/\sqrt{3\gamma}$. Similarly, by setting $n=3$ in (7) we obtain

$$a_3 = \frac{\gamma}{36\pi\omega}a_1^3\int_0^{2\pi/\omega}\cos^2 3\omega t\, dt = \frac{\gamma a_1^3}{36\omega^2} = \frac{a_1}{27}. \qquad (26)$$

Thus, the solution approximated by the first two harmonics is

$$x(t) = \frac{2\omega}{\sqrt{3\gamma}}\left(\cos\omega t + \frac{1}{27}\cos 3\omega t\right). \qquad (27)$$

By imposing the initial condition $(2)_1$ on (27) we obtain for the angular frequency and the period

$$\omega = \left(27\sqrt{3\gamma}/56\right)x_0, \qquad (28)$$

$$T_1 = 2\pi/\omega = \frac{7.524}{\sqrt{\gamma}}\frac{1}{x_0} = 1.014\,T. \qquad (29)$$

Equations (27) and (28) give the approximate solution

$$x(t) = x_0\left(\frac{27}{28}\cos\omega t + \frac{1}{28}\cos 3\omega t\right). \qquad (30)$$

(e) For the ansatz (12) we write

$$x^3 \approx a_1^3 \cos^3 \omega t + 3a_1^2 a_3 \cos^2 \omega t \cos 3\omega t$$
$$= \tfrac{3}{4}\left(a_1^3 + a_1^2 a_3\right)\cos \omega t + \left(\tfrac{1}{4}a_1^3 + \tfrac{3}{2}a_1^2 a_3\right)\cos 3\omega t + \tfrac{3}{4}a_1^2 a_3 \cos 5\omega t \,. \quad (31)$$

In the first step above we have neglected terms in $a_1 a_3^2$ and a_3^3, and in the second step we have used the identities (11) and (15). By substituting (31) in (7) and setting $n = 1, 3$ and 5 in turn, and using (25), we obtain the three equations:

$$a_1 = \frac{3\gamma}{4\omega^2}\left(a_1^3 + a_1^2 a_3\right), \qquad a_3 = \frac{\gamma}{9\omega^2}\left(\tfrac{1}{4}a_1^3 + \tfrac{3}{2}a_1^2 a_3\right), \qquad a_5 = \frac{3\gamma}{100\omega^2}\, a_1^2 a_3 \,. \quad (32)$$

The ratio $(32)_2 \div (32)_1$ yields a quadratic equation

$$27\,(a_3/a_1)^2 + 21\,(a_3/a_1) - 1 = 0 \,. \quad (33)$$

One root is

$$a_3/a_1 \approx 0.04501 \,. \quad (34)$$

(We ignore the second root ≈ -1.85 because it gives an imaginary value for a_1 in $(32)_1$.) From $(32)_1$ and (34) we have

$$a_1 \approx 0.9782 \frac{2\omega}{\sqrt{3\gamma}} \,. \quad (35)$$

Equations $(32)_3$, (34) and (35) yield

$$a_5/a_1 \approx \frac{3\gamma}{100\omega^2}\, 4.501 \times 10^{-2} a_1^2 \approx 0.001723 \,. \quad (36)$$

Thus, the solution approximated by the first three harmonics of (5) is

$$x(t) = 0.9782 \frac{2\omega}{\sqrt{3\gamma}}\left(\cos \omega t + 0.04501 \cos 3\omega t + 0.001723 \cos 5\omega t\right). \quad (37)$$

The initial condition $(2)_1$ applied to (37) requires

$$\omega = 0.4883 \sqrt{3\gamma}\, x_0 \,, \quad (38)$$

and so

$$T_2 = \frac{7.4287}{\sqrt{\gamma}} \frac{1}{x_0} = 1.002\, T \,. \quad (39)$$

Equations (37) and (38) give the approximate solution

$$x(t) = x_0 \left(0.9553 \cos \omega t + 0.04300 \cos 3\omega t + 0.001646 \cos 5\omega t\right). \quad (40)$$

Comments

(i) According to (3), the period T is inversely proportional to x_0 (this is unlike simple harmonic motion where the period is independent of amplitude).

(ii) The approximate solutions obtained above are compared with numerical solutions in the following question.

Question 13.6

(a) Rewrite the equation of motion $\ddot{x}+\gamma x^3 = 0$ of Question 13.5 in the dimensionless form
$$\frac{d^2 X}{d\tau^2} + 2B^2(\tfrac{1}{4}, \tfrac{1}{2}) X^3 \approx \frac{d^2 X}{d\tau^2} + 55.00\, X^3 = 0, \tag{1}$$
where $X = x/x_0$, $\tau = t/T$ and T is given by (3) of Question 13.5.

(b) Use the approximation (40) of Question 13.5 and its derivatives to plot graphs of $X(\tau)$, $\dot{X}(\tau)$ and $\ddot{X}(\tau)$ for $0 \leq \tau \leq 1$ and the initial conditions $X = 1.0$ and $\dot{X} = dX/d\tau = 0$ at $\tau = 0$. On the same graph, use points to represent values obtained from a numerical solution of (1).

(c) With a step size $\Delta\tau = 0.001$, create a data file from the numerical solutions for $X(\tau)$ obtained in (b). Next, apply *Mathematica*'s `FindFit` function to this data and calculate the first three Fourier coefficients a_1, a_3 and a_5. Compare these with the approximations in (d) and (e) of Question 13.5 (i.e. the first and second Fourier approximations).

(d) Plot a graph that shows the relative error of both these Fourier approximations for $X(\tau)$ in the first cycle of the motion.

Solution

(a) By expressing the equation of motion in terms of X and τ, and using (3) of Question 13.5, we obtain (1).

(b) The graphs and the *Mathematica* notebook used to obtain them are shown below.

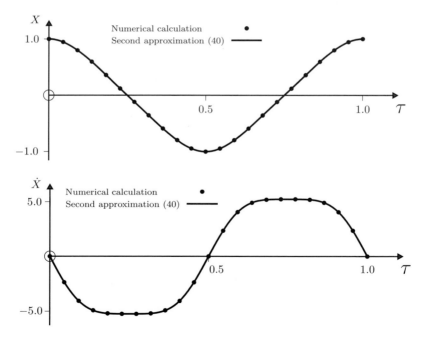

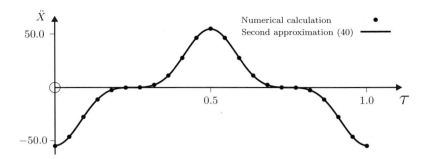

(c) The second cell in the *Mathematica* notebook below uses the data file `FourierDat` to calculate the Fourier coefficients. The values obtained are shown in the following table, where the first and second approximations refer to the values in (30) and (40) of Question 13.5.

	a_1/x_0	a_3/x_0	a_5/x_0
First approximation	0.9643	0.03571	0
Second approximation	0.9553	0.04300	0.001646
Numerical value	0.9550	0.04305	0.001861

(d) The following graph shows the fractional error $(X^*(\tau) - X(\tau))/X(\tau)$, where $X^*(\tau)$ represents either the first or second approximation. The error in the second Fourier approximation is less than 0.3%.

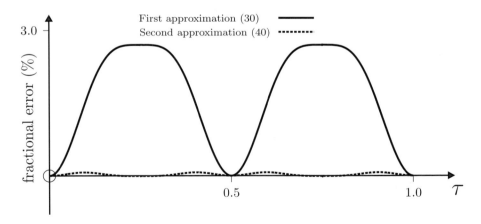

```
In[1]:= τmax = 1.0;

       Sol = NDSolve[{X''[τ] + 2Beta[1/4, 1/2]^2 X[τ]^3 == 0, X[0] == 1.0,
              X'[0] == 0}, {X[τ], X'[τ], X''[τ]}, {τ, 0, τmax}];
```

```
In[2]:= FourierDat = Table[{τ, First[X[τ]/.Sol]}, {τ, 0, τmax, τmax/100}];

       Coeff = FindFit[FourierDat, a1 Cos[2π τ] + a3 Cos[6π τ] + a5 Cos[10π τ],
         {a1, a3, a5}, τ]

       gr1 = Plot[Evaluate[a1 Cos[2π τ] + a3 Cos[6π τ] + a5 Cos[10π τ]/.Coeff],
         {τ, 0, τmax}, PlotStyle → {Red, Thin}];  gr2 = ListPlot[FourierDat];

       Show[gr1, gr2]
```

Question 13.7

Consider the equation of motion of the Duffing oscillator:

$$\ddot{x} + \alpha x + \gamma x^3 = 0, \tag{1}$$

where α and γ are positive constants. The initial conditions are

$$x = x_0 \quad \text{and} \quad \dot{x} = 0 \quad \text{at } t = 0. \tag{2}$$

(a) Show that the period of the oscillations is

$$T(\epsilon) = \frac{2}{\pi} T_0 \int_0^1 \frac{du}{\sqrt{1 - u^2 + \epsilon(1 - u^4)}}, \tag{3}$$

where $T_0 = 2\pi/\sqrt{\alpha}$ is the period when $\gamma = 0$ and ϵ is the dimensionless quantity

$$\epsilon = \gamma x_0^2 / 2\alpha. \tag{4}$$

(Note: in the numerical work that follows, it is convenient to express (3) as

$$\frac{T(\epsilon)}{T_0} = \frac{2}{\pi\sqrt{1+\epsilon}} F\left(\frac{-\epsilon}{1+\epsilon}\right), \tag{5}$$

where F is a complete elliptic integral of the first kind, as defined in *Mathematica*.)

(b) Extend the Fourier analysis of Question 13.5 to obtain approximate solutions for $x(t)$ and $T(\epsilon)$. Plot a graph of $T(\epsilon)/T_0$ versus ϵ for $0 \leq \epsilon \leq 10$ and compare it with the numerical values obtained from (5) using *Mathematica*. Plot also the fractional error in $T(\epsilon)$ versus ϵ.

Solution

(a) Use the formula (see Question 5.17)

$$T = \int_{-x_0}^{x_0} \sqrt{\frac{2m}{V(x_0) - V(x)}}\, dx. \tag{6}$$

Here, $V(x) = \frac{1}{2} m \alpha x^2 + \frac{1}{4} m \gamma x^4$, and the substitution $u = x/x_0$ yields (3).

(b) The relevant results from Question 13.5 are the Fourier expansion

$$x(t) = \sum_{n=0}^{\infty} a_{2n+1} \cos(2n+1)\omega t, \qquad (7)$$

the formula for the Fourier coefficients

$$a_n = -\frac{1}{\pi n^2 \omega} \int_0^{2\pi/\omega} \ddot{x} \cos n\omega t \, dt, \qquad (8)$$

and the orthonormality relation

$$\frac{\omega}{\pi} \int_0^{2\pi/\omega} \cos m\omega t \cos n\omega t \, dt = \delta_{mn}. \qquad (9)$$

Equations (1), (2)$_2$ and (9) allow us to express (8) as

$$a_n = \frac{\alpha}{n^2 \omega^2} a_n + \frac{\gamma}{\pi n^2 \omega} \int_0^{2\pi/\omega} x^3(t) \cos n\omega t \, dt \qquad (n = 1, 3, 5\ldots). \qquad (10)$$

The last term in (10) has been discussed in Question 13.5. Thus we have:

1. The ansatz $x = a_1 \cos \omega t$ in (10) yields two equations,

$$(\omega^2 - \alpha)a_1 = 3\gamma a_1^3/4, \qquad (9\omega^2 - \alpha)a_3 = \gamma a_1^3/4, \qquad (11)$$

and their solutions for ω^2 and a_3 in terms of a_1 are

$$\omega^2 = \alpha + 3\gamma a_1^2/4, \qquad \frac{a_3}{a_1} = \frac{\gamma}{32\alpha + 27\gamma a_1^2} a_1^2. \qquad (12)$$

Equations (12), together with the initial condition

$$a_1 + a_3 = x_0, \qquad (13)$$

determine a_1, a_3 and ω. We obtain approximate solutions as follows. Because $\alpha > 0$ it follows from (12)$_2$ that $a_3 < a_1/27$. So, as a first approximation, we set $a_1 = x_0$ in the right-hand side of (12)$_2$. Then,

$$\frac{a_3}{a_1} = \frac{\epsilon}{16 + 27\epsilon}, \qquad (14)$$

where ϵ is given by (4). A better approximation for a_1 follows from (13) and (14):

$$a_1 = \frac{16 + 27\epsilon}{16 + 28\epsilon} x_0, \qquad (15)$$

and this gives for the frequency in (12)$_1$

$$\omega = \sqrt{\alpha + \frac{3\alpha\epsilon}{2}\left(\frac{16 + 27\epsilon}{16 + 28\epsilon}\right)^2}. \qquad (16)$$

Thus, based on the first two harmonics we have the following approximations:

$$\frac{x(t)}{x_0} = \left(\frac{16+27\epsilon}{16+28\epsilon}\right)\cos\omega t + \left(\frac{\epsilon}{16+28\epsilon}\right)\cos 3\omega t, \qquad (17)$$

$$\frac{T_1(\epsilon)}{T_0} = \left[1 + \frac{3\epsilon}{2}\left(\frac{16+27\epsilon}{16+28\epsilon}\right)^2\right]^{-1/2}, \qquad (18)$$

where $T_0 = 2\pi/\sqrt{\alpha}$. Note that (14)–(16) have the desired property that in the limit $\alpha \to 0$ (i.e. $\epsilon \to \infty$) they reduce to the exact solutions to (12) and (13) that were obtained in Question 13.5. The graph of (18) is shown below, together with numerical values of (5) calculated using *Mathematica*'s EllipticK function.

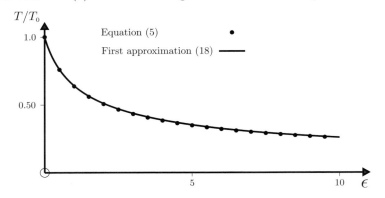

2. The ansatz: $x(t) = a_1 \cos\omega t + a_3 \cos 3\omega t$ in (10) yields three equations

$$(\omega^2 - \alpha)a_1 = \tfrac{3}{4}\gamma(a_1^3 + a_1^2 a_3) \qquad (19)$$

$$(9\omega^2 - \alpha)a_3 = \gamma(\tfrac{1}{4}a_1^3 + \tfrac{3}{2}a_1^2 a_3) \qquad (20)$$

$$(25\omega^2 - \alpha)a_5 = \tfrac{3}{4}\gamma a_1^2 a_3. \qquad (21)$$

Equation (19) expresses ω^2 in terms of a_1 and a_3/a_1:

$$\omega^2 = \alpha + \tfrac{3}{4}\gamma(1 + a_3/a_1)a_1^2. \qquad (22)$$

If this is substituted in (20) we obtain an equation for a_3/a_1 in terms of a_1^2:

$$\left\{32\alpha + 27\gamma\left(1 + \frac{a_3}{a_1}\right)a_1^2\right\}\frac{a_3}{a_1} = \gamma a_1^2\left(1 + \frac{6a_3}{a_1}\right). \qquad (23)$$

The Fourier coefficients must satisfy the initial condition

$$a_1 + a_3 + a_5 = x_0. \qquad (24)$$

We obtain an approximate solution to (23) by first neglecting a_3 and a_5 in (24). Thus, we set $a_1^2 = x_0^2$ in (23) to obtain the quadratic equation

$$27\epsilon(a_3/a_1)^2 + (16 + 21\epsilon)(a_3/a_1) - \epsilon = 0. \tag{25}$$

The solution that remains finite when $\epsilon \to 0$ (i.e. $\gamma \to 0$) is

$$\frac{a_3}{a_1} = G(\epsilon) = \frac{1}{54\epsilon}\left\{\sqrt{108\epsilon^2 + (16 + 21\epsilon)^2} - (16 + 21\epsilon)\right\}. \tag{26}$$

Similarly, from (21), (22) and (26) with $a_1^2 = x_0^2$ we have

$$\frac{a_5}{a_1} = \frac{\epsilon G(\epsilon)}{16 + 25\epsilon + 25\epsilon G(\epsilon)}. \tag{27}$$

Finally, we use (26) and (27) in (24) and solve for a_1:

$$\frac{a_1}{x_0} = H(\epsilon) = \frac{16 + 25\epsilon\{1 + G(\epsilon)\}}{16\{1 + G(\epsilon)\} + 25\epsilon\{1 + G^2(\epsilon)\} + 51\epsilon G(\epsilon)}. \tag{28}$$

Thus, the solution approximated by the first three harmonics is

$$\frac{x(t)}{x_0} = H(\epsilon)\left\{\cos\omega t + G(\epsilon)\cos 3\omega t + \frac{\epsilon G(\epsilon)}{16 + 25\epsilon\{1 + G(\epsilon)\}}\cos 5\omega t\right\}. \tag{29}$$

Here, ω is given by (22), (26) and (28) as

$$\omega = \sqrt{\alpha\left[1 + \tfrac{3}{2}\epsilon\{1 + G(\epsilon)\}H^2(\epsilon)\right]}, \tag{30}$$

and so the period is

$$\frac{T_2(\epsilon)}{T_0} = \frac{1}{\sqrt{1 + \tfrac{3}{2}\epsilon\{1 + G(\epsilon)\}H^2(\epsilon)}}. \tag{31}$$

The graph of (31) is not shown because it is almost coincident with the graph of $T_1(\epsilon)$ above. Instead, a plot of the fractional error $(T^* - T)/T$ versus ϵ is given below, where T^* represents either $T_1(\epsilon)$ or $T_2(\epsilon)$ and T is given by (3).

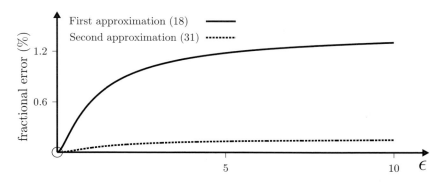

Comment

The Duffing oscillator has both mechanical and electrical analogues. A damped, driven Duffing oscillator is considered in Questions 13.11 and 13.12.

Question 13.8

(a) Rewrite the equation of motion $\ddot{x} + \alpha x + \gamma x^3 = 0$ of Question 13.7 in the dimensionless form

$$\frac{d^2 X}{d\tau^2} + \frac{16}{1+\epsilon} F^2\left(\frac{-\epsilon}{1+\epsilon}\right)\{X + 2\epsilon X^3\} = 0, \tag{1}$$

where $X = x/x_0$, $\tau = t/T$, F is a complete elliptic integral of the first kind (as defined in *Mathematica*) and ϵ and T are given by (4) and (5) of Question 13.7.

(b) Obtain numerical solutions to (1) for $0 \leq \epsilon \leq 10$, and with initial conditions $X = 1.0$ and $\dot{X} = dX/d\tau = 0$ at $\tau = 0$. Use *Mathematica*'s FindFit function to calculate the Fourier coefficients $a_1(\epsilon)$, $a_3(\epsilon)$ and $a_5(\epsilon)$ of these numerical solutions, and show these as a discrete set of points on a graph. On each graph also plot the approximate $a_i(\epsilon)$ contained in (17) and (29) of Question 13.7.

(c) Take $\epsilon = 0.9$, and use the second approximation (29) of Question 13.7, and its derivatives, to plot graphs of $X(\tau)$, $\dot{X}(\tau)$ and $\ddot{X}(\tau)$, where $\tau = \omega t/2\pi$, in the interval $0 \leq \tau \leq 1$. On the same graph, use points to represent values obtained from the numerical solution of the equation of motion.

Solution

(a) By expressing the equation of motion in terms of X and τ, and using (4) and (5) of Question 13.7 we obtain (1).

(b) Replace cell 1 in the *Mathematica* notebook of Question 13.6 with the following:

```
In[1]:= ε = 0.9; τmax = 1.0;
    Sol = NDSolve[{X''[τ] + 16/(1+ε) (X[τ] + 2ε X[τ]^3) (EllipticK[-ε/(1+ε)])^2 == 0,
        X[0] == 1.0, X'[0] == 0}, {X[τ], X'[τ], X''[τ]}, {τ, 0, τmax}];
```

In the following figures the horizontal dotted lines show the asymptotic values of the approximate Fourier coefficients in (29) of Question 13.7 for $\epsilon \to \infty$.

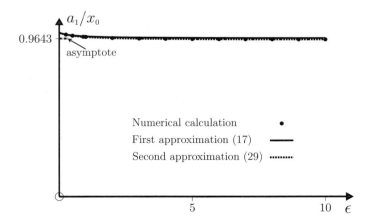

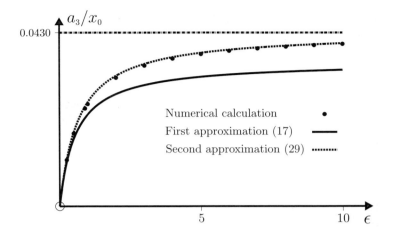
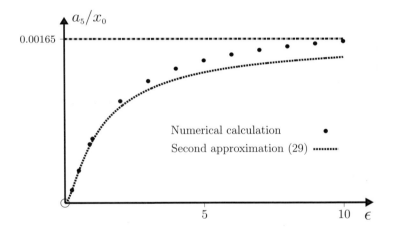
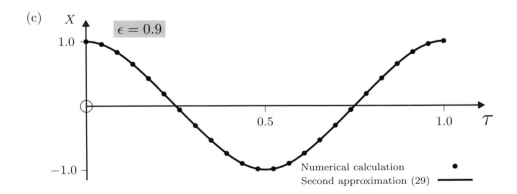

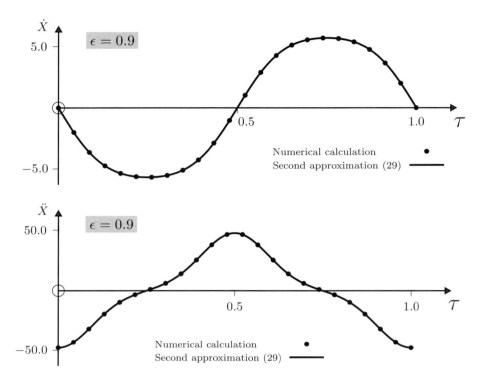

Comment

The numerical solutions for $X(\tau)$ approach their asymptotic (large ϵ) behaviour already at quite low values of ϵ.

Question 13.9

Consider the equation of motion
$$\ddot{x} + \alpha x - \beta x^2 = 0, \tag{1}$$
where α and β are positive constants, and the initial conditions are
$$x = x_0 \; (>0) \quad \text{and} \quad \dot{x} = 0 \quad \text{at } t = 0. \tag{2}$$

(a) Show that the period of bounded motion is given by
$$T(\epsilon) = \frac{T_0}{\pi} \int_{u(\epsilon)}^{1} \frac{du}{\sqrt{1 - u^2 + \tfrac{2}{3}\epsilon(u^3 - 1)}}, \tag{3}$$
where $T_0 = 2\pi/\sqrt{\alpha}$ is the period when $\beta = 0$ and ϵ is the dimensionless quantity
$$\epsilon = \beta x_0/\alpha < 1, \tag{4}$$

and
$$u(\epsilon) = -\frac{3-2\epsilon}{4\epsilon}\left(\sqrt{1+\frac{8\epsilon}{3-2\epsilon}}-1\right). \tag{5}$$

(b) Extend the Fourier analysis of Question 13.5 to obtain approximate solutions for bounded motion $x(t)$ when $\epsilon \ll 1$, based on 1. the first two harmonics and 2. the first three harmonics. Compare the results for the corresponding approximate periods $T(\epsilon)$ with (3) by plotting graphs of $T(\epsilon)$ versus ϵ.

Solution

(a) For the asymmetric potential $V(x) = \frac{1}{2}m\alpha x^2 - \frac{1}{3}m\beta x^3$ the period is
$$T = \int_{x_1}^{x_0} \sqrt{\frac{2m}{V(x_0) - V(x)}}\, dx, \tag{6}$$

where x_1 is a solution to $V(x) = V(x_0)$. By making the substitution $u = x/x_0$, we can express (6) as
$$T(\epsilon) = \frac{2}{\sqrt{\alpha}}\int_{u(\epsilon)}^{1}\frac{du}{\sqrt{1-u^2+\frac{2}{3}\epsilon(u^3-1)}}, \tag{7}$$

where $u(\epsilon)$ is a solution to
$$\tfrac{2}{3}\epsilon u^3 - u^2 + (1-\tfrac{2}{3}\epsilon) = 0. \tag{8}$$

One solution of (8) is $u = 1$ (corresponding to the initial point $x = x_0$) and the other is therefore a solution of the quadratic equation
$$2\epsilon u^2 - (3-2\epsilon)u - (3-2\epsilon) = 0. \tag{9}$$

The root that remains finite when $\epsilon \to 0$ is (5). This root is real if $\epsilon < \frac{3}{2}$, which is automatically satisfied because of the condition (4) for bounded motion (x_0 less than the coordinate α/β at which $V(x)$ has a local maximum).

(b) The Fourier analysis of (1) is similar to that in Question 13.5 and we give just an outline. The Fourier expansion for the bounded (periodic) solutions of (1) that satisfies the initial condition $v(0) = 0$ is
$$x(t) = \tfrac{1}{2}a_0 + \sum_{n=1}^{\infty} a_n \cos n\omega t, \tag{10}$$

where
$$a_n = \frac{\omega}{\pi}\int_0^{2\pi/\omega} x(t)\cos n\omega t\, dt$$
$$= -\frac{\beta\omega}{\pi(n^2\omega^2-\alpha)}\int_0^{2\pi/\omega} x^2(t)\cos n\omega t\, dt. \tag{11}$$

Note that because $V(x)$ is asymmetric, the expansion (10) contains both even and odd harmonics (and not just odd harmonics as is the case for a symmetric potential, see Question 13.5).

1. We start by making the approximation $x(t) = \frac{1}{2}a_0 + a_1 \cos\omega t$ in (11). That is,

$$x^2(t) = \tfrac{1}{4}a_0^2 + \tfrac{1}{2}a_1^2 + a_0 a_1 \cos\omega t + \tfrac{1}{2}a_1^2 \cos 2\omega t. \tag{12}$$

Now

$$\int_0^{2\pi/\omega} \cos m\omega t \cos n\omega t \, dt = \frac{\pi}{\omega}\delta_{mn}, \tag{13}$$

and it therefore follows that with the approximation (12) the only non-zero coefficients are a_0, a_1 and a_2, and that these satisfy the three equations

$$a_0 = (\beta/\alpha)(\tfrac{1}{2}a_0^2 + a_1^2), \qquad a_1 = -\frac{\beta}{\omega^2 - \alpha}a_0 a_1, \qquad a_2 = \frac{\beta}{4\omega^2 - \alpha}\tfrac{1}{2}a_1^2. \tag{14}$$

When $\epsilon \ll 1$, the oscillator is almost harmonic. Then, $a_1 \approx x_0$ and $a_0 \ll x_0$. So, $(14)_1$ yields the estimate

$$a_0 \approx \epsilon x_0, \tag{15}$$

and then $(14)_2$ gives

$$\omega^2 \approx \alpha(1 - \epsilon^2). \tag{16}$$

From $(14)_3$ and (16) we have

$$a_2 \approx -\tfrac{1}{6}\epsilon x_0. \tag{17}$$

An improved value for a_1 is obtained by imposing the initial condition

$$\tfrac{1}{2}a_0 + a_1 + a_2 = x_0. \tag{18}$$

From (15), (17) and (18) we find

$$a_1 \approx (1 - \tfrac{1}{3}\epsilon)x_0. \tag{19}$$

Thus, for $\epsilon \ll 1$ the solution based on the first two harmonics is

$$\frac{x(t)}{x_0} = \tfrac{1}{2}\epsilon + (1 - \tfrac{1}{3}\epsilon)\cos\omega t - \tfrac{1}{6}\epsilon\cos 2\omega t, \tag{20}$$

where

$$\omega = \sqrt{\alpha(1 - \epsilon^2)}. \tag{21}$$

The period is

$$T_1(\epsilon) = \frac{T_0}{\sqrt{1 - \epsilon^2}}. \tag{22}$$

2. For the next approximation, $x(t) = \tfrac{1}{2}a_0 + a_1 \cos\omega t + a_2 \cos 2\omega t$, we can express $x^2(t) = \tfrac{1}{4}a_0^2 + \tfrac{1}{2}a_1^2 + (a_0 a_1 + a_1 a_2)\cos\omega t + (\tfrac{1}{2}a_1^2 + a_0 a_2)\cos 2\omega t + a_1 a_2 \cos 3\omega t$, and (11) and (13) yield

$$a_0 = (\beta/\alpha)(\tfrac{1}{2}a_0^2 + a_1^2), \qquad a_1 = \frac{-\beta}{\omega^2 - \alpha}(a_0 a_1 + a_1 a_2)$$

$$a_2 = \frac{-\beta}{4\omega^2 - \alpha}(\tfrac{1}{2}a_1^2 + a_0 a_2), \qquad a_3 = \frac{-\beta}{9\omega^2 - \alpha}a_1 a_2. \tag{23}$$

The initial condition $(2)_1$ requires

$$\tfrac{1}{2}a_0 + a_1 + a_2 + a_3 = x_0. \tag{24}$$

For $\epsilon \ll 1$ we can obtain approximate solutions to these equations by starting with $a_1 \approx x_0$ and iterating. Up to the leading terms in ϵ the results are

$$a_0 \approx \epsilon x_0, \quad a_1 \approx (1 - \tfrac{1}{3}\epsilon)x_0, \quad a_2 \approx -\tfrac{1}{6}\epsilon x_0, \quad a_3 \approx \tfrac{1}{48}\epsilon^2 x_0, \tag{25}$$

$$\omega^2 \approx \alpha(1 - \tfrac{5}{6}\epsilon^2). \tag{26}$$

Thus, the solution based on the first three harmonics is

$$\frac{x(t)}{x_0} = \tfrac{1}{2}\epsilon + (1 - \tfrac{1}{3}\epsilon)\cos\omega t - \tfrac{1}{6}\epsilon\cos 2\omega t + \tfrac{1}{48}\epsilon^2 \cos 3\omega t, \tag{27}$$

and the period is

$$T_2(\epsilon) = \frac{T_0}{\sqrt{1 - \tfrac{5}{6}\epsilon^2}}. \tag{28}$$

The Fourier coefficients in (25) are compared with numerical values in the next question. A comparison of the periods (22) and (28) with a numerical integration of (3) is shown below, together with an improved approximation given in (30).

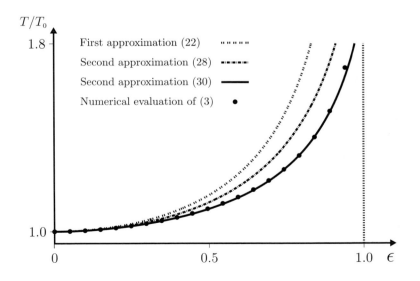

Comments

(i) Higher-order corrections to (25) can be obtained by further iteration of (23) and (24). For example, to order ϵ^3 one finds

$$\left.\begin{array}{ll} a_0 = (\epsilon - \frac{2}{3}\epsilon^2 + \frac{11}{18}\epsilon^3)x_0\,, & a_1 = (1 - \frac{1}{3}\epsilon + \frac{29}{144}\epsilon^2 - \frac{59}{432}\epsilon^3)x_0 \\ a_2 = (-\frac{1}{6}\epsilon + \frac{1}{9}\epsilon^2 - \frac{4}{27}\epsilon^3)x_0\,, & a_3 \approx \frac{1}{48}\epsilon^2\, x_0\,. \end{array}\right\} \quad (29)$$

We have not evaluated the cubic terms in the small coefficient a_3.

(ii) By using these expressions for a_0 and a_2 in $(23)_2$ we obtain a better estimate of ω and hence the period:

$$T_2(\epsilon) = \frac{T_0}{\sqrt{1 - \frac{5}{6}\epsilon^2(1 - \frac{2}{3}\epsilon + \frac{5}{9}\epsilon^2)}}\,. \qquad (30)$$

The fractional error $(T^* - T)/T$ versus ϵ, where T^* is either (28) or (30) and T is given by (3), is plotted below.

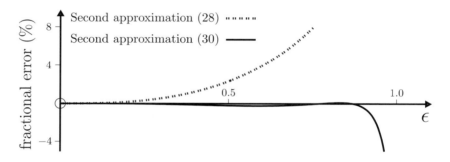

Comment

The error in (30) is less than 0.5% for $\epsilon \lesssim 0.9$.

Question 13.10

(a) Rewrite the equation of motion $\ddot{x} + \alpha x - \beta x^2 = 0$ of Question 13.9 in the dimensionless form

$$\frac{d^2 X}{d\tau^2} + 4\bigl[X - \epsilon X^2\bigr] I^2(\epsilon) = 0\,, \qquad (1)$$

where $X = x/x_0$, $\tau = t/T$ and $T = T_0 I(\epsilon)$ is given by (3) of Question 13.9.

(b) Find numerical solutions of (1) for $0.0001 \leq \epsilon \leq 0.99$ and initial conditions $X = 1.0$ and $\dot{X} = dX/d\tau = 0$ at $\tau = 0$. For each of these solutions, use *Mathematica*'s FindFit function to calculate the Fourier coefficients a_0, a_1, a_2 and a_3 in the expansion (10) of Question 13.9. Plot graphs of these coefficients as a function of ϵ using a set of discrete points. On each graph, plot also the approximations (25) and (29) of Question 13.9.

(c) Plot graphs of $X(\tau)$, $\dot{X}(\tau)$ and $\ddot{X}(\tau)$ in the interval $0 \leq \tau \leq 1$ and for $\epsilon = 0.3$ and $\epsilon = 0.6$, corresponding to the approximations (27) and (29) of Question 13.9. As before, also include points representing numerical solutions of the equation of motion.

Solution

(a) By expressing the equation of motion in terms of X and τ, and using (3) and (4) of Question 13.9 we obtain (1).

(b) Replace cell 1 in the *Mathematica* notebook of Question 13.6 with the following:

```
In[1]:= X''[τ] + 4(X[τ] - ε X[τ]²)(Chop[NIntegrate[ 1/√((2/3)ε(u³-1) - u² + 1) ,
        {u, -(3-2ε)/(4ε) (√(1 + 8ε/(3-2ε)) - 1), 1}], 10⁻⁶])² == 0
```

and suitably modify the `FindFit` function to calculate the required Fourier coefficients. This yields the following graphs:

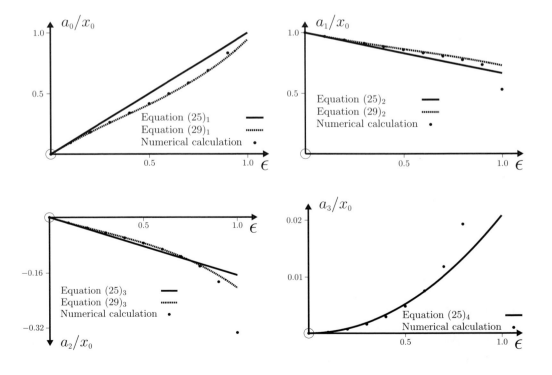

(c)

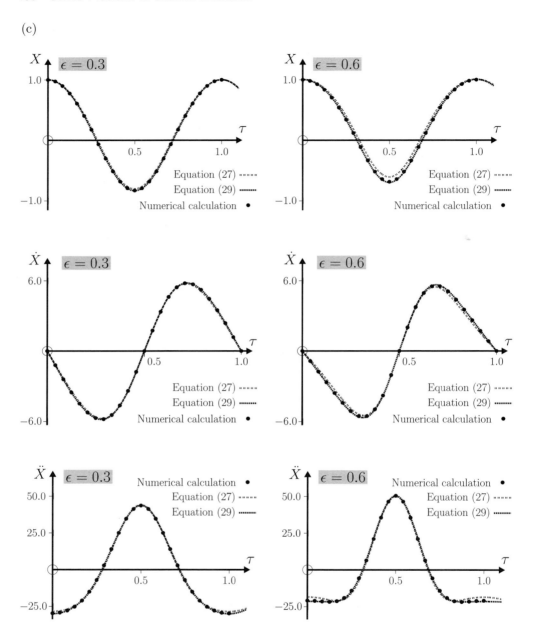

Question 13.11

Suppose a damped, driven harmonic oscillator is perturbed by a cubic force $-m\gamma x^3$. That is, by including a term γx^3 in the equation of motion (2) of Question 4.7.

(a) Write down the equation of motion and look for an approximate steady-state solution
$$x(t) = a(\omega)\cos(\omega t - \vartheta(\omega)). \tag{1}$$
Show that this requires
$$\tan\vartheta = \frac{\omega}{Q(1-\omega^2+a^2\mathrm{sign}\gamma)} \tag{2}$$
and the following relations between ω^2 and a:
$$\omega_\pm^2 = 1 + a^2\mathrm{sign}\gamma - \frac{1}{2Q^2} \pm \frac{\sqrt{1-4Q^2-4Q^2a^2\mathrm{sign}\gamma+4Q^4(F_0/ma)^2}}{2Q^2}, \tag{3}$$
where $Q = \frac{1}{2}\omega_0\tau$. In (3), ω, a and F_0/m represent the dimensionless quantities
$$\frac{\omega}{\omega_0},\quad \frac{\sqrt{3|\gamma|}}{2\omega_0}a,\quad \frac{\sqrt{3|\gamma|}}{2\omega_0^3}\frac{F_0}{m}. \tag{4}$$

(b) Suppose $\gamma > 0$ (a 'hard' system). Deduce that:
1. Equation (3) has no real values if $a > a_{\max}$, where
$$a_{\max}^2 = \frac{1}{2}\sqrt{\left(1-\frac{1}{4Q^2}\right)^2 + 4Q^2\left(\frac{F_0}{m}\right)^2} - \frac{1}{2}\left(1-\frac{1}{4Q^2}\right). \tag{5}$$

2. $\omega_-^2(a) < 0$ if $a < a_{\min}$, where $a_{\min}$ is the largest real root of
$$a^6 + 2a^4 + a^2 - (F_0/m)^2 = 0. \tag{6}$$

3. $\omega_\pm^2(a) > 0$ if $a_{\min} < a < a_{\max}$.

(c) Write a *Mathematica* notebook to calculate $a(\omega)$ and $\vartheta(\omega)$ for $\gamma > 0$, $Q = 4$ and $F_0/m = 0.4$. Plot the graphs of $a(\omega)$ and $\vartheta(\omega)$ versus ω for $0 \le \omega \le 2.5$. Repeat for $Q = 8$ and $F_0/m = 0.4$. (Hint: Use (5) and (6) to find $a_{\max}$ and $a_{\min}$. Then use (2) and (3) to calculate $a(\omega)$ and $\vartheta(\omega)$ for $a \le a_{\max}$.)

Solution

(a) The equation of motion is that of a damped, driven Duffing oscillator:
$$\ddot{x} + (2/\tau)\dot{x} + \omega_0^2 x + \gamma x^3 = (F_0/m)\cos\omega t. \tag{7}$$
By substituting the trial solution (1) into (7) we obtain
$$(\omega_0^2-\omega^2)a\cos(\omega t-\vartheta) - (2\omega/\tau)a\sin(\omega t-\vartheta) + \gamma a^3\cos^3(\omega t-\vartheta) = (F_0/m)\cos\omega t. \tag{8}$$
With the aid of the identities

$$\left.\begin{array}{l}\cos\omega t = \cos\vartheta \cos(\omega t - \vartheta) - \sin\vartheta \sin(\omega t - \vartheta) \\ \cos^3(\omega t - \vartheta) = \tfrac{3}{4}\cos(\omega t - \vartheta) + \tfrac{1}{4}\cos 3(\omega t - \vartheta)\end{array}\right\} \quad (9)$$

we rewrite (8) as

$$\{(\omega_0^2 - \omega^2)a + \tfrac{3}{4}\gamma a^3 - (F_0/m)\cos\vartheta\}\cos(\omega t - \vartheta) + \{-(2\omega/\tau)a$$
$$+ (F_0/m)\sin\vartheta\}\sin(\omega t - \vartheta) + \tfrac{1}{4}\gamma a^3\cos 3(\omega t - \vartheta) = 0. \quad (10)$$

Now, the approximation (1) represents the first term (the fundamental) of a Fourier series with harmonics $\cos(3\omega t - \phi)$, etc. Thus, the last term in (10) plays a role in the higher harmonics and can be ignored for the fundamental. Then, for (10) to hold for all t requires that each of the coefficients in curly brackets be zero:

$$(F_0/m)\cos\vartheta = (\omega_0^2 - \omega^2)a + \tfrac{3}{4}\gamma a^3 \quad (11)$$

$$(F_0/m)\sin\vartheta = (2\omega/\tau)a. \quad (12)$$

We wish to solve these two equations for $\vartheta(\omega)$ and $a(\omega)$ (alternatively, $\omega(a)$). To simplify the calculations we first express (11) and (12) in terms of the dimensionless quantities

$$\overline{\omega} = \frac{\omega}{\omega_0}, \qquad \overline{a} = \frac{\sqrt{3|\gamma|}}{2\omega_0}a, \qquad \overline{\frac{F_0}{m}} = \frac{\sqrt{3|\gamma|}}{2\omega_0^3}\frac{F_0}{m}, \quad (13)$$

and $Q = \tfrac{1}{2}\omega_0\tau$. Then

$$\overline{(F_0/m)}\cos\vartheta = (1 - \overline{\omega}^2)\overline{a} + \overline{a}^3\mathrm{sign}\gamma \quad (14)$$

$$\overline{(F_0/m)}\sin\vartheta = \overline{\omega}\,\overline{a}/Q. \quad (15)$$

The term $\overline{a}^3\mathrm{sign}\gamma$ in (14) represents the effect of the anharmonic (cubic) force. The ratio (15)÷(14) yields (2).[‡] Also, by squaring and adding (14) and (15) we have

$$\{(1 - \omega^2)a + a^3\mathrm{sign}\gamma\}^2 + \omega^2 a^2/Q^2 = (F_0/m)^2. \quad (16)$$

This can be regarded either as a cubic equation for $a^2(\omega)$ or as a quadratic equation for $\omega^2(a)$. The latter is easier to solve and the roots are (3).

(b) **1.** If $\gamma > 0$ then the quantity under the square root in (3) becomes negative for $a > a_{\max}$, where $a_{\max}$ is the positive root of (5). Thus, there are no real values of $\omega_\pm^2(a)$ for $a > a_{\max}$. At $a = a_{\max}$, $\omega_+ = \omega_-$ is real.

2. ω_-^2 in (3) is zero when (6) is satisfied. For $a < a_{\min}$, ω_-^2 becomes negative: only the upper sign in (3) yields a real $\omega(a)$.

3. For $a_{\min} < a < a_{\max}$ both signs in (3) yield positive values for ω^2. Thus, the graph of ω versus a is double-valued for $a_{\min} < a < a_{\max}$ and single-valued for $a < a_{\min}$ (see also the graphs below). As $a \to 0$, $\omega_+ \to \infty$.

[‡]For simplicity we omit the bar on ω, a and F_0/m: in the following these are understood to be dimensionless.

(c) The *Mathematica* notebook is given after the comments, and it yields the following graphs:

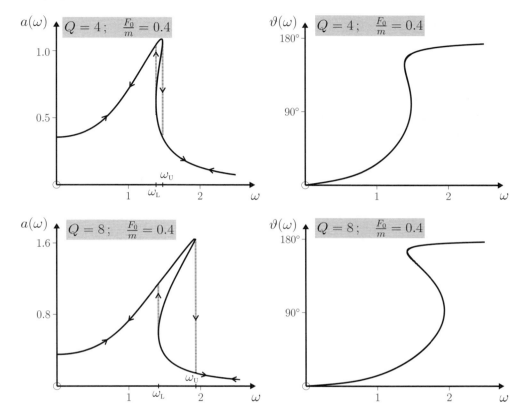

Comments

(i) We see that between the frequencies marked ω_L and ω_U in the diagrams, the amplitude a is a triple-valued function of ω. Associated with this are the phenomena of amplitude jumps and hysteresis. If ω is increased from below ω_L then $a(\omega)$ follows the upper branch of the curve (that is, $\omega_-(a)$) until $\omega = \omega_U$ is reached. Here, the amplitude jumps to the lower branch (that is, $\omega_+(a)$) and then it follows this branch with further increase in ω. The result is a sudden large decrease in the amplitude at ω_U. Conversely, if ω is decreased from above ω_U then $a(\omega)$ follows the lower branch of the curve until $\omega = \omega_L$, where there is a jump to the upper branch. These amplitude jumps and hysteresis can occur in oscillating mechanical and electrical systems.

(ii) The damped, driven Duffing oscillator has other interesting properties such as harmonic resonances (see below), bifurcation and chaos.[4]

---

[4] See, for example, V. D. Barger and M. G. Olsson, *Classical mechanics*. New York: McGraw-Hill, 2nd edn, 1995. Chap. 11.

```
In[1]:= Q = 8.0; α = 0.4;  (* THIS IS DIMENSIONLESS F₀/m *)

root = Flatten[{ToRules[N[Reduce[{x⁶ + 2x⁴ + x² - α² == 0}, x, Reals]]]}];

x1 = x/.First[root]; x2 = x/.Last[root]; amin = Max[x1, x2];

amax = √((1/2)√((1 - 1/(4Q²))² + 4Q²α²) - (1/2)(1 - 1/(4Q²))); astep = (amax - amin)/10000;

f[a_] := √(1 - 4Q² - 4Q²a² + 4Q⁴(α/a)²)/(2Q²);

ω1[a_] := Chop[√(1 + a² - 1/(2Q²) + f[a]), 10⁻⁸];

ω2[a_] := Chop[√(1 + a² - 1/(2Q²) - f[a]), 10⁻⁸];

ParametricPlot[{{ω1[a], a}, {ω2[a], a}}, {a, amin/5, amax}]

φ1[a_] := Chop[ω1[a]/(Q(1 - ω1[a]² + a²)), 10⁻⁸];

φ2[a_] := Chop[ω2[a]/(Q(1 - ω2[a]² + a²)), 10⁻⁸];

θ1[a_] := 180 + (180/π) ArcTan[φ1[a]]/; φ1[a] < 0

θ1[a_] := (180/π) ArcTan[φ1[a]]/; φ1[a] > 0;    θ2[a_] := (180/π) ArcTan[φ2[a]];

ParametricPlot[{{ω1[a], θ1[a]}, {ω2[a], θ2[a]}}, {a, amin/5, amax},
AspectRatio → 0.8]
```

Question 13.12

Write a *Mathematica* notebook to find numerical solutions $x(t)$ for the driven Duffing oscillator using the dimensionless variables introduced in (13) of the previous question.

(a) Plot graphs of the amplitude $a(\omega)$ versus ω for $\gamma > 0$ and $\omega_{\min} \leq \omega \leq \omega_{\max}$. Take $\omega_{\min} = 0.1$, $\omega_{\max} = 2.5$, $F_0/m = 0.4$ and $Q = 4$. Repeat for $Q = 8$. (Hint: To observe hysteresis, sweep the frequency from $\omega_{\min}$ to $\omega_{\max}$ using a step size $\delta\omega = (\omega_{\max} - \omega_{\min})/1000$. At each frequency calculate $x(t)$ for $0 \leq t \leq 40T$, where $T = 2\pi/\omega$. Wait (for a time of, say, $30T$) for the transient to decay. Then, calculate the average value of $a(\omega)$ in the interval $30T \leq t \leq 40T$. Store the values of ω, $a(\omega)$ and then repeat for the next value of ω until $\omega_{\max}$ is reached. Repeat the above, sweeping downwards in frequency from $\omega_{\max}$ to $\omega_{\min}$ in steps $-\delta\omega$. For both increasing and decreasing ω use the initial conditions: $\dot{x}(0) = 0$ and $x(0) = 0$ if $\omega = \omega_{\min}$, otherwise $x(0) = a(\omega)$ obtained from the previous calculation.)

(b) Compare these numerical solutions with the approximate $a(\omega)$ of Question 13.11.

Solution

(a) In terms of the dimensionless variables of Question 13.11, the equation of motion is $\ddot{x} + \dot{x}/Q + x + (4/3)x^3 = (F_0/m)\cos\omega t$. The notebook below produces the following graphs.

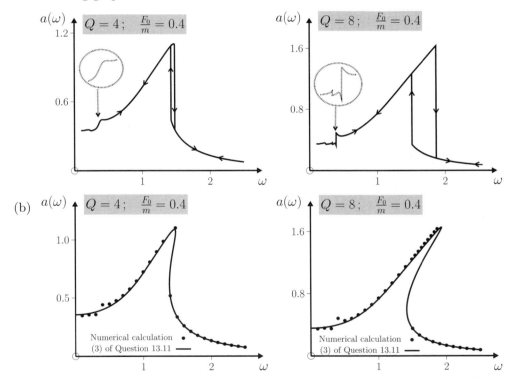

Comments

(i) The numerical calculations confirm the hysteresis found in the approximate analysis of Question 13.11. Agreement with the numerical values can be improved by including the next harmonic, $\cos(3\omega t - \phi)$, in (1) of Question 13.11.

(ii) In addition to the primary resonance, a small resonance peak is discernible at lower ω. This is known as a harmonic resonance. Additional harmonic resonances become evident at larger values of the amplitude F_0 of the driving force.[4]

```
In[1]:= ω0 = 1; Q = 8; α = 4/10; ωstart = 1/10; ωstop = 31/10; ωstep = (ωstop - ωstart)/2400;

amp = 0; ymax = 0; list1 = {}; list2 = {};

(* INCREASING FREQUENCY *)

Do[T = 2π/ω;
  Sol = NDSolve[{x''[t] + 1/Q x'[t] + x[t] + 4/3 x[t]^3 == α Cos[ω/ω0 t],
      x[0] == Floor[1000 amp]/1000, x'[0] == 0}, x[t], {t, 0, 40 T},
    WorkingPrecision → 18, MaxSteps → 100000];
  list = Table[First[x[t] /. Sol], {t, 30T, 40T, T/1000}];
  n = Length[list]; i = 2; count = 0; amp = 0; While[i < n,
    a = Abs[list[[i - 1]]]; b = Abs[list[[i]]]; c = Abs[list[[i + 1]]];
    If[a < b && c < b, amp = amp + b; count = count + 1; i = i + 1, i = i + 1]];
  amp = amp/count; list1 = Append[list1, {N[ω], amp}];
  If[amp > ymax, ymax = amp], {ω, ωstart, ωstop, ωstep}];

gr1 = ListLinePlot[list1, PlotRange → {{0, ωstop}, {0, ymax}}];

(* DECREASING FREQUENCY *)

Do[T = 2π/ω;
  Sol = NDSolve[{x''[t] + 1/Q x'[t] + x[t] + 4/3 x[t]^3 == α Cos[ω/ω0 t],
      x[0] == Floor[1000 amp]/1000, x'[0] == 0}, x[t], {t, 0, 40 T},
    WorkingPrecision → 18, MaxSteps → 100000];
  list = Table[First[x[t] /. Sol], {t, 30T, 40T, T/1000}];
  n = Length[list]; i = 2; count = 0; amp = 0; While[i < n,
    a = Abs[list[[i - 1]]]; b = Abs[list[[i]]]; c = Abs[list[[i + 1]]];
    If[a < b && c < b, amp = amp + b; count = count + 1; i = i + 1, i = i + 1]];
  amp = amp/count; list2 = Append[list2, {N[ω], amp}];
  If[amp > ymax, ymax = amp], {ω, ωstop, ωstart, -ωstep}];

gr2 = ListLinePlot[list2, PlotRange → {{0, ωstop}, {0, ymax}}];

Show[gr1, gr2]
```

Question 13.13

A particle of mass m is subject to a one-dimensional restoring force $F_r = -kx$ and a quadratic frictional force $F_f = -\beta v^2$, where k and β are positive constants. Obtain numerical solutions for the position $x(t)$ up to $t = 100\,\text{s}$, for $m = 1.0\,\text{kg}$, $\beta = 0.125\,\text{kg}\,\text{m}^{-1}$, $k = 1.0\,\text{N}\,\text{m}^{-1}$ and the initial conditions $x_0 = 1.0\,\text{m}$, $v_0 = 0$.

Solution

Write $F_f = -\beta v|v|$ to account for the change in direction of this force every half-cycle. The equation of motion is

$$m\ddot{x} + \beta v|v| + kx = 0. \tag{1}$$

The *Mathematica* notebook used to solve (1) and the graph of $x(t)$ versus t are:

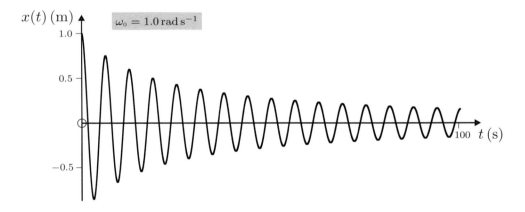

Comments

(i) Values of the half-periods $\frac{1}{2}T_n$ of the nth half-cycle, obtained from intercepts $x(t) = 0$, are tabulated below for the first 30 half-cycles. We see that $\frac{1}{2}T_n$ changes very slowly, and mainly during the first few cycles; thereafter it is nearly constant and close to the value $\pi/\omega_0 = \pi\,\text{s}$. Thus, the angular frequency of the oscillator is little different from its natural frequency $\omega_0 = 1\,\text{rad}\,\text{s}^{-1}$, in agreement with theoretical analysis.[5]

---

[5] N. N. Bogolyubov and Y. A. Mitropolskii, *Asymptotic methods in the theory of nonlinear oscillations.* New York: Gordon and Breach, 1961.

(ii) Experiments show that for a simple pendulum in air the frictional force is usually a combination of linear and quadratic terms, whose respective contributions can be determined by a fit to the experimental data.[6] Pendula have been constructed that demonstrate three types of damping (constant, linear and quadratic).[7,8] Experiments on oscillatory motion of viscously damped spheres show the complexity of the actual damping force for oscillatory motion, including the role of so-called history and added-mass terms.[9]

n	$\frac{1}{2}T_n$ (s)	n	$\frac{1}{2}T_n$ (s)	n	$\frac{1}{2}T_n$ (s)	n	$\frac{1}{2}T_n$ (s)	n	$\frac{1}{2}T_n$ (s)
1	3.142500	7	3.141856	13	3.141716	19	3.141664	25	3.141639
2	3.142287	8	3.141819	14	3.141704	20	3.141658	26	3.141636
3	3.142142	9	3.141790	15	3.141693	21	3.141654	27	3.141633
4	3.142037	10	3.141766	16	3.141685	22	3.141649	28	3.141631
5	3.141960	11	3.141747	17	3.141677	23	3.141646	29	3.141629
6	3.141901	12	3.141730	18	3.141670	24	3.141642	30	3.141627

Question 13.14

A particle (bead) of mass m is constrained to slide without friction on a circular loop of radius R placed in a vertical plane. A spring with force constant k and equilibrium length $\frac{1}{2}R$ is attached to the particle and to a fixed point P that is a distance $\frac{1}{2}R$ above the centre O of the loop. The position of the particle is given by the angle $\theta(t)$ and the instantaneous length of the spring is $\ell(\theta)$.

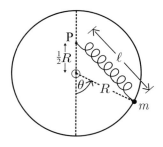

(a) Show that the potential energy of the system is

$$V(\theta) = mgR\left\{1 - \cos\theta + \frac{3\alpha}{8}\left(\sqrt{5 + 4\cos\theta} - 1\right)^2\right\}, \tag{1}$$

where $\alpha = k/k_c$ and $k_c = 3mg/R$.

(b) Determine the points of stable and unstable equilibrium of the particle.
(c) Plot graphs of $V(\theta)/mgR$ versus θ (for $-\pi \leq \theta \leq \pi$) that illustrate the various possibilities for this function.

---

[6] M. Bacon and D. D. Nguyen, "Real-world damping of a physical pendulum," European Journal of Physics, vol. 26, pp. 651–655, 2005.
[7] L. F. C. Zonetti et al., "A demonstration of dry and viscous damping of an oscillating pendulum," European Journal of Physics, vol. 20, pp. 85–88, 1999.
[8] X. Wang, C. Schmitt, and M. Payne, "Oscillations with three damping effects," European Journal of Physics, vol. 23, pp. 155–164, 2002.
[9] P. Alexander and E. Indelicato, "A semi-empirical approach to a viscously damped oscillating sphere," European Journal of Physics, vol. 26, pp. 1–10, 2005.

Solution

(a) The gravitational potential energy of the bead is $V_g(\theta) = mgR(1 - \cos\theta)$ and its elastic potential energy is $V_e(\theta) = \frac{1}{2}k(\ell - \frac{1}{2}R)^2$. By applying the cosine formula to the triangle in the diagram we have $\ell^2 = (\frac{1}{2}R)^2 + R^2 - R^2\cos(\pi - \theta) = (\frac{5}{4} + \cos\theta)R^2$. The total potential energy $V_g + V_e$ is therefore

$$V(\theta) = mgR\left\{1 - \cos\theta + \frac{3\alpha}{8}\left(\sqrt{5 + 4\cos\theta} - 1\right)^2\right\}, \tag{2}$$

which is (1). (The reason for including a factor 3 in α will become apparent below.)

(b) Points of equilibrium are given by the roots of $V' \equiv dV/d\theta = 0$. That is,

$$\sin\theta - \frac{3\alpha}{2}\left\{1 - (5 + 4\cos\theta)^{-1/2}\right\}\sin\theta = 0. \tag{3}$$

There are always at least three roots to (3) in $[-\pi, \pi]$, namely $\theta_1 = 0$ and $\theta_2 = \pm\pi$. It is not difficult to show that $V''(0) = (1 - \alpha)mgR$: so there is a minimum at $\theta = 0$ if $\alpha < 1$ and a maximum if $\alpha > 1$. Also, $V''(\pm\pi) = -mgR$ and so there always are maxima at $\theta = \pm\pi$. According to (3), there is the possibility of two more roots, $\pm\theta_3$, which are the solutions to

$$\sqrt{5 + 4\cos\theta} = 3\alpha/(3\alpha - 2). \tag{4}$$

Now, the left-hand side of (4) is restricted to values from 1 to 3, and these values are attained by the right-hand side only if $\alpha > 1$. Thus, the roots $\pm\theta_3$ exist only if $\alpha > 1$. One can show that

$$V''(\theta_3) = \frac{(\alpha - 1)(3\alpha - 1)(2\alpha - 1)}{\alpha^2(3\alpha - 2)}mgR, \tag{5}$$

which is positive for $\alpha > 1$, and therefore the turning points at $\pm\theta_3$ are always minima. To summarize:

1. $\theta = \pm\pi$ are always points of unstable equilibrium.
2. $\theta = 0$ is a point of stable (unstable) equilibrium if $\alpha < 1$ ($\alpha > 1$).
3. For $\alpha > 1$ there are also two points of stable equilibrium located at the solutions $\pm\theta_3$ to (4). For large values of α the solutions to (4) approach $\pm\pi$.

(c) The three curves plotted below show the range of behaviour for $V(\theta)$. Note that from (1), $V(\theta)/mgR = \frac{3}{2}\alpha$ at $\theta = 0$, and 2 at $\theta = \pm\pi$.

Comments

(i) When the force constant k is increased through the critical value $k_c = 3mg/R$ a notable change occurs in the mechanics of this simple system: the point of stable equilibrium shifts from $\theta = 0$ to $\theta = \pm\theta_3$, given by the solutions to (4) – a particle in equilibrium can be located at either θ_3 or $-\theta_3$.

(ii) This is an example of a widespread phenomenon known as 'spontaneous symmetry breaking'. Here, the word 'spontaneous' refers to the fact that the above change is not accompanied by any alteration in the symmetry of the loop or surroundings (source of the gravitational field).

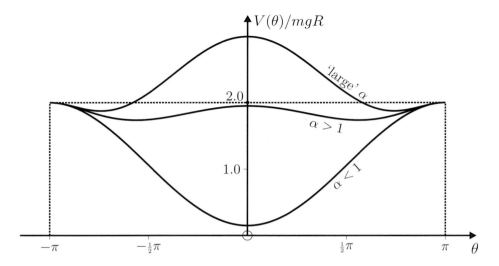

(iii) In fact, a rather detailed formal analogy can be drawn between this mechanical system and a ferroelectric undergoing a Landau second-order phase transition.[10] Here, θ_3 is analogous to an order parameter and $1/k$ to absolute temperature. For example, when k is just above k_c an expansion of (4) yields

$$\theta_3 = 3\sqrt{2}\sqrt{1 - k_c/k}. \tag{6}$$

By comparison, for a ferroelectric the order parameter (the polarization in zero electric field) just below the critical temperature T_c varies as $\sqrt{1 - T/T_c}$. The numerical solution to (4) shown in the following graph is similar to a plot of order parameter versus T/T_c. Further discussion of this analogy and the important topics of critical exponents and universality classes can be found in Ref. [10].

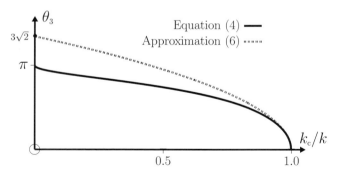

(iv) There are other simple mechanical systems that exhibit spontaneous symmetry breaking and an analogy with phase transitions (see Question 10.14). They all involve a competition between two (or more) forces – gravity and elasticity in the above example.

---

[10] J. R. Drugowich de Felício and O. Hipólito, "Spontaneous symmetry breaking in a simple mechanical model," American Journal of Physics, vol. 53, pp. 690–693, 1985.

Question 13.15

(a) Show that for the particle in Question 13.14 the equation of motion is

$$\frac{d^2\theta}{d\tau^2} + \left\{1 - \frac{3\alpha}{2}\left(1 - (5 + 4\cos\theta)^{-1/2}\right)\right\}\sin\theta = 0, \qquad (1)$$

where $\tau = \sqrt{g/R}\,t$ is a dimensionless time.

(b) Write a *Mathematica* notebook to find numerical solutions of (1) for $0 \leq \tau \leq 100$ using **1.** $\theta_o = 179.0°$, $\dot{\theta}_o = 0$, $\alpha = 0.1$ and **2.** $\theta_o = 99.0°$, $\dot{\theta}_o = 0$, $\alpha = 1.1$. Plot the graphs of $\theta(\tau)$ in this range.

(c) Use *Mathematica*'s Manipulate function to animate the above notebook. Hence simulate the particle's motion for the four cases below and with $\dot{\theta}_o = 0$. Describe and explain what is observed.

  1. $\theta_o = 19.0°$, $\alpha = 0.1$.   **2.** $\theta_o = 47.0°$, $\alpha = 1.1$.
  3. $\theta_o = 171.0°$, $\alpha = 50.0$.   **4.** $\theta_o = 170.5°$, $\alpha = 50.0$.

Solution

(a) The component F_θ of the net force tangential to the loop is $-R^{-1}dV/d\theta$, where $V(\theta)$ is given by (1) of Question 13.14. Newton's second law, $F_\theta = mR\ddot{\theta}$, yields the equation of motion (1).

(b) The notebook given below produces the following graphs:

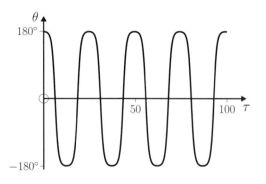
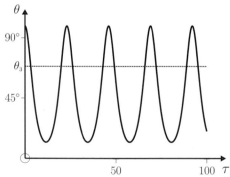

(c) **1.** Here, the point of stable equilibrium is at $\theta = 0$. The initial amplitude θ_o is 'small' and the oscillations are approximately harmonic (in contrast with the first graph for (b) above).

  2. The point of stable equilibrium is now at $\theta_3 = 68.8°$ (see (4) of Question 13.14) and the oscillations are again approximately harmonic about this point (in contrast to the second graph in (b)).

3. For this 'large' α the points of stable equilibrium are close to the vertical ($\theta_3 = 173.3°$). The amplitude of the oscillations is 2.3° and they occur in a shallow potential well. The total energy of the system is small enough that the particle does not cross the point of unstable equilibrium at $\theta = \pi$.
4. The small increase of 0.5° in the initial amplitude for case 3. is sufficient for the particle to cross the potential hump at $\theta = \pi$, and it oscillates symmetrically about this point.

```
In[1]:= α = 0.1; R = 1.0; τmax = 100.0; If[α > 1, θ3 = ArcCos[(-9α² + 15α - 5)/(3α - 2)²], θ3 = 0];

       θ0 = 19 π/180; θ0dot = 0;

       Sol = NDSolve[{θ''[τ] + (1 - 3α/2 (1 - 1/√(5 + 4 Cos[θ[τ]]))) Sin[θ[τ]] == 0,
              θ[0] == θ0, θ'[0] == θ0dot}, {θ[τ], θ'[τ], θ''[τ]}, {τ, 0, τmax},
              MaxSteps → 100000];

       X[t_] := R Sin[θ[τ]]/.Sol/.τ → t; Y[t_] := -R Cos[θ[τ]]/.Sol/.τ → t;

       data = Table[{First[Evaluate[180/π θ[τ]/.Sol]]}, {τ, 0, τmax, τmax/2000}];

       ymin = Min[Flatten[data]]; ymax = Max[Flatten[data]];

       Plot[{Evaluate[180/π θ[τ]/.Sol]}, {τ, 0, τmax}, PlotRange → {ymin, ymax}]

       traj[t_] := Graphics[{Circle[{0, 0}, R], PointSize[0.0125], Black,
              Point[{0, R/2}], Green, Point[{R Sin[θ3], -R Cos[θ3]}],
              PointSize[0.025], Blue, Point[{First[X[t]], First[Y[t]]}],
              Purple, Line[{{0, R/2}, {First[X[t]], First[Y[t]]}}],
              {PointSize[0.0075], Point[{0, 0}]}}]

       Manipulate[traj[t], {t, 0, τmax}]
```

Question 13.16

Consider the system of two particles, a spring and a rotating hoop discussed in Questions 10.14 and 10.15. The position of the particles in a frame rotating with the hoop is specified by their dimensionless z-coordinate Z, which satisfies the non-linear equation of motion (see (4) of Question 10.15)

$$\frac{d^2 Z}{d\tau^2} + \frac{Z}{1-Z^2}\left(\frac{dZ}{d\tau}\right)^2 + 4\pi^2\left\{1 - (1-\epsilon)\frac{\omega^2}{\omega_c^2}\right\}Z(Z^2 - 1) + 4\pi^2\epsilon Z\sqrt{1-Z^2} = 0. \quad (1)$$

Here, $\epsilon = r_0/R \ (\leq 1)$ is the ratio of the equilibrium length $2r_0$ of the spring to the diameter $2R$ of the hoop; $Z = z/R$ and so $-1 \leq Z \leq 1$; $\tau = \omega_0 t/2\pi$ (where $\omega_0 = \sqrt{2k/m}$) is a dimensionless time; and $\omega_c = \omega_0\sqrt{1-\epsilon}$ is the critical angular velocity discussed in Question 10.14. In general, Z is a function of τ, ω/ω_c, ϵ and the initial conditions Z_0 and $\dot{Z}_0$ at $\tau = 0$. Furthermore, the points of equilibrium are at (see (7) and (16) of Question 10.14)

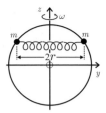

$$Z_\omega = 0 \quad \text{and} \quad Z_\omega = \pm\sqrt{1 - \epsilon^2\{1 - (\omega^2/\omega_c^2)(1-\epsilon)\}^{-2}}. \tag{2}$$

In the calculations that follow we take $\epsilon = \tfrac{1}{2}$ and $\dot{Z}_0 = 0$.

(a) Write a *Mathematica* notebook to find numerical solutions to (1) for $Z(\tau)$ and $\dot{Z}(\tau)$, in which the user specifies values for ϵ, ω/ω_c, Z_0 and $\dot{Z}_0$. Consider the following cases: **1.** $\omega/\omega_c = 0.8$, $Z_0 = 0.030$; **2.** $\omega/\omega_c = 0.8$, $Z_0 = 0.700$; **3.** $\omega/\omega_c = 0.8$, $Z_0 = 0.880$; **4.** $\omega/\omega_c = 0.8$, $Z_0 = 0.883$; **5.** $\omega/\omega_c = 1.2$, $Z_0 = 0.100$; **6.** $\omega/\omega_c = 1.2$, $Z_0 = 0.900$.

(b) For each of these cases, plot the graphs of $Z(\tau)$ for $0 \leq \tau \leq 10$, and comment on the results. Also, plot the phase portraits (that is, $\dot{Z}(\tau)$ versus $Z(\tau)$) for $\omega/\omega_c = 0.8$ and $\omega/\omega_c = 1.2$.

(c) Use *Mathematica*'s Manipulate function to extend the above notebook and produce a dynamic display of the rotating hoop and constrained particles relative to the laboratory frame. Comment on the behaviour of the system for the following cases: **1.** $\omega/\omega_c = 0.8$, $Z_0 = 0.677$ and $Z_0 = 0$; **2.** $\omega/\omega_c = 0.8$, $Z_0 = 0.777$ and $Z_0 = 0.100$; **3.** $\omega/\omega_c = 0.8$, $Z_0 = 0.883$; **4.** $\omega/\omega_c = 1.2$, $Z_0 = 0.1$.

Solution

(a) The *Mathematica* notebook is:

```
In[1]:= ε = 1/2;  ωOverωc = 4/5;  τmax = 100.0;  Z0 = 883/1000;  Z0dot = 0;

(* FindRoot[(1 - ε)^2 - (1 - ε)(ωOverωc)^2 - (√(1 - Z^2) - ε)^2 + (1 - ε)
           (ωOverωc)^2 (1 - Z^2) == 0, {Z, 0.883}] *)

Sol = NDSolve[{Z''[τ] + (Z[τ] Z'[τ]^2)/(1 - Z[τ]^2) + 4 π^2 (1 - (1 - ε)(ωOverωc)^2)
       Z[τ] (Z[τ]^2 - 1) + 4 π^2 ε Z[τ] √(1 - Z[τ]^2) == 0, Z[0] == Z0, Z'[0] == Z0dot},
       {Z[τ], Z'[τ], Z''[τ]}, {τ, 0, τmax}, MaxSteps → 100000,
       AccuracyGoal → 400, WorkingPrecision → 32];

Plot[{Evaluate[{Z[τ]}/.Sol]}, {τ, 0, τmax},
     PlotPoints → 1000, AspectRatio → 1]

ParametricPlot[{Evaluate[{Z'[τ], Z[τ]}/.Sol]}, {τ, 0, τmax},
     PlotRange → {{-1.5, 1.5}, {-1.5, 1.5}}, PlotPoints → 1000]
```

494 Solved Problems in Classical Mechanics

(b) This notebook yields the following diagrams and phase portraits:

1. $Z_0 = 0.03$; $\dot{Z}_0 = 0$; $\omega/\omega_c = 0.8$; $\epsilon = 0.5$

2. $Z_0 = 0.7$; $\dot{Z}_0 = 0$; $\omega/\omega_c = 0.8$; $\epsilon = 0.5$

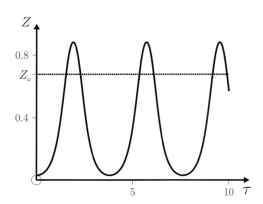
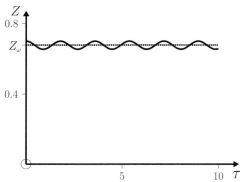

3. $Z_0 = 0.880$; $\dot{Z}_0 = 0$; $\omega/\omega_c = 0.8$; $\epsilon = 0.5$

4. $Z_0 = 0.883$; $\dot{Z}_0 = 0$; $\omega/\omega_c = 0.8$; $\epsilon = 0.5$

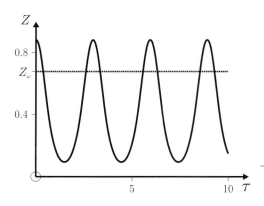
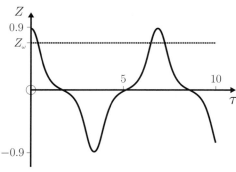

5. $Z_0 = 0.1$; $\dot{Z}_0 = 0$; $\omega/\omega_c = 1.2$; $\epsilon = 0.5$

6. $Z_0 = 0.9$; $\dot{Z}_0 = 0$; $\omega/\omega_c = 1.2$; $\epsilon = 0.5$

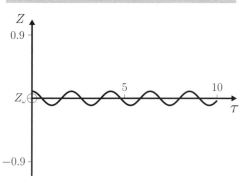
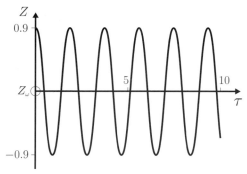

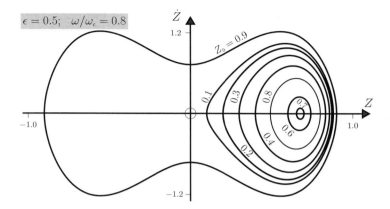

In the previous diagram, the trajectories for $Z_0 = 0.5$ and 0.8 are almost identical and so only the latter is plotted. For cases 1. – 4. the point of stable equilibrium in $(2)_2$ is $Z_\omega = 0.677$, and this is indicated by the horizontal line on the graphs for $Z(\tau)$. In 1. , where $Z_0 = 0.03$, the particles start far from Z_ω, and consequently they perform large-amplitude, anharmonic oscillations. In 2. , where $Z_0 = 0.7$, the initial displacement from Z_ω is smaller and the particles perform nearly harmonic oscillations. Further increase of Z_0 produces again large-amplitude oscillations (case 3. , where $Z_0 = 0.880$). There is a critical value of Z_0, above which the energy of the particles exceeds the maximum $V_e(0)$ in the effective potential at $Z = 0$, and the particles oscillate across the equator: according to (9) of Question 10.14 this critical value is a root of

$$(1-\epsilon)^2 - (1-\epsilon)(\omega/\omega_c)^2 = (\sqrt{1-Z^2} - \epsilon)^2 - (1-\epsilon)(\omega/\omega_c)^2(1-Z^2). \quad (3)$$

Mathematica's `FindRoot` function with $\epsilon = 0.5$ and $\omega/\omega_c = 0.8$ gives $Z = 0.882353$. This is exceeded in case 4. , where $Z_0 = 0.883$. In 5. and 6. , where $\omega/\omega_c > 1$, the point of stable equilibrium is $Z_\omega = 0$, and the graphs above show the expected small- and large-amplitude oscillations about the equator.

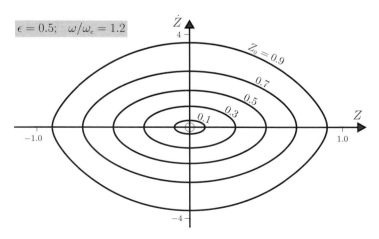

496 *Solved Problems in Classical Mechanics*

(c) We first express the equation of the rotating hoop in terms of a parameter u and in dimensionless form:
$$\left.\begin{aligned} X_{\rm h}(u,\tau) &= \cos u \sin 2\pi\tau \\ Y_{\rm h}(u,\tau) &= \cos u \cos 2\pi\tau \\ Z_{\rm h}(u,\tau) &= \sin u \, . \end{aligned}\right\} \quad (4)$$

Equation (4) and *Mathematica*'s `ParametricPlot3D` function are used to produce the graphics for the hoop. The coordinates $X_i(\tau)$, $Y_i(\tau)$ and $Z_i(\tau)$ of the particles in the laboratory frame are given, in terms of $Z(\tau)$, by

$$\begin{pmatrix} X_i(\tau) \\ Y_i(\tau) \\ Z_i(\tau) \end{pmatrix} = \begin{pmatrix} \cos 2\pi\tau & \sin 2\pi\tau & 0 \\ -\sin 2\pi\tau & \cos 2\pi\tau & 0 \\ 0 & 0 & 1 \end{pmatrix} \begin{pmatrix} 0 \\ (-1)^{i-1}\sqrt{1-Z^2(\tau)} \\ Z(\tau) \end{pmatrix}, \quad (5)$$

for $i = 1, 2$. *Mathematica*'s `Graphics3D` function uses the coordinates given by (5) to locate the particles on the hoop. The modified notebook is shown below.

```
In[1]:= ε = 1/2;  ωOverωc = 99/100;  τmax = 1000.0;  Z0 = 883/1000;  Z0dot = 0;

Sol = NDSolve[{Z''[t] + (Z[t] Z'[t]^2)/(1 - Z[t]^2) + 4 π^2 (1 - (1 - ε) (ωOverωc)^2) Z[t]
    (Z[t]^2 - 1) + 4 π^2 ε Z[t] √(1 - Z[t]^2) == 0, Z[0] == Z0, Z'[0] == Z0dot},
    {Z[t], Z'[t], Z''[t]}, {t, 0, τmax}, MaxSteps → 100000]; (*,
    AccuracyGoal → 400, WorkingPrecision → 32]; *)

X[τ_] := 0; Y[τ_] := √(1 - (First[Z[t] /. Sol /. t → τ])^2);
X1[τ_] := Y[τ] Sin[2π τ]; Y1[τ_] := Y[τ] Cos[2π τ];
Z1[τ_] := First[Z[t] /. Sol /. t → τ];
X2[τ_] := -X1[τ]; Y2[τ_] := -Y1[τ]; Z2[τ_] := Z1[τ];

hoop[u_, τ_] := ParametricPlot3D[{Cos[u] Sin[2π τ], Cos[u]
    Cos[2π τ], Sin[u]}, {u, 0, 2π}, PlotRange →
    {{-2, 2}, {-2, 2}, {-2, 2}}, Axes → {False, False, False},
    Boxed → False, ViewVertical → {1, 0, 0}, ViewPoint → {-20, 0, 0}]

particles[τ_] := Graphics3D[{PointSize[0.025], Red, Point[{X1[τ],
    Y1[τ], Z1[τ]}], Green, Point[{X2[τ], Y2[τ], Z2[τ]}], Purple,
    Line[{{X1[τ], Y1[τ], Z1[τ]}, {X2[τ], Y2[τ], Z2[τ]}}],
    {PointSize[0.0125], Point[{0, 0, 0}]}}]

Manipulate[Show[hoop[u, τ], particles[τ]], {τ, 0, τmax}]
```

1. From (2), $Z_0 = 0.677$ and 0 are points of equilibrium. The particles remain at rest relative to the rotating hoop. **2.** Changing Z_0 to 0.777 and 0.100 produces small-amplitude, harmonic and large-amplitude, anharmonic oscillations, respectively, such as those described in (b) above. **3.** The particles have sufficient energy to oscillate across the equator. **4.** Here, $\omega > \omega_c$ and the oscillations are about the equator.

Comment

It is important to set a suitable `AccuracyGoal` and `WorkingPrecision` in the *Mathematica* notebook. Failure to do so can produce arithmetical errors, resulting in unphysical behaviour. For example, oscillations across the equator by particles with energy just less than $V_e(0)$, where V_e is the effective potential discussed in Question 10.14. Furthermore, in plotting the phase trajectories, the `PlotPoints` setting must be increased beyond the default value, otherwise it is possible to obtain phase-space plots that exhibit features usually associated with chaotic behaviour. This can result in incorrect conclusions being drawn: see, for example, Figure 9 and the accompanying discussion in Ref. [11].

Question 13.17

Consider a rigid simple pendulum with linear drag and harmonic driving force applied tangential to the path. The equation of motion is (see Questions 4.3 and 4.7)

$$\frac{d^2\theta}{d\tau^2} + \frac{1}{Q}\frac{d\theta}{d\tau} + \sin\theta = A\cos(\omega\tau/\omega_0). \tag{1}$$

Here ω_0 is the natural frequency, ω is the driving frequency, $\tau = \omega_0 t$ is a dimensionless time,[‡] $Q = m\omega_0/\alpha$ is the quality factor and A is a constant. Suppose the system is started twice from rest at slightly different initial positions θ_0 and $\theta_0 + \epsilon$. Write a *Mathematica* notebook to calculate $\Delta\theta(\tau) = |\theta_1(\tau) - \theta_2(\tau)|$, where $\theta_1(\tau)$ and $\theta_2(\tau)$ are the respective solutions. **1.** Take $Q = 4$, $\omega/\omega_0 = 1/2$, $A = 1/2$, $\theta_0 = 0$, $\dot{\theta}_0 = 0$ and $\epsilon = 0.000001°$. Plot $\Delta\theta(\tau)$ on a logarithmic scale for $0 \leq \tau \leq 120$. **2.** Repeat the above, changing A to $5/2$, and plot $\Delta\theta(\tau)$ on the same axes. (Use increased precision in the notebook to minimize round-off errors.)

Solution

The notebook and resulting plot of $\Delta\theta(\tau)$ are

[‡]This use of the symbol τ should not be confused with the relaxation time $2m/\alpha$ of the oscillator introduced in Chapter 4.

---

[11] F. Ochoa and J. Clavijo, "Bead, hoop and spring as a classical spontaneous symmetry breaking problem," European Journal of Physics, vol. 27, pp. 1277–1288, 2006.

498 *Solved Problems in Classical Mechanics*

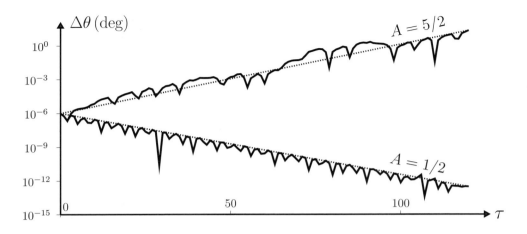

```
In[1]:= τmax = 200;   ω0 = 1;   ω = 5/10;   Q = 4;   φ = 0;   theta0 = 0;   ε = π/180 10^-6;

       θ0 = theta0 π/180;   θ0dot = 0;

       DataTable = {};   counter = 0;   NumPrec = 32;

       Δθ[τ_] := Abs[180/π (Evaluate[θ1[τ]/.Sol1] - Evaluate[θ2[τ]/.Sol2])];

       Do[If[counter == 0, A = 50/100, A = 250/100];
         Sol1 = NDSolve[{θ1''[τ] + θ1'[τ]/Q + Sin[θ1[τ]] == A Sin[ω/ω0 τ + φ],
           θ1[0] == θ0, θ1'[0] == θ0dot}, {θ1[τ], θ1'[τ], θ1''[τ]}, {τ, 0, τmax},
           MaxSteps → 100000, WorkingPrecision → NumPrec];
         Sol2 = NDSolve[{θ2''[τ] + θ2'[τ]/Q + Sin[θ2[τ]] == A Sin[ω/ω0 τ + φ],
           θ2[0] == θ0 + ε, θ2'[0] == θ0dot}, {θ2[τ], θ2'[τ], θ2''[τ]}, {τ, 0, τmax},
           MaxSteps → 100000, WorkingPrecision → NumPrec];
         DataTable = Join[DataTable, Table[{τ, First[Δθ[τ]]}, {τ, 0τmax, τmax}]],
         {counter, 0, 1}];

       ListLogPlot[{DataTable}, PlotRange → {{0, 120}, {0, 10^3}}, Joined → False]
```

Comments

(i) These results illustrate an interesting and important feature of this system. For $A = \frac{1}{2}$ the small uncertainty ϵ in θ_0 has a negligible effect on the long-term behaviour of the pendulum. However, for $A = \frac{5}{2}$ the quantity $\Delta\theta$ increases by seven orders of magnitude in the interval shown: $\theta(\tau)$ is very sensitive to the initial conditions. In this case the motion is said to be chaotic. The study of chaos is an important branch of physics. In the following comments we mention briefly some developments in this field, and some distinguishing features of chaotic systems.

(ii) Towards the end of the nineteenth century, Poincaré discovered irregular patterns of behaviour, indicative of chaos, in the gravitational three-body problem (see also Question 11.7). However, it turned out that for some time "Poincaré's fundamental studies had only a minimum impact. His investigations were mathematically so novel and sophisticated that they were not widely read or rapidly accepted. Furthermore, the three-body problem was no longer at the center of interest. It was considered a rather 'far-out' problem, so that peculiar and non-intuitive behaviour did not cause a great deal of interest, let alone excitement. The equations of motion of the three-body problem, which formed the starting point of Poincaré's considerations, are a set of non-linear ordinary differential equations. At the beginning of this century, no general theory existed to handle such equations. . . . It was hard, even impossible to obtain general conclusions. Thus the chaotic behaviour of the three-body problem might well be an idiosyncracy of just that problem without general significance. In any case, the physics community paid very little attention to Poincaré's results."[12] Later – around 1960 – a period of rapid development in the study of chaos began. Dresden has suggested three reasons for this: "Perhaps most important were the advances in computer technology. . . . The second, related, factor was the realization that many systems described by quite simple equations nevertheless could show behaviour indistinguishable from random behaviour and so could exhibit chaos. Finally, a combination of computational, analytical, and abstract mathematical methods led to a much deeper understanding of non-linear equations."[12]

(iii) In 1959, E. N. Lorenz used a computer to study the numerical solutions of a set of non-linear differential equations he had derived to describe the evolution of weather systems. He noticed that these solutions possessed extreme sensitivity to the initial conditions. Lorenz's research "marked the beginning of a remarkable story which involved a number of individuals working largely in isolation, who all came across different aspects of what we would now call chaotic behaviour. The systems studied included dripping taps, electronic circuits, turbulence, population dynamics, fractal geometry, the stability of Jupiter's great red spot, the dynamics of stellar orbits and the tumbling of Hyperion, one of the satellites of Saturn."[13]

(iv) The non-linear equations that describe chaotic systems are deterministic in the usual sense: the initial conditions specify a unique solution. "However, it is now clear that non-linear deterministic systems typically behave in such a way that, even with the most powerful computers available, it would be impossible to predict their state for a very long time. This interesting, and apparently contradictory, behaviour (*deterministic* yet *unpredictable*) is possible because the solutions for

---

[12] M. Dresden, "Chaos: A new paradigm – or science by public relations?," The Physics Teacher, vol. 30, pp. 74–80, 1992.

[13] M. Longair, *Theoretical concepts in physics (An alternative view of theoretical reasoning in physics)*, p. 182. Cambridge: Cambridge University Press, 2nd edn, 2003.

a non-linear system can depend very sensitively on the initial conditions."[14] The term 'deterministic chaos' is widely used to mean chaotic motion of dynamical systems whose time evolution is obtained from a knowledge of its previous history.[15] It has been mentioned that "the shock which was associated with the discovery of deterministic chaos has therefore been compared to that which spread when it was found that quantum mechanics only allows statistical predictions."[15]

(v) If prediction becomes impossible, "it is evident that a chaotic system can resemble a stochastic system (a system subject to random external forces). However, the source of the irregularity is quite different. For chaos, the irregularity is part of the intrinsic dynamics of the system, not unpredictable outside influences."[16]

(vi) For chaotic behaviour of a system, its equation of motion must be non-linear, and it must possess at least three independent dynamical variables.[16] For the driven pendulum discussed above, these variables are θ, $d\theta/d\tau$ and $\phi = \omega\tau/\omega_0$.[16]

(vii) The two dotted lines in the preceeding plot of $\Delta\theta(\tau)$ show, respectively, either an exponential increase or an exponential decrease in the uncertainty $\Delta\theta$ with time. These rates "of orbital divergence or convergence, called Lyapunov exponents ... are clearly of fundamental importance in studying chaos. Positive Lyapunov exponents indicate orbital divergence and chaos, and set the time scale on which state prediction is possible. Negative Lyapunov exponents set the time scale on which transients or perturbations of the system's state will decay."[17]

Question 13.18

Consider a compass needle that rotates freely in a uniform, oscillatory magnetic field $\mathbf{B} = \mathbf{B}_0 \cos\omega t$ aligned perpendicular to its pivot. The needle has magnetic dipole moment $\mathbf{m}$ and moment of inertia I about a perpendicular axis through its centre.

(a) Show that the dimensionless equation of motion is

$$\frac{d^2\theta}{d\tau^2} + \cos(\omega\tau/\omega_0)\sin\theta = 0, \tag{1}$$

where θ is the angle between $\mathbf{m}$ and $\mathbf{B}_0$, $\omega_0^2 = mB_0/I$ and $\tau = \omega_0 t$.

(b) Suppose the system is started twice from rest at slightly different initial positions θ_0 and $\theta_0 + \epsilon$. Write a *Mathematica* notebook to obtain numerical solutions of (1) for $\theta_0 = 30°$, $\dot\theta_0 = 0$ and **1.** $\epsilon = 0.000001°$; $\lambda = 2$; and **2.** $\epsilon = 1°$; $\lambda = \frac{1}{2}$, where $\lambda = \sqrt{2}\omega_0/\omega$. In **1.** plot the total angle turned through, $\theta(\tau)$, for $0 \leq \tau \leq 100\pi$. In **2.** plot $\theta(\tau)$ up to $\tau = 50$. Also, plot the corresponding Poincaré sections, using a sampling frequency $\omega_s = \omega$ (see Question 4.15).

[14] S. De Souza-Machado, R. W. Rollins, D. T. Jacobs, and J. L. Hartman, "Studying chaotic systems using microcomputer simulations and Lyapunov exponents," American Journal of Physics, vol. 58, pp. 321–329, 1990.
[15] H. G. Schuster, *Deterministic chaos*, pp. 1–5. Frankfurt: Physik-Verlag, 1984.
[16] G. L. Baker and J. P. Gollub, *Chaotic dynamics (an introduction)*. Cambridge: Cambridge University Press, 1990.
[17] A. Wolf, "Quantifying chaos with Lyapunov exponents," in *Chaos* (A. V. Holden, ed.), Princeton, Princeton University Press, p. 273, 1986.

Solution

(a) The needle experiences a restoring torque $\mathbf{m} \times \mathbf{B}$ and so the equation of motion is (see Question 12.4)

$$I\frac{d^2\theta}{dt^2} = -mB\sin\theta = -mB_0\cos\omega t\sin\theta, \qquad (2)$$

which is (1).

(b) The *Mathematica* notebook is listed below. In both Poincaré sections, the angular velocity $\dot{\theta} = d\theta/d\tau$ is given in degrees. Also, for clarity of presentation, some of the data in the Poincaré section for case 1. have been omitted.

1. $\theta_0 = 30°$, $\dot{\theta}_0 = 0$, $\epsilon = 0.000001°$ and $\lambda = 2$

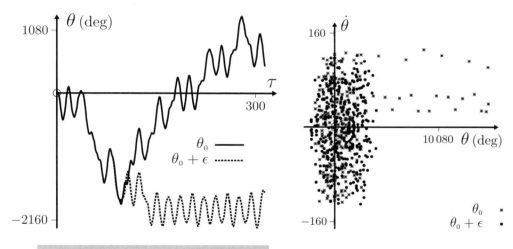

2. $\theta_0 = 30°$, $\dot{\theta}_0 = 0$, $\epsilon = 1°$ and $\lambda = \frac{1}{2}$

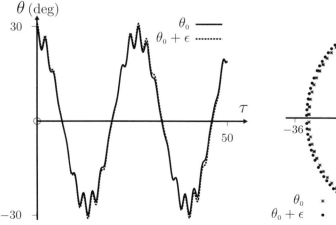
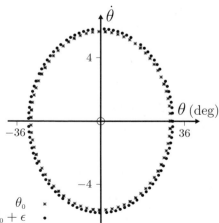

Comments

(i) For $\lambda > 1$ the system exhibits chaotic behaviour.[18] Chaos is evident in the Poincaré section for case 1. above, which illustrates a diffuse scatter of points covering large areas of the phase space. Here, the initial 'uncertainty' ϵ in θ_0 is $0.000001°$ and the two trajectories soon diverge. Contrast this with case 2. , where the behaviour of the system is insensitive to the initial value of θ even though ϵ is now one million times larger.

(ii) For $\lambda \ll 1$ the motion of the needle is approximately sinusoidal: the compass needle becomes phase locked and rotates with the frequency of the magnetic field.[18] As λ increases, the time taken to "reach phase locking becomes longer and longer until a critical value is reached at which the time becomes infinite and the motion becomes chaotic. In this state, the magnet typically makes many revolutions in one direction before unpredictably reversing."[18]

(iii) Various features characteristic of chaotic systems, such as bifurcation and period doubling have been observed in experiments on a compass needle in an alternating magnetic field.[18,19]

```
In[1]:= τmax = 65π;   ω0 = 1;   ω = 2√2;   theta0 = 30;   τstep = τmax/100000;

t = 0;   X1 = -40;   Y1 = -6;   X2 = 40;   Y2 = 6;   ε = 1;

(* τmax = 1000π;   ω0 = 1;   ω = √2/2;   theta0 = 30;   τstep = τmax/100000;

t = 0;   X1 = -5000;   Y1 = -200;   X2 = 20000;   Y2 = 200;   ε = 1.10^-6;  *)

θ0 = theta0 π/180;   θ0dot = 0;   ε = ε π/180;

Sol1 = NDSolve[{θ1''[τ] + Cos[ω/ω0 τ] Sin[θ1[τ]] == 0, θ1[0] == θ0,
    θ1'[0] == θ0dot}, {θ1[τ], θ1'[τ], θ1''[τ]}, {τ, 0, τmax},
    MaxSteps → 200000];

Sol2 = NDSolve[{θ2''[τ] + Cos[ω/ω0 τ] Sin[θ2[τ]] == 0, θ2[0] == θ0 + ε,
    θ2'[0] == θ0dot}, {θ2[τ], θ2'[τ], θ2''[τ]}, {τ, 0, τmax},
    MaxSteps → 200000];

MyTable1 = {180/π {First[θ1[τ]/.Sol1/.τ → t], First[θ1'[τ]/.Sol1/.τ → t]}};
```

---

[18] K. Briggs, "Simple experiments in chaotic dynamics," American Journal of Physics, vol. 55, pp. 1083–1089, 1987.

[19] H. Meissner and G. Schmidt, "A simple experiment for studying the transition from order to chaos," American Journal of Physics, vol. 54, pp. 800–804, 1986.

```
In[2]:= MyTable2 = { 180/π {First[θ2[τ]/.Sol2/.τ → t], First[θ2'[τ]/.Sol2/.τ → t]}};

       While[t < τmax, ang1 = 180/π First[θ1[τ]/.Sol1/.τ → t];
         vel1 = 180/π First[θ1'[τ]/.Sol1/.τ → t];
         MyTable1 = Append[MyTable1, {ang1, vel1}];
         ang2 = 180/π First[θ2[τ]/.Sol2/.τ → t];
         vel2 = 180/π First[θ2'[τ]/.Sol2/.τ → t];
         MyTable2 = Append[MyTable2, {ang2, vel2}]; t = t + 2π/ω ]

       ListLinePlot[{MyTable1, MyTable2}, PlotRange → {{X2, X1},
         {Y2, Y1}}, Joined → False, Frame → True, FrameLabel → {"θ",
         "θdot", "", ""}, PlotStyle → {PointSize[0.0075]}]

       Plot[{{ 180/π Evaluate[θ1[τ]/.Sol1]},
         { 180/π Evaluate[θ2[τ]/.Sol2]}}, {τ, 0, τmax/4.5},
         AspectRatio → 0.5, PlotPoints → 200]
```

Question 13.19

Two smooth planes inclined at angles α ($\leq \frac{1}{4}\pi$) and $\pi - \alpha$ to the horizontal are joined to form a symmetrical wedge. Cartesian axes are oriented with the y-axis vertical and z-axis along the join. A particle released from rest above the left plane at (x_0, y_0) falls under gravity and makes elastic collisions with the two faces of the wedge. The motion is confined to the xy-plane.

(a) Suppose $\mathbf{u} = (u_x, u_y)$ and $\mathbf{u}'$ are the velocities immediately before and after a collision of the particle with the wedge. Then $\theta = \cos^{-1}(\hat{\mathbf{x}} \cdot \mathbf{u}'/u')$ is the corresponding angle of projection after the collision. For the first impact, $\theta = \frac{1}{2}\pi - 2\alpha$. Show that for subsequent impacts the various values of θ are:

$$\theta = \begin{cases} \pi - 2\alpha + \phi & (u_y > 0; u_x < 0) \\ \pi - 2\alpha - \phi & (u_y < 0; u_x < 0) \\ -2\alpha + \phi & (u_y < 0; u_x > 0) \end{cases} \quad \begin{array}{ll} 2\alpha - \phi & (u_y > 0; u_x > 0) \\ \pi + 2\alpha - \phi & (u_y < 0; u_x < 0) \\ 2\alpha + \phi & (u_y < 0; u_x > 0), \end{array} \quad (1)$$

left plane / right plane

where $\phi = \tan^{-1}|u_y/u_x|$.

(b) Use (5) of Question 7.1 and the values of θ in (1) to write a *Mathematica* notebook that calculates the trajectory $y(x)$ of the particle for a specified number n of bounces. (Hint: Recall that for elastic collisions $u = u'$.) Take $x_0 = -1$ m, $y_0 = 2$ m, $\alpha = 30°$ and $n = 7$. Plot a graph of y versus x.

(c) Plot the graphs of $x(t)$ and $y(t)$ for $n = 10$.

Solution

(a) For elastic collisions on a smooth plane surface the 'angle of incidence' of the particle equals the 'angle of reflection' (see Chapter 6). This, together with simple geometry, yields the values of θ listed in (1).

(b) The trajectory obtained from the notebook below is:

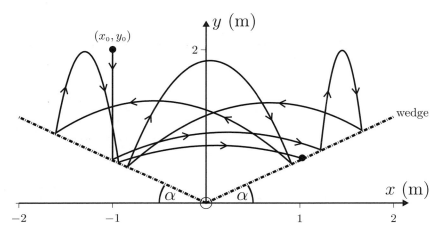

(c) We obtain the graphs:

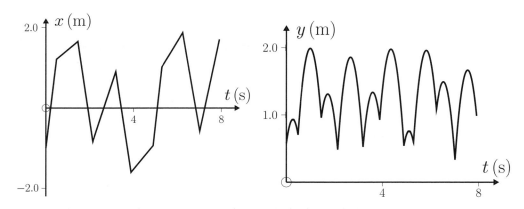

Comments

(i) Phase-space plots illustrate the rich behaviour of this non-linear system, which is sometimes referred to as 'gravitational billiards'.[20]

---

[20] H. J. Korsch and H. J. Jodl, *Chaos*, pp. 67–88. Berlin: Springer, 1999.

(ii) For this system the "motion cannot converge to a well defined sustained motion because there are *no* attractors in frictionless, conservative systems. As a result, the nature of all motion strongly depends on the initial conditions and the total energy. Regular motion corresponds to certain sets of initial conditions, while chaotic motion corresponds to other sets. The initial conditions that lead to chaotic motion form chaotic bands that, contrary to chaotic attractors, are plane-filling objects."[21]

```
In[1]:= α = 30 π/180 ;  g = 9.8;  x0 = -1.0;  y0 = 2.0;  n = 10;  R = 5/4 Max[Abs[x0], y0];
  If[x0 < 0, If[α ≤ π/4, θ1 = π/2 - 2α, θ1 = -π/2 + 2α], θ1 = π/2 + 2α];
  If[y0 < Abs[x0 Tan[α]], Interrupt[]]
  x1 = x0; y1 = Abs[x1] Tan[α]; t1 = √(2(y0 - y1)/g) ;
  gr[0] = ParametricPlot[{x1, y0 - g/2 t²}, {t, -t1, 0}, PlotRange → {{-R, R},
      {0, R}}, Axes → True, PlotStyle → Dashed]; list = {gr[0]};
  Do[
    v1 = √(2g(y0 - y1)); Sola = Solve[{x1 + v1 Cos[θ1] t == x,
        y == Abs[x1] Tan[α] + v1 Sin[θ1] t - g/2 t², y == x Tan[α]}, {x, y, t}];
    Solb = Solve[{x1 + v1 Cos[θ1] t == x, y == Abs[x1] Tan[α] + v1 Sin[θ1] t - g/2 t²,
        y == -x Tan[α]}, {x, y, t}];
    xa = x/.Sola; ya = y/.Sola; ta = Chop[t/.Sola, 10⁻⁴];
    xb = x/.Solb; yb = y/.Solb; tb = Chop[t/.Solb, 10⁻⁴];
    Do[
      If[ya[[j]] > 0 && ta[[j]] > 0, x2 = xa[[j]]; y2 = ya[[j]]; t2 = ta[[j]];,
      If[yb[[j]] > 0 && tb[[j]] > 0, x2 = xb[[j]]; y2 = yb[[j]]; t2 = tb[[j]];]],
      {j, 1, 2}]; vy = v1 Sin[θ1] - g t2; vx = v1 Cos[θ1]; φ2 = ArcTan[Abs[vy/vx]];
    If[x2 > 0, If[vy < 0, If[vx < 0, θ2 = π + 2α - φ2, θ2 = 2α + φ2],
        If[vx < 0, Beep[], θ2 = 2α - φ2]], If[vy < 0, If[vx < 0, θ2 = π - 2α - φ2,
        θ2 = -2α + φ2], If[vx < 0, θ2 = π - 2α + φ2, Beep[]]]];
    X[t_] := x1 + v1 Cos[θ1] t; Y[t_] := Abs[x1] Tan[α] + v1 Sin[θ1] t - g/2 t²;
    gr[i] = ParametricPlot[{X[t], Y[t]}, {t, 0, t2}, PlotRange → {{-R, R},
        {0, R}}, Axes → True, PlotPoints → 100]; list = Append[list, gr[i]];
    x1 = x2; y1 = y2; t1 = t2; θ1 = θ2; v1 = √(2g(y0 - y1));
    , {i, 1, n}]
  Wedge = ParametricPlot[{{x, x Tan[α]}, {x, -x Tan[α]}}, {x, -5, 5},
      PlotRange → {{-R, R}, {0, R}}, PlotStyle → {Thickness[0.0075]},
      Axes → True, PlotPoints → 100]; Show[Wedge, list]
```

---

[21] T. Tél and M. Gruiz, *Chaotic Dynamics*, pp. 13–15. Cambridge: Cambridge University Press, 2006.

Question 13.20

Consider a simple pendulum consisting of a mass m attached to one end of a massless, rigid rod of length ℓ. The pivot point is subject to a vertical oscillation $y_0 = a\sin\omega t$, where a and ω are constants.

(a) Show that the equation of motion for oscillations of the pendulum in a vertical plane is
$$\ddot\theta + \left(\omega_0^2 - (a/\ell)\omega^2 \sin\omega t\right)\sin\theta = 0, \tag{1}$$
where θ is the angular coordinate of m measured counter-clockwise from the downward vertical direction and $\omega_0 = \sqrt{g/\ell}$ is the natural frequency. Do this in two ways: 1. in an inertial frame, and 2. in an accelerated (non-inertial) frame whose origin is located at the pivot point.

(b) Suppose the pivot is subject to a horizontal oscillation $b\sin\omega t$ rather than a vertical one. Deduce the equation of motion of the pendulum.

(c) Extend the equations of motion in (a) and (b) to include the effect of a linear frictional force $-\alpha\dot\theta$ on m. Express the result in terms of the quality factor $Q = m\ell\omega_0/\alpha$ for a weakly damped oscillator (see Question 4.10).

(d) Extend (1) to apply to a compound pendulum of mass m and moment of inertia I_C about its CM.

Solution

(a) 1. Let F_x and F_y denote the external forces acting on the rod at the pivot P. Oxy is an inertial frame, and P oscillates about O along the y-axis. The CM of the pendulum is located at the position (x, y) of m. The equation of motion of the CM has components
$$m\ddot x = F_x, \qquad m\ddot y = F_y - mg. \tag{2}$$

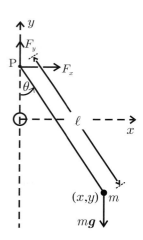

The equation of motion for rotation about the CM is
$$\begin{aligned}I_C\ddot\theta &= -F_x\ell\cos\theta - F_y\ell\sin\theta \\ &= -m\ell\ddot x\cos\theta - m\ell(g+\ddot y)\sin\theta,\end{aligned} \tag{3}$$
where I_C is the moment of inertia about the CM. Now, $x = \ell\sin\theta$ and $y = y_0 - \ell\cos\theta$, where $y_0 = a\sin\omega t$ is the position of P relative to O. So
$$\left.\begin{aligned}\ddot x &= \ell\ddot\theta\cos\theta - \ell\dot\theta^2\sin\theta \\ \ddot y &= \ddot y_0 + \ell\ddot\theta\sin\theta + \ell\dot\theta^2\cos\theta.\end{aligned}\right\} \tag{4}$$

Equations (3) and (4) show that
$$I_C\ddot\theta = -m\ell^2\ddot\theta - m\ell(g+\ddot y_0)\sin\theta. \tag{5}$$

Here, $I_C = 0$ because the mass of the rod is negligible compared to the mass m of the bob. Also, $y_o = a \sin \omega t$. Thus, (5) yields (1).

2. This question can be solved most simply by referring the motion to a non-inertial frame with origin at P – that is, at $y_o \hat{\mathbf{y}}$ relative to O. In such a frame the gravitational force $m\mathbf{g} = -mg\hat{\mathbf{y}}$ is modified by a translational force $-m\ddot{y}_o\hat{\mathbf{y}}$ acting on m (see Chapters 1 and 14), and the effective force is

$$\mathbf{F} = -m(g + \ddot{y}_o)\hat{\mathbf{y}}. \tag{6}$$

Therefore, in the equation of motion $\ddot{\theta} + (g/\ell)\sin\theta = 0$ for a simple pendulum (see Question 4.3) one should replace g with $g + \ddot{y}_o$. This yields (1).

(b) If the pivot is subject to a horizontal oscillation $x_o = b\sin\omega t$, rather than a vertical one, then we should set $x = x_o + \ell\sin\theta$ and $y = -\ell\cos\theta$ in (3) and so obtain

$$\ddot{\theta} + \omega_0^2 \sin\theta - (b/\ell)\omega^2 \sin\omega t \cos\theta = 0. \tag{7}$$

Alternatively, use an effective force $\mathbf{F} = -mg\hat{\mathbf{y}} - m\ddot{y}_o\hat{\mathbf{x}}$.

(c) A frictional force $-\alpha\dot{\theta}$ acting on m will produce an additional term $-\ell\alpha\dot{\theta}$ on the right-hand sides of (3) and (5). Thus, (1) and (7) become, respectively,

$$\ddot{\theta} + (\omega_0/Q)\dot{\theta} + \left[\omega_0^2 - (a/\ell)\omega^2 \sin\omega t\right]\sin\theta = 0, \tag{8}$$

and

$$\ddot{\theta} + (\omega_0/Q)\dot{\theta} + \omega_0^2 \sin\theta - (b/\ell)\omega^2 \sin\omega t \cos\theta = 0. \tag{9}$$

(d) Equation (5) will apply to a compound pendulum of mass m if we replace ℓ by ℓ_C, the distance of the CM from P. So, the equation of motion becomes

$$\left[1 + I_C/m\ell_C^2\right]\ddot{\theta} + \left[\omega_0^2 - (a/\ell_C)\omega^2 \sin\omega t\right]\sin\theta = 0. \tag{10}$$

Comments

(i) The above system, consisting of an ordinary rigid pendulum whose pivot point is driven sinusoidally, possesses a remarkable variety of motions for such a seemingly simple object. (Many references to this topic are provided in Ref. [22].) Among these are parametric resonance, chaos, bifurcation, and induced stability. The latter occurs in vertical driving, and it is an example of a response that is counter-intuitive: when the ratio $a\omega/\ell\omega_0$ exceeds a critical value, the upward (unstable) equilibrium position of the pendulum becomes stable – the pendulum can perform oscillatory motion about the vertical inverted position and, if it is damped, it will come to rest in this position. This peculiar behaviour, which was discovered about one hundred years ago, is studied in the next two questions.

(ii) It is evident from (6) that an oscillation $y_o(t)$ of the pivot point P is equivalent to a modulation $\ddot{y}_o$ of the constant gravitational acceleration g.

---

[22] E. I. Butikov, "On the dynamic stabilization of an inverted pendulum," American Journal of Physics, vol. 69, pp. 755–768, 2001.

(iii) The non-linear equation (1) cannot be solved analytically, although for small oscillations about either the downward or upward equilibrium positions it reduces to Mathieu's equation, and the solution is given in terms of a doubly periodic infinite series.[23] In Question 13.21 we consider numerical solutions to the equations of motion (8) and (9).

Question 13.21

(a) Use the dimensionless time $\tau = \omega_0 t$ to express the equations of motion (8) and (9) of Question 13.20 in the dimensionless forms

$$\frac{d^2\theta}{d\tau^2} + \frac{1}{Q}\frac{d\theta}{d\tau} + \left[1 - \frac{a}{\ell}\frac{\omega^2}{\omega_0^2}\sin\frac{\omega}{\omega_0}\tau\right]\sin\theta = 0 \tag{1}$$

and

$$\frac{d^2\theta}{d\tau^2} + \frac{1}{Q}\frac{d\theta}{d\tau} + \left[1 - \frac{b}{\ell}\frac{\omega^2}{\omega_0^2}\sin\frac{\omega}{\omega_0}\tau\right]\cos\theta = 0. \tag{2}$$

(b) Suppose the pivot oscillates vertically and that $a/\ell = 0.1$, $\omega/\omega_0 = 20.0$, $\theta_0 = 170°$ and $\dot\theta_0 = 0$. Write a *Mathematica* notebook to solve (1) for $\theta(\tau)$. Plot graphs of the angular position $\phi = \pi - \theta$, measured from the upward vertical, versus τ for $Q = \infty$ (an undamped pendulum) and $Q = 10$, for $0 \leq \tau \leq 15$.
(c) Suppose the pivot oscillates horizontally and that $b/\ell = 0.1$, $\omega/\omega_0 = 20.0$, $\theta_0 = 80°$ and $\dot\theta_0 = 0$. Modify the notebook in (b) and plot graphs of $\theta(\tau)$ versus τ for $Q = \infty$ and $Q = 10$.
(d) Animate the motion of the pendulum using *Mathematica*'s Manipulate function.

Solution

(a) Equations (1) and (2) follow by dividing (8) and (9) of Question 13.20 by ω_0^2.
(b) and (c) See the notebook given in (d) below. On the following figures we have also plotted the oscillation of the pivot.

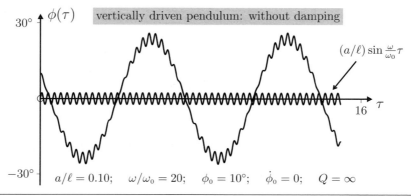

[23] F. M. Phelps, III and J. H. Hunter, Jr., "An analytical solution of the inverted pendulum," *American Journal of Physics*, vol. 33, pp. 285–295, 1965.

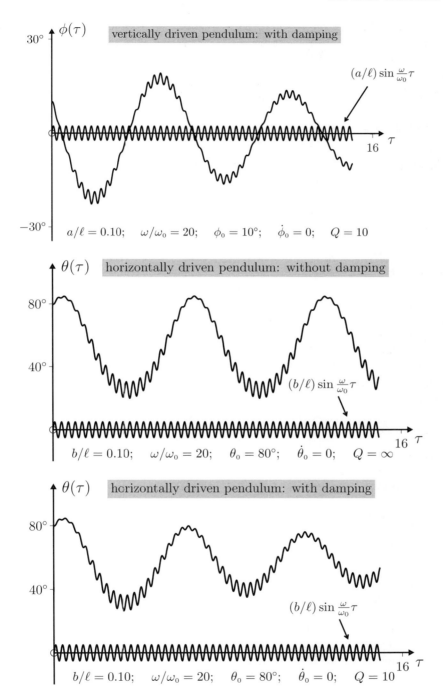

(d) Use the following notebook:

```
In[1]:= τmax = 15.0;  ω0 = 1.0;  ω = 20.0;  Q = 10;  a = 0.1;  λ = 1;  theta0 = 170;

(* b = 0.1;   theta0 = 80;   (* horizontally - driven pivot *) *)

θ0 = theta0 π/180;   θ0dot = 0;

Sol = NDSolve[{θ''[τ] + θ'[τ]/Q + (1 - a/λ ω²/ω0² Sin[ω/ω0 τ]) Sin[θ[τ]] == 0,
    θ[0] == θ0, θ'[0] == θ0dot}, {θ[τ], θ'[τ], θ''[τ]}, {τ, 0, τmax},
    MaxSteps → 100000];

(* Sol = NDSolve[{θ''[τ] + θ'[τ]/Q + Sin[θ[τ]] - b/λ ω²/ω0² Sin[ω/ω0 τ] Cos[θ[τ]] == 0,
    θ[0] == θ0, θ'[0] == θ0dot}, {θ[τ], θ'[τ], θ''[τ]},
    {τ, 0, τmax}, MaxSteps → 100000];  (* horizontally - driven pivot *) *)

Plot[{{180 - 180/π Evaluate[θ[τ]/.Sol]}, 15a Sin[ω/ω0 τ]},
    {τ, 0, τmax}, AspectRatio → 0.5]

pendulum1[t_] := Graphics[{Circle[{0, 0}, 0.025], PointSize[0.01], Black,
    Point[{{0, a/λ Sin[ω/ω0 τ]}}/.Sol/.τ → t}], PointSize[0.05], Green,
    Point[{{Sin[θ[τ]], a/λ Sin[ω/ω0 τ] - Cos[θ[τ]]}}/.Sol/.τ → t}], Black,
    Line[{First[{0, a/λ Sin[ω/ω0 τ]}/.Sol/.τ → t],
        First[{Sin[θ[τ]], (a/λ Sin[ω/ω0 τ] - Cos[θ[τ]])}/.Sol/.τ → t]}],
    Dashed, Line[{{0, 1}, {0, -1}}], Dashed, Line[{{-1, 0}, {1, 0}}]},
    PlotRange → {{-1.2, 1.2}, {-1.2, 1.2}}, AspectRatio → 1]

Manipulate[Show[pendulum1[t]], {t, 0, τmax}]
```

Comments

(i) A simple pendulum possesses two equilibrium positions: the downward position $\theta = 0$, which is stable, and the inverted position $\phi = \pi - \theta = 0$, which is unstable (see Question 5.14). We see from the first pair of figures above that a vertically driven pendulum can oscillate about $\phi = 0$. In particular, if the pendulum is damped it eventually comes to rest in the inverted position.

(ii) This property, where vertical oscillations of the pivot can cause the position of unstable equilibrium to become stable, is known as induced stability. It was evidently first studied (for small oscillations $\phi(t)$) by Stephenson in 1908[24] and

---

[24] A. Stephenson, "On an induced stability," Philosophical Magazine, vol. 15, pp. 233–236, 1908.

later in more detail by Kapitza and others (see references in Refs. [22] and [25].)
(iii) For the horizontally driven pendulum the second pair of figures above show that a lateral position of equilibrium is induced below the horizontal position $\theta = \tfrac{1}{2}\pi$.
(iv) It is evident from the figures that the motion consists of a slow oscillation plus a small, superimposed rapid vibration at the frequency ω of the forced oscillation of the pivot. This feature is present for small, rapid oscillations of the pivot ($a \ll \ell$ and $\omega \gg \omega_0$), and it provides a useful clue for performing an approximate analysis of the motion (see Questions 13.22 and 13.23).[22]

Question 13.22

Consider the vertically driven simple pendulum of Question 13.20 in the limit of small driving amplitude a and large driving frequency ω:

$$a/\ell \ll 1, \qquad \omega/\omega_0 \gg 1. \tag{1}$$

Neglect friction. Motivated by the numerical solutions of Question 13.21, express the angular position $\theta(t)$ as the sum of a 'slow' component θ_s and a small 'rapid' component θ_r:

$$\theta = \theta_s + \theta_r \qquad (|\theta_r| \ll 1). \tag{2}$$

(a) Show that

$$\theta_r \approx -\frac{a}{\ell}\sin\omega t \sin\theta_s. \tag{3}$$

(Hint: Consider a non-inertial frame that oscillates with the pivot and note that θ_r is an oscillatory response of frequency ω to the torque exerted by the translational force[‡] in this frame.)

(b) Evaluate the total torque $\boldsymbol{\Gamma}$ about the oscillating pivot and show that its value averaged over the period $2\pi/\omega$ of the rapid oscillation is

$$\langle \boldsymbol{\Gamma} \rangle = -m(g\ell + \tfrac{1}{2}a^2\omega^2\cos\theta_s)\sin\theta_s \hat{\mathbf{z}}. \tag{4}$$

(c) Consider oscillations of the pendulum about the vertical position $\theta = \pi$. Express (4) in terms of the supplementary angle $\phi = \pi - \theta$ measured from the upward vertical, and deduce that the maximum deviation $\phi_{\max}$ that can occur in such oscillations is given by

$$\cos\phi_{\max} = 2\left(\frac{\ell}{a}\frac{\omega_0}{\omega}\right)^2, \tag{5}$$

where $\omega_0 = \sqrt{g/\ell}$. Hence, obtain the condition for induced stability (that is, for oscillations of an inverted pendulum).

(d) Use (4) to obtain the effective potential $V_e(\theta_s)$ that governs the slow oscillations of the pendulum. Sketch the graphs of $V_e(\theta_s)$ versus θ_s for $a\omega/\ell\omega_0 = 1/\sqrt{2}$ and $5/\sqrt{2}$, and use them to analyze the possible motions of the pendulum.

[‡]See Chapter 1 and Question 14.11.

---

[25] A. B. Pippard, "The inverted pendulum," European Journal of Physics, vol. 8, pp. 203–206, 1987.

Solution

(a) The torque about the pivot P exerted by the translational force $-m\ddot{y}_0 \hat{\mathbf{y}}$ is

$$\mathbf{\Gamma}_{\text{tr}} = -m\ell \ddot{y}_0 \sin\theta \, \hat{\mathbf{z}} = m\ell a\omega^2 \sin\omega t \sin\theta \, \hat{\mathbf{z}}. \tag{6}$$

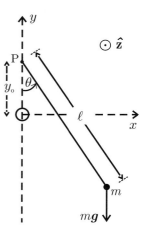

We look for a solution for the rapid variation θ_r in (2) that is associated with $\mathbf{\Gamma}_{\text{tr}}$: so we consider the equation of motion

$$m\ell^2 \ddot{\theta}_r = m\ell a\omega^2 \sin\omega t \sin\theta. \tag{7}$$

We are interested in a solution θ_r that is small and oscillates with frequency ω. So, we approximate θ in (7) by θ_s and write

$$\ddot{\theta}_r = (a\omega^2/\ell) \sin\omega t \sin\theta_s. \tag{8}$$

Here, the slow variable θ_s can be treated as a constant, and therefore integration of (8) yields the desired solution (3). Note that θ_r oscillates with the rapid frequency ω and that $|\theta_r|$ is small because of $(1)_1$. Also, θ_r is proportional to $\sin\theta_s$ and so it vanishes where θ_s is zero. These features are in agreement with the numerical results in Question 13.21. An alternative derivation of (3) is given in Comment (ii) below.

(b) The total torque $\mathbf{\Gamma}$ about P is the sum of the gravitational torque $\mathbf{\Gamma}_g = -mg\ell \sin\theta \, \hat{\mathbf{z}}$ and $\mathbf{\Gamma}_{\text{tr}}$ in (6):

$$\mathbf{\Gamma} = (-mg\ell \sin\theta + m\ell a\omega^2 \sin\omega t \sin\theta)\hat{\mathbf{z}}. \tag{9}$$

Here

$$\sin\theta = \sin(\theta_s + \theta_r) \approx \sin\theta_s + \theta_r \cos\theta_s. \tag{10}$$

With θ_r given by (3), equations (9) and (10) yield

$$\mathbf{\Gamma} \approx \left[-mg\ell\left(1 - \frac{a}{\ell}\sin\omega t \cos\theta_s\right) + m\ell a\omega^2\left(\sin\omega t - \frac{a}{\ell}\sin^2\omega t \cos\theta_s\right)\right]\sin\theta_s \, \hat{\mathbf{z}}. \tag{11}$$

In averaging (11) over the period $2\pi/\omega$ of the rapid oscillation we can treat θ_s as a constant. Also, $\langle\sin\omega t\rangle = 0$ and $\langle\sin^2\omega t\rangle = \frac{1}{2}$. Therefore

$$\langle\mathbf{\Gamma}\rangle = -m(g\ell + \tfrac{1}{2}a^2\omega^2\cos\theta_s)\sin\theta_s \, \hat{\mathbf{z}}. \tag{12}$$

The first term in (12) is the average $\langle\mathbf{\Gamma}_g\rangle$ of the gravitational torque and the second term is the average $\langle\mathbf{\Gamma}_{\text{tr}}\rangle$ of the torque due to the translational force.

(c) It is clear from (12) that if $\theta_s < \frac{1}{2}\pi$ then both $\langle\mathbf{\Gamma}_g\rangle$ and $\langle\mathbf{\Gamma}_{\text{tr}}\rangle$ tend to turn the pendulum downward: the pendulum can perform oscillations about the point of stable equilibrium at $\theta = 0$. However, if $\theta_s > \frac{1}{2}\pi$ then these torques oppose each other: $\langle\mathbf{\Gamma}_{\text{tr}}\rangle$ tends to turn the pendulum upward. To analyze this case it is helpful to introduce the supplementary angle $\phi = \pi - \theta_s$ measured from the upward vertical and write (12) as

$$\langle \mathbf{\Gamma} \rangle = -m(g\ell - \tfrac{1}{2}a^2\omega^2 \cos\phi)\sin\phi \,\hat{\mathbf{z}}. \tag{13}$$

It follows that oscillations about the upward vertical ($\phi = 0$) can occur up to a maximum deflection $\phi_{\max}$ where

$$\cos\phi_{\max} = 2\frac{g\ell}{a^2\omega^2} = 2\left(\frac{\ell}{a}\frac{\omega_0}{\omega}\right)^2. \tag{14}$$

These oscillations of an inverted pendulum are possible if the right-hand side of (14) is less than 1. That is, if

$$\frac{a}{\ell}\frac{\omega}{\omega_0} > \sqrt{2}. \tag{15}$$

Thus, induced stability (stable equilibrium at $\phi = 0$) occurs when the ratio $a\omega/\ell\omega_0$ exceeds the critical value $\sqrt{2}$. According to (14), for values of $a\omega/\ell\omega_0$ slightly above $\sqrt{2}$ an inverted pendulum can perform only small oscillations, whereas for $a\omega/\ell\omega_0 \gg 1$ these inverted oscillations can extend to almost the horizontal position ($\phi_{\max} \approx \tfrac{1}{2}\pi$):

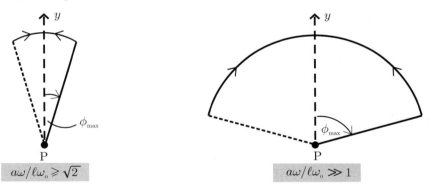

(d) The effective potential $V_e(\theta_s)$ for the slow oscillations is related to the average torque (4) by

$$\langle \mathbf{\Gamma} \rangle = -(dV_e/d\theta_s)\,\hat{\mathbf{z}}. \tag{16}$$

Therefore, choosing the zero of potential at $\theta_s = 0$ we have

$$V_e(\theta_s) = m\ell^2\omega_0^2\left[1 - \cos\theta_s + \frac{1}{8}\left(\frac{a\omega}{\ell\omega_0}\right)^2(1 - \cos 2\theta_s)\right]. \tag{17}$$

From the derivatives $dV_e/d\theta_s$ and $d^2V_e/d\theta_s^2$ we see that V_e has a minimum at $\theta_s = 0$, and when $a\omega/\ell\omega_0 > \sqrt{2}$ it also has minima at $\theta_s = \pm\pi$, otherwise it has maxima there. The graphs for $a\omega/\ell\omega_0 = 1/\sqrt{2}$ and $5/\sqrt{2}$ illustrate this:

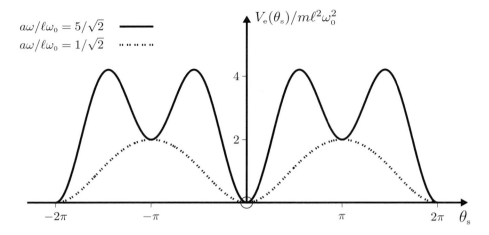

We conclude that if $a\omega/\ell\omega_0 < \sqrt{2}$ there is just one point of stable equilibrium (at $\theta_s = 0$) about which the pendulum can perform oscillatory motion. However, if $a\omega/\ell\omega_0 > \sqrt{2}$, then $\theta_s = \pi$ (or, equivalently, $\theta_s = -\pi$) is an additional (induced) point of stable equilibrium, about which oscillations of an inverted pendulum are possible, and to which damped oscillations will tend.

Comments

(i) The equation of motion of the slow oscillations is $m\ell^2 \ddot{\theta}_s \hat{\mathbf{z}} = \langle \mathbf{\Gamma} \rangle$, where $\langle \mathbf{\Gamma} \rangle$ is given by (4). That is,

$$\ddot{\theta}_s + \omega_0^2 \sin\theta_s + (a^2\omega^2/2\ell^2)\cos\theta_s \sin\theta_s = 0. \tag{18}$$

Thus, for small oscillations about the lower equilibrium position ($|\theta_s| \ll 1$) this equation is harmonic with angular frequency

$$\omega_L = \omega_0\sqrt{1 + a^2\omega^2/2\ell^2\omega_0^2}\,; \tag{19}$$

and for small oscillations about the upper equilibrium position ($\theta_s = \phi + \pi$, where $|\phi| \ll 1$) the angular frequency is

$$\omega_U = \omega_0\sqrt{a^2\omega^2/2\ell^2\omega_0^2 - 1}\,, \tag{20}$$

which exists if $a\omega/\ell\omega_0 > \sqrt{2}$, in agreement with (15).

(ii) The effective potential (17) can also be obtained from the equation of motion (1) of Question 13.20. With the superposition (2) and the approximation (10) we can write the following coupled equations of motion for θ_s and θ_r:

$$\ddot{\theta}_s + \omega_0^2 \sin\theta_s - (a\omega^2/\ell)\,\theta_r \cos\theta_s \sin\omega t = 0 \tag{21}$$

$$\ddot{\theta}_r + \omega_0^2 \theta_r \cos\theta_s - (a\omega^2/\ell) \sin\theta_s \sin\omega t = 0. \tag{22}$$

Because θ_s is slowly varying compared to θ_r, the solution to (22) can be approximated as $\theta_r = A\sin\omega t$, where

$$A = -\frac{(a\omega^2/\ell)\sin\theta_s}{\omega^2 - \omega_0^2\cos\theta_s} \approx -\frac{a}{\ell}\sin\theta_s. \tag{23}$$

Thus θ_r is given by (3). (The second term in (22), which is absent from (7), does not contribute materially to the solution.) From (21) and (3) we have

$$\ddot{\theta}_s + \omega_0^2\sin\theta_s + (a\omega/\ell)^2\cos\theta_s\sin\theta_s\sin^2\omega t = 0. \tag{24}$$

This reduces to (18) when $\sin^2\omega t$ is replaced by its average value over the driving period $2\pi/\omega$. Equation (18) and the relation $m\ell^2\ddot{\theta}_s = -dV_e/d\theta_s$ yield (17).

(iii) The analysis given in parts (a) to (d) above is valid only in the limits (1). Other motions occur when (1) is not satisfied, including parametric resonance and chaotic modes, which can be either oscillatory (θ bounded), or rotational (θ unbounded), or a combination of the two.[22]

Question 13.23

Extend the analysis of the previous question to a simple pendulum whose pivot is driven horizontally according to $x_0 = b\sin\omega t$ in the limits $\omega/\omega_0 \gg 1$ and $b/\ell \ll 1$.

Solution

The calculations are similar to those above for a vertically driven pendulum, and we present a brief outline. The translational force $-m\ddot{x}_0\hat{\mathbf{x}}$ exerts a torque

$$\boldsymbol{\Gamma}_{\text{tr}} = -m\ell\ddot{x}_0\cos\theta\,\hat{\mathbf{z}} = m\ell b\omega^2\sin\omega t\cos\theta\,\hat{\mathbf{z}}. \tag{1}$$

The equation of motion of the small, rapid oscillation θ_r in the superposition $\theta = \theta_s + \theta_r$ is therefore

$$\ddot{\theta}_r = (b\omega^2/\ell)\sin\omega t\cos\theta_s. \tag{2}$$

Thus, $\theta_r = (-b/\ell)\sin\omega t\cos\theta_s$, and with $\cos\theta = \cos(\theta_s+\theta_r) \approx \cos\theta_s - \theta_r\sin\theta_s$ in (1) we find an average torque

$$\langle\boldsymbol{\Gamma}_{\text{tr}}\rangle = \tfrac{1}{2}mb^2\omega^2\cos\theta_s\sin\theta_s\,\hat{\mathbf{z}}. \tag{3}$$

Therefore, the average value of the total torque is

$$\langle\boldsymbol{\Gamma}\rangle = -m(g\ell - \tfrac{1}{2}b^2\omega^2\cos\theta_s)\sin\theta_s\,\hat{\mathbf{z}}. \tag{4}$$

Consequently, an induced equilibrium position at θ_s given by

$$\cos\theta_s = 2g\ell/b^2\omega^2 = 2(\ell\omega_0/b\omega)^2 \tag{5}$$

exists if

$$\frac{b}{\ell}\frac{\omega}{\omega_0} > \sqrt{2}. \tag{6}$$

The effective potential $V_e(\theta_s)$ for the slow oscillations is related to (4) by $\langle \mathbf{\Gamma} \rangle = -(dV_e/d\theta_s)\,\hat{\mathbf{z}}$. The potential, which is zero at $\theta_s = 0$, is therefore

$$V_e(\theta_s) = m\ell^2\omega_0^2\left[1 - \cos\theta_s - \frac{1}{8}\left(\frac{b\omega}{\ell\omega_0}\right)^2(1 - \cos 2\theta_s)\right]. \qquad (7)$$

This differs from the corresponding potential for a vertically driven pendulum (see (17) of Question 13.22) only in the sign of the last term in the square brackets. The potential (7) has maxima at $\theta_s = \pm\pi$. If $b\omega/\ell\omega_0 > \sqrt{2}$ it has a maximum at $\theta_s = 0$ and minima at $\theta_s = \cos^{-1} 2(\ell\omega_0/b\omega)^2$:

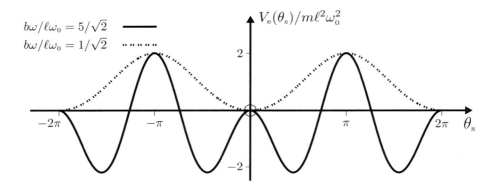

We conclude that if $b\omega/\ell\omega_0 < \sqrt{2}$ there is just one point of stable equilibrium, the vertically downward position $\theta_s = 0$. But, if $b\omega/\ell\omega_0 > \sqrt{2}$, this point becomes unstable and instead a point of stable equilibrium is induced at a θ_s between 0 and $\frac{1}{2}\pi$, and at a symmetric point between 0 and $-\frac{1}{2}\pi$. For $b\omega/\ell\omega_0 \gg 1$ these two points of stable equilibrium approach the horizontal ($\theta_s = \pm\frac{1}{2}\pi$): a damped pendulum that is driven horizontally with a sufficiently large amplitude b and/or frequency ω will eventually perform small, rapid oscillations[‡] (of amplitude b/ℓ and angular frequency ω) about an almost horizontal position.

Comment

The following curve is a plot of the induced equilibrium position θ_s given by the approximation (5). The points are numerical results obtained by averaging asymptotic numerical solutions to the dimensionless equation of motion of a damped pendulum (see Question 13.21),

$$\frac{d^2\theta}{d\tau^2} + \frac{1}{Q}\frac{d\theta}{d\tau} + \left[1 - \frac{b}{\ell}\frac{\omega^2}{\omega_0^2}\sin\frac{\omega}{\omega_0}\tau\right]\cos\theta = 0, \qquad (8)$$

[‡]The reader should keep in mind that $\theta_s = \theta - \theta_r$ represents the 'smoothed' motion of the pendulum: it is equal to the value of θ averaged over the period $2\pi/\omega$ of the rapid oscillations, because $\langle\theta_r\rangle = 0$.

over the (dimensionless) period $2\pi\omega_0/\omega$ of the rapid oscillations. We have used $Q = 10$, $\omega/\omega_0 = 80$ and increasing values of b/ℓ, starting at $b/\ell = 1.416/80$, which is just above the critical value in (6). The approximate theory is seen to be rather accurate. In fact, there is good agreement with numerical values at even smaller values of ω/ω_0 and

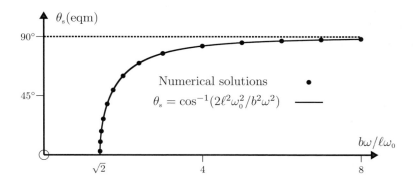

larger starting values of b/ℓ, particularly if $b\omega/\ell\omega_0$ is not too close to the critical value $\sqrt{2}$ (see the two tables below, where ω/ω_0 decreases and b/ℓ increases as one reads from left to right). The conditions $\omega/\omega_0 \gg 1$ and $b/\ell \ll 1$ for the validity of the theory are evidently not particularly onerous in practice.

$$\frac{b}{\ell}\frac{\omega}{\omega_0} = 1.416 \quad \text{and} \quad \theta_{s,\text{theory}} = 4.070°$$

ω/ω_0	1000	500	100	80	60	40	20	10	5
$100b/\ell$	0.1416	0.2832	1.416	1.770	2.360	3.540	7.080	14.16	28.32
$\theta_{s,\text{num}}$	4.075°	4.077°	4.096°	4.114°	4.151°	4.251°	4.749°	6.358°	10.554°

$$\frac{b}{\ell}\frac{\omega}{\omega_0} = 1.750 \quad \text{and} \quad \theta_{s,\text{theory}} = 49.227°$$

ω/ω_0	500	100	80	60	40	20	10	5
$100b/\ell$	0.350	1.750	2.188	2.917	4.375	8.750	17.500	35.000
$\theta_{s,\text{num}}$	49.227°	49.229°	49.230°	49.232°	49.238°	49.270°	49.397°	49.901°

14
Translation and rotation of the reference frame

In this chapter we consider a topic that has been discussed at least from the time of Galileo and Newton, and which centres on the question: what if we change the frame of reference in some way? At first sight this question may seem of no special significance. However, the harvest turns out to be surprisingly rich and it contains (and points to) some of the most interesting and profound results in physics.

Broadly speaking, two types of question arise regarding the effects (if any) of a change in the frame of reference, namely 1. effects on physical quantities and 2. effects on the equations of physics. As Newton realized, this involves one in questions concerning the nature of space and time, and it leads – after a subtle change in emphasis (the relativity principle) – to Einstein's theory of space-time.

In the following we consider physical quantities and relations in a frame S' that differs from an inertial frame S by a translation, or a rotation, or both. These frames have Cartesian coordinate axes $O'x'y'z'$ and $Oxyz$, and the position vector of the origin O' of S' relative to O is denoted by **D**. When no rotation is involved the corresponding axes of S' and S are always taken to be parallel.

Questions in which **D** is constant deal with origin independence/dependence of physical quantities and equations, and with conservation of momentum and its connection with symmetry (see Questions 14.1–14.7). When **D** is time dependent one can treat topics such as the Galilean transformation, invariance of the equation of motion, the translational force and applications (see Questions 14.8–14.15).

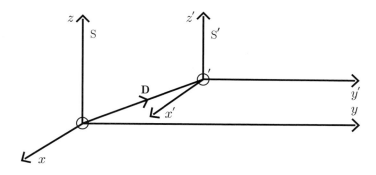

A fixed rotation of S' relative to S enables one to discuss vectors (in general, tensors) and orthogonal transformations, invariance of the equations of physics under such transformations, conservation of angular momentum and symmetry (see Questions 14.16–14.19). Examples in which S' rotates with angular velocity $\boldsymbol{\omega}$ relative to S deal

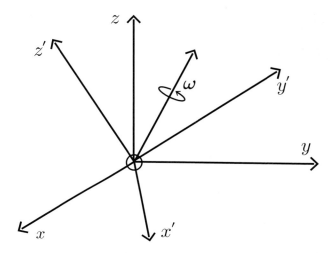

with the centrifugal, Coriolis, and azimuthal (Euler) forces and some applications (see Questions 14.20–14.30). If the rotation is accompanied by translation then both $\mathbf{D}$ and $\boldsymbol{\omega}$ are non-zero, as illustrated on page 8.

Newtonian (or Galilean) relativity – that is, absolute space, time, and mass – is assumed in this chapter. The next chapter touches on the topic of relative space-time.

Question 14.1

State what is meant by 1. an origin-dependent physical quantity, and 2. an origin-independent physical quantity.

Solution

Let O denote the origin of the coordinate axes relative to which the position vector $\mathbf{r}$ is specified. Let A denote a theoretical expression (or numerical value) for a physical quantity relative to O; and let A' be its expression, or value, relative to a different origin O'. Then, A is termed origin dependent if (for at least some choices of O')

$$A' \neq A, \qquad (1)$$

and origin independent if (for all choices of O')

$$A' = A. \qquad (2)$$

Comments

(i) So, the origin dependence of a quantity is determined simply by comparing its theoretical expression (or numerical value) relative to coordinate origins O and O' that are separated by a constant displacement $\mathbf{D}$, so that $\mathbf{r} = \mathbf{r}' + \mathbf{D}$.

(ii) The value of an origin-dependent quantity has meaning only with respect to a stated coordinate origin. Some examples are considered below.

(iii) In general, A represents the components of a tensor (scalar, vector, ...). In the above we have supposed that the corresponding axes of $Oxyz$ and $O'x'y'z'$ are parallel so that we are not concerned with the effect of rotation on these components (see Question 14.16).

(iv) Usually O and O' are arbitrary, but in certain cases they may be somewhat constrained, for example by convergence of a series expansion.

(v) In practice, the condition for origin independence is used to check the validity of a theoretical result for a quantity that is expected to be origin independent. For example, after calculating such a quantity (magnetic susceptibility), Van Vleck remarked: "The reader has possibly wondered what point should be used as the origin for computing $\mathbf{r}$... . This choice is immaterial as (the expression) is invariant of the origin."[1]

(vi) Conversely, the condition for origin independence can sometimes be used to deduce theoretical expressions to within dimensionless factors. In this regard it is somewhat analogous to dimensional analysis.

Question 14.2

Determine the effect of a change in coordinate origin on the following quantities: (a) The position and displacement vectors $\mathbf{r}$ and $d\mathbf{r}$. (b) The velocity $\mathbf{v}$ and acceleration $\mathbf{a}$. (c) Relative mass (see Question 2.6). (d) The momentum $\mathbf{p}$ and angular momentum $\mathbf{L}$ of a particle. (e) The spin angular momentum $\mathbf{L}_s$, orbital angular momentum $\mathbf{L}_o$, and total angular momentum $\mathbf{L}$ of a system of particles (see Question 11.3).

Solution

(a) The position vectors $\mathbf{r}$ and $\mathbf{r}'$ of a particle relative to coordinate origins O and O' are related by $\mathbf{r} = \mathbf{r}' + \mathbf{D}$. So, $\mathbf{r}$ is an origin-dependent vector. But, $d\mathbf{r} = d(\mathbf{r}' + \mathbf{D}) = d\mathbf{r}'$ is origin independent, as therefore is any finite displacement vector.

(b) $\mathbf{v} = d\mathbf{r}/dt$ is therefore also origin independent, as is $\mathbf{a} = d\mathbf{v}/dt$.

(c) An operational definition of relative mass was given in terms of a ratio of accelerations in Question 2.6. Therefore, relative mass is origin independent, and

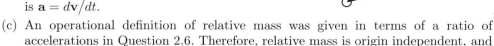

---

[1] J. H. Van Vleck, *The theory of electric and magnetic susceptibilities*, p. 276. Oxford: Clarendon Press, 1932.

so is the mass m in Newtonian mechanics (the standard mass – for example, the kilogram – is an arbitrary, origin-independent quantity).

(d) So $\mathbf{p} = m\mathbf{v}$ is origin independent, but $\mathbf{L}$ is origin dependent:
$$\mathbf{L} = \mathbf{r} \times \mathbf{p} = (\mathbf{r}' + \mathbf{D}) \times \mathbf{p}' = \mathbf{L}' + \mathbf{D} \times \mathbf{p}. \tag{1}$$

(e) The various angular momenta of a system of particles are (see Question 11.3):
$$\mathbf{L}_s = \sum_i (\mathbf{r}_i - \mathbf{R}) \times \mathbf{p}_i = \sum_i (\mathbf{r}'_i + \mathbf{D} - \{\mathbf{R}' + \mathbf{D}\}) \times \mathbf{p}'_i$$
$$= \sum_i (\mathbf{r}'_i - \mathbf{R}') \times \mathbf{p}'_i = \mathbf{L}'_s. \tag{2}$$

$$\mathbf{L}_o = \mathbf{R} \times \mathbf{P} = (\mathbf{R}' + \mathbf{D}) \times \mathbf{P}' = \mathbf{L}'_o + \mathbf{D} \times \mathbf{P}, \tag{3}$$

where $\mathbf{P}\ (=\mathbf{P}')$ is the total momentum. So $\mathbf{L}_s$ is origin independent, whereas $\mathbf{L}_o$, and also the total angular momentum $\mathbf{L} = \mathbf{L}_o + \mathbf{L}_s$, are origin dependent.

Question 14.3

Consider a set of Cartesian axes $Ox_1x_2x_3$ and the inertia tensor of a rigid, continuous body relative to these axes:
$$I_{\alpha\beta} = \int_V (\delta_{\alpha\beta} x_\gamma x_\gamma - x_\alpha x_\beta) \rho(\mathbf{r})\, dV \qquad (\alpha, \beta = 1, 2 \text{ or } 3), \tag{1}$$

where V and ρ are the volume and density of the body, and the repeated index γ implies summation from 1 to 3 (see (18) of Question 12.6). Let $O'x'_1x'_2x'_3$ be a second set of axes parallel to $Ox_1x_2x_3$ and with O' displaced from O by a constant vector $\mathbf{D} = (D_1, D_2, D_3)$. Show that the inertia tensor relative to $O'x'_1x'_2x'_3$ has components
$$I'_{\alpha\beta} = I_{\alpha\beta} + M(\delta_{\alpha\beta} D^2 - D_\alpha D_\beta) + M(R_\alpha D_\beta + R_\beta D_\alpha - 2\delta_{\alpha\beta} R_\gamma D_\gamma), \tag{2}$$

where $D^2 = D_1^2 + D_2^2 + D_3^2$, $\mathbf{R}$ is the position vector of the CM relative to O, and M is the mass of the body,
$$M = \int_V \rho(\mathbf{r})\, dV. \tag{3}$$

Solution

We start with (1) expressed in terms of the primed coordinates:
$$I'_{\alpha\beta} = \int_V (\delta_{\alpha\beta} x'_\gamma x'_\gamma - x'_\alpha x'_\beta) \rho(\mathbf{r}')\, dV', \tag{4}$$

where $x'_\alpha = x_\alpha - D_\alpha$ and $dV' = dx'_1 dx'_2 dx'_3$. Because the D_α are constants it is clear that $dV' = dV$, and also $\rho(\mathbf{r}') = dm'/dV' = dm/dV = \rho(\mathbf{r})$. So, (4) can be written

$$I'_{\alpha\beta} = \int_V \{\delta_{\alpha\beta}(x_\gamma - D_\gamma)(x_\gamma - D_\gamma) - (x_\alpha - D_\alpha)(x_\beta - D_\beta)\}\rho(\mathbf{r})\,dV. \qquad (5)$$

Expanding (5), then using (1) and (3) and rearranging terms gives

$$I'_{\alpha\beta} = I_{\alpha\beta} + M(\delta_{\alpha\beta}D^2 - D_\alpha D_\beta) + \int_V (D_\beta x_\alpha + D_\alpha x_\beta - 2\delta_{\alpha\beta}D_\gamma x_\gamma)\rho(\mathbf{r})\,dV. \qquad (6)$$

Now

$$\int_V x_\alpha \rho(\mathbf{r})\,dV = MR_\alpha, \qquad (7)$$

and so (6) reduces to (2).

Comments

(i) It is clear from (2) that, in general, the inertia tensor is origin dependent.
(ii) If O is at the CM, then $R_\alpha = 0$ ($\alpha = 1, 2$ or 3) and (2) simplifies to

$$I'_{\alpha\beta} = I_{\alpha\beta} + M(\delta_{\alpha\beta}D^2 - D_\alpha D_\beta). \qquad (8)$$

Equation (8) is the parallel-axis theorem for the inertia tensor (see Question 12.7).
(iii) It is not difficult to show that (2) and (8) are valid also for a discrete mass distribution (a system of particles), though the notation is more cumbersome (the components of the position vectors of the particles are denoted $x_{i\alpha}$, where $i = 1, 2, \cdots, N$ labels the particle and $\alpha = 1, 2, 3$ labels the component).

Question 14.4

The components of the electric dipole moment μ_α and the electric quadrupole moment $q_{\alpha\beta}$ of a continuous distribution of charge are defined by

$$\mu_\alpha = \int_V x_\alpha \rho(\mathbf{r})\,dV \qquad (1)$$

$$q_{\alpha\beta} = \int_V x_\alpha x_\beta \rho(\mathbf{r})\,dV, \qquad (2)$$

where V and ρ are the volume and charge density of the distribution, and $\alpha, \beta = 1, 2,$ or 3. Determine the change in these moments when the coordinate origin is shifted by $\mathbf{D} = (D_1, D_2, D_3)$.

Solution

Relative to a coordinate origin O' the dipole moment is

$$\mu'_\alpha = \int_V x'_\alpha \rho(\mathbf{r'})\,dV', \qquad (3)$$

where $x'_\alpha = x_\alpha - D_\alpha$ and $dV' = dx'_1 dx'_2 dx'_3$. Because the D_α are constant, $dV' = dV$ and $\rho(\mathbf{r}') = \rho(\mathbf{r})$. So (3) can be written

$$\mu'_\alpha = \int_V (x_\alpha - D_\alpha)\rho(\mathbf{r})\, dV = \mu_\alpha - qD_\alpha,\qquad(4)$$

where

$$q = \int_V \rho(\mathbf{r})\, dV \qquad(5)$$

is the total charge of the distribution. Similarly, for the quadrupole moment we have

$$q'_{\alpha\beta} = \int_V x'_\alpha x'_\beta \rho(\mathbf{r}')\, dV' = \int_V (x_\alpha - D_\alpha)(x_\beta - D_\beta)\rho(\mathbf{r})\, dV$$
$$= q_{\alpha\beta} - \mu_\alpha D_\beta - \mu_\beta D_\alpha + qD_\alpha D_\beta. \qquad(6)$$

Equations (4) and (6) are the desired results for the origin dependences of μ_α and $q_{\alpha\beta}$.

Comments

(i) Equations (4) and (6) are valid also for a discrete distribution of charge.
(ii) It follows from (4) that the dipole moment μ_α is origin independent only if the charge distribution is neutral ($q = 0$).
(iii) Similarly, from (6) the quadrupole moment $q_{\alpha\beta}$ is origin independent only if the total charge q and the dipole moment μ_α are both zero. So, the electric quadrupole moment of a dipolar molecule (one with $\mu_\alpha \neq 0$) is origin dependent. An ingenious experimental technique exists to measure this observable, based on a theory by Buckingham and Longuet-Higgins of the birefringence induced in a gas by an electric field gradient.[2,3]
(iv) These conclusions regarding origin dependence of μ_α and $q_{\alpha\beta}$ can be generalized: only the leading non-vanishing electric multipole moment is origin independent.

Question 14.5

Show that the equation of motion $\mathbf{F} = d\mathbf{p}/dt$ is origin independent (that is, it is valid for any choice of the coordinate origin of an inertial frame).

Solution

This follows directly because $\mathbf{p}' = \mathbf{p}$ (see Question 14.2). Therefore, the rates of change $d\mathbf{p}/dt$ and $d\mathbf{p}'/dt$ are the same, as are the associated forces $\mathbf{F}$ and $\mathbf{F}'$.

---

[2] A. D. Buckingham and H. C. Longuet-Higgins, "The quadrupole moments of dipolar molecules," Molecular Physics, vol. 14, pp. 63–72, 1968.
[3] A. D. Buckingham, R. L. Disch, and D. A. Dunmur, "The quadrupole moments of some simple molecules," Journal of the American Chemical Society, vol. 90, pp. 3104–3107, 1968.

Comments

(i) This result was known already to Newton (see Question 14.9). Its extension to frames that are translating relative to each other with constant velocity (where the separation $\mathbf{D}$ of the coordinate origins O and O' is not constant) occurred in two stages. First, for Newton's absolute space and time (see Question 14.9), and then for relative space-time (special relativity – see Chapter 15).

(ii) In the above, the vectors $\mathbf{p}$ and $\mathbf{F}$ entering the equation of motion are both invariant with respect to the coordinate origin. In other instances the vectors in an equation of physics can be origin dependent but the equation is, nevertheless, invariant (see below).

Question 14.6

Show that the equation

$$\dot{\mathbf{L}} = \mathbf{\Gamma} \tag{1}$$

for the rate of change of the total angular momentum of a system of particles (see Question 11.2) is origin independent. Here, $\mathbf{\Gamma}$ is the total torque on the particles due to external forces.

Solution

In terms of the separation $\mathbf{D}$ of the origins of the two frames, the position vectors of the ith particle are related by $\mathbf{r}_i = \mathbf{r}'_i + \mathbf{D}$. Also $\mathbf{p}_i = \mathbf{p}'_i$ (see Question 14.2). So

$$\mathbf{L} = \sum_i \mathbf{r}_i \times \mathbf{p}_i = \sum_i (\mathbf{r}'_i + \mathbf{D}) \times \mathbf{p}'_i = \mathbf{L}' + \mathbf{D} \times \mathbf{P}, \tag{2}$$

where $\mathbf{P} = \sum_i \mathbf{p}_i$ is the total momentum. Then

$$\dot{\mathbf{L}} = \dot{\mathbf{L}}' + \mathbf{D} \times \mathbf{F}, \tag{3}$$

because $\mathbf{D}$ is constant and $\dot{\mathbf{P}} = \mathbf{F}$, the total external force acting on the particles. Also,

$$\mathbf{\Gamma} = \sum_i \mathbf{r}_i \times \mathbf{F}_i = \sum_i (\mathbf{r}'_i + \mathbf{D}) \times \mathbf{F}'_i = \mathbf{\Gamma}' + \mathbf{D} \times \mathbf{F}, \tag{4}$$

because the external force $\mathbf{F}_i$ acting on each particle is origin independent. According to (1), (3) and (4), $\dot{\mathbf{L}}' = \mathbf{\Gamma}'$.

Comment

Even though the vectors $\dot{\mathbf{L}}$ and $\mathbf{\Gamma}$ depend on the choice of coordinate origin, the equality (1) between them is origin independent.

Question 14.7

(a) What are the conditions for the Lagrangian of a system of particles to be translationally invariant (that is, unchanged by a displacement of the entire system)?
(b) Use this invariance to deduce the law of conservation of momentum for a system of particles.
(c) Deduce also the third law of motion for the interaction between two particles.

Solution

(a) The frame of reference should be inertial (which ensures that the space is homogeneous – see Chapter 1) and the system should be isolated (or closed – meaning that it does not interact with anything external).
(b) An infinitesimal, constant displacement $\delta \mathbf{r}_i = \mathbf{D}$ of each particle leaves the velocity $\mathbf{v}_i$ unaltered and changes the Lagrangian $L(\mathbf{r}_i, \mathbf{v}_i)$ by[‡]

$$\delta L = \sum_i \frac{\partial L}{\partial \mathbf{r}_i} \cdot \delta \mathbf{r}_i = \mathbf{D} \cdot \sum_i \frac{\partial L}{\partial \mathbf{r}_i}. \qquad (1)$$

Translational invariance of L means $\delta L = 0$ for arbitrary $\mathbf{D}$, and therefore

$$\sum_i \frac{\partial L}{\partial \mathbf{r}_i} = 0. \qquad (2)$$

According to Lagrange's equations (see Chapter 1) this means

$$\frac{d}{dt} \sum_i \frac{\partial L}{\partial \mathbf{v}_i} = 0. \qquad (3)$$

But $\partial L / \partial \mathbf{v}_i = \mathbf{p}_i$, the momentum of the ith particle, and so (3) can be written

$$\frac{d\mathbf{P}}{dt} = 0, \qquad (4)$$

where $\mathbf{P} = \sum_i \mathbf{p}_i$ is the total momentum of the system. Equation (4) is the law of conservation of momentum for a closed system of particles in an inertial frame.

(c) According to (2) the total force on the system is zero:

$$\sum_i \mathbf{F}_i = 0, \qquad (5)$$

where $\mathbf{F}_i = \partial L / \partial \mathbf{r}_i$ is the force on the ith particle. In an isolated system this force is entirely due to interparticle interactions:

$$\mathbf{F}_i = \sum_j \mathbf{F}_{ji} \qquad (j \neq i), \qquad (6)$$

[‡]We use the following notation: $\partial L / \partial \mathbf{r}$ denotes the vector $(\partial L / \partial x, \partial L / \partial y, \partial L / \partial z)$, and similarly for $\partial L / \partial \mathbf{v}$. This avoids the use of double indices in (1)–(3).

and therefore
$$\sum_i \sum_j \mathbf{F}_{ji} = 0 \quad (j \neq i). \tag{7}$$

For the interaction of two particles, (7) yields Newton's third law:
$$\mathbf{F}_{12} + \mathbf{F}_{21} = 0. \tag{8}$$

Comments

(i) This example is a particular case of the general connection between symmetry properties and conservation laws. The homogeneity of space in an inertial frame means that spatial translation of an isolated system is a symmetry transformation, and the consequence of this is conservation of momentum. We have previously seen that homogeneity of time in an inertial frame means that translation in time is a symmetry transformation for an isolated system (see Question 11.33), and correspondingly the energy is conserved. The connection between rotational symmetry and conservation of angular momentum is treated in Question 14.18.

(ii) If the system is not isolated then the particles will also experience external forces: in general, the invariance of L with respect to displacements along the three coordinate axes will be broken and none of the components of **P** will be conserved. In particular cases there may, nevertheless, be partial invariance, and consequently conservation of the corresponding component(s): for example, invariance of L with respect to displacement along the x-axis means that P_x is conserved, etc.

(iii) In an inertial frame the Lagrangian for an isolated system of particles interacting via two-body central potentials is
$$\mathsf{L} = \sum_i \tfrac{1}{2} m_i \mathbf{v}_i^2 - \tfrac{1}{2} \sum_i \sum_j V_{ij}(|\mathbf{r}_i - \mathbf{r}_j|), \tag{9}$$

and the translational invariance is obvious. External interactions that add terms of the form $V_i(\mathbf{r}_i)$ to (9) will, in general, break this invariance, as will use of a non-inertial frame (see Questions 14.11 and 14.22).

Question 14.8

Suppose S' is a frame that is translating with constant velocity **v** relative to an inertial frame S. Let $(\mathbf{r}, t)$ and $(\mathbf{r}', t')$ be the space and time coordinates of the same event relative to S and S'. Give a brief discussion of the basis for the Galilean transformation connecting $(\mathbf{r}', t')$ with $(\mathbf{r}, t)$.

Solution

Suppose first that S' is at rest relative to S. Then
$$\mathbf{r}' = \mathbf{r} - \mathbf{D} \quad \text{and} \quad t' = t, \tag{1}$$

where $\mathbf{D}$ is a constant. Equation (1)$_1$ is a property of the assumed Euclidean nature of space (see Chapter 1), and (1)$_2$ asserts that a universal time exists in a given frame. The Galilean transformation results from the assumption that equations (1) are valid also when S$'$ is translating relative to S with constant velocity $\mathbf{v}$: that is, when $\mathbf{D} = \mathbf{v}t + \mathbf{D}_0$. Usually one sets $\mathbf{D}_0 = 0$ for convenience. (This means that O and O' coincide at $t = t' = 0$.) Then

$$\mathbf{r}' = \mathbf{r} - \mathbf{v}t \quad \text{and} \quad t' = t. \tag{2}$$

The Galilean transformation (2) is the basis for Newtonian relativity.

Comments

(i) The transformation (2) seems intuitively reasonable. For example, it yields the velocity transformation

$$\dot{\mathbf{r}}' = \dot{\mathbf{r}} - \mathbf{v}. \tag{3}$$

(ii) However, (2) is based on the assumption of absolute space and absolute time (see Question 15.1). Despite attracting much criticism,[4] about two hundred and fifty years elapsed before Newton's relativity was replaced by Einstein's theory of special relativity, following work by Fitzgerald, Lorentz and Poincaré. This showed that the Galilean transformation is a good approximation at low speeds (see Question 15.3).

(iii) When $\mathbf{v}$ is along one of the coordinate axes, say the x-axis, (2) reduces to the Galilean transformation for the 'standard configuration':

$$x' = x - vt, \quad y' = y, \quad z' = z, \quad t' = t. \tag{4}$$

Question 14.9

Show that the equation of motion $\mathbf{F} = m\mathbf{a}$ is invariant under the Galilean transformation (2) of the previous question.

Solution

According to the Galilean transformation, acceleration is the same in all frames that are translating relative to each other with constant velocity:

$$\mathbf{a}' = \frac{d^2\mathbf{r}'}{dt^2} = \frac{d^2}{dt^2}(\mathbf{r} - \mathbf{v}t) = \frac{d^2\mathbf{r}}{dt^2} = \mathbf{a}. \tag{1}$$

Also, relative mass (as determined by a ratio of accelerations – see Question 2.6), and therefore mass, is the same in these frames.[‡] So,

[‡]In general, in Newtonian physics it is assumed that the mass of an object is an invariant.

---

[4] See, for example, A. Danto and S. Morgenbesser, *Philosophy of science*. New York: Meridian, 1960.

$$m'\mathbf{a}' = m\mathbf{a} = \mathbf{F}. \qquad (2)$$

Thus the equation of motion holds – with the same values of acceleration, mass and force – in all frames that move with constant velocity relative to an inertial frame.

Comments

(i) Therefore, if one inertial frame exists then infinitely many exist.

(ii) Invariance under a Galilean transformation is clear for any force that depends on the difference $\mathbf{r}_i - \mathbf{r}_j$ between the simultaneous positions of particles: e.g. the gravitational and electrostatic forces between particles vary as $(\mathbf{r}_i - \mathbf{r}_j)/|\mathbf{r}_i - \mathbf{r}_j|^3$, and so $\mathbf{F}'_{ij} = \mathbf{F}_{ij}$. More generally, the equations of motion for a system of N particles interacting via two-body central potentials,

$$m_i \ddot{\mathbf{r}}_i = -\sum_{j=1}^{N} \boldsymbol{\nabla}_i V_{ij}(|\mathbf{r}_i - \mathbf{r}_j|) \qquad (i = 1, 2, \cdots, N \text{ and } j \neq i), \qquad (3)$$

are invariant (provided $m'_i = m_i$).

(iii) The result (2) – that Newton's equation of motion is valid in all inertial frames – is the earliest example of what came to be known as the relativity principle. As formulated by Einstein, it states that the laws of physics are equally valid in all inertial frames, and it is one of the fundamental principles of physics. Rindler[5] has emphasized that "Einstein's principle is really a *metaprinciple*: it puts constraints on *all* the laws of physics."

(iv) In particular, the relativity principle requires that the equations of physics have the same mathematical form in all inertial frames. This is known as form invariance (or covariance). In the above example it means that if $\mathbf{F} = m\mathbf{a}$ in S, then in S' one should have the same form, namely $\mathbf{F}' = m'\mathbf{a}'$. (The invariance of $\mathbf{a}$, m and $\mathbf{F}$ is peculiar to this example. In other instances the quantities involved – such as electric and magnetic fields – are different in S and S'.) The relativity principle is applied to the space-time transformation in Questions 14.10 and 15.1.

(v) When the Galilean transformation was applied to the theory of the electromagnetic field, it was found that the relativity principle is not completely satisfied.[†] This indicates that the Galilean transformation is an approximation. It was replaced by the Lorentz transformation (and consequently a new theory of space and time) in Einstein's theory of special relativity (see Question 15.1).

(vi) Penrose[6] points out that Newton "was certainly well aware that his dynamical laws are invariant under change to a uniformly moving frame (as well as under shift

---

[†]Gauss's law for the magnetic field and Faraday's law of electromagnetic induction are form invariant under the Galilean transformation, but Gauss's law for the electric field and Ampère's law are not.

---

[5]   W. Rindler, *Introduction to special relativity*, p. 2. Oxford: Oxford University Press, 1982.
[6]   R. Penrose, "Newton, quantum theory and reality," in *Experimental gravitation from Newton's Principia to Einstein's general relativity* (S. W. Hawking and W. Israel, eds.), Cambridge, Cambridge University Press, pp. 17 – 49, 1987.

of origin and under rotation)" and asks "what evidence is there that Newton at any time shared Einstein's *conviction* that physics *must* be invariant under change to uniform motion?" Penrose mentions that in writings prior to the *Principia*, Newton had "proposed to base his mechanics on five (or six) fundamental laws rather than the three that have come down to us through *Principia*. Law 4 was actually a clear statement of the (Galilean) relativity principle! Newton was well aware that these laws were not independent of one another, and for *Principia* he settled on the three that are now familiar to us. What is remarkable – apparently supporting the viewpoint tentatively put forward in this article – is that Newton had at one time, indeed, seriously contemplated using the relativity principle *as a fundamental principle* (despite what his views on 'absolute space' may or may not have been)!"

(vii) Apparently, Huyghens also identified the relativity principle "as something deeper in mechanics than a mere property of Newton's laws."[7]

(viii) The relativity principle is an example of a result that retains its validity even when the theory from which it emerged is found to be approximate.

Question 14.10

Is the Galilean transformation of Question 14.8 consistent with the relativity principle discussed in Comments (iii) and (iv) above?

Solution

The Galilean transformation

$$\mathbf{r}' = \mathbf{r} - \mathbf{v}t, \qquad t' = t \qquad (1)$$

changes space and time coordinates $(\mathbf{r}, \mathbf{t})$ into coordinates $(\mathbf{r}', t')$. The inverse transformation

$$\mathbf{r} = \mathbf{r}' + \mathbf{v}t, \qquad t = t' \qquad (2)$$

has the same form as (1) and is therefore consistent with the relativity principle. (Note the change in the sign of $\mathbf{v}$ in (2) which reflects the fact that while S′ moves with velocity $\mathbf{v}$ relative to S, the latter moves with velocity $-\mathbf{v}$ relative to S′.)

Comments

(i) In general, the following rule encapsulates the requirement that the relativity principle imposes on the space and time transformation between inertial frames:

$$\left.\begin{array}{l}\text{The inverse transformation is obtained from the direct transformation}\\ \text{by priming the unprimed coordinates, unpriming the primed}\\ \text{coordinates, and replacing } \mathbf{v} \text{ with } -\mathbf{v}.\end{array}\right\} \quad (3)$$

---

[7] W. Rindler, *Essential relativity*. New York: Springer, 2nd edn, 1977. Chap. 2.

(ii) This raises an important question: is the Galilean transformation (1) the most general linear, homogeneous transformation consistent with (3)? The reader may wish to attempt an answer to this (using, for simplicity, the standard configuration where S' moves along the x-axis). A solution is given in Question 15.1.

Question 14.11

Let S' be a frame that is accelerating without rotation relative to an inertial frame S. Assume Newtonian relativity and show that relative to S' the equation of motion of a particle of constant mass m is

$$m\frac{d^2\mathbf{r}'}{dt^2} = \mathbf{F} - m\frac{d^2\mathbf{D}}{dt^2}, \qquad (1)$$

where $\mathbf{F}$ is the force acting in S and $\mathbf{D}(t)$ is the position vector of the origin O' relative to O. Do this in two ways: (a) directly, and (b) using Lagrange's equations.

Solution

(a) In Newtonian relativity $\mathbf{r}'(t') = \mathbf{r}(t) - \mathbf{D}(t)$ and $t' = t$. These imply the acceleration addition formula

$$\frac{d^2\mathbf{r}'}{dt^2} = \frac{d^2\mathbf{r}}{dt^2} - \frac{d^2\mathbf{D}}{dt^2}. \qquad (2)$$

Also, mass is assumed to be absolute ($m'=m$). Then, because S is inertial and m is constant we have $md^2\mathbf{r}/dt^2 = \mathbf{F}$, and so (2) yields (1).

(b) The Lagrangian of the particle in S is

$$\mathsf{L} = \tfrac{1}{2}m\mathbf{v}^2 - V(\mathbf{r}), \qquad (3)$$

where $\mathbf{v} = d\mathbf{r}/dt$ is the velocity of the particle relative to S and $V(\mathbf{r})$ is the potential. Now $\mathbf{v} = \mathbf{v}' + \dot{\mathbf{D}}$, where $\mathbf{v}' = d\mathbf{r}'/dt$ is the velocity relative to S'. So (3) gives for the Lagrangian in S':

$$\mathsf{L}' = \tfrac{1}{2}m\mathbf{v}'^2 + m\mathbf{v}' \cdot \dot{\mathbf{D}} + \tfrac{1}{2}m\dot{\mathbf{D}}^2 - V(\mathbf{r}'), \qquad (4)$$

where we have again assumed $m' = m$. Now use the identity

$$\mathbf{v}' \cdot \dot{\mathbf{D}} = \frac{d}{dt}(\mathbf{r}' \cdot \dot{\mathbf{D}}) - \mathbf{r}' \cdot \ddot{\mathbf{D}}, \qquad (5)$$

and recall that a derivative with respect to time (such as the first term in (5)) can be omitted from a Lagrangian because it does not affect Lagrange's equations (see Question 4.16). Similarly, $\dot{\mathbf{D}}^2(t)$ in (4) can be written as a time derivative and therefore omitted. So (4) becomes

$$\mathsf{L}' = \tfrac{1}{2}m\mathbf{v}'^2 - m\mathbf{r}' \cdot \ddot{\mathbf{D}} - V(\mathbf{r}'). \qquad (6)$$

If m is constant, then (1) follows from (6) and the Lagrange equations in S':

$$\frac{d}{dt}\frac{\partial \mathsf{L}'}{\partial \mathbf{v}'} - \frac{\partial \mathsf{L}'}{\partial \mathbf{r}'} = 0. \qquad (7)$$

Comments

(i) Equation (1) is the extension of Newton's equation of motion to a frame that is translating in an arbitrary manner relative to an inertial frame.

(ii) The accelerated frame S' is non-inertial because relative to it a free particle ($\mathbf{F} = 0$) does not have constant velocity.

(iii) The term $-m\ddot{\mathbf{D}}$ in (1) is a particular type of 'fictitious force' known as the translational force (see Chapter 1). The translational force has been used in Question 10.4 (relative motion in the two-body problem) and Question 13.20 (the pivot-driven pendulum).

(iv) The Lagrangian and the equation of motion are not form invariant under transformation to an accelerating frame – see (6) and (1). And the Lagrangian (6) is not translationally invariant even if $V = 0$ (space is not homogeneous in an accelerating frame).

Question 14.12

Consider oscillations of a rigid simple pendulum whose pivot point is accelerating vertically at a constant rate A in a uniform gravitational field $\mathbf{g}$.

(a) Show that the angular frequency of small oscillations is

$$\omega = \sqrt{|g - A|/\ell}, \tag{1}$$

where ℓ is the length of the pendulum, g (> 0) is the magnitude of the gravitational acceleration, and A is positive if the acceleration of the pivot is 'downward' (that is, along $\mathbf{g}$).

(b) Discuss the result (1).

Solution

(a) Let S' be a frame that is accelerating with the pendulum. In S' the gravitational force $m\mathbf{g}$ on the bob is modified by the translational force $-m\mathbf{A}$ to yield an effective force

$$\mathbf{F}_e = m(\mathbf{g} - \mathbf{A}), \tag{2}$$

where $\mathbf{A} = \ddot{\mathbf{D}}$ is the acceleration of S' relative to an inertial frame (see Question 14.11). It follows that in the usual equation of motion, $\ddot{\theta} + g\theta/\ell = 0$, for small oscillations of a simple pendulum in an inertial frame, (see Question 4.3) we should replace g with $|g - A|$, where $A > 0$ if $\mathbf{A}$ is downwards (parallel to $\mathbf{g}$). Thus, we obtain (1).

(b) **1.** A downward acceleration ($A > 0$) diminishes the effect of gravity on the pendulum, and in free fall ($A = g$) its effect disappears.

2. If $A > g$ then the effective gravity is reversed, and the pendulum oscillates about an inverted position ($\theta = \pi$).

3. An upward acceleration ($A < 0$) increases the effective gravity and hence the frequency of the oscillations.

4. A pendulum accelerating upward at a rate $A = -g$ in the absence of gravity oscillates with the same frequency as an unaccelerated pendulum in a gravitational field **g**.

We emphasize that strictly uniform gravitational fields do not exist, and it is only locally (in small regions of space) that the uniform quantity $-m\mathbf{A}$ can simulate a non-uniform gravitational field $m\mathbf{g}$. For a pendulum, restriction to a small region of space can be satisfied by driving the pivot sinusoidally (see Question 13.20), which simulates a modulated gravitational field.

Comments

(i) In the above we have assumed the weak equivalence principle (equality of the inertial and gravitational masses $m^{\rm I}$ and $m^{\rm G}$ – see Questions 2.4 and 2.5). Otherwise, (1) and (2) should read

$$\mathbf{F}_e = m^{\rm G}\mathbf{g} - m^{\rm I}\mathbf{A}, \qquad \omega = \sqrt{|(m^{\rm G}/m^{\rm I})g - A|}. \qquad (3)$$

(ii) If the weak equivalence principle holds then in mechanics one cannot distinguish between a uniform gravitational acceleration **g** and an acceleration $\mathbf{A} = -\mathbf{g}$ of the frame (relative to inertial space) in the absence of gravity. The generalization of this statement beyond mechanics, to the rest of physics, is known as Einstein's equivalence principle.

(iii) Einstein has described the importance of accelerated frames to his thinking: "The breakthrough came suddenly one day. I was sitting on a chair in my patent office in Bern. Suddenly a thought struck me: If a man falls freely, he would not feel his weight. I was taken aback. This simple thought experiment made a deep impression on me. This led me to the theory of gravity. I continued my thought: A falling man is accelerated. Then what he feels and judges is happening in the accelerated frame of reference. I decided to extend the theory of relativity to the reference frame with acceleration. I felt that in doing so I could solve the problem of gravity at the same time. A falling man does not feel his weight because in his reference frame there is a new gravitational field which cancels the gravitational field due to the Earth."[8]

Question 14.13

At the centre C of the Earth there is a balance between the inward gravitational force $m\mathbf{g}_C$ exerted by the Sun and the outward translational force $-m\mathbf{g}_C$ due to the radial acceleration $\mathbf{g}_C$ of the Earth towards the Sun.[‡] At points away from C, and

[‡]That is, the Earth is in free fall towards the Sun, and according to the equivalence principle it is not possible to observe effects due to the Sun's field in an Earth-fixed system at C.

---

[8] A. Einstein, "How I created the theory of relativity," Physics Today, vol. 35, pp. 45–47, August 1982.

fixed relative to the Earth, this cancellation is incomplete (because the Sun's field is non-uniform) and so there is a residual effective force $\mathbf{F}'$. Evaluate $\mathbf{F}'$ to first order in r/R_s, where r is the distance of the field point from C, and R_s is the distance from the centre of the Sun to C. Express the result in terms of radial and transverse components.

Solution

In the diagram, P is an arbitrary Earth-fixed point with $r/R_s \ll 1$. The effective force $\mathbf{F}'$ on a particle of mass m at P is the sum of the Sun's gravitational force and the translational force due to free fall of P at the rate $\mathbf{g}_C$:

$$\mathbf{F}' = -GM_s m \left\{ \frac{\mathbf{R}_s + \mathbf{r}}{|\mathbf{R}_s + \mathbf{r}|^3} - \frac{\mathbf{R}_s}{R_s^3} \right\} \approx -G \frac{M_s m}{R_s^3} \left\{ \mathbf{r} - 3 \frac{\mathbf{r} \cdot \mathbf{R}_s}{R_s^2} \mathbf{R}_s \right\} \quad (1)$$

to first order in r/R_s. Note that $\mathbf{F}' = 0$ at $\mathbf{r} = 0$ as required. To express (1) in terms of

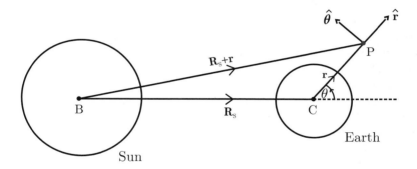

radial and transverse components, convert it to plane polar coordinates using

$$\mathbf{r} = r\hat{\mathbf{r}} \quad \text{and} \quad \mathbf{R}_s = \hat{\mathbf{r}} R_s \cos\theta - \hat{\boldsymbol{\theta}} R_s \sin\theta. \quad (2)$$

This yields the components

$$F'_r = G \frac{M_s m}{R_s^3} r(3\cos^2\theta - 1) \quad (3)$$

$$F'_\theta = -G \frac{M_s m}{R_s^3} r(3\cos\theta \sin\theta). \quad (4)$$

Comments

(i) These residual forces are due to the non-uniformity of the Sun's field and the uniformity of the translational force. They are responsible for oceanic tides and are referred to as tidal forces. There are, of course, also tidal forces due to the

Moon. They are given by (3) and (4) with M_S replaced by M_M (the Moon's mass) and R_S by R_M (the distance between the centres of the Earth and the Moon). So

$$\frac{F'(\text{Moon})}{F'(\text{Sun})} = \frac{M_M}{M_S}\left(\frac{R_S}{R_M}\right)^3 \approx \frac{7.3 \times 10^{22}}{2 \times 10^{30}}\left(\frac{1.5 \times 10^{10}}{3.8 \times 10^7}\right)^3 \approx 2.2. \quad (5)$$

The Moon's effect on tides is about twice that of the Sun. This is a result of the non-uniformity of the Moon's field at the Earth being larger than that of the Sun.

(ii) The ratio of the tidal force of the Moon to the weight $mg = GM_E m/\mathcal{R}_E^2$ of an object at the Earth's surface ($\mathcal{R}_E$ is the Earth's radius) is of order

$$\frac{F'}{mg} \approx \frac{M_M}{M_E}\left(\frac{\mathcal{R}_E}{R_M}\right)^3 \approx \frac{1}{81}\left(\frac{1}{60}\right)^3 \approx 6 \times 10^{-8}. \quad (6)$$

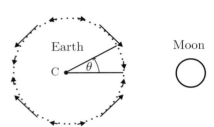

Despite the smallness of this ratio, the effect on the oceans is noticeable for two reasons. First, because the horizontal component F'_θ moves water relative to the Earth's surface: this force is depicted in the diagram and it is clear that it causes 'bulges' on the sides of the Earth facing towards and away from the Moon. Secondly, the effect of these bulges is time dependent: the two tidal bulges move round the Earth with the Moon and the result is two high tides and two low tides each day.

(iii) The maximum height H_M of these tides, for an ocean of uniform depth covering the entire planet, can be shown to be[9]

$$H_M = \frac{3}{2}\frac{M_M}{M_E}\left(\frac{\mathcal{R}_E}{R_M}\right)^3 \mathcal{R}_E, \quad (7)$$

with a similar expression for H_S (due to the Sun). These give $H_M \approx 0.54\,\text{m}$ and $H_S \approx 0.24\,\text{m}$, which provide an estimate for the tidal range in mid-ocean and the value $(H_M + H_S)/(H_M - H_S) \approx 2.6$ for the ratio of spring tide to neap tide. Further discussion of this model is given in Ref. [9]. In reality, the Earth's tides are complicated, being strongly influenced by topography and other factors.

(iv) Long ago, when the Moon was much closer to Earth, the tides it produced were about a hundred to a thousand times larger than today, and they occurred more frequently because of the faster spin of the young Earth. It is thought that the scouring action of these great surges onto land was a factor in the origin of life through its effect on the chemical composition of the oceans.

(v) In the limit $\mathcal{R}(\text{planet})/R(\text{orbit}) \to 0$ the residual forces disappear; the region over which the equivalence principle applies covers the entire planet and not just the neighbourhood of its centre.

---

[9] E. I. Butikov, "A dynamical picture of the oceanic tides," American Journal of Physics, vol. 70, pp. 1001–1011, 2002.

Question 14.14

During take-off an aircraft accelerates horizontally in a straight line at a rate A. A small bob of mass m is suspended on a string attached to the roof of the cabin, and a hydrogen balloon (total mass m) is tethered to the floor by a string. For each, determine the tension in the string and the equilibrium angle θ between the string and the vertical.

Solution

We work in the reference frame of the aircraft. In this accelerated frame the effective gravitational acceleration is $\mathbf{g}_e = \mathbf{g} - \mathbf{A}$ (see Questions 14.11 and 14.12). In terms of the coordinates shown in the diagram:

$$\mathbf{g}_e = -A\hat{\mathbf{x}} - g\hat{\mathbf{y}} \qquad (A, g > 0). \tag{1}$$

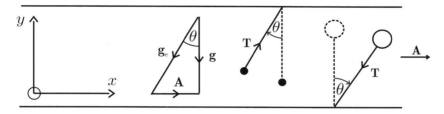

1. For the suspended mass the condition for equilibrium in this frame is $\mathbf{T} + m\mathbf{g}_e = 0$, where $\mathbf{T}$ is the tension in the string. That is,

$$\mathbf{T} = mA\hat{\mathbf{x}} + mg\hat{\mathbf{y}}. \tag{2}$$

So, the magnitude of the tension and the angle of inclination are given by

$$T = m\sqrt{A^2 + g^2}, \qquad \theta = \tan^{-1}(A/g). \tag{3}$$

2. For the tethered balloon in the same frame there are three forces: the tension $\mathbf{T}$, the weight $m\mathbf{g}_e$, and the upthrust $-m_a\mathbf{g}_e$ given by Archimedes's principle and the equivalence principle. Here, $m_a (> m)$ is the mass of air displaced. So, the condition for equilibrium is $\mathbf{T} + (m - m_a)\mathbf{g}_e = 0$. That is,

$$\mathbf{T} = -(m_a - m)A\hat{\mathbf{x}} - (m_a - m)g\hat{\mathbf{y}}, \tag{4}$$

and hence

$$T = (m_a - m)\sqrt{A^2 + g^2}, \qquad \theta = \tan^{-1}(A/g). \tag{5}$$

Comments

(i) In both cases the equilibrium orientation of the string is along $\mathbf{g}_e$. For the bob, $\mathbf{T}$ and $\mathbf{g}_e$ are anti-parallel; for the balloon they are parallel. As a result, the bob is displaced towards the rear of the aircraft as one expects. But, the response of the balloon is counter-intuitive: it is displaced towards the front of the aircraft.

(ii) The above calculations in the accelerated frame, including the use of Archimedes's principle in this frame, are based on the equivalence principle.

Question 14.15

According to Hubble's law, galaxies recede from the Earth with a velocity that is proportional to their distance. Determine the form of Hubble's law relative to an observer in some other galaxy. (Assume Newtonian relativity.)

Solution

Let O be an observer on Earth, and O' an observer in some other galaxy, located at position $\mathbf{D}(t)$ relative to O. Both observers measure the velocity of a galaxy G that is located at $\mathbf{r}(t)$ relative to O and $\mathbf{r}'(t')$ relative to O'. We know that the measurements made by O satisfy Hubble's law:

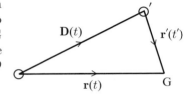

$$\frac{d\mathbf{r}}{dt} = H_o \mathbf{r}, \qquad \frac{d\mathbf{D}}{dt} = H_o \mathbf{D}, \qquad (1)$$

where H_o is a constant. The question is: what is the velocity $d\mathbf{r}'/dt'$ of G as measured by O'? In Newtonian relativity $t' = t$ and $\mathbf{r}' = \mathbf{r} - \mathbf{D}$, and the answer is immediate:

$$\frac{d\mathbf{r}'}{dt'} = \frac{d}{dt}(\mathbf{r} - \mathbf{D}) = H_o \mathbf{r} - H_o \mathbf{D} = H_o \mathbf{r}'. \qquad (2)$$

That is, Hubble's law applies for all observers.

Comments

(i) This result brings to mind a picture of an expanding balloon (all points recede from any point on the balloon) rather than the idea of a preferred point from which all others recede.

(ii) Hubble's law was formulated in 1929, based on earlier measurements of galactic velocities obtained from red shifts. The Hubble constant H_o is an important cosmological parameter. For example, a rough estimate of the age of the universe (post the big bang) is given by H_o^{-1}, and current measurements of H_o provide a value of about 13.5 billion years.

(iii) Recent measurements of the red shifts of supernovae have yielded an astonishing result: the expansion of the universe is accelerating, and this acceleration started about 5 to 7 billion years ago. It is thought that the acceleration is caused by some gravitationally repulsive substance, and that perhaps three quarters of the energy density of the universe is due to this substance. The total amount of matter in the universe (ordinary plus 'dark' – see Question 11.20) accounts for only one-quarter of the energy; the rest is of unknown origin and is referred to as 'dark energy'. Evidently, the future development of the universe hinges on the competition between these attractive and repulsive constituents.

Question 14.16

Let frames S and S' have a common origin O and Cartesian axes $Ox_1x_2x_3$ and $Ox'_1x'_2x'_3$, where the latter are obtained from the former by rotation about an axis through O.

(a) Show that the components of the position vector $\mathbf{r}$ relative to S' and S are related by
$$x'_i = a_{ij}x_j \qquad (i = 1, 2, 3), \tag{1}$$
where the repeated index j implies summation from 1 to 3, and the nine coefficients a_{ij} should be defined.

(b) Prove that the a_{ij} must satisfy the orthogonality relations
$$a_{ir}a_{jr} = \delta_{ij}, \qquad a_{ri}a_{rj} = \delta_{ij}, \tag{2}$$
where δ_{ij} is the Kronecker delta symbol ($\delta_{ij} = 1$ if $i = j$; $\delta_{ij} = 0$ if $i \neq j$). How many independent a_{ij} are there in general?

(c) Illustrate the above for the special case of a positive (anti-clockwise) rotation about Ox_3.

(d) What is the generalization of (1) to an arbitrary vector $\mathbf{A}$?

Solution

(a) Each component x_j of $\mathbf{r}$ may be projected onto the axis x'_i and then added to give
$$x'_i = x_1 \cos\theta_{i1} + x_2 \cos\theta_{i2} + x_3 \cos\theta_{i3} = a_{ij}x_j, \tag{3}$$
where θ_{ij} is the angle between x'_i and x_j, and $a_{ij} = \cos\theta_{ij}$ is the direction cosine of x'_i relative to x_j.

(b) The unit vectors of S' relative to S are $\mathbf{n}'_i = (a_{i1}, a_{i2}, a_{i3})$; and the unit vectors of S relative to S' are $\mathbf{n}_i = (a_{1i}, a_{2i}, a_{3i})$. Therefore, the orthonormality conditions for unit vectors, namely $\mathbf{n}'_i \mathbf{n}'_j = \delta_{ij}$ and $\mathbf{n}_i \mathbf{n}_j = \delta_{ij}$, require (2). Equations (2) provide six relations, thereby reducing the number of independent a_{ij} to three.

(c) We have $a_{11} = a_{22} = \cos\theta$; $a_{12} = \cos(90-\theta) = \sin\theta$; $a_{21} = \cos(90+\theta) = -\sin\theta$; $a_{13} = a_{23} = 0$; $a_{33} = 1$. Equation (1) can be written in matrix form as

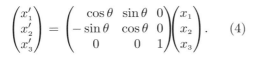

(4)

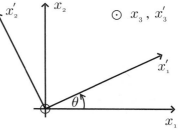

(d) By definition, the components of a vector $\mathbf{A} = (A_1, A_2, A_3)$ transform under rotation in the same way as do the components of $\mathbf{r}$. That is,

$$A'_i = a_{ij} A_j. \tag{5}$$

According to (5) and $(2)_2$, the inverse relation is

$$A_j = a_{ij} A'_i. \tag{6}$$

Comments

(i) The above can be generalized to Cartesian tensors of any rank. A vector is a tensor of rank 1: it has three components A_i that transform under rotations according to (5). A second-rank tensor has nine components T_{ij} that transform according to

$$T'_{ij} = a_{ir} a_{js} T_{rs}, \tag{7}$$

and so on for higher rank. A scalar S is a tensor of zero rank and it is unchanged by rotation: $S' = S$.

(ii) A number of useful results follow from this. For example, from (5) and $(2)_2$ we have

$$A'_i A'_i = a_{ij} a_{ik} A_j A_k = \delta_{jk} A_j A_k = A_j A_j. \tag{8}$$

Thus the magnitude of a vector (for example, the distance between two points) is unchanged by rotation – it is a scalar. Similarly, $A_i B_i$ is a scalar. Also,

$$\frac{\partial}{\partial x'_i} = \frac{\partial x_j}{\partial x'_i} \frac{\partial}{\partial x_j} = a_{ij} \frac{\partial}{\partial x_j}. \tag{9}$$

Thus $\boldsymbol{\nabla} = (\partial/\partial x_1, \partial/\partial x_2, \partial/\partial x_3)$ is a vector, and $\nabla^2 = \boldsymbol{\nabla} \cdot \boldsymbol{\nabla}$ is a scalar, as is the divergence $\partial A_i/\partial x_i$.

(iii) Tensors in three-dimensional space are referred to as three-tensors to distinguish them from those in higher dimensions, particularly the four-tensors encountered in special relativity (see Question 15.7).

Question 14.17

Show that the equation of motion $\mathbf{F} = d\mathbf{p}/dt$ is invariant under rotation of the coordinate system (i.e. that it is valid for any orientation of the axes of an inertial frame).

Solution

In an inertial frame S, $F_j = dp_j/dt$. By definition, $\mathbf{F}$ and $\mathbf{p}$ are vectors. So, in a frame S' obtained from S by rotation about an axis through O, they have components

$$F'_i = a_{ij} F_j \quad \text{and} \quad p'_i = a_{ij} p_j, \tag{1}$$

where the a_{ij} are independent of time. Therefore

$$F'_i - \frac{d}{dt}p'_i = a_{ij}(F_j - \frac{d}{dt}p_j) = 0, \qquad (2)$$

and the equation of motion holds in S'.

Comments

(i) This result was known to Newton (see Question 14.9). It is a consequence of the transformations (1) being linear and homogeneous. The components of **F** and **ṗ** change when the axes are rotated, but the relationship between them does not.

(ii) The same conclusion applies to any three-tensor equation. For example, if

$$A_i = B_i, \qquad (3)$$

then

$$A'_i - B'_i = a_{ij}(A_j - B_j) = 0. \qquad (4)$$

That is, (3) transforms into

$$A'_i = B'_i. \qquad (5)$$

Again, the components of **A** and **B** change but the relationship between them does not – one says that the terms of (3) are covariant. In general, tensor equations consist of linear relations among covariant quantities and therefore they are invariant under rotations.

Question 14.18

(a) What are the conditions for the Lagrangian of a system of particles to be rotationally invariant (that is, unchanged by a rotation of the entire system about any axis)?

(b) Use this invariance to deduce the law of conservation of angular momentum for a system of particles.

Solution

(a) The frame of reference should be inertial (which ensures that the space is isotropic – see Chapter 1) and the system should be isolated.

(b) When the system is rotated through an infinitesimal angle $\delta\boldsymbol{\theta}$ about some axis, both the position vectors $\mathbf{r}_i$ and the velocity vectors $\mathbf{v}_i$ of the particles change according to (see Question 12.3)

$$\delta\mathbf{r}_i = \delta\boldsymbol{\theta} \times \mathbf{r}_i, \qquad \delta\mathbf{v}_i = \delta\boldsymbol{\theta} \times \mathbf{v}_i. \qquad (1)$$

The resulting change in the Lagrangian $\mathsf{L}(\mathbf{r}_i, \mathbf{v}_i)$ is

$$\delta\mathsf{L} = \sum_i \left(\frac{\partial\mathsf{L}}{\partial\mathbf{r}_i} \cdot \delta\mathbf{r}_i + \frac{\partial\mathsf{L}}{\partial\mathbf{v}_i} \cdot \delta\mathbf{v}_i \right). \qquad (2)$$

Here, we are using the same notation as in Question 14.7. By substituting (1) in (2) and using the relation $\partial L/\partial \mathbf{v}_i = \mathbf{p}_i$ and the Lagrange equations $\partial L/\partial \mathbf{r}_i = \dot{\mathbf{p}}_i$ (see Chapter 1) we obtain

$$\delta L = \sum_i \{\dot{\mathbf{p}}_i \cdot (\delta\boldsymbol{\theta} \times \mathbf{r}_i) + \mathbf{p}_i \cdot (\delta\boldsymbol{\theta} \times \dot{\mathbf{r}}_i)\} = \delta\boldsymbol{\theta} \cdot \sum_i (\mathbf{r}_i \times \dot{\mathbf{p}}_i + \dot{\mathbf{r}}_i \times \mathbf{p}_i)$$

$$= \delta\boldsymbol{\theta} \cdot \frac{d}{dt}\sum_i (\mathbf{r}_i \times \mathbf{p}_i). \qquad (3)$$

(In the second step we have used the identity $\mathbf{a} \cdot (\mathbf{b} \times \mathbf{c}) = \mathbf{b} \cdot (\mathbf{c} \times \mathbf{a})$.) Rotational invariance of L means $\delta L = 0$ for arbitrary $\delta\boldsymbol{\theta}$ and so it follows from (3) that

$$\frac{d\mathbf{L}}{dt} = 0, \qquad (4)$$

where $\mathbf{L} = \sum_i \mathbf{r}_i \times \mathbf{p}_i$ is the total angular momentum. Equation (4) is the law of conservation of angular momentum for a closed system of particles in an inertial frame.

Comments

(i) This is a further example of the general connection between symmetry and conservation laws. The isotropy of space in an inertial frame means that rotation of an isolated system is a symmetry transformation, and the consequence is conservation of angular momentum. Previous examples dealt with translation in space (see Question 14.7) and translation in time (see Question 11.33), and the corresponding laws of conservation of momentum and energy.

(ii) If the system is not isolated then, in general, the rotational invariance of **L** will be broken and none of the components of **L** will be conserved. In special cases there may be partial invariance and consequently conservation of the corresponding component(s) of **L**: for example, invariance with respect to rotations about the z-axis means that L_z is conserved, where L_z is evaluated relative to an origin on the z-axis.

Question 14.19

(a) Reconcile the analysis of conservation of angular momentum in terms of Newton's equation of motion and central interparticle forces (see Question 11.2) with that based on Lagrange's equations and rotational invariance (see Question 14.18).[10]
(b) Illustrate your answer by considering the two-particle Lagrangian[10]

$$L = \tfrac{1}{2}m_1 \mathbf{v}_1^2 + \tfrac{1}{2}m_2 \mathbf{v}_2^2 - V(|\mathbf{r}_1 - \mathbf{r}_2|) + b\mathbf{v}_1 \cdot \mathbf{v}_2, \qquad (1)$$

where b is a constant.

---

[10] B. Podolsky, "Conservation of angular momentum," American Journal of Physics, vol. 34, pp. 42–45, 1966.

Solution

(a) The Newtonian analysis deals with the angular momentum

$$\mathbf{L}_{\text{N}} = \sum_i \mathbf{r}_i \times m_i \mathbf{v}_i, \tag{2}$$

and establishes conservation of $\mathbf{L}_{\text{N}}$ for an isolated system with central interparticle forces. The Lagrangian treatment involves the canonical angular momentum

$$\mathbf{L}_{\text{c}} = \sum_i \mathbf{r}_i \times \frac{\partial \mathsf{L}}{\partial \mathbf{v}_i}, \tag{3}$$

and establishes conservation of $\mathbf{L}_{\text{c}}$ for isolated, rotationally invariant systems. Now, $\mathbf{L}_{\text{c}} = \mathbf{L}_{\text{N}}$ only when the canonical and Newtonian momenta are equal:

$$\frac{\partial \mathsf{L}}{\partial \mathbf{v}_i} = m_i \mathbf{v}_i \quad (i = 1, 2, \cdots, N). \tag{4}$$

That is, when

$$\mathsf{L}(\mathbf{r}_i, \mathbf{v}_i) = \sum_i \tfrac{1}{2} m_i \mathbf{v}_i^2 - V(\mathbf{r}_1, \mathbf{r}_2, \cdots, \mathbf{r}_N). \tag{5}$$

If this L is to be rotationally invariant then V can depend only on the magnitude of the particle separations:

$$V = \tfrac{1}{2} \sum_i \sum_j V_{ij}(|\mathbf{r}_i - \mathbf{r}_j|) \quad (i \neq j). \tag{6}$$

The force $\mathbf{F}_i$ on particle i is a sum of interparticle forces

$$\mathbf{F}_i = \frac{\partial \mathsf{L}}{\partial \mathbf{r}_i} = -\sum_j \frac{\mathbf{r}_i - \mathbf{r}_j}{|\mathbf{r}_i - \mathbf{r}_j|} \frac{dV_{ij}}{dr_{ij}} \quad (j \neq i), \tag{7}$$

where $r_{ij} = |\mathbf{r}_i - \mathbf{r}_j|$. That is, the interparticle forces are central. So, the Newtonian and Lagrangian analyses of conservation of angular momentum in inertial frames agree when the canonical and Newtonian momenta are equal. When these momenta are unequal, the angular momenta (2) and (3) are not the same. If there is rotational invariance then it is (3) that is conserved, and not (2).

(b) The two-particle Lagrangian (1) illustrates this conclusion. The canonical momenta $\mathbf{p}_i = \partial \mathsf{L}/\partial \mathbf{v}_i$ are

$$\mathbf{p}_1 = m_1 \mathbf{v}_1 + b \mathbf{v}_2, \qquad \mathbf{p}_2 = m_2 \mathbf{v}_2 + b \mathbf{v}_1. \tag{8}$$

It is apparent that $\mathbf{p}_i \neq m_i \mathbf{v}_i$ if $b \neq 0$. The Lagrange equations yield the equations of motion

$$\left. \begin{array}{l} \dfrac{d}{dt}(m_1 \mathbf{v}_1 + b \mathbf{v}_2) = -\hat{\mathbf{r}} \dfrac{dV(\mathbf{r})}{dr} = \mathbf{F}_1 \\[2ex] \dfrac{d}{dt}(m_2 \mathbf{v}_2 + b \mathbf{v}_1) = \hat{\mathbf{r}} \dfrac{dV(\mathbf{r})}{dr} = \mathbf{F}_2 = -\mathbf{F}_1, \end{array} \right\} \tag{9}$$

where $\mathbf{r} = \mathbf{r}_1 - \mathbf{r}_2$. The rate of change of the canonical angular momentum $\mathbf{L}_c = \mathbf{r}_1 \times \mathbf{p}_1 + \mathbf{r}_2 \times \mathbf{p}_2$ is

$$\dot{\mathbf{L}}_c = \mathbf{v}_1 \times \mathbf{p}_1 + \mathbf{r}_1 \times \dot{\mathbf{p}}_1 + \mathbf{v}_2 \times \mathbf{p}_2 + \mathbf{r}_2 \times \dot{\mathbf{p}}_2$$
$$= b\mathbf{v}_1 \times \mathbf{v}_2 + \mathbf{r}_1 \times \mathbf{F}_1 + b\mathbf{v}_2 \times \mathbf{v}_1 + \mathbf{r}_2 \times \mathbf{F}_2$$
$$= (\mathbf{r}_1 - \mathbf{r}_2) \times \mathbf{F}_1 = 0, \tag{10}$$

because $\mathbf{F}_1$ in (9) is along $\mathbf{r}_1 - \mathbf{r}_2$. Thus, the canonical angular momentum is conserved, as one expects because the Lagrangian (1) is rotationally invariant. By contrast, $\mathbf{L}_N = \mathbf{r}_1 \times m_1 \mathbf{v}_1 + \mathbf{r}_2 \times m_2 \mathbf{v}_2$ is not conserved. To see this, first note that from (9) the accelerations of the particles are

$$\dot{\mathbf{v}}_1 = \frac{m_2 + b}{m_1 m_2 - b^2} \mathbf{F}_1, \qquad \dot{\mathbf{v}}_2 = -\frac{m_1 + b}{m_1 m_2 - b^2} \mathbf{F}_1. \tag{11}$$

Consequently,

$$\dot{\mathbf{L}}_N = \mathbf{r}_1 \times m_1 \dot{\mathbf{v}}_1 + \mathbf{r}_2 \times m_2 \dot{\mathbf{v}}_2 = \frac{m_1(m_2 + b)\mathbf{r}_1 - m_2(m_1 + b)\mathbf{r}_2}{m_1 m_2 - b^2} \times \mathbf{F}_1, \tag{12}$$

which is not zero if $b \neq 0$ and $m_1 \neq m_2$.

Question 14.20

Reference frames S and S' share a common origin O, and S' rotates about an axis through O with angular velocity $\boldsymbol{\omega}$ relative to S.

(a) Deduce the velocity transformation formula

$$\left(\frac{d\mathbf{r}}{dt}\right)_S = \left(\frac{d\mathbf{r}}{dt}\right)_{S'} + \boldsymbol{\omega} \times \mathbf{r} \tag{1}$$

connecting velocities relative to S and S'. What is the generalization of (1) to an arbitrary differentiable vector $\mathbf{A}(t)$?

(b) Use the above results to deduce the acceleration transformation formula

$$\left(\frac{d^2\mathbf{r}}{dt^2}\right)_S = \left(\frac{d^2\mathbf{r}}{dt^2}\right)_{S'} + \boldsymbol{\omega} \times (\boldsymbol{\omega} \times \mathbf{r}) + 2\boldsymbol{\omega} \times \left(\frac{d\mathbf{r}}{dt}\right)_{S'} + \frac{d\boldsymbol{\omega}}{dt} \times \mathbf{r}. \tag{2}$$

(c) Suppose S is inertial. Use (2) to write down the equation of motion for a particle of constant mass m relative to S'.

Solution

(a) Consider the position vector expressed in terms of the coordinates of S':

$$\mathbf{r} = x'\hat{\mathbf{x}}' + y'\hat{\mathbf{y}}' + z'\hat{\mathbf{z}}'. \tag{3}$$

The velocity relative to S′ is

$$\left(\frac{d\mathbf{r}}{dt}\right)_{S'} = \frac{dx'}{dt}\hat{\mathbf{x}}' + \frac{dy'}{dt}\hat{\mathbf{y}}' + \frac{dz'}{dt}\hat{\mathbf{z}}'. \tag{4}$$

Relative to S both the components and the unit vectors in (3) vary, and so

$$\left(\frac{d\mathbf{r}}{dt}\right)_{S} = \frac{dx'}{dt}\hat{\mathbf{x}}' + \frac{dy'}{dt}\hat{\mathbf{y}}' + \frac{dz'}{dt}\hat{\mathbf{z}}' + x'\frac{d\hat{\mathbf{x}}'}{dt} + y'\frac{d\hat{\mathbf{y}}'}{dt} + z'\frac{d\hat{\mathbf{z}}'}{dt}. \tag{5}$$

The frame S′ is rigid and therefore (see Question 12.3)

$$\frac{d\hat{\mathbf{x}}'}{dt} = \boldsymbol{\omega} \times \hat{\mathbf{x}}', \qquad \frac{d\hat{\mathbf{y}}'}{dt} = \boldsymbol{\omega} \times \hat{\mathbf{y}}', \qquad \frac{d\hat{\mathbf{z}}'}{dt} = \boldsymbol{\omega} \times \hat{\mathbf{z}}'. \tag{6}$$

Equations (3)–(6) yield (1). For a vector $\mathbf{A} = A_{x'}\hat{\mathbf{x}}' + A_{y'}\hat{\mathbf{y}}' + A_{z'}\hat{\mathbf{z}}'$, the same reasoning yields the generalization of (1):

$$\left(\frac{d\mathbf{A}}{dt}\right)_{S} = \left(\frac{d\mathbf{A}}{dt}\right)_{S'} + \boldsymbol{\omega} \times \mathbf{A}. \tag{7}$$

(b) By differentiating (1) with respect to t and using (7) we have

$$\left(\frac{d^2\mathbf{r}}{dt^2}\right)_{S} = \left\{\left(\frac{d}{dt}\right)_{S'} + \boldsymbol{\omega}\times\right\}\left(\frac{d\mathbf{r}}{dt}\right)_{S'} + \left[\left\{\left(\frac{d}{dt}\right)_{S'} + \boldsymbol{\omega}\times\right\}\boldsymbol{\omega}\right] \times \mathbf{r} + \boldsymbol{\omega} \times \left\{\left(\frac{d}{dt}\right)_{S'} + \boldsymbol{\omega}\times\right\}\mathbf{r}, \tag{8}$$

which simplifies to (2) because $\boldsymbol{\omega} \times \boldsymbol{\omega} = 0$. In (2) we have dropped the label S′ on $\dot{\boldsymbol{\omega}}$ because, according to (7), $\dot{\boldsymbol{\omega}}_S = \dot{\boldsymbol{\omega}}_{S'}$.

(c) The transformations (1) and (2) apply to any two frames. If S is inertial and m is constant, then $(d^2\mathbf{r}/dt^2)_S = \mathbf{F}/m$. Use of this in (2) and a rearrangement of terms yields the equation of motion relative to a rotating frame S′:

$$m\frac{d^2\mathbf{r}}{dt^2} = \mathbf{F} - m\boldsymbol{\omega} \times (\boldsymbol{\omega} \times \mathbf{r}) - 2m\boldsymbol{\omega} \times \frac{d\mathbf{r}}{dt} - m\frac{d\boldsymbol{\omega}}{dt} \times \mathbf{r}. \tag{9}$$

Here, it is understood that the acceleration $\ddot{\mathbf{r}}$ and velocity $\dot{\mathbf{r}}$ are relative to S′.

Comment

The above results are based on Newtonian relativity – absolute space, time and mass. The analysis is readily extended to frames that are translating as well as rotating – see below.

Question 14.21

Extend the analysis of Question 14.20 to include translation of S′ relative to S.

Solution

Here, $\mathbf{r} = \mathbf{r}' + \mathbf{D}$, where $\mathbf{D}$ is the position vector of the origin O' of S' relative to S. Then, $\dot{\mathbf{r}}_S = \dot{\mathbf{r}}'_S + \dot{\mathbf{D}}$ and $\dot{\mathbf{r}}'_S = \dot{\mathbf{r}}'_{S'} + \boldsymbol{\omega} \times \mathbf{r}'$, and the velocity transformation formula is

$$\left(\frac{d\mathbf{r}}{dt}\right)_S = \left(\frac{d\mathbf{r}'}{dt}\right)_{S'} + \boldsymbol{\omega} \times \mathbf{r}' + \frac{d\mathbf{D}}{dt}. \tag{1}$$

Now differentiate (1) with respect to t. For the derivatives of the first two terms on the right-hand side of (1) we repeat the step in (8) of Question 14.20. The derivative of $d\mathbf{D}/dt$ is $d^2\mathbf{D}/dt^2$. Thus, we obtain the acceleration transformation formula

$$\left(\frac{d^2\mathbf{r}}{dt^2}\right)_S = \left(\frac{d^2\mathbf{r}'}{dt^2}\right)_{S'} + \boldsymbol{\omega} \times (\boldsymbol{\omega} \times \mathbf{r}') + 2\boldsymbol{\omega} \times \left(\frac{d\mathbf{r}'}{dt}\right)_{S'} + \frac{d\boldsymbol{\omega}}{dt} \times \mathbf{r}' + \frac{d^2\mathbf{D}}{dt^2}. \tag{2}$$

Equations (1) and (2) are the extensions, to include translation, of the transformations (1) and (2) of Questions 14.20. If S is inertial then $(d^2\mathbf{r}/dt^2)_S = \mathbf{F}/m$ and (2) yields the equation of motion

$$m\frac{d^2\mathbf{r}'}{dt^2} = \mathbf{F} - m\boldsymbol{\omega} \times (\boldsymbol{\omega} \times \mathbf{r}') - 2m\boldsymbol{\omega} \times \frac{d\mathbf{r}'}{dt} - m\frac{d\boldsymbol{\omega}}{dt} \times \mathbf{r}' - m\frac{d^2\mathbf{D}}{dt^2}. \tag{3}$$

To simplify the notation, we have again omitted the subscript S' on $\ddot{\mathbf{r}}'$ and $\dot{\mathbf{r}}'$.

Comments

(i) Equation (3) shows how Newton's second law $m\ddot{\mathbf{r}} = \mathbf{F}$ is modified due to rotation and translation of the frame relative to an inertial frame. In general, there are four additional forces:

1. Centrifugal force $\quad\quad \mathbf{F}_{cf} = -m\boldsymbol{\omega} \times (\boldsymbol{\omega} \times \mathbf{r}'),$ (4)

2. Coriolis force $\quad\quad \mathbf{F}_{Cor} = -2m\boldsymbol{\omega} \times \dfrac{d\mathbf{r}'}{dt},$ (5)

3. Azimuthal force $\quad\quad \mathbf{F}_{az} = -m\dfrac{d\boldsymbol{\omega}}{dt} \times \mathbf{r}',$ (6)

4. Translational force $\quad\quad \mathbf{F}_{tr} = -m\dfrac{d^2\mathbf{D}}{dt^2}.$ (7)

(ii) The Coriolis and centrifugal forces are important in a number of phenomena. For example, the Coriolis force is responsible for the circulation of winds around regions of low and high pressure on Earth; for the deflection of trade winds; for the deflection of the trajectories of long-range projectiles; for the rotation of the plane of oscillation of a simple pendulum on Earth (the Foucault pendulum); for the oscillations of a gyrocompass; and for the asymmetry in the precession of a gyroscope. The centrifugal force is responsible for the flattening of the Earth; for the variations in the direction and magnitude of the gravitational acceleration $\mathbf{g}$; and for the 'centrifugal barrier' in the two-body problem. Some of these have been encountered previously (see Chapter 8 and Questions 10.4, 10.14, 10.17, and 12.25–12.27), and others are analyzed below.

(iii) The analysis leading to (3) is largely kinematical in nature: dynamics enters only when $\ddot{\mathbf{r}}$ is replaced by $\mathbf{F}/m$. An alternative approach, which uses Lagrange's equations in a non-inertial frame, is given in Question 14.22.

Question 14.22

Use Lagrange's equations to obtain the equation of motion (9) in Question 14.20 for motion of a particle in a rotating frame.

Solution

First, express the Lagrangian $\mathsf{L} = \frac{1}{2}m\mathbf{v}_\mathrm{S}^2 - V(\mathbf{r})$ in S in terms of the velocity $\mathbf{v}_{\mathrm{S}'} = (d\mathbf{r}/dt)_{\mathrm{S}'} = \mathbf{v}_\mathrm{S} - \boldsymbol{\omega} \times \mathbf{r}$ relative to S':

$$\mathsf{L}' = \tfrac{1}{2}m\mathbf{v}_{\mathrm{S}'}^2 + m\mathbf{v}_{\mathrm{S}'} \cdot (\boldsymbol{\omega} \times \mathbf{r}) + \tfrac{1}{2}m(\boldsymbol{\omega} \times \mathbf{r})^2 - V(\mathbf{r}). \tag{1}$$

Then, use the Lagrange equations in S':

$$\frac{d}{dt}\frac{\partial \mathsf{L}'}{\partial \mathbf{v}_{\mathrm{S}'}} - \frac{\partial \mathsf{L}'}{\partial \mathbf{r}} = 0. \tag{2}$$

From (1) we have

$$\frac{\partial \mathsf{L}'}{\partial \mathbf{v}_{\mathrm{S}'}} = m\mathbf{v}_{\mathrm{S}'} + m\boldsymbol{\omega} \times \mathbf{r}. \tag{3}$$

To evaluate $\partial \mathsf{L}'/\partial \mathbf{r}$, first use the vector identities

$$\mathbf{v}_{\mathrm{S}'} \cdot (\boldsymbol{\omega} \times \mathbf{r}) = \mathbf{r} \cdot (\mathbf{v}_{\mathrm{S}'} \times \boldsymbol{\omega}) \quad \text{and} \quad (\boldsymbol{\omega} \times \mathbf{r})^2 = (\boldsymbol{\omega} \cdot \boldsymbol{\omega})(\mathbf{r} \cdot \mathbf{r}) - (\boldsymbol{\omega} \cdot \mathbf{r})(\boldsymbol{\omega} \cdot \mathbf{r}). \tag{4}$$

Then

$$\frac{\partial \mathsf{L}'}{\partial \mathbf{r}} = m\mathbf{v}_{\mathrm{S}'} \times \boldsymbol{\omega} + m(\boldsymbol{\omega} \cdot \boldsymbol{\omega})\mathbf{r} - m\boldsymbol{\omega}(\boldsymbol{\omega} \cdot \mathbf{r}) - \frac{\partial V}{\partial \mathbf{r}}$$

$$= m\mathbf{v}_{\mathrm{S}'} \times \boldsymbol{\omega} - m\boldsymbol{\omega} \times (\boldsymbol{\omega} \times \mathbf{r}) - \frac{\partial V}{\partial \mathbf{r}}. \tag{5}$$

If S is inertial then $\partial V/\partial \mathbf{r} = -\mathbf{F}$ and (2)–(5) yield

$$m\dot{\mathbf{v}}_{\mathrm{S}'} + m\dot{\boldsymbol{\omega}} \times \mathbf{r} + m\boldsymbol{\omega} \times \mathbf{v}_{\mathrm{S}'} - m\mathbf{v}_{\mathrm{S}'} \times \boldsymbol{\omega} + m\boldsymbol{\omega} \times (\boldsymbol{\omega} \times \mathbf{r}) - \mathbf{F} = 0. \tag{6}$$

This is the same as the equation of motion (9) in Question 14.20, where $\mathbf{v}_{\mathrm{S}'}$ is abbreviated as $d\mathbf{r}/dt$.

Comments

(i) The above is readily extended to include the effect of translation of S' relative to S (see Question 14.11), and it yields (3) of Question 14.21.

(ii) The Lagrangian (1) is not translationally or rotationally invariant, even if $V = 0$ (space is neither homogeneous nor isotropic in the rotating frame). Also, the Lagrangian and the equation of motion (6) are not form invariant, when compared with their counterparts in an inertial frame.

Question 14.23

Determine the equation of motion, relative to a non-inertial (rotating and translating) frame, of the centre of mass of a system of particles subject to external and interparticle forces.

Solution

We use the same notation as in Question 11.1. The equation of motion of the ith particle relative to the non-inertial frame is (see Question 14.21)

$$m_i \frac{d^2 \mathbf{r}_i}{dt^2} = \sum_j \mathbf{F}_{ji} + \mathbf{F}_i^{(e)} - m_i \boldsymbol{\omega} \times (\boldsymbol{\omega} \times \mathbf{r}_i) - 2 m_i \boldsymbol{\omega} \times \frac{d \mathbf{r}_i}{dt} - m_i \frac{d \boldsymbol{\omega}}{dt} \times \mathbf{r}_i - m_i \frac{d^2 \mathbf{D}}{dt^2}. \quad (1)$$

(Here, we have simplified the notation by omitting the prime on $\mathbf{r}_i$.) By summing this over all i, and recalling that the m_i are constant, we can write

$$\frac{d^2}{dt^2} \sum_i m_i \mathbf{r}_i = \sum_i \sum_j \mathbf{F}_{ji} + \sum_i \mathbf{F}_i^{(e)} - \boldsymbol{\omega} \times \left(\boldsymbol{\omega} \times \sum_i m_i \mathbf{r}_i \right)$$

$$- 2\boldsymbol{\omega} \times \frac{d}{dt} \sum_i m_i \mathbf{r}_i - \frac{d\boldsymbol{\omega}}{dt} \times \sum_i m_i \mathbf{r}_i - \left(\sum_i m_i \right) \frac{d^2 \mathbf{D}}{dt^2}. \quad (2)$$

The double sum in (2) is zero because $\mathbf{F}_{ji} = -\mathbf{F}_{ij}$ (see Question 11.1). And $\sum_i \mathbf{F}_i^{(e)} = \mathbf{F}^{(e)}$ is the total external force acting on the system. Also, $\sum_i m_i \mathbf{r}_i = M\mathbf{R}$, where $M = \sum_i m_i$ is the total mass and $\mathbf{R}$ is the position vector of the CM relative to the non-inertial frame. So, (2) yields

$$M\ddot{\mathbf{R}} = \mathbf{F}^{(e)} - M\boldsymbol{\omega} \times (\boldsymbol{\omega} \times \mathbf{R}) - 2M\boldsymbol{\omega} \times \dot{\mathbf{R}} - M\dot{\boldsymbol{\omega}} \times \mathbf{R} - M\ddot{\mathbf{D}}. \quad (3)$$

Comment

In an inertial frame the equation of motion of the CM is $M\ddot{\mathbf{R}} = \mathbf{F}^{(e)}$. Equation (3) is its extension to motion in a non-inertial (rotating and translating) frame. We conclude that in any reference frame the CM motion is a single-particle problem: the interparticle forces $\mathbf{F}_{ij}$ play no role and the four additional forces (centrifugal, Coriolis, azimuthal and translational) in (3) are those of a particle of mass M located at $\mathbf{R}$ and moving with velocity $\dot{\mathbf{R}}$ relative to the non-inertial frame.

Question 14.24

A long straight wire rotates in free space with constant angular velocity $\boldsymbol{\omega}$ about a perpendicular axis through its midpoint O. This axis is fixed in inertial space. A bead of mass m slides on the wire. The coefficient of kinetic friction between the bead and the wire is μ. The bead is initially a distance x_0 from O and moving with speed v_0 along the wire. Calculate its position $x(t)$ on the wire in terms of ω, μ, x_0 and v_0.

Solution

The problem is one-dimensional in a rotating frame whose x-axis (say) is along the wire. In this frame the forces acting on the bead are the centrifugal force $\mathbf{F}_{\text{cf}}$ along the wire, the Coriolis force $\mathbf{F}_{\text{Cor}}$ perpendicular to the wire, the reaction $\mathbf{N}$ of the wire (which balances the Coriolis force), and the frictional force $\mathbf{F}_{\text{f}}$ along the wire. The equation of motion along Ox is

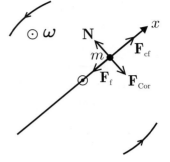

$$m\ddot{x} = m\omega^2 x - 2m\mu\omega\dot{x}. \qquad (1)$$

By considering a solution $x = e^{qt}$, where q is a constant, we readily find that the general solution to (1) is

$$x(t) = e^{-\mu\omega t}\left(A_+ e^{\sqrt{1+\mu^2}\,\omega t} + A_- e^{-\sqrt{1+\mu^2}\,\omega t}\right), \qquad (2)$$

where $A_{\pm}$ are arbitrary constants. By differentiating (2) we obtain the velocity of the bead along the wire:

$$v(t) = -\mu\omega x(t) + \sqrt{1+\mu^2}\,\omega e^{-\mu\omega t}\left(A_+ e^{\sqrt{1+\mu^2}\,\omega t} - A_- e^{-\sqrt{1+\mu^2}\,\omega t}\right). \qquad (3)$$

It follows from (2) and (3) that the initial conditions $x(0) = x_0$ and $v(0) = v_0$ require

$$A_{\pm} = \frac{1}{2}\left(x_0 \pm \frac{v_0 + \mu\omega x_0}{\omega\sqrt{1+\mu^2}}\right), \qquad (4)$$

and therefore

$$x(t) = e^{-\mu\omega t}\left(x_0 \cosh\sqrt{1+\mu^2}\,\omega t + \frac{v_0 + \mu\omega x_0}{\omega\sqrt{1+\mu^2}}\sinh\sqrt{1+\mu^2}\,\omega t\right). \qquad (5)$$

Comment

According to (5) there is a critical value of the initial velocity v_0, namely

$$v_{0c} = -(\mu + \sqrt{1+\mu^2}\,)\omega x_0, \qquad (6)$$

for which the motion is bounded:

$$x(t) = x_0 e^{-(\mu+\sqrt{1+\mu^2})\omega t}. \qquad (7)$$

(Note that $v_{0c} < 0$, meaning the particle is projected towards O.) A particle having $v_0 = v_{0c}$ comes to rest at $x = 0$ on the axis of rotation, which is a point of unstable equilibrium. For any other value of v_0 the motion is unbounded – along the positive x-axis if $v_0 > v_{0c}$, and along the negative x-axis if $v_0 < v_{0c}$. This is illustrated below in terms of the characteristic time $\tau = \omega^{-1}(1+\mu^2)^{-1/2}$.

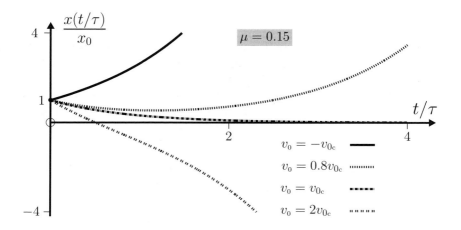

Question 14.25

Solve the problem of a charged particle moving in a uniform magnetostatic field **B** by transforming to a suitable rotating frame.

Solution

The equation of motion in a frame S′ that is rotating with constant angular velocity $\boldsymbol{\omega}$ relative to an inertial frame S can be written (see (7) of Question 14.20)

$$m\left(\frac{d\mathbf{v}}{dt}\right)_{S'} = m\left(\frac{d\mathbf{v}}{dt}\right)_{S} - m\boldsymbol{\omega} \times \mathbf{v} = q\mathbf{v} \times \mathbf{B} - m\boldsymbol{\omega} \times \mathbf{v}, \qquad (1)$$

which is zero if

$$\boldsymbol{\omega} = -q\mathbf{B}/m. \qquad (2)$$

Then **v** is constant in S′. The simplest description of the motion occurs relative to a frame S′ in which the particle is at rest. The corresponding trajectory in S is a circle in a plane perpendicular to **B**, and the particle moves with constant speed $v_\perp = R|\boldsymbol{\omega}|$ where R is the distance of the particle from the axis of rotation. The trajectory is also simple if the particle has velocity $v_0 \hat{\mathbf{B}}$ in S′: the corresponding trajectory in S is a helix with axis along **B**, and of constant pitch $2\pi v_0/|\boldsymbol{\omega}|$, and radius $v_\perp/|\boldsymbol{\omega}|$. Other choices of the velocity in S′ produce distorted helices.

Comment

In this example there is an advantage to working in a suitable non-inertial frame (cf. the solution to Question 7.17). Other instances are given elsewhere in this chapter.

Question 14.26

(a) A charged particle moves under the combined effects of an electrostatic field $\mathbf{E}(\mathbf{r})$ and a weak, uniform magnetostatic field $\mathbf{B}$. Show that the effect of $\mathbf{B}$ can be removed by transforming to a suitable rotating frame.

(b) Apply this result to bounded motion of a charge q in the electrostatic field of a fixed charge q' perturbed by the field $\mathbf{B}$.

Solution

(a) In a frame rotating with constant angular velocity $\boldsymbol{\omega}$ relative to an inertial frame, the equation of motion is

$$m\ddot{\mathbf{r}} = q\mathbf{E} + q\dot{\mathbf{r}} \times \mathbf{B} - 2m\boldsymbol{\omega} \times \dot{\mathbf{r}} - m\boldsymbol{\omega} \times (\boldsymbol{\omega} \times \mathbf{r}). \tag{1}$$

If we choose

$$\boldsymbol{\omega} = -q\mathbf{B}/2m \tag{2}$$

then the Coriolis force $-2m\boldsymbol{\omega} \times \dot{\mathbf{r}}$ in (1) cancels the Lorentz force $q\dot{\mathbf{r}} \times \mathbf{B}$. Suppose also that $\mathbf{B}$ is sufficiently weak that in (1) the centrifugal force is negligible compared to the electrostatic force; that is,

$$B \ll \sqrt{|4m\mathbf{E}(\mathbf{r})/qr|}. \tag{3}$$

Then, (1) reduces to

$$m\ddot{\mathbf{r}} = q\mathbf{E}. \tag{4}$$

(b) Here, $\mathbf{E} = q'\hat{\mathbf{r}}/4\pi\epsilon_0 r^2$ and the bounded solutions to (4) are elliptical orbits with q' at a focus of the ellipse (see Question 8.9). These are the trajectories in the rotating frame. In an inertial frame which shares a common origin with the rotating frame, the ellipses precess slowly about $\mathbf{B}$ with the angular velocity (2). In particular, if $\mathbf{B}$ is perpendicular to the plane of the ellipse then the trajectory forms a rosette pattern, such as that depicted in Question 8.15. The condition (3) for $\mathbf{B}$ to be weak is

$$B \ll \sqrt{|mq'/\epsilon_0 qr^3|}, \tag{5}$$

which means that ω is small compared to the average angular speed of the particle in its elliptical orbit.

Comments

(i) The transformation from (1) to (4) is known as Larmor's theorem. It is a consequence of the similarity between the Coriolis and Lorentz forces, with the angular velocity $\boldsymbol{\omega}$ being analogous to the magnetic field $\mathbf{B}$. The slow precession about $\mathbf{B}$ is known as the Larmor effect, and $\omega_{\mathrm{L}} = qB/2m$ is the Larmor frequency.

(ii) The analogy between electromagnetic and inertial forces applies elsewhere; for example, to the gyrocompass and magnetic compass (see Question 12.27), to the magnetic and mechanical Hall effects,[11] and to the dynamics of systems of particles and rigid bodies in non-inertial frames.[12]

(iii) For the hydrogen atom in its ground state, the condition (5) requires $B \ll 10^4$ T.
(iv) If **B** is not weak, the motion is more complicated and exhibits chaos.
(v) The Larmor effect changes electronic energy levels in atoms and gives rise to the Zeeman effect – dependence of spectral frequencies on an applied magnetic field.

Question 14.27

A spherically symmetric planet has radius R, mass M, and rotates with constant angular velocity $\boldsymbol{\Omega}$ about an axis through its centre.

(a) Consider motion of a particle of mass m close to the surface of the planet. Construct an equation of motion relative to a frame fixed on the planet (a laboratory frame) based on the approximation that the centrifugal and gravitational forces are constant and can be combined in an effective gravitational force.
(b) Evaluate and analyze the effective gravitational acceleration.

Solution

(a) The coordinate system $Oxyz$ is fixed on the surface of the planet, with x-axis pointing East, y-axis North, and z-axis vertical (here, we have in mind the example of the Earth). **R** is the position vector of O relative to the centre C of the planet, and **r** is the position vector of m relative to $Oxyz$.

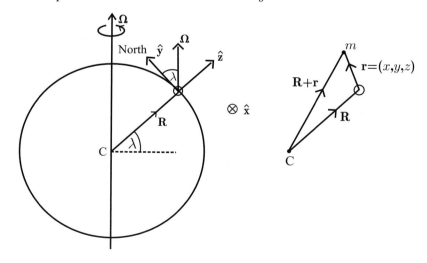

---

[11] B. L. Johnson, "Inertial forces and the Hall effect," American Journal of Physics, vol. 68, pp. 649–653, 2000.
[12] G. A. Moreno and R. O. Barrachina, "A velocity-dependent potential of a rigid body in a rotating frame," American Journal of Physics, vol. 76, pp. 1146–1149, 2008.

The equation of motion relative to this frame is (see Question 14.20)

$$m\frac{d^2\mathbf{r}}{dt^2} = \mathbf{F}_{\text{ng}} - \frac{GMm}{|\mathbf{R}+\mathbf{r}|^3}(\mathbf{R}+\mathbf{r}) - m\boldsymbol{\Omega}\times(\boldsymbol{\Omega}\times(\mathbf{R}+\mathbf{r})) - 2m\boldsymbol{\Omega}\times\frac{d\mathbf{r}}{dt}, \quad (1)$$

where $\mathbf{F}_{\text{ng}}$ is the total non-gravitational force on the particle. It is convenient to combine the gravitational and centrifugal forces in (1) into a term $m\mathbf{g}$, where

$$\mathbf{g} = -\frac{GM}{|\mathbf{R}+\mathbf{r}|^3}(\mathbf{R}+\mathbf{r}) - \boldsymbol{\Omega}\times(\boldsymbol{\Omega}\times(\mathbf{R}+\mathbf{r})) \quad (2)$$

is an effective gravitational acceleration, and to write (1) as

$$m\frac{d^2\mathbf{r}}{dt^2} = \mathbf{F}_{\text{ng}} + m\mathbf{g} - 2m\boldsymbol{\Omega}\times\frac{d\mathbf{r}}{dt}. \quad (3)$$

For motions where $r \ll R$ it is customary to assume that $\mathbf{r}$ can be neglected in (2). That is, $\mathbf{g}$ is approximated by a constant vector equal to its expression at the origin O of the laboratory frame:

$$\mathbf{g} = -g_0\hat{\mathbf{z}} - \boldsymbol{\Omega}\times(\boldsymbol{\Omega}\times\mathbf{R}), \quad \text{where } g_0 = GM/R^2. \quad (4)$$

(b) In the laboratory frame:

$$\boldsymbol{\Omega} = \Omega_y\hat{\mathbf{y}} + \Omega_z\hat{\mathbf{z}}, \qquad \mathbf{R} = R\hat{\mathbf{z}}. \quad (5)$$

The components of $\boldsymbol{\Omega}$ are given in terms of the latitude λ by

$$\Omega_y = \Omega\cos\lambda, \qquad \Omega_z = \gamma\Omega\sin\lambda, \quad (6)$$

where λ is positive, and $\gamma = +1$ in the northern hemisphere; $\gamma = -1$ in the southern hemisphere. Equations (4)–(6) yield

$$\mathbf{g} = -\gamma R\Omega^2\cos\lambda\sin\lambda\,\hat{\mathbf{y}} - (g_0 - R\Omega^2\cos^2\lambda)\hat{\mathbf{z}}. \quad (7)$$

Thus, $\mathbf{g}$ (and hence a plumb line) deviates from the downward vertical $-\hat{\mathbf{z}}$ by an amount ϵ given by

$$\tan\epsilon = \frac{g_y}{g_z} = \frac{R\Omega^2\cos\lambda\sin\lambda}{g_0 - R\Omega^2\cos^2\lambda}. \quad (8)$$

Northern hemisphere

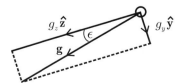

Southern hemisphere

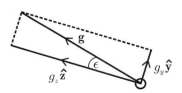

ϵ is zero at the equator ($\lambda = 0$) and the poles ($\lambda = 90°$) and has a maximum value at $\lambda = 45°$. For the Earth, $M = 6.00 \times 10^{24}$ kg, $R = 6.37 \times 10^6$ m, $\Omega = 7.29 \times 10^{-5}$ rad s^{-1} and $\epsilon_{\max} \approx 0.1°$. According to (7), the centrifugal force produces a difference between g at the equator and at the poles amounting to $R\Omega^2$. For the Earth this is approximately 3.4×10^{-2} m s^{-2}, which is smaller than the measured value $\approx 5.2 \times 10^{-2}$ m s^{-2}. This discrepancy is associated with departures from a spherically symmetric model of the Earth.

Comments

(i) Although it is in widespread use, there is no general criterion for the validity of the approximation (4); instead, "The quantitative validity of any approximation must be tested case by case and component by component, by comparing the magnitudes and/or the estimated effects of the terms neglected and the terms retained. This lesson has too often not been followed in a number of standard treatments."[13] We comment on these in the next three questions.

(ii) Reference [13] contains a wealth of information on diverse aspects of dynamics relative to a laboratory frame on Earth. The authors conclude that "A selective review of the twentieth century physics literature on motion relative to the earth demonstrates that errors and omissions abound."

Question 14.28

Consider a freely falling particle that is released from rest (relative to the Earth) at a low altitude.

(a) Use the equation of motion (3) in Question 14.27 to determine an approximate trajectory (valid to first order in Ω) relative to a laboratory frame on Earth. (Neglect air resistance.)

(b) Determine the deviation from the vertical of the point of impact with the ground.

Solution

(a) If non-gravitational forces (such as air resistance) are neglected then the equation of motion relative to the Earth is

$$\ddot{\mathbf{r}} = \mathbf{g} - 2\mathbf{\Omega} \times \dot{\mathbf{r}}. \tag{1}$$

If we neglect terms of order Ω^2 in $\mathbf{g}$, and also changes with altitude, then $\mathbf{g} = \mathbf{g}_0 = -GM\hat{\mathbf{z}}/R^2$ is a constant and (1) can be integrated. With $\mathbf{r} = \mathbf{r}_0$ and $\mathbf{v} = 0$ at $t = 0$ we have

$$\dot{\mathbf{r}} = \mathbf{g}_0 t - 2\mathbf{\Omega} \times (\mathbf{r} - \mathbf{r}_0). \tag{2}$$

---

[13] M. Tiersten and H. Soodak, "Dropped objects and other motions relative to the noninertial earth," American Journal of Physics, vol. 68, pp. 129–142, 2000.

Equation (2) cannot be integrated. Instead we substitute (2) in (1) and neglect a term of order Ω^2. The resulting equation,

$$\ddot{\mathbf{r}} = \mathbf{g}_o - 2\boldsymbol{\Omega} \times \mathbf{g}_o t, \qquad (3)$$

can be integrated twice to yield

$$\mathbf{r} = \mathbf{r}_o + \tfrac{1}{2}\mathbf{g}_o t^2 - \tfrac{1}{3}\boldsymbol{\Omega} \times \mathbf{g}_o t^3. \qquad (4)$$

(b) In terms of the laboratory coordinates specified in Question 14.27, $\mathbf{r}_o = H\hat{\mathbf{z}}$, $\mathbf{g}_o = -g_o\hat{\mathbf{z}}$ and $\boldsymbol{\Omega} \times \mathbf{g}_o = -g_o\Omega_y\hat{\mathbf{x}}$. So, the components of (4) are

$$x = \tfrac{1}{3}g_o\Omega\cos\lambda\, t^3, \qquad y = 0 \quad\text{and}\quad z = H - \tfrac{1}{2}g_o t^2. \qquad (5)$$

Thus, the time taken to reach the ground is $t = \sqrt{2H/g_o}$ and the deflection is

$$x = \frac{2\Omega}{3}\sqrt{\frac{2H^3}{g_o}}\cos\lambda. \qquad (6)$$

This is positive and so the deflection is to the East in both hemispheres.

Comments

(i) To first order in Ω the eastward deflection is entirely due to the Coriolis force. For an object falling from $H = 100$ m, and with $g_o = 9.8\,\mathrm{m\,s^{-2}}$, (6) gives $x = 0.022\cos\lambda$ m.

λ	0°	45°	90°
x (mm)	22	16	0

(ii) A reader may, at first, find the eastward direction of deflection surprising because the Earth rotates towards the East. A simple explanation was given by Newton "based on the observation that, with respect to inertial space, the dropped object, being further from earth's spin axis, has a larger eastward initial velocity than the plumb-bob, and will land eastward of the bob."[13]

(iii) Tiersten and Soodak[13] have calculated the terms of order Ω^2 in the trajectory. They find that:
- ☞ the vertical component $z = H - \tfrac{1}{2}g_o t^2$ is accurate to second order in Ω;
- ☞ the corrections to (6) are of order H/R; and
- ☞ there is a deflection towards the equator, whose value relative to a plumb line is given by

$$y = -4(\Omega^2 H^2/g_o)\cos\lambda\sin\lambda. \qquad (7)$$

Both the Coriolis force and the non-uniformity of $\mathbf{g}$ contribute to (7), with the latter contribution being about five times the former. Thus, the assumption of constant $\mathbf{g}$ will produce an error here amounting to about a factor of six.

(iv) The problem of the deviation of the path of a freely falling object from a plumb line has a long history, and satisfactory agreement between theory and experiment is still lacking.[13]

Question 14.29

Determine (to first order in Ω) the effect of the Earth's rotation on small oscillations of a simple pendulum. Do this in two ways:

(a) by transforming to a frame that is rotating relative to a laboratory frame, and
(b) by solving the equation of motion in a laboratory frame.

Solution

We use the same coordinate system as in Question 14.27. The pivot is on the z-axis, a distance ℓ above O, where ℓ is the length of the pendulum. For small oscillations (x and y both $\ll \ell$) the vertical displacement of the bob $z \approx (x^2 + y^2)/2\ell$ is of second order and therefore negligible. The equation of motion is (see (3) of Question 14.27)

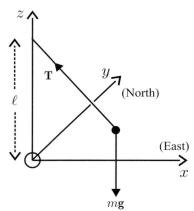

$$m\ddot{\mathbf{r}} = \mathbf{F}_{\mathrm{ng}} + m\mathbf{g}_\mathrm{o} - 2m\boldsymbol{\Omega} \times \dot{\mathbf{r}}. \qquad (1)$$

Here, $\mathbf{F}_{\mathrm{ng}} = \mathbf{T}$, the tension in the string, and we have approximated $\mathbf{g}$ by its value at O. (Changes in the centrifugal force are of second order in Ω, and changes in the gravitational force are of second order in x and y.) The motion is essentially two-dimensional in the xy-plane, and the vertical component of (1) is $\mathbf{T} = -m\mathbf{g}_\mathrm{o}$. The horizontal component of $\mathbf{T}$ is $-mg_\mathrm{o}\boldsymbol{\rho}/\ell = -m\omega_0^2 \boldsymbol{\rho}$, where $\boldsymbol{\rho} = (x, y)$ is the position vector of the bob and $\omega_0 = \sqrt{g_0/\ell}$ is the angular frequency of the oscillations when $\Omega = 0$. Thus, the horizontal component of (1) is

$$\ddot{\boldsymbol{\rho}} = -\omega_0^2 \boldsymbol{\rho} - 2\boldsymbol{\Omega} \times \dot{\boldsymbol{\rho}}. \qquad (2)$$

To this order the effect of the Earth's rotation is given by the Coriolis term in (2).

(a) Now, $\boldsymbol{\Omega} = (0, \Omega_y, \Omega_z)$ and so $\boldsymbol{\Omega} \times \dot{\boldsymbol{\rho}} = \Omega_z \hat{\mathbf{z}} \times \dot{\boldsymbol{\rho}}$. Thus, (2) can be written

$$\ddot{\boldsymbol{\rho}} = -\omega_0^2 \boldsymbol{\rho} - 2\Omega_z \hat{\mathbf{z}} \times \dot{\boldsymbol{\rho}}. \qquad (3)$$

The Coriolis term in (3) can be removed by transforming to a frame S' rotating with constant angular velocity

$$\boldsymbol{\omega} = -\Omega_z \hat{\mathbf{z}} = -\gamma \Omega \sin \lambda \, \hat{\mathbf{z}}, \qquad (4)$$

where $\gamma = +1$ in the northern hemisphere; $\gamma = -1$ in the southern hemisphere (see (2) of Question 14.20 with terms in Ω^2 neglected). In S' the pendulum oscillates in a vertical plane with angular frequency ω_0; in the laboratory frame this plane rotates slowly (precesses) about the vertical with angular frequency $\boldsymbol{\omega}$. This precession is clockwise when viewed from above in the northern hemisphere, and anti-clockwise in the southern hemisphere.

(b) Equation (2) has components
$$(\ddot{x}, \ddot{y}) = -\omega_0^2(x, y) + 2\Omega_z(\dot{y}, -\dot{x}). \tag{5}$$

Thus, the quantity $\eta = x + iy$ satisfies the differential equation
$$\ddot{\eta} + 2i\Omega_z \dot{\eta} + \omega_0^2 \eta = 0. \tag{6}$$

The general solution to (6) is
$$\eta = e^{-i\Omega_z t}\left(Ae^{i\sqrt{\omega_0^2+\Omega_z^2}\,t} + Be^{-i\sqrt{\omega_0^2+\Omega_z^2}\,t}\right). \tag{7}$$

The solution that satisfies the initial condition $x = x_0$, $y = 0$ at $t = 0$ is
$$\eta = x_0 e^{-i\Omega_z t} \cos\sqrt{\omega_0^2 + \Omega_z^2}\,t \approx x_0 e^{-i\Omega_z t} \cos\omega_0 t \tag{8}$$

because $\Omega_z \ll \omega_0$. That is,
$$x = x_0 \cos\omega_0 t \cos\Omega_z t, \qquad y = -x_0 \cos\omega_0 t \sin\Omega_z t. \tag{9}$$

This is an oscillation $x_0 \cos\omega_0 t$ in a vertical plane that rotates about the z-axis with angular velocity $-\Omega_z \hat{\mathbf{z}}$. The motion is the same as that found in (a).

Comments

(i) The period of precession, $T = 2\pi/\Omega \sin\lambda = 24/\sin\lambda$ in hours, increases from 24 h at the poles, to 33.7 h at $\lambda = 45°$, and becomes infinite (i.e. no precession) at the Equator.

(ii) The precessing pendulum is known as Foucault's pendulum, after Jean Foucault who first demonstrated the effect in a series of experiments in 1851, culminating in the use of a 28-kg cannon ball suspended by a 65-m long wire in the Pantheon in Paris. His experiment provides a terrestrial demonstration of the Earth's rotation, and it is a popular exhibit in science museums and other public places. In principle, other effects could be used for this demonstration, such as the deflection of a falling object or a projectile. The essential advantage of the pendulum (and also the gyroscope, see Question 12.25) is the cumulative nature of the effect. An interesting account of Foucault's work on the pendulum has been given by Crane.[14] Foucault was an accomplished experimentalist who invented the gyroscope, performed an accurate measurement (at the time) of the speed of light, discovered eddy currents, and worked on many practical devices. Foucault's pendulum was selected in a poll as one of the 'most beautiful experiments in physics'.[15]

(iii) Usually, the Foucault pendulum consists of a massive ball suspended by a long wire. However, in recent years considerable progress has been made in eliminating

---

[14] H. R. Crane, "The Foucault pendulum as a murder weapon and a physicist's delight," The Physics Teacher, vol. 28, pp. 264–269, 1990.

[15] R. P. Crease, "The most beautiful experiment," Physics World, vol. 15, pp. 19–20, September 2002.

the perturbing effects that plague more compact pendulums – to such an extent that the effect has been observed with pendulums as short as 15 cm, and a 70-cm pendulum has been reported to be in continuous operation for over ten years, precessing to within 2% of the theoretical rate.[16]

(iv) If non-uniformity of the gravitational and centrifugal forces – that is, terms due to non-zero **r** in (2) of Question 14.27 – is included, then (3) is modified to read

$$\ddot{\boldsymbol{\rho}} = -(\omega_0^2 - \Omega_z^2 + g_0/R)\boldsymbol{\rho} + \Omega_y^2 x\hat{\mathbf{x}} - 2\Omega_z \hat{\mathbf{z}} \times \dot{\boldsymbol{\rho}}. \tag{10}$$

It is apparent that the additional terms are negligible (for example, $g_0/R = \ell\omega_0^2/R \ll \omega_0^2$) and the main contribution to non-inertiality is the Coriolis term in (10). In the following question the opposite is true.

Question 14.30

Obtain an approximate equation of motion for free motion of a particle on a horizontal frictionless plane in the laboratory frame. (Hint: Refer to Question 14.29.)

Solution

Here, the components of the non-gravitational force $\mathbf{F}_{\text{ng}}$ in the plane of the table are zero. So, the equation of motion can be obtained by setting $\omega_0 = 0$ in (10) of Question 14.29. With $g_0/R = \omega_p^2$ we have

$$\ddot{\boldsymbol{\rho}} = -(\omega_p^2 - \Omega_z^2)\boldsymbol{\rho} + \Omega_y^2 x\hat{\mathbf{x}} - 2\Omega_z \hat{\mathbf{z}} \times \dot{\boldsymbol{\rho}}. \tag{1}$$

Comment

It is often claimed in the literature that this problem can be solved by making the uniform **g** approximation (4) of Question 14.27. That is, by neglecting non-uniformity of the gravitational and centrifugal forces, so that

$$\ddot{\boldsymbol{\rho}} = -2\Omega_z \hat{\mathbf{z}} \times \dot{\boldsymbol{\rho}}. \tag{2}$$

However, the analysis given in Ref. [13] shows that this approximation is not valid – in fact, it would be more accurate to neglect the Coriolis term in (1). These authors point out that the solutions to (1) and (2) are very different: according to (1) the motion is an oscillation of angular frequency ω_p (period $\approx 5000\,\text{s}$) that precesses at a rate Ω_z (period $\geq 86\,400\,\text{s}$); whereas, according to (2) the particle performs a uniform circular gyration if $\dot{\boldsymbol{\rho}}_0 \neq 0$, and remains at rest if $\dot{\boldsymbol{\rho}}_0 = 0$.

---

[16] H. R. Crane, "Foucault's pendulum 'wall clock'," American Journal of Physics, vol. 63, pp. 33–39, 1995.

15

The relativity principle and some of its consequences

Einstein's relativity principle is an assertion that the laws of physics are equally valid in all inertial frames – that is, these laws have the same mathematical form in all inertial frames. This principle is the basis for the theory of special relativity, yet students often have difficulty understanding its significance and power. It is with this in mind that we have devised the following short set of questions.

If one uses the relativity principle to go beyond the absolute space and time of Newtonian relativity, then a startling result soon emerges: the theory admits the possibility of a universal speed, and shows that particles travelling with this speed are peculiar – their speed relative to us is independent of whether we move towards or away from them (see Questions 15.1 to 15.3).

Following Rindler[1] we denote the universal speed by V and refer to the space-time transformation allowed by the relativity principle as the V^2-Lorentz transformation.[‡] Because of our focus on the relativity principle, many of the following questions involve V and the V^2-Lorentz transformation. The rationale behind this approach is emphasized in the comments to Question 15.2. Readers who have in mind also the theory of electromagnetism and the abundant experimental evidence concerning V can, at any stage, make the identification $V = c$ (the speed of light in vacuum).

We hope that our examples will indicate to the reader that "Einstein's principle is really a *metaprinciple*: it puts constraints on *all* the laws of physics ... (and) is a beautiful example of the power of pure thought to leap ahead of the empirical frontier."[2] For further study we recommend the books by Rindler[1,2] and Barton.[3]

Question 15.1

Consider two inertial frames S and S′ in the standard configuration, with S′ moving relative to S at constant velocity v along the x-axis (see Question 14.8 and the diagram

[‡]This notation should not be confused with our previous use of V for a potential or a volume.

---

[1]   W. Rindler, *Essential relativity*. New York: Springer, 2nd edn, 1977. Chap. 2.
[2]   W. Rindler, *Introduction to special relativity*, p. 2. Oxford: Oxford University Press, 1982.
[3]   G. Barton, *Introduction to the relativity principle*. Chichester: Wiley, 1999.

below). Let (x, t) and (x', t') be the space and time coordinates of the same event relative to S and S'. Show that the most general linear, homogeneous transformation between (x, t) and (x', t') that is consistent with the relativity principle (see Question 14.10) is

$$x' = \frac{1}{\sqrt{1 - v^2/V^2}}(x - vt) \tag{1}$$

$$t' = \frac{1}{\sqrt{1 - v^2/V^2}}\left(t - \frac{v}{V^2}x\right), \tag{2}$$

where $V^2 = V^2(v)$ is an even function of v.

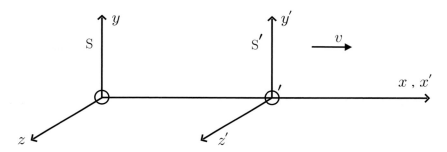

Solution

The analysis involves four steps:

1. For the standard configuration depicted above, O and O' coincide at $t = t' = 0$ (see also Question 14.8). Therefore, $x = vt$ implies $x' = 0$. This, together with linearity, means x' is proportional to $x - vt$:

$$x' = \gamma(v)(x - vt), \tag{3}$$

where the coefficient $\gamma(v)$ is independent of x and t.

2. Now, apply the relativity principle to (3). That is, unprime x', prime x and t, and replace v with $-v$ (see (3) of Question 14.10). Then

$$x = \gamma(-v)(x' + vt'). \tag{4}$$

3. From (3), $\gamma(v)$ is equal to the ratio[†] $O'x'/Ox$ at $t = 0$, and this cannot depend on the sign of v because space is isotropic in an inertial frame (see Chapter 1). So

$$\gamma(-v) = \gamma(v). \tag{5}$$

[†] In general, $\gamma(v) = \Delta x'/\Delta x$, where $\Delta x'$ is the length of a rod at rest along the x'-axis and Δx is the length of that rod in S, the endpoints of Δx being measured at the same instant t.

4. Now, use (3) to eliminate x' from (4), and then solve for t' in terms of t and x:

$$t' = \gamma(v)\left[t - \frac{\gamma^2(v) - 1}{v^2\gamma^2(v)}\, vx\right]. \qquad (6)$$

Equations (3) and (6) are the linear transformations allowed by the relativity principle and isotropy. They contain one unknown coefficient, the even function $\gamma(v)$. For further analysis it is helpful to write (3) and (6) in terms of a related unknown, V^2, defined in terms of γ^2 and v by the inverse of the coefficient of vx in (6),

$$V^2(v) \equiv \frac{\gamma^2(v)}{\gamma^2(v) - 1}\, v^2, \qquad (7)$$

and in terms of which

$$\gamma(v) = \frac{1}{\sqrt{1 - v^2/V^2}}. \qquad (8)$$

(In going from (7) to (8) a possible minus sign is excluded because $\gamma(0) = 1$.) Equations (3) and (6)–(8) yield (1) and (2). Note that V has the dimensions of velocity.

Comments

(i) The transformations (1) and (2) are known as the V^2-Lorentz transformation.[1]

(ii) If $V^2 = \infty$ (that is, $\gamma = 1$) they reduce to the Galilean transformation of absolute space and time (see Question 14.8): $\Delta x' = \Delta x$ and $t' = t$. But, if V^2 is finite, then (1) and (2) describe relative space-time (space and time intervals are different in different inertial frames).

(iii) The restriction to linear transformations is associated with the restriction to inertial frames, although the connection requires some discussion. We have also supposed that the velocity of S relative to S' is $-v$, where v is the velocity of S' relative to S. Again, this seemingly mild result (known as reciprocity) requires elaboration. Linearity follows from the homogeneity and isotropy of space and the homogeneity of time in an inertial frame (see Chapter 1); reciprocity requires these and the relativity principle.[1,2,4,5]

(iv) In the above we have not mentioned the coordinates perpendicular to the direction of motion. It is easily shown that these are unchanged:

$$y' = y, \qquad z' = z. \qquad (9)$$

For example, with $y' = \alpha(v)y$ the relativity principle and reciprocity require $\alpha^2(v) = 1$ and therefore, since $\alpha \to 1$ as $v \to 0$, the relevant root is $\alpha(v) = 1$.

(v) We emphasize the importance of the coefficient $\gamma(v)$ introduced in (3) (equivalently, the quantity $V^2(v)$ defined in (7)): it allows one to depart from Newton's restriction to absolute space and absolute time – see (6) – and it pervades all of relativistic physics. The next question deals with some properties of V.

---

[4] A. R. Lee and T. M. Kalotas, "Response to Comments on 'Lorentz transformations from the first postulate'," American Journal of Physics, vol. 44, pp. 1000–1002, 1976.

[5] V. Berzi and V. Gorini, "Reciprocity principle and Lorentz transformations," Journal of Mathematical Physics, vol. 10, pp. 1518–1524, 1969.

Question 15.2

(a) Show that the quantity $V^2(v)$ that appears in the V^2-Lorentz transformation of Question 15.1 is a universal constant. (Hint: Consider successive V^2-Lorentz transformations from S to S' to S''.)

(b) Deduce that $V^2 > 0$.

Solution

(a) Let S'' be an inertial frame in the standard configuration with S' and moving with velocity u relative to S'. According to Question 15.1 the space-time coordinates of S'' and S' are related by a V^2-Lorentz transformation:

$$x'' = \gamma(u)\left(x' - ut'\right), \qquad t'' = \gamma(u)\left(t' - \frac{u}{V^2(u)}x'\right). \tag{1}$$

Also, x' and t' are given in terms of x and t by the V^2-Lorentz transformation (1) and (2) of Question 15.1. It follows that (x'', t'') are related to (x, t) by

$$x'' = \gamma(u)\gamma(v)\left[\left(1 + \frac{uv}{V^2(v)}\right)x - (u+v)t\right] \tag{2}$$

$$t'' = \gamma(u)\gamma(v)\left[\left(1 + \frac{uv}{V^2(u)}\right)t - \left(\frac{u}{V^2(u)} + \frac{v}{V^2(v)}\right)x\right]. \tag{3}$$

But, (x'', t'') and (x, t) are related by a V^2-Lorentz transformation from S to S'':

$$x'' = \gamma(w)(x - wt), \qquad t'' = \gamma(w)\left(t - \frac{w}{V^2(w)}x\right), \tag{4}$$

where w is the velocity of S'' relative to S. Equations (4) require equality of the coefficient of x in (2) with the coefficient of t in (3). That is,

$$V^2(u) = V^2(v). \tag{5}$$

But u and v are arbitrary. Thus, $V^2(v)$ must be independent of v: it is the same for all inertial observers and is therefore a universal constant.

(b) By equating the ratio of the coefficients of t and x in $(4)_1$ with the ratio of those in (2) we obtain a velocity addition formula:

$$w = \frac{u+v}{1 + uv/V^2}. \tag{6}$$

Now, on physical grounds we require of (6) that for positive u and v the resultant w must be positive. (Velocities in the same direction cannot add to produce a velocity in the opposite direction.[6]) But, if $V^2 < 0$, then $w < 0$ when $uv > -V^2$. Therefore, only positive values of V^2 are physically acceptable. (Negative values of V^2 produce other unphysical consequences such as negative values of γ and violation of causality.[1])

---

[6] A. R. Lee and T. M. Kalotas, "Lorentz transformation from the first postulate," American Journal of Physics, vol. 43, pp. 434–437, 1975.

Comments

(i) The condition $V^2 > 0$ means that V is a real quantity. Also, because γ is necessarily real, it follows from (8) of Question 15.1 that $v^2 < V^2$: the relative speed of inertial frames is bounded above by $|V|$. (No such bound is implied if V^2 were negative.)

(ii) The foregoing derivation and analysis of the V^2-Lorentz transformation shows "that the Lorentz transformations are attainable up to an unspecified universal constant $[V]$, without recourse to any specific physical phenomena apart from those underlying the very existence of classical inertial frames."[4] The prevailing tendency to base the theory instead on propagation of light has been criticized particularly by Lévy-Leblond: "By establishing special relativity on a property of the speed of light, one seems to link this theory to a restricted class of natural phenomena, namely, electromagnetic radiations. However, ... special relativity up to now seems to rule *all* classes of natural phenomena, whether they depend on electromagnetic, weak, strong, or even gravitational interactions. This theory does not derive from the use of electromagnetic signals for synchronizing clocks, for example ...; quite the contrary, it is the validity of the theory which constrains electromagnetic signals to have their specific propagation properties. We believe that special relativity at the present time stands as a universal theory describing the structure of a common space-time arena in which all fundamental processes take place ... and electromagnetic interactions here have no privilege other than a historic and anthropocentric one."[7]

(iii) The only possible space-time transformations between inertial frames are the V^2-Lorentz transformation or its special limit, the Galilean transformation (corresponding to $V = \infty$). "The Lorentz case is characterized by a parameter with the dimensions of a velocity which is a universal constant associated with the very structure of space-time."[7]

Question 15.3

(a) Let $\mathbf{u} = (u_x, u_y, u_z)$ and $\mathbf{u'} = (u'_x, u'_y, u'_z)$ be the velocities of a particle relative to inertial frames S and S' in the standard configuration. Use the V^2-Lorentz transformation to obtain the velocity transformation

$$u'_x = \frac{u_x - v}{1 - u_x v/V^2}, \quad u'_y = \frac{u_y}{\gamma(1 - u_x v/V^2)}, \quad u'_z = \frac{u_z}{\gamma(1 - u_x v/V^2)}. \quad (1)$$

Comment on (1) with regard to form invariance (see Question 14.9).

(b) Deduce that if a particle moves with speed V relative to an inertial frame, then its speed is V relative to all inertial frames. (Hint: Use the inverse of (1) to obtain a relation between $V^2 - u^2$ and $V^2 - u'^2$.)

---

[7] J. M. Lévy-Leblond, "One more derivation of the Lorentz transformation," American Journal of Physics, vol. 44, pp. 271–277, 1976.

Solution

(a) By definition, the velocities are

$$\mathbf{u}' = (dx'/dt', dy'/dt', dz'/dt'), \qquad \mathbf{u} = (dx/dt, dy/dt, dz/dt), \qquad (2)$$

where the differentials are given by the V^2-Lorentz transformation as

$$dx' = \gamma(dx - v\,dt), \quad dy' = dy, \quad dz' = dz, \quad dt' = \gamma(dt - v\,dx/V^2). \qquad (3)$$

Equations (1) follow directly from (2) and (3). The inverses of (1) are

$$u_x = \frac{u'_x + v}{1 + u'_x v/V^2}, \qquad u_y = \frac{u'_y}{\gamma(1 + u'_x v/V^2)}, \qquad u_z = \frac{u'_z}{\gamma(1 + u'_x v/V^2)}. \qquad (4)$$

Comparison of (1) and (4) shows that they are related by the rule: interchange primed and unprimed quantities, and replace v with $-v$. Therefore, they satisfy the form invariance required by the relativity principle (see Questions 14.9 and 14.10).

(b) With $u^2 = u_x^2 + u_y^2 + u_z^2$, and from (4) we have[1]

$$\gamma^2\left(1 + u'_x \frac{v}{V^2}\right)^2 (V^2 - u^2) = \gamma^2\left(1 + u'_x \frac{v}{V^2}\right)^2 V^2 - \left[\gamma^2(u'_x + v)^2 + u'^{\,2}_y + u'^{\,2}_z\right]$$

$$= V^2 - u'^2.$$

That is,

$$V^2 - u^2 = V^2(V^2 - v^2)(V^2 + \mathbf{u}' \cdot \mathbf{v})^{-2}(V^2 - u'^2), \qquad (5)$$

where we have replaced $u'_x v$ with $\mathbf{u}' \cdot \mathbf{v}$, thereby generalizing the result to relative motion of S and S' along any direction. It follows from (5) that if $u' = V$ then $u = V$, for arbitrary inertial frames S and S'.

Comments

(i) Such behaviour is outside our experience, which concerns objects that can be pursued, caught and overtaken.

(ii) Because $v^2 < V^2$ (see Question 15.2), it follows from (5) that $u' < V$ implies $u < V$. Thus, we can distinguish two types of inhabitant of the universe: those moving with speed less than V and those with speed equal to V (relative to an inertial frame). The former can be pursued and even overtaken – and they cannot move with speeds equal to or greater than V relative to any inertial frame. (Actually, because $u' > V$ implies $u > V$, there is a third possible inhabitant, which moves with speed greater than V relative to all inertial frames.)

(iii) Further development of this theory shows that V is the speed of any particle that has zero mass. To within the accuracy of existing experiments, photons (light quanta) have zero mass and probably also neutrinos and gravitons. All existing measurements in physics are consistent with the identification $V = c$, the speed of light in vacuum.

(iv) If it should one day turn out that the photon (and the neutrino, etc.) all have a small but non-zero mass, then V would not be the speed of any actual particle. This "would not, as such, shake in any way the validity of special relativity. It would, however, nullify all of its derivations which are based on the invariance of the photon velocity."[7]

(v) The Galilean transformation applies when $v \ll c$, and Newtonian physics is a good approximation for the large class of phenomena where particle speeds are small compared to c.

Question 15.4

Consider the wave equation

$$\left(\frac{\partial^2}{\partial x^2} + \frac{\partial^2}{\partial y^2} + \frac{\partial^2}{\partial z^2} - \frac{1}{c^2}\frac{\partial^2}{\partial t^2}\right)\phi(\mathbf{r}, t) = 0, \tag{1}$$

where ϕ is a scalar function and c is the speed of light in vacuum. Determine how (1) transforms under the V^2-Lorentz transformation of Question 15.1. Deduce the value of V for which (1) is form invariant.

Solution

For the transformation

$$x' = \gamma(x - vt), \qquad y' = y, \qquad z' = z, \qquad t' = \gamma(t - vx/V^2) \tag{2}$$

we have

$$\left.\begin{array}{l} \dfrac{\partial}{\partial x} = \gamma\dfrac{\partial}{\partial x'} - \gamma\dfrac{v}{V^2}\dfrac{\partial}{\partial t'} \\[2mm] \dfrac{\partial}{\partial t} = -\gamma v\dfrac{\partial}{\partial x'} + \gamma\dfrac{\partial}{\partial t'}, \end{array}\right\} \tag{3}$$

and $\partial/\partial y = \partial/\partial y'$, $\partial/\partial z = \partial/\partial z'$. Also, because ϕ is a scalar under (2), we have $\phi'(\mathbf{r}', t') = \phi(\mathbf{r}, t)$. So, (1) transforms into

$$\left[\frac{V^2}{V^2 - v^2}\frac{c^2 - v^2}{c^2}\frac{\partial^2}{\partial x'^2} + \frac{\partial^2}{\partial y'^2} + \frac{\partial^2}{\partial z'^2} - \frac{1}{V^2 - v^2}\left(\frac{V^2}{c^2} - \frac{v^2}{V^2}\right)\frac{\partial^2}{\partial t'^2}\right.$$

$$\left. + 2v\frac{V^2}{V^2 - v^2}\left(\frac{1}{c^2} - \frac{1}{V^2}\right)\frac{\partial^2}{\partial t'\partial x'}\right]\phi'(\mathbf{r}', t') = 0. \tag{4}$$

In general, this is not form invariant because of the term in $\partial^2/\partial t'\partial x'$ and the coefficients of $\partial^2/\partial x'^2$ and $\partial^2/\partial t'^2$. However, if $V = c$ then (2) becomes form invariant:

$$\left(\frac{\partial^2}{\partial x'^2} + \frac{\partial^2}{\partial y'^2} + \frac{\partial^2}{\partial z'^2} - \frac{1}{c^2}\frac{\partial^2}{\partial t'^2}\right)\phi'(\mathbf{r}', t') = 0. \tag{5}$$

Comments

(i) This is a long-hand (first principles) calculation of the invariance of (1) under a V^2-Lorentz transformation. There is a powerful formalism based on four-vectors that enables one to determine, by inspection, the invariance of (1) under Lorentz transformation: the differential operator acting on ϕ is a four-scalar when $V = c$ (see Question 15.7).

(ii) Identification of c with V means that c is a universal constant (the same in all inertial frames – see Question 15.2).

(iii) For the Galilean transformation ($V = \infty$), (4) is

$$\left(\frac{c^2 - v^2}{c^2}\frac{\partial^2}{\partial x'^2} + \frac{\partial^2}{\partial y'^2} + \frac{\partial^2}{\partial z'^2} - \frac{1}{c^2}\frac{\partial^2}{\partial t'^2} + \frac{2v}{c^2}\frac{\partial^2}{\partial t'\partial x'}\right)\phi'(\mathbf{r}', t') = 0, \quad (6)$$

which is not form invariant.

(iv) The operator in brackets in (1) is the same as that which appears in the equation for propagation of electromagnetic waves in vacuum. This indicates that form invariance will require $V = c = (1/\sqrt{\epsilon_0 \mu_0})$, the speed of light in vacuum (see also Question 15.9).

Question 15.5

(a) Express the V^2-Lorentz transformation in a form connecting a space and time interval $(\Delta\mathbf{r}, \Delta t)$ relative to S with the corresponding interval $(\Delta\mathbf{r}', \Delta t')$ relative to S'.

(b) Hence, deduce the phenomena of **1.** length contraction, and **2.** time dilation.

Solution

(a) Let $(\mathbf{r}_1, t_1)$ and $(\mathbf{r}_2, t_2)$ be the space and time coordinates of two events relative to S, and $(\mathbf{r}'_1, t'_1)$ and $(\mathbf{r}'_2, t'_2)$ be the corresponding coordinates relative to S'. The respective intervals are $(\Delta\mathbf{r}, \Delta t) = (\mathbf{r}_2 - \mathbf{r}_1, t_2 - t_1)$ and $(\Delta\mathbf{r}', \Delta t') = (\mathbf{r}'_2 - \mathbf{r}'_1, t'_2 - t'_1)$. The various coordinates are related by the V^2-Lorentz transformation of Question 15.1, and therefore so are the intervals:

$$\Delta x' = \gamma(\Delta x - v\Delta t), \quad \Delta y' = \Delta y, \quad \Delta z' = \Delta z, \quad \Delta t' = \gamma\left(\Delta t - \frac{v\Delta x}{V^2}\right). \quad (1)$$

(b) **1.** Let $\Delta x' = \ell_0$ be the length of a rod at rest in S'. Its length ℓ relative to S is defined as the difference $\Delta x = x_2 - x_1$, where x_1 and x_2 are measured at the same instant t. From (1)$_1$ with $\Delta t = 0$ we have the formula for length contraction:

$$\ell_0 = \gamma\ell. \quad (2)$$

2. Consider the inverse transformation to (1)$_4$, namely

$$\Delta t = \gamma\left(\Delta t' + \frac{v\Delta x'}{V^2}\right), \quad (3)$$

and let $\Delta t' = \Delta \tau$ be the time interval between two events occurring at the same position in S'. From (3) with $\Delta x' = 0$ we have the formula for time dilation:

$$\Delta t = \gamma \Delta \tau. \qquad (4)$$

Comments

(i) According to (2), $\ell = \sqrt{1-v^2/V^2}\,\ell_0 < \ell_0$: the length ℓ of a moving rod is always less than the so-called proper‡ length (or rest length).

(ii) According to (4), $\Delta \tau = \sqrt{1-v^2/V^2}\,\Delta t$: a time interval is always least in the rest frame of a clock. A *clock* translating with constant speed v relative to an inertial frame S runs slows by a factor $\sqrt{1-v^2/V^2}$ relative to the *clocks* of S. $\Delta \tau$ is known as the proper time interval.

(iii) The frame S' in which measurements are made on an object or clock at rest is known as the proper (or rest) frame. There is no preferred (absolute) rest frame.

(iv) In (2), the length ℓ (which is measured by an observer in motion relative to the rod) is known as the improper length. Similarly, Δt in (4) is called the improper time interval.

Question 15.6

Show that, in terms of the coordinates

$$x_1 = x, \qquad x_2 = y, \qquad x_3 = z, \qquad x_4 = iVt \qquad (1)$$

($i = \sqrt{-1}$), the V^2-Lorentz transformation of Question 15.1 can be expressed as an orthogonal transformation in four dimensions:

$$x'_\mu = a_{\mu\nu} x_\nu, \qquad (2)$$

where the coefficients $a_{\mu\nu}$ satisfy the orthogonality relations

$$a_{\mu\nu} a_{\mu\lambda} = \delta_{\nu\lambda}, \qquad a_{\nu\mu} a_{\lambda\mu} = \delta_{\nu\lambda}. \qquad (3)$$

(Here, and in the following, it is understood that Greek indices have values from 1 to 4, and that a repeated index, such as ν in (2), implies summation from 1 to 4.)

Solution

In terms of the notation (1) and because V is the same in all inertial frames, the V^2-Lorentz transformation can be expressed as

$$x'_1 = \gamma(x_1 + ivx_4/V), \qquad x'_2 = x_2, \qquad x'_3 = x_3, \qquad x'_4 = \gamma(x_4 - ivx_1/V), \qquad (4)$$

‡The word 'proper' derives from the Latin *proprius*, meaning 'own' or 'belonging to oneself'; and so it is a helpful adjective for evoking a measurement done in the rest frame of an object.

where $\gamma = \dfrac{1}{\sqrt{1 - v^2/V^2}}$ and $x'_4 = iV't' = iVt'$. That is,

$$\begin{pmatrix} x'_1 \\ x'_2 \\ x'_3 \\ x'_4 \end{pmatrix} = \begin{pmatrix} \gamma & 0 & 0 & i\gamma v/V \\ 0 & 1 & 0 & 0 \\ 0 & 0 & 1 & 0 \\ -i\gamma v/V & 0 & 0 & \gamma \end{pmatrix} \begin{pmatrix} x_1 \\ x_2 \\ x_3 \\ x_4 \end{pmatrix}. \tag{5}$$

Equations (4) are the desired relation (2) with coefficients $a_{\mu\nu}$ given by the elements of the 4×4 matrix in (5). It follows from (4) that the quantity $x_1^2 + x_2^2 + x_3^2 + x_4^2$ is an invariant (the same for all inertial observers):

$$x'^2_\mu = x_\mu^2. \tag{6}$$

The orthogonality relations (3) are a consequence of the condition (6):

1. By substituting (2) in (6) we have $a_{\mu\nu} a_{\mu\lambda} x_\nu x_\lambda = x_\mu^2$, for all values of the x_ν and hence $(3)_1$.

2. Multiplication of (2) by $a_{\mu\lambda}$ and use of $(3)_1$ gives

$$a_{\mu\lambda} x'_\mu = a_{\mu\lambda} a_{\mu\nu} x_\nu = x_\lambda. \tag{7}$$

Thus we have the inverse transformation

$$x_\mu = a_{\nu\mu} x'_\nu. \tag{8}$$

From (8) we have $x_\mu^2 = a_{\nu\mu} a_{\lambda\mu} x'_\nu x'_\lambda$, and hence $(3)_2$.

Comments

(i) The inclusion of the imaginary number i in x_4 enables one to work in a complex Cartesian space known as Minkowski space. The alternative is to work in a real space where $x_4 = Vt$, and consequently $x_\mu^2 = x_1^2 + x_2^2 + x_3^2 - x_4^2$. Such a space is Riemannian. We have chosen the former for our discussion of special relativity because "the formulas in complex Minkowski space are usually particularly simple and neat, without the encumbrances of metric tensors or the (here) artificial distinction between covariant and contravariant quantities. It also permits natural extensions from our experience with ordinary three-dimensional space. ... Most of the equations look the same in either space; in any case, conversion from one space to the other is a simple matter."[8]

(ii) The coefficients $a_{\mu\nu}$ contained in (5) are for frames S and S' that are in the standard configuration. However, the transformation (2) and the invariance (6) are valid for arbitrary direction of the velocity $\mathbf{v}$ of S' relative to S.

(iii) Clearly, (6) is also valid for infinitesimal intervals:

$$dx'^2_\mu = dx_\mu^2. \tag{9}$$

---

[8] H. Goldstein, *Classical mechanics*, p. 293. Reading: Addison-Wesley, 1980. Chap. 7.

Question 15.7

Question 14.17 details the form invariance of three-dimensional tensor equations under rotation of the coordinate system. Based on Question 15.6, and reasoning by analogy, outline a mathematical framework that enables one to treat form invariance of physical laws under a V^2-Lorentz transformation.

Solution

We first recall some properties of three-tensors (see Questions 14.16 and 14.17). These are defined by the way they transform under rotation of the coordinate system – that is, under a linear, homogeneous, orthogonal transformation. The prototype three-tensor is the position vector **r** whose three components transform according to $x'_i = a_{ij} x_j$. In general, the components of three-tensors transform according to $A'_i = a_{ij} A_j$, $T'_{ij} = a_{ir} a_{js} T_{rs}$, etc. Three-tensor equations consist of linear relations among tensors of the same rank, and form invariance of the equations is achieved by covariance of these tensors.

The extension of these ideas to quantities that are four-tensors under the V^2-Lorentz transformation is simple. The prototype four-vector is the position vector $x_\mu = (\mathbf{r}, iVt)$ of a point in Minkowski space, and we have seen that its components transform according to the linear, homogeneous, orthogonal transformation $x'_\mu = a_{\mu\nu} x_\nu$. In general, a four-vector is a set of four quantities A_μ for which

$$A'_\mu = a_{\mu\nu} A_\nu \,; \tag{1}$$

a second-rank four-tensor comprises sixteen quantities $A_{\mu\nu}$ for which

$$A'_{\mu\nu} = a_{\mu\rho} a_{\nu\sigma} A_{\rho\sigma} \,, \tag{2}$$

and so on. A four-scalar S is a tensor of rank zero and it is unchanged by the transformation: $S' = S$. A four-tensor equation is a linear relation among tensors of the same rank. For example, the four-vector equation

$$A_\mu = B_\mu \,. \tag{3}$$

Then

$$A'_\mu - B'_\mu = a_{\mu\nu}(A_\nu - B_\nu) = 0 \,. \tag{4}$$

That is, (3) transforms into the form-invariant result

$$A'_\mu = B'_\mu \,. \tag{5}$$

Similarly for four-tensor equations of any rank. Form invariance of a four-tensor equation under the V^2-Lorentz transformation is achieved by covariance of the terms (tensors) in that equation. Thus, the relativity principle imposes a condition on the equations of physics: they should be four-tensor equations. This means that kinematical quantities (velocity and acceleration) and dynamical quantities (such as momentum and force) have to be reformulated in terms of four-vectors.

Comments

(i) Several important results for four-tensors can simply be written down based on the analogy with three-tensors and results in Question 14.17. For example: $A_\mu B_\mu$ is a four-scalar; the four-gradient with components $\partial/\partial x_\mu$ is a four-vector; the four-divergence $\partial A_\mu/\partial x_\mu$ is a four-scalar, as is $\partial^2/\partial x_\mu^2$.

(ii) Note that the orthogonality relations are crucial in obtaining these results. For example,
$$A'_\mu B'_\mu = a_{\mu\nu} a_{\mu\lambda} A_\nu B_\lambda = \delta_{\nu\lambda} A_\nu B_\lambda = A_\nu B_\nu. \tag{6}$$

(iii) In particular, $A'^2_\mu = A^2_\mu$: the length of a four-vector is unchanged by a V^2-Lorentz transformation. (This is analogous to the invariance of the length of a three-vector under rotation.)

(iv) The scalar property of
$$\frac{\partial^2}{\partial x_\mu^2} \equiv \nabla^2 - \frac{1}{V^2}\frac{\partial^2}{\partial t^2} \tag{7}$$

means that form invariance of the wave equation in Question 15.4, when $V = c$, can be decided by inspection.

(v) In the following we will, for brevity, refer to A_μ as a four-vector, rather than as the components of a four-vector.

Question 15.8

Consider a particle moving with velocity $\mathbf{u}(t)$ relative to an inertial frame S. Let $d\tau$ be an infinitesimal proper time interval; that is, a time interval measured in the instantaneous proper frame (rest frame) S′ of the particle.

(a) Show that $d\tau$ is an invariant (a four-scalar). (Hint: Consider dx'^2_μ.)
(b) Show that $d\tau$ is related to the corresponding improper time interval dt measured in S by
$$d\tau = \sqrt{1 - u^2(t)/V^2}\, dt. \tag{1}$$

Solution

(a)
$$dx'^2_\mu = dx'^2_1 + dx'^2_2 + dx'^2_3 - V^2 d\tau^2 = -V^2 d\tau^2 \tag{2}$$

because $d\tau$ is measured in S′, where $dx'_i = 0$. Now, V^2 and dx'^2_μ are invariants (see Questions 15.2 and 15.6), and therefore so is $d\tau$.

(b)
$$dx^2_\mu = dx^2_1 + dx^2_2 + dx^2_3 - V^2 dt^2. \tag{3}$$

Equating (2) and (3), and setting $dx^2_1 + dx^2_2 + dx^2_3 = u^2 dt^2$, gives (1).

Comments

(i) Equation (1) describes time dilation: $dt > d\tau$. (See Question 15.5.)
(ii) The invariance of $d\tau$ means that it is independent of the choice of inertial frame S relative to which $u(t)$ in (1) is measured. Consequently, events on the trajectory can be assigned a unique proper time (relative to an initial value τ_i), independent of S:

$$\tau = \tau_i + \int_{t_i}^{t} \sqrt{1 - u^2(t)/V^2}\, dt. \tag{4}$$

Question 15.9

(a) Explain why the expressions

$$U_\mu = \frac{dx_\mu}{d\tau} \quad \text{and} \quad A_\mu = \frac{dU_\mu}{d\tau} \tag{1}$$

are suitable definitions for the four-velocity and four-acceleration of a particle.
(b) Define the four-momentum P_μ.
(c) Evaluate U_μ^2 and P_μ^2.

Solution

(a) An acceptable definition of the four-velocity should provide a four-vector whose first three components tend to the familiar Newtonian velocity $\mathbf{u} = d\mathbf{r}/dt$ in the limit $u \ll V$. This is satisfied by $(1)_1$: $dx_\mu = (d\mathbf{r}, iV\,dt)$ is a four-vector and $d\tau = \sqrt{1 - u^2/V^2}\, dt$ is a four-scalar; consequently, their ratio is a four-vector. Similarly, A_μ given by $(1)_2$ is a four-vector with the desired Newtonian limit $d^2\mathbf{r}/dt^2$ for its first three components.

(b) The four-momentum of a particle is defined by

$$P_\mu = m_0 U_\mu. \tag{2}$$

Here m_0 is the mass of the particle, as measured in an inertial frame in which the particle is at rest.[‡] This is presumed to be a unique attribute of the particle – thus, m_0 is a scalar under the V^2-Lorentz transformation. Equation (2) defines a four-vector whose first three components have the desired Newtonian limit $\mathbf{p} = m_0 \mathbf{u}$.

(c) From Question 15.8, $dx_\mu^2 = -V^2 d\tau^2$ and therefore

$$U_\mu^2 = -V^2, \qquad P_\mu^2 = -m_0^2 V^2. \tag{3}$$

Comments

(i) U_μ is a hybrid of the four-vector $dx_\mu = (d\mathbf{r}, iV\,dt)$ defined relative to an inertial frame S and the four-scalar $d\tau = \sqrt{1 - u^2/V^2}\, dt$ defined relative to the instantaneous rest frame S'. In terms of the velocity $\mathbf{u} = d\mathbf{r}/dt$ relative to S:

[‡] In the literature, m_0 is often referred to as the rest mass, or proper mass.

$$U_\mu = \gamma(u)(\mathbf{u},\, iV)\,, \tag{4}$$

where $\gamma(u) = (1 - u^2/V^2)^{-1/2}$. From (4) it is obvious that $U_\mu^2 = -V^2$.

(ii) This is a suitable point at which to consider the equation of motion of a particle. The Newtonian form

$$\frac{d}{dt}(m_o \mathbf{u}) = \mathbf{F} \tag{5}$$

is invariant under the Galilean transformation (see Question 14.9) but not under the V^2-Lorentz transformation. To generalize (5), so that it satisfies the relativity principle when V is finite, requires a classical theory of the force $\mathbf{F}$ that is invariant under the V^2-Lorentz transformation. The only candidate for this is the Lorentz force

$$\mathbf{F} = q(\mathbf{E} + \mathbf{u} \times \mathbf{B}) \tag{6}$$

acting on a charged particle in an electromagnetic field. The theory of the transformation of this field is beyond the scope of our book, and we simply state the results. First, covariance of the theory of the electromagnetic field requires

$$V = c\,, \tag{7}$$

where $c = (\epsilon_0 \mu_0)^{-1/2}$ is the speed of light in vacuum. (It is therefore at this point that an experimental value for V enters the discussion. An intimation of (7) is contained in Question 15.4.) Secondly, the equation of motion that satisfies the relativity principle when $V = c$ is[8,9]

$$\frac{d}{dt}(\gamma m_o \mathbf{u}) = \mathbf{F}, \quad \text{where} \quad \gamma(u) = \frac{1}{\sqrt{1 - u^2/c^2}}, \tag{8}$$

and $\mathbf{F}$ is given by (6). The modification of Newton's theory is contained in the factor $\gamma(u)$ in (8), and in the limit $u/c \ll 1$ equation (8) tends to (5). The V^2-Lorentz transformation with $V = c$ is known as the Lorentz transformation.

(iii) Equation (8) comprises the first three components of a four-vector equation of motion that we can write as

$$\frac{dP_\mu}{d\tau} = K_\mu\,, \tag{9}$$

where

$$K_\mu = \bigl(\gamma(u)\,\mathbf{F},\, K_4\bigr) \tag{10}$$

is the four-force. The question arises: what is the fourth component of (9), and what is its physical significance. If m_o is constant, the answer is contained in (9) and the fact that $U_\mu^2 = -c^2$ is a constant (see Question 15.10).

---

[9]   J. D. Jackson, *Classical Electrodynamics*, Chaps. 11 and 12. New York: Wiley, 1962.

Question 15.10

Consider a particle whose mass m_0 is constant (independent of time).

(a) Show that the fourth component of the equation of motion (9) of Question 15.9 is

$$\frac{d}{dt}\frac{m_0 c^2}{\sqrt{1-u^2/c^2}} = \mathbf{F}\cdot\mathbf{u}. \tag{1}$$

(Hint: First use (9) above to evaluate the fourth component K_4 of the four-force.)

(b) Deduce that the kinetic energy of the particle is given by

$$K = \left(\frac{1}{\sqrt{1-u^2/c^2}} - 1\right) m_0 c^2. \tag{2}$$

Solution

(a) For constant m_0 the equation of motion (9) of Question 15.9 is

$$m_0 \frac{dU_\mu}{d\tau} = K_\mu. \tag{3}$$

Multiply this by U_μ. Then

$$K_\mu U_\mu = m_0 U_\mu \frac{dU_\mu}{d\tau} = \tfrac{1}{2} m_0 \frac{d}{d\tau} U_\mu^2 = 0, \tag{4}$$

because $U_\mu^2 = -c^2$ is constant. From (4), and (4), (7) and (10) of Question 15.9, we have

$$K_4 = (i/c)\,\gamma(u)\, \mathbf{F}\cdot\mathbf{u}. \tag{5}$$

Thus the fourth component of (3) is (1).

(b) According to (1), the work done on the particle during a displacement $d\mathbf{r}$ is

$$\mathbf{F}\cdot d\mathbf{r} = d\left(\frac{m_0 c^2}{\sqrt{1-u^2/c^2}}\right). \tag{6}$$

The kinetic energy $K(u)$ of the particle is defined by (see Question 5.1)

$$dK = \mathbf{F}\cdot d\mathbf{r} \quad \text{with} \quad K(0) = 0. \tag{7}$$

The solution to (6) and (7) is (2).

Comments

(i) Equation (2) reduces to the Newtonian expression $K = \tfrac{1}{2} m_0 u^2$ when $u \ll c$.

(ii) The four-momentum $P_\mu = \gamma m_0(\mathbf{u},\, ic)$ can be written

$$P_\mu = (\mathbf{p},\, iE/c), \tag{8}$$

where

$$\mathbf{p} = \gamma m_0 \mathbf{u} \tag{9}$$

is the relativistic momentum and

$$E = \gamma m_0 c^2 = m_0 c^2 + K \tag{10}$$

is the sum of the proper (or rest mass) energy $m_0 c^2$ and the kinetic energy K. For a free particle E is the total energy[‡], and (8) and the relation $P_\mu^2 = -m_0^2 c^2$ (see Question 15.9) yield the relativistic energy–momentum relation

$$E^2 = \mathbf{p}^2 c^2 + m_0^2 c^4. \tag{11}$$

(iii) When $\mathbf{F} = 0$ the four-force K_μ is zero and the components of the four-momentum (8) are constants. That is, $\mathbf{p}$ and E are conserved.

(iv) The mass m_0 is constant for a charged particle in an electromagnetic field, but not in general (see Questions 15.11 and 15.12).

Question 15.11

Show that a necessary and sufficient condition for the mass m_0 of a particle to be constant is orthogonality of the four-force and the four-velocity:

$$K_\mu U_\mu = 0. \tag{1}$$

Solution

Multiply the equation of motion $d(m_0 U_\mu)/d\tau = K_\mu$ by U_μ. Then,

$$K_\mu U_\mu = U_\mu \frac{d}{d\tau}(m_0 U_\mu) = \tfrac{1}{2} m_0 \frac{d}{d\tau} U_\mu^2 + U_\mu^2 \frac{dm_0}{d\tau} = -c^2 \frac{dm_0}{d\tau}, \tag{2}$$

because $U_\mu^2 = -c^2$ (see Question 15.9). According to (2), a constant m_0 implies (1) and vice versa.

Comments

(i) The condition (1) is satisfied for a charged particle in an electromagnetic field and hence m_0 is constant. (Here, $K_\mu \propto U_\nu F_{\mu\nu}$, where $F_{\mu\nu}$ is anti-symmetric in μ and ν. Therefore (1) holds.[9])

[‡]But not for a particle in a potential – see, for example, (1)$_1$ of Question 15.16 and (3) of Question 15.17.

(ii) In general, forces can be classified as 'pure' if $K_\mu U_\mu = 0$ (and hence m_o is constant) and 'impure' otherwise. Any force can be divided into pure and impure parts, $\mathcal{K}_\mu$ and $\mathcal{I}_\mu$, according to

$$K_\mu = \mathcal{K}_\mu + \mathcal{I}_\mu, \tag{3}$$

where

$$\mathcal{K}_\mu = K_\mu + c^{-2}(K_\nu U_\nu)U_\mu, \qquad \mathcal{I}_\mu = -c^{-2}(K_\nu U_\nu)U_\mu. \tag{4}$$

It follows from (2) and (4)$_2$ that

$$\mathcal{I}_\mu = U_\mu \frac{dm_o}{d\tau}, \tag{5}$$

and hence

$$\mathcal{K}_\mu = m_o \frac{dU_\mu}{d\tau}. \tag{6}$$

Pure forces change U_μ, impure forces change m_o. An example of an impure force is given next.

Question 15.12

Consider a particle in a field

$$K_\mu = -\frac{\partial \phi}{\partial x_\mu}, \tag{1}$$

where $\phi(x_\nu)$ is a four-scalar. Show that the mass m_o depends on ϕ according to

$$m_o(\phi) = m_o(0) + \phi/c^2. \tag{2}$$

Solution

From (2) of Question 15.11 and (1) we have

$$-c^2 \frac{dm_o}{d\tau} = -\frac{\partial \phi}{\partial x_\mu} \frac{dx_\mu}{d\tau} = -\frac{d\phi}{d\tau}. \tag{3}$$

That is,

$$\frac{d}{d\tau}(m_o - \phi/c^2) = 0. \tag{4}$$

The solution to (4) is (2).

Comment

Use of (3) in (2) of Question 15.11 shows that the expression for K_4 is modified to read

$$K_4 = \frac{i}{c}\left(\gamma \mathbf{F} \cdot \mathbf{u} + \frac{d\phi}{dt}\right), \tag{5}$$

and as a result the fourth component of the equation of motion becomes

$$\frac{d}{dt}\left(\frac{m_0 c^2}{\sqrt{1 - u^2/c^2}}\right) = \mathbf{F} \cdot \mathbf{u} + \frac{1}{\gamma}\frac{d\phi}{dt}. \tag{6}$$

Question 15.13

Consider one-dimensional motion of a particle of constant mass m_0 subject to a constant force $F(> 0)$ in an inertial frame S. The initial conditions are $x = 0$ and $\dot{x} = 0$ at $t = 0$.

(a) Determine the velocity $u(t)$ and the position $x(t)$ of the particle relative to S at time t in terms of F, m_0 and c.
(b) Determine also $t(\tau)$, $u(\tau)$ and $x(\tau)$, where τ is the proper time.
(c) Using suitable dimensionless variables, plot the graphs of $u(t)$, $x(t)$ and $t(\tau)$. On each graph also show the corresponding Newtonian limit.

Solution

(a) The equation of motion in S is $\bigl($see (8) of Question 15.9$\bigr)$:

$$\frac{d}{dt}\frac{u}{\sqrt{1 - u^2/c^2}} = g, \tag{1}$$

where $u = \dot{x}$ and $g = F/m_0$ is a constant. By integrating (1) with respect to t and setting $u = 0$ at $t = 0$ we have

$$\frac{u}{\sqrt{1 - u^2/c^2}} = gt. \tag{2}$$

For motion along the positive x-axis the relevant solution to (2) is

$$u(t) = \frac{gt}{\sqrt{1 + (gt/c)^2}}. \tag{3}$$

By integrating (3) with respect to t and setting $x = 0$ at $t = 0$ we have

$$x(t) = \frac{c^2}{g}\left(\sqrt{1 + (gt/c)^2} - 1\right). \tag{4}$$

(b) The elapsed proper time is (see Question 15.8)

$$\tau = \int_0^t \sqrt{1 - u^2/c^2}\, dt = \int_0^t \frac{dt}{\sqrt{1 + (gt/c)^2}} = \frac{c}{g} \sinh^{-1} \frac{gt}{c}. \tag{5}$$

Thus, the time in S is given in terms of the proper time by

$$t(\tau) = \frac{c}{g} \sinh \frac{g\tau}{c}. \tag{6}$$

Equations (3), (4) and (6) yield

$$u(\tau) = c \tanh \frac{g\tau}{c}, \qquad x(\tau) = \frac{c^2}{g}\left(\cosh \frac{g\tau}{c} - 1\right). \tag{7}$$

(c) For $gt/c \ll 1$, (3) and (4) reduce to the non-relativistic (Newtonian) expressions $u = gt$ and $x = \tfrac{1}{2}gt^2$. For $gt/c \gg 1$, $u \to c$ and $x \to ct - c^2/g$. To plot the graphs of (3) and (4) we use the dimensionless quantities $\bar{u} = u/c$, $\bar{x} = gx/c^2$ and $\bar{t} = gt/c$. In terms of these, (3) and (4) are

$$\bar{u} = \bar{t}(1 + \bar{t}^{\,2})^{-1/2}, \qquad \bar{x} = (1 + \bar{t}^{\,2})^{1/2} - 1. \tag{8}$$

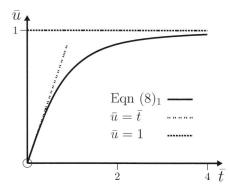
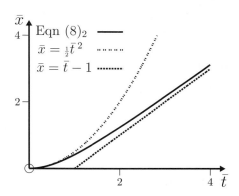

On the diagrams we have also shown the Newtonian expressions $\bar{u} = \bar{t}$ and $\bar{x} = \tfrac{1}{2}\bar{t}^{\,2}$, which are reasonable approximations only for the initial motion $\bar{t} \ll 1$ (that is, $t \ll c/g$). The limiting forms of (6) are $t \approx \tau$ (if $g\tau/c \ll 1$) and $t \approx (c/2g)e^{g\tau/c}$ (if $g\tau/c \gg 1$). In dimensionless units (6) is $\bar{t} = \sinh \bar{\tau}$, and this is shown in the diagram together with the Newtonian relation $\bar{t} = \bar{\tau}$ (absolute time). For large values of t the proper time τ is much less than t.

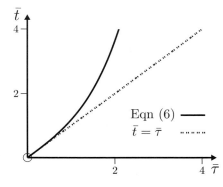

Comments

(i) The equation of motion (1) is equivalent to the statement: 'the particle undergoes constant proper acceleration g.' To see this, consider first the four-acceleration (see $(1)_2$ of Question 15.9) in the initial frame S:

$$A_\mu = \frac{1}{\sqrt{1-u^2/c^2}} \frac{d}{dt} \frac{1}{\sqrt{1-u^2/c^2}} (u, 0, 0, ic). \qquad (9)$$

A short calculation shows that $A_\mu^2 = (1 - u^2/c^2)^{-3} (du/dt)^2$. On the other hand, according to the above statement, in the instantaneous rest frame $A_\mu = (g, 0, 0, 0)$ and so $A_\mu^2 = g^2$. By equating these two expressions for A_μ^2 we obtain (1).

(ii) It follows that for a space ship accelerating in this way, an astronaut of mass m_0 would experience a weight equal to $m_0 g$.

(iii) Consider travel in a space ship for which $c/g = 1$ Earth year, and therefore $c^2/g = 1$ light year. (This requires $g \approx 9.5 \,\mathrm{m\,s^{-2}}$, and so, with regard to their weight, the astronauts would feel quite comfortable.) The following table is obtained from (4) and (7). It gives values of x and u/c (the distance travelled and the speed attained, both relative to the starting frame S) in terms of t (the time in S) and the proper time τ. We see that for a journey across the Milky Way ($x \sim 10^5$ light years), τ is considerably less than t, and (in principle) travel amounting to billions of light years relative to S is possible during the lifetime of an astronaut.

x (light years)	u/c	t (yr)	τ (yr)
10^{-1}	0.417	0.458	0.444
10^0	0.866	1.732	1.317
10^1	$1-4\times 10^{-3}$	10.95	3.089
10^5	$1-5\times 10^{-11}$	10^5	12.21
10^{10}	$1-5\times 10^{-21}$	10^{10}	23.72

(iv) Experimental tests of the theory for motion in a constant force have been performed using charged particles accelerated by a uniform electrostatic field. For these, $F = qE$ and $g = qE/m_0$. The measurements are in good agreement with the relativistic theory and confirm that, to within experimental error, the limiting value of the speed is c.[10]

Question 15.14

A one-dimensional restoring force $F = -kx$ (k is a positive constant) acts on a particle of constant mass m_0 in an inertial frame S. The particle starts from rest at $x = A$.

---

[10] W. Bertozzi, "Speed and kinetic energy of relativistic electrons," American Journal of Physics, vol. 32, pp. 551–555, 1964.

(a) Show that its velocity relative to S is given by

$$\frac{dx}{dt} = \pm c\sqrt{1 - \{1 + (\omega^2/2c^2)(A^2 - x^2)\}^{-2}}, \qquad (1)$$

where $\omega = \sqrt{k/m_o}$ is the angular frequency in the non-relativistic limit.

(b) Show that the period of the oscillation is given by

$$T(\epsilon) = \frac{2\epsilon T_o}{\pi} \int_0^1 \frac{dX}{\sqrt{1 - \{1 + \tfrac{1}{2}\epsilon^2(1 - X^2)\}^{-2}}}, \qquad (2)$$

where $T_o = 2\pi/\omega$ is the non-relativistic period and $\epsilon = \omega A/c$. Determine the non-relativistic limits of (1) and (2).

(c) Write a *Mathematica* notebook to calculate u/c and $T(\epsilon)/T_o$ from (1) and (2). Plot the graphs of 1. u/c versus x/A for $\epsilon = 0.1, 0.3, 0.6, 1.0, 2.0, 3.0$ and 10; and 2. $T(\epsilon)/T_o$ versus ϵ up to $\epsilon = 10$.

Solution

(a) The equation of motion is (see Question 15.9)

$$\frac{d}{dt}(\gamma m_o u) = -kx, \qquad (3)$$

where $\gamma = (1 - u^2/c^2)^{-1/2}$. If m_o is constant, (3) can be written as

$$\frac{1}{(1 - u^2/c^2)^{3/2}} \frac{du}{dt} = -\omega^2 x. \qquad (4)$$

The relation $du/dt = u\, du/dx$ enables us to put (4) in the separated form

$$\frac{u\, du}{(1 - u^2/c^2)^{3/2}} = -\omega^2 x\, dx, \qquad (5)$$

and therefore

$$\int_0^u \frac{u\, du}{(1 - u^2/c^2)^{3/2}} = -\omega^2 \int_A^x x\, dx. \qquad (6)$$

That is,

$$(1 - u^2/c^2)^{-1/2} = 1 + (\omega^2/2c^2)(A^2 - x^2). \qquad (7)$$

The solutions to (7) for $u = dx/dt$ are (1), where the positive (negative) sign refers to motion along the positive (negative) x-axis.

(b) By rearranging and integrating (1) we have

$$c\int_0^{\frac{1}{4}T} dt = \int_0^A \frac{dx}{\sqrt{1-\{1+(\omega^2/2c^2)(A^2-x^2)\}^{-2}}}, \qquad (8)$$

and hence (2). In the non-relativistic limit $\epsilon = \omega A/c \ll 1$ and the expansion of (1) to order ϵ^3 is

$$u(x) = \pm\omega\sqrt{A^2-x^2}\left[1 - \frac{3}{8}\frac{\omega^2 A^2}{c^2}\left(1-\frac{x^2}{A^2}\right)\right]. \qquad (9)$$

The factor multiplying the square brackets is just the familiar Newtonian result (5) of Question 4.1. Similarly, (2) becomes

$$T = \frac{2T_0}{\pi}\int_0^1 \frac{dX}{\sqrt{1-X^2}} = T_0. \qquad (10)$$

(c) The *Mathematica* notebook and graphs of (1) and (2) are:

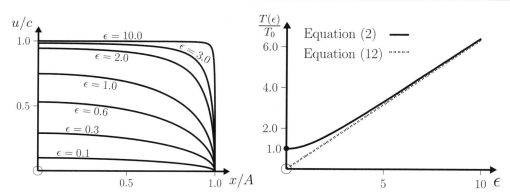

Comments

(i) $\epsilon^2 = kA^2/m_0c^2$ is twice the maximum potential energy ÷ the rest-mass energy.
(ii) In the relativistic limit $\epsilon \gg 1$ and for x not too close to A, the speed (1) is

$$u(x) \approx c(1-2\epsilon^{-4}). \qquad (11)$$

Equation (11) represents the plateaus that stretch to x/A just below 1 in graphs of u/c versus x/A, and then plummet to zero at $x/A = 1$. In the limit $\epsilon \to \infty$,

$u(x)$ resembles a step function. This behaviour is evident in the first diagram above. Here $u(x) \approx c$ in most of the interval, and so

$$T \approx \frac{4A}{c} = \frac{2\epsilon}{\pi} T_0. \tag{12}$$

The asymptote (12) is indicated by a dotted line in the second diagram. This diagram also shows the Newtonian limit $T \to T_0$ as $\epsilon \to 1$.

(iii) According to the fourth component of the equation of motion (see Question 15.10),

$$\frac{d}{dt}(\gamma m_0 c^2) = -kx\frac{dx}{dt} = -\frac{d}{dt}(\tfrac{1}{2}kx^2), \tag{13}$$

and therefore the energy $E = \gamma m_0 c^2 + \tfrac{1}{2}kx^2$ is conserved. This provides an alternative way of obtaining (7) and hence (1).

Question 15.15

(a) Show that the equation of motion for a particle of mass m_0 and charge q in a magnetic field **B** can be expressed as

$$\frac{m_0}{\sqrt{1 - u^2/c^2}} \frac{d\mathbf{u}}{dt} = q\mathbf{u} \times \mathbf{B}. \tag{1}$$

(b) Determine the trajectory for motion in a uniform magnetostatic field $\mathbf{B} = (0, 0, B)$. (Hint: Compare (1) with the non-relativistic equation in Question 7.17.)

Solution

(a) The equation of motion is (see Question 15.9)

$$\frac{d}{dt}\left(\frac{m_0 \mathbf{u}}{\sqrt{1 - u^2/c^2}}\right) = \mathbf{F} = q\mathbf{u} \times \mathbf{B}. \tag{2}$$

For motion in an electromagnetic field m_0 is a constant (see Question 15.11). Also, **F** is perpendicular to **u**, and so $\mathbf{F} \cdot \mathbf{u} = 0$. It follows from (1) of Question 15.10 that the speed u of the particle is constant. Therefore, (2) reduces to (1).

(b) Equation (1) is just the non-relativistic equation of motion (4) of Question 7.17 with the constant mass m replaced by the constant $m_0(1 - u^2/c^2)^{-1/2}$. It follows that we can obtain the trajectory for (1) by making this substitution in the non-relativistic trajectory (2) of Question 7.17. The solution corresponding to the initial conditions[‡] $\mathbf{r}_0 = 0$ and $\mathbf{u}_0 = (u_1(0), u_2(0), u_3(0))$ is therefore

[‡]In Question 7.17, **u** denotes the velocity at $t = 0$; here it represents the velocity at time t.

$$\mathbf{r}(t) = \left(\frac{u_1(0)}{\omega}\sin\omega t + \frac{u_2(0)}{\omega}(1-\cos\omega t),\right.$$
$$\left.\frac{u_1(0)}{\omega}(\cos\omega t - 1) + \frac{u_2(0)}{\omega}\sin\omega t,\; u_3(0)t\right), \quad (3)$$

where

$$\omega = \sqrt{1 - \frac{u^2}{c^2}}\,\frac{qB}{m_0}. \quad (4)$$

Comments

(i) We see that the modification to the non-relativistic motion consists entirely of the time-independent factor $\sqrt{1 - u^2/c^2}$ in (4). The relativistic angular frequency ω differs from the cyclotron frequency (the non-relativistic magnetic resonance frequency) qB/m_0 by this factor, and $\omega \to 0$ as $u \to c$.

(ii) The interpretation of (3) is the same as for the non-relativistic motion. With a suitable choice of axes (see Question 7.17), the trajectory is a helix of constant pitch with axis parallel to the z-axis and centred at $(u_2(0), -u_1(0))/\omega$. The radius and pitch of the helix are given by

$$R = \sqrt{u_1^2(0) + u_2^2(0)}/\omega, \qquad D = 2\pi u_3(0)/\omega. \quad (5)$$

Both R and D are time independent, and both become large as $u \to c$.

(iii) There is abundant experimental support for these results, coming mainly from the operation of modern particle accelerators. For example, the relativistic dependence of ω on u in (4) limits the functioning of a cyclotron to about 20 MeV for protons, and this dependence has to be incorporated in the design of machines such as the synchrocyclotron and synchrotron that function at much higher energies.

Question 15.16

Consider a particle of mass m_0 subject to an attractive inverse-square force $\mathbf{F} = -k\mathbf{r}/r^3$, where k is a positive constant (the relativistic Coulomb problem). Assume m_0 is constant.
(a) Show that the energy and angular momentum,

$$E = \gamma m_0 c^2 - k/r, \qquad \mathbf{L} = \mathbf{r} \times \mathbf{p}, \quad (1)$$

are conserved.
(b) Determine the polar equation of the trajectory in terms of k, c, L and E. (Hint: Adapt the method of Questions 8.7 and 8.8 for the non-relativistic motion.)

Solution

(a) If m_0 is constant, the relativistic equation of motion is

$$\frac{d}{d\tau}\bigl(\gamma m_0 \mathbf{u},\; (i/c)\gamma m_0 c^2\bigr) = \gamma\bigl(\mathbf{F},\; (i/c)\mathbf{F}\cdot\mathbf{u}\bigr), \quad (2)$$

where $\mathbf{u} = d\mathbf{r}/dt$ and $d\tau = dt/\gamma$ (see Questions 15.8, 15.9 and 15.10). That is,

$$\frac{d}{dt}(\gamma m_0 \mathbf{u}) = -\frac{k}{r^3}\mathbf{r}, \qquad \frac{d}{dt}(\gamma m_0 c^2) = -\frac{k}{r^2}\frac{dr}{dt}. \tag{3}$$

From $(3)_2$ and $(1)_1$ it is apparent that $dE/dt = 0$. Also, from $(1)_2$ with $\mathbf{p} = \gamma m_0 \mathbf{u}$ we have

$$\frac{d\mathbf{L}}{dt} = \mathbf{u} \times \gamma m_0 \mathbf{u} + \mathbf{r} \times \frac{d}{dt}(\gamma m_0 \mathbf{u}) = \mathbf{r} \times (-k/r^3)\mathbf{r} = 0. \tag{4}$$

(b) Just as in the non-relativistic case, conservation of $\mathbf{L}$ means that the motion is confined to a plane defined by the initial position and velocity vectors $\mathbf{r}_0$ and $\mathbf{u}_0$. Therefore, plane polar coordinates r and θ can be used (see Chapter 8). In terms of these

$$L = \gamma m_0 r^2 \frac{d\theta}{dt} = m_0 r^2 \frac{d\theta}{d\tau}. \tag{5}$$

Also, the invariance of dx_μ^2 can be written (see (2) and (3) of Question 15.8)

$$-c^2 d\tau^2 = d\mathbf{r}^2 - c^2 dt^2 = dr^2 + r^2 d\theta^2 - c^2 dt^2. \tag{6}$$

Therefore

$$\left(\frac{dr}{d\tau}\right)^2 + r^2\left(\frac{d\theta}{d\tau}\right)^2 - c^2\gamma^2 = -c^2. \tag{7}$$

By writing $(dr/d\tau) = (dr/d\theta)(d\theta/d\tau)$ and using $(1)_1$ and (5) to eliminate γ and $d\theta/d\tau$, respectively, from (7), we find

$$\frac{L^2}{m_0^2 r^4}\left(\frac{dr}{d\theta}\right)^2 + \frac{L^2}{m_0^2 r^2} - \frac{1}{m_0^2 c^2}\left(E + \frac{k}{r}\right)^2 = -c^2. \tag{8}$$

This differential equation can be converted to a standard form by setting $r = 1/w$ and then differentiating both sides of (8) with respect to θ:

$$\frac{d^2 w}{d\theta^2} + \lambda^2 w = \frac{kE}{c^2 L^2}, \tag{9}$$

where

$$\lambda^2 = 1 - k^2/c^2 L^2. \tag{10}$$

Consider first the case $\lambda^2 > 0$. Then, the general solution to (9) is $w(\theta) = (kE/\lambda^2 c^2 L^2) + A\cos\lambda(\theta - \theta_0)$, where A and θ_0 are arbitrary constants and $0 < \lambda \leq 1$. Thus, we have the polar equation of the trajectory:

$$r(\theta) = \frac{1}{(kE/\lambda^2 c^2 L^2) + A\cos\lambda(\theta - \theta_0)}. \tag{11}$$

The value of A^2 is determined by the requirement that (8) be satisfied. By substituting (11) in (8) we find, after some calculation, that

$$A^2 = \frac{m_0^2 c^2}{\lambda^4 L^2}\left(\frac{E^2}{m_0^2 c^4} - \lambda^2\right). \tag{12}$$

If $\lambda^2 < 0$ (that is, $L < k/c$) then we replace λ^2 with $-|\lambda|^2$ in (9). The solution is

$$r(\theta) = \frac{1}{(kE/\lambda^2 c^2 L^2) + A\cosh|\lambda|(\theta - \theta_0)}. \tag{13}$$

582 *Solved Problems in Classical Mechanics*

Comments

(i) The condition for bounded motion is evident from $(1)_1$ and the fact that $\gamma > 1$. We see that r will always be finite provided $E < m_o c^2$. We know that for non-relativistic motion ($\lambda \to 1_-$) bounded trajectories are ellipses (see Question 8.9). For $0 < \lambda^2 < 1$ it is clear from (11) that the perihelion (the point of closest approach to the origin) precesses. In the non-relativistic limit ($\lambda = 1$) the position vector $\mathbf{r}$ rotates through π (clockwise, say) in going from perihelion to aphelion, whereas for relativistic motion ($\lambda < 1$) $\mathbf{r}$ must rotate by more than π. Thus, the relativistic trajectory is an ellipse whose perihelion advances relative to the inertial frame (or remains stationary in an appropriate rotating frame). This is shown in the figure below, where both the precession and the sense in which the orbit is traversed are clockwise (or anti-clockwise if all arrows are reversed). The angular displacement of successive perihelia is

$$\Delta\theta = (2\pi/\lambda) - 2\pi = 2\pi\left\{(1 - k^2/c^2 L^2)^{-1/2} - 1\right\}. \tag{14}$$

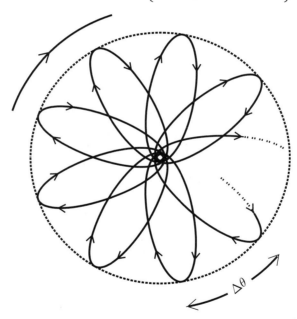

(ii) For circular (and almost circular) orbits the angular momentum and equation of motion are $L = rp = r\gamma m_o u$ and $\gamma m_o u^2/r = k/r^2$. Thus, $k/L = u$ and

$$\lambda = \sqrt{1 - u^2/c^2}. \tag{15}$$

Also, $(1)_1$ becomes

$$E = \gamma m_o c^2 (1 - u^2/c^2). \tag{16}$$

It follows from (14) and (15) that almost circular, non-relativistic orbits precess by $\Delta\theta \approx \pi(u/c)^2$.

(iii) The trajectory (13) for $\lambda^2 < 0$ is a spiral into the centre of force at the origin.
(iv) An important application of the relativistic theory was made by Sommerfeld in 1916. Using Bohr's theory he was able to extend Bohr's non-relativistic result to obtain the relativistic energy levels of the hydrogen-like atom. It later turned out that Sommerfeld's formula is in exact agreement with that given by relativistic quantum mechanics (the Dirac equation). An interesting analysis and resolution of this so-called 'Sommerfeld Puzzle' has been provided by Biedenharn.[11]

Question 15.17

Extend the analysis of Question 15.16 to a particle of constant mass m_0 moving in a spherically symmetric potential $V(r)$. Specifically, show that the energy and angular momentum are conserved, and obtain the differential equation for the inverse of the polar trajectory, $w(\theta) = 1/r(\theta)$. (That is, obtain the relativistic extension to (1) of Question 8.8.)

Solution

The force is
$$\mathbf{F} = F(r)\hat{\mathbf{r}}, \qquad \text{where } F(r) = -dV(r)/dr, \tag{1}$$

and the corresponding equations of motion for constant m_0 are
$$\frac{d}{dt}(\gamma m_0 \mathbf{u}) = \frac{1}{r} F(r)\mathbf{r}, \qquad \frac{d}{dt}(\gamma m_0 c^2) = F(r)\frac{dr}{dt}, \tag{2}$$

where $\mathbf{u} = d\mathbf{r}/dt$. The total energy
$$E = \gamma m_0 c^2 + V(r) \tag{3}$$

is conserved:
$$\frac{dE}{dt} = \frac{d}{dt}(\gamma m_0 c^2) + \frac{dV}{dt} = F(r)\frac{dr}{dt} + \frac{dr}{dt}\frac{dV}{dr} = 0, \tag{4}$$

and so is the angular momentum $\mathbf{L} = \mathbf{r} \times \gamma m_0 \mathbf{u}$:
$$\frac{d\mathbf{L}}{dt} = \mathbf{u} \times \gamma m_0 \mathbf{u} + \mathbf{r} \times \frac{d}{dt}(\gamma m_0 \mathbf{u}) = \mathbf{r} \times \frac{1}{r} F(r)\mathbf{r} = 0. \tag{5}$$

The analysis leading to (8) of Question 15.16 is valid for any spherically symmetric potential. Thus, we can replace the Coulomb potential $-k/r$ in this equation by $V(r)$:
$$\frac{L^2}{m_0^2 r^4}\left(\frac{dr}{d\theta}\right)^2 + \frac{L^2}{m_0^2 r^2} - \frac{1}{m_0^2 c^2}\{E - V(r)\}^2 = -c^2. \tag{6}$$

---

[11] L. C. Biedenharn, "The 'Sommerfeld Puzzle' revisited and resolved," Foundations of Physics, vol. 13, pp. 13–34, 1983.

In (6) we make the substitution $r = 1/w$ and then differentiate (6) with respect to θ. This yields

$$\frac{d^2w}{d\theta^2} + w = -\frac{1}{c^2L^2}\frac{1}{w^2}\left\{E - V\left(\frac{1}{w}\right)\right\}F\left(\frac{1}{w}\right), \qquad (7)$$

which is the relativistic extension to (1) of Question 8.8.

Comments

(i) In the non-relativistic limit $(E \approx m_o c^2 \gg |V|)$, (7) reduces to (1) of Question 8.8.

(ii) For a Coulomb potential $(V = -kw$ and $F = -kw^2)$, equation (7) has the simple form given in (9) of Question 15.16.

(iii) But, in general (7) is not a simple differential equation. Even the isotropic harmonic oscillator $(V = -k/2w^2$ and $F = -k/w)$ poses the challenging equation

$$\frac{d^2w}{d\theta^2} + w = \frac{kE}{c^2L^2}\frac{1}{w^3} - \frac{k^2}{2c^2L^2}\frac{1}{w^5}, \qquad (8)$$

whose non-relativistic limit,

$$\frac{d^2w}{d\theta^2} + w = \frac{m_o k}{L^2}\frac{1}{w^3}, \qquad (9)$$

was solved in Question 8.10.

Question 15.18

A rocket having initial mass M is accelerated in a straight line in free space by exhausting material at a constant speed v_e relative to the rocket. Let u be the speed of the rocket relative to its initial rest frame S when its rest mass has decreased to m. Show that

$$\frac{u}{c} = \frac{1 - (m/M)^{2v_e/c}}{1 + (m/M)^{2v_e/c}}. \qquad (1)$$

(Hint: Use conservation of energy and momentum in the instantaneous rest frame S'.)

Solution

Consider the rocket at proper time τ when its rest mass is m and its speed relative to S is u. At this instant the rocket is instantaneously at rest in the inertial frame S'. At

Relative to S' at $\tau + d\tau$:

$v_e \quad dm_g \qquad m+dm \quad du'$

proper time $\tau+d\tau$ its rest mass is $m+dm$ (where $dm < 0$) and it has acquired velocity du' (directed to the right) relative to S'; exhaust 'gases' of rest mass dm_g (> 0) have been expelled with velocity v_e (directed to the left) relative to S'. The energy and momentum in S' at time τ are mc^2 and 0, respectively, and conservation of energy and momentum relative to S' during the interval $d\tau$ require

$$\frac{(m+dm)c^2}{\sqrt{1-du'^2/c^2}} + \frac{dm_g c^2}{\sqrt{1-v_e^2/c^2}} = mc^2 \tag{2}$$

and

$$\frac{(m+dm)du'}{\sqrt{1-du'^2/c^2}} - \frac{dm_g v_e}{\sqrt{1-v_e^2/c^2}} = 0. \tag{3}$$

To first order in small quantities these equations simplify to

$$dm_g = -\sqrt{1-v_e^2/c^2}\, dm, \qquad m\, du' = dm_g v_e (1-v_e^2/c^2)^{-1/2}. \tag{4}$$

Therefore

$$m\, du' = -v_e\, dm. \tag{5}$$

We can eliminate du' from (5) in favour of du (the velocity increment relative to S) by using the velocity transformation between S and S' (see Question 15.3). The velocity $u+du$ relative to S is the combination of the velocity u of S' relative to S and the velocity du' of the rocket relative to S':

$$u+du = \frac{u+du'}{1+u\, du'/c^2} \approx u + (1-u^2/c^2)du'. \tag{6}$$

Equations (5) and (6) yield a differential equation for $u(m)$:

$$\frac{du}{1-u^2/c^2} = -v_e \frac{dm}{m}. \tag{7}$$

Integration of the left-hand side of (9) between 0 and u, and the right-hand side between M and m gives

$$\tfrac{1}{2}c\ln\left(\frac{1+u/c}{1-u/c}\right) = -v_e \ln\frac{m}{M}, \tag{8}$$

and hence (1).

Comments

(i) The limiting form of (1) in the non-relativistic limit $u/c \to 0$ is independent of c. To see this, write (8) as

$$\frac{m}{M} = \left(\frac{1-u/c}{1+u/c}\right)^{c/2v_e} = \left[\left(\frac{1-\alpha^{-1}}{1+\alpha^{-1}}\right)^\alpha\right]^{u/2v_e}, \tag{9}$$

where $\alpha = c/u$. The quantity in square brackets tends to e^{-2} as $\alpha \to \infty$ and therefore $u \to v_e \ln M/m$, which is independent of c and the same as the Newtonian result in Question 11.26.

(ii) Graphs of u/c versus v_e/c obtained from (1) for various values of m/M are shown. In general, u/c is largest (for given m/M) when $v_e = c$. Also shown are graphs of u/c versus m/M for various values of v_e/c. In these, $u/c \to 1$ as $m/M \to 0$.

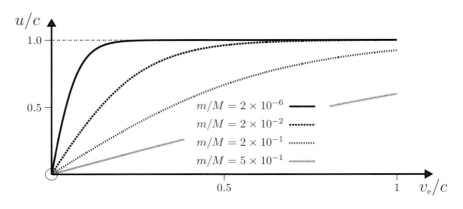

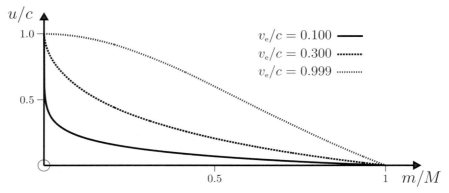

Question 15.19

With the analysis of Questions 15.1 and 15.2 in mind, comment on the question: 'Why did it take so long to improve on Newtonian relativity – that is, to discover the V^2-Lorentz transformation?'

Solution

Question 15.1 shows how directly and simply one can obtain this transformation, based on: 1. A willingness to forego absolute space by writing $x' = \gamma(v)(x - vt)$, where $\gamma(v)$ is an even function and not necessarily equal to unity. 2. The relativity principle, which requires the crucial form invariance $x = \gamma(v)(x' + vt')$. From these two equations it follows that time is relative – see (6) of Question 15.1. That is, relative space + relativity principle $\Rightarrow$ relative time. When put this way it seems surprising that it took about two hundred and fifty years to reveal the Lorentz transformation as a richer alternative to the Galilean transformation. After all, there was much criticism of absolute space and time (albeit with a philosophical bent)[12] and, later, concern with intricate properties of a mysterious 'ether', which was believed necessary for propagation of electromagnetic waves.[1] These endeavours involved far more (and fruitless) labour than the two steps detailed above. The main impediment was probably the conceptual leap involved in raising the relativity principle from the status of a curiosity in Newtonian dynamics to a principle that also constrains the space-time transformation, as in step 2. above. By contrast, the associated mathematical development is not arduous, compared especially to mathematical achievements during these centuries.

If Newton had carried out his original intention and presented the relativity principle as a fundamental law, or if Huyghens had been more insistent on the significance of this principle, (see Question 14.9) it may have triggered a very different development of the theory. A latter-day Lagrange could well have obtained the V^2-Lorentz transformation, as in Question 15.1, and either he or others (such as Gauss or Hamilton) could have begun to extract the startling consequences of this result, including a tentative identification of the speed of light in vacuum with the universal (frame-independent) speed V.

Perhaps this chronology has been (or will be) followed by some other developing civilization in the universe, so that their theory of space-time for inertial frames developed independently of other physical theories – notably, the theory of classical electromagnetism. And their Michelson–Morley experiment confirmed an anticipated result and impressed on them the power of inductive reasoning: that extension of the notion of form invariance (manifest in the non-relativistic limit) to the space-time transformation (and beyond) could produce such deep results.

As a further example of such reasoning we mention the extension of the Planck–Einstein relations ($E = \hbar\omega$ and $\mathbf{p} = \hbar\mathbf{k}$) for light quanta (photons) to apply to all particles. For a non-relativistic particle of mass m in a potential V, this implies the dispersion relation $\hbar\omega = \hbar^2\mathbf{k}^2/2m + V$ and so, by inspection, the time-dependent Schrödinger equation of quantum mechanics.[13]

> *To know what questions may reasonably be asked*
> *is already a great and necessary proof of sagacity*
> *and insight. (Immanuel Kant)*

---

[12] See, for example, A. Danto and S. Morgenbesser, *Philosophy of science*. New York: Meridian, 1960.
[13] I. G. Main, *Vibrations and waves in physics*. Cambridge: Cambridge University Press, 3rd edn, 1993. Chap. 15.

Appendix

The following questions provide suitable material for project work in computational physics. They require computer calculations in their solution, or for analysis of the solution, or both. Those incorporating the interactive Manipulate function, which enables one to observe and control motion on a computer screen, are highlighted below.

1. One dimensional motion of a particle subject to a time-harmonic force (Question 3.3).
2. One-dimensional motion in an attractive inverse-square field (Question 3.5).
3. Free fall of a sphere through an atmosphere of varying density (Question 3.14).
4. Damped linear oscillator without driving and with driving (Questions 4.5 to 4.7).
5. Phase plot and Poincaré section of a damped, driven oscillator (Question 4.15).
6. Periods of some non-linear oscillations (Question 5.18).
7. Field lines and equipotentials for a non-uniform **B** field (Question 5.26).
8. Animation of the trajectory of a ball bouncing across a horizontal surface (Question 6.5).
9. The variable-length (Lorentz) pendulum (Question 6.18).
10. Projectile and target (Question 7.4).
11. Projectile with quadratic drag and variable atmospheric density (Questions 7.8 and 7.9).
12. Sliding on a rough sphere (Question 7.12).
13. Sliding on a smooth wire (Questions 7.13 to 7.15).
14. Trajectory in a non-uniform **B** field (Question 7.23 and 7.24).
15. Lissajous figures for an anisotropic harmonic oscillator (Question 7.26).
16. Energy diagrams for the Yukawa potential (Question 8.6).
17. Trajectories in an inverse-cube force (Question 8.11).
18. Trajectories in a perturbed Coulomb potential (Question 8.12).
19. Time-dependent polar coordinates for motion in a Coulomb potential (Questions 8.18 and 8.19).
20. Rocket motion near the Earth (Question 8.20).
21. Animation of trajectories for the gravitational two-body problem (Questions 10.9 and 10.10).
22. One-dimensional coupled oscillator (Question 10.12).
23. Two interacting particles on a hoop: effective potential, and symmetry breaking, etc. (Question 10.14).

24. Two interacting particles on a hoop: dynamical properties and dynamic display (Question 13.16).
25. Two interacting charges moving in a plane perpendicular to a uniform **B** field (Questions 10.16 to 10.19).
26. Earth's orbit with a time-dependent gravitational constant $G(t)$ (Question 10.20).
27. Gravitational three-body problem: equilateral triangle solutions and their animation (Question 11.6).
28. Restricted three-body problem: solutions and their animation for the heliocentric, geocentric and centre-of-mass frames (Question 11.7).
29. Restricted three-body problem: effective potential and Lagrange points (Question 11.8).
30. Restricted three-body problem: bounded motion around a Lagrange point, and its animation (Question 11.9).
31. Equilibrium configurations of discrete and continuous arrays (Questions 11.12 and 11.13).
32. Rotation curves for a galaxy with a halo (Question 11.20).
33. Rocket take-off from Earth (Question 11.28).
34. Variable-mass oscillator (Question 11.30).
35. Dynamics of a falling/sliding rod (Question 12.15).
36. Dynamics of a spinning golf ball (Question 12.29).
37. Dynamics of a spinning tennis ball (Question 12.30).
38. Fourier approximations for $\ddot{x} + \gamma x^3 = 0$ (Questions 13.5 and 13.6).
39. Fourier approximations for $\ddot{x} + \alpha x + \gamma x^3 = 0$ (Questions 13.7 and 13.8).
40. Fourier approximations for $\ddot{x} + \alpha x - \beta x^2 = 0$ (Questions 13.9 and 13.10).
41. Dynamics of a damped, driven Duffing oscillator (Questions 13.11 and 13.12).
42. Oscillator with quadratic drag (Question 13.13).
43. Bead, loop and spring: effective potential and symmetry breaking (Question 13.14).
44. Bead, loop and spring: motion and its animation (Question 13.15).
45. The driven rigid pendulum and chaos (Question 13.17).
46. Chaos of a compass needle in an oscillatory **B** field (Question 13.18).
47. Ball bouncing in a wedge: 'gravitational billiards' (Question 13.19).
48. Pivot-driven pendulum. Vertical driving. The inverted pendulum (Questions 13.20 to 13.22).
49. Pivot-driven pendulum. Horizontal driving (Questions 13.20 and 13.23).
50. Relativistic motion in a uniform field (Question 15.13).
51. Relativistic one-dimensional harmonic oscillator (Question 15.14).
52. Relativistic Coulomb problem (Question 15.16).
53. Relativistic rocket (Question 15.18).

Index

Abraham–Lorentz equation, 58
absolute space, 527
absolute time, 527
accelerated frame
   Einstein on, 532
   suspended bob in, 535
   tethered balloon in, 535
acceleration, 2
   effective gravitational, 47
   gravitational, 14, 18, 24, 26, 31, 45, 55,
        157, 168, 330, 434
   in plane polar coordinates, 217
   origin independence of, 520
   radial, 217
   transformation, 542, 544
   transverse, 217
accidental degeneracy, 276
action integral, 6
age of universe, 536
Aharonov–Bohm effect, 122
air-suspension gyroscope, 438–443
Ampère's law, 126, 528
amplitude, 22, 63, 103
   of damped oscillations, 86
   of forced oscillations, 73
amplitude jumps, 483
amplitude-dependent frequency, 458–485
angular frequency, 63, 66, 108, 303, 531
angular momentum, 127, 144
   about centre of mass, 329
   absolute, 330
   canonical, 541–542
   conservation of, 145, 146, 290, 291, 328,
        424, 540, 583
   in polar coordinates, 218
   'mixed' definition of, 145
   Newtonian, 541–542
   of rigid body, 403–408
   of two-body system, 292
   orbital, 330
   rate of change of, 291, 328
   relative, 330
   spin, 329, 424, 436
   total, 290, 316, 328, 432
angular velocity, 8, 431
   critical, 309
   of a rigid body, 402
   of Earth, 441, 444
anharmonic oscillations, 454–517

anharmonic oscillator, 103, 105, 456
anti-gravity glider, 19
aphelion, 245, 266, 582
Archimedes's principle
   in accelerated frame, 535, 536
Aristotle, 318
attractor, 90
axis of symmetry, 408, 431
azimuthal force, 9, 544–546

backspin, 448
Barton, 557
bead, loop and spring, 488–492
beads
   interacting on a hoop, 307–313, 492–497
beat frequency, 76
Bernoulli
   solution for inverse-cube force, 239
Bertrand's theorem, 251, 284
Bessel functions, 152
Biedenharn, 583
big bang, 536
billiard table, 12, 134
black hole, 377
boat crossing a stream, 172–173
Bohr, 200
Bohr magneton, 29
Bohr's theory, 200
bound motion, 33, 43, 63, 64, 105–107, 189,
        223, 252, 315, 455, 547
broken symmetry, 312
Buckingham, 523
bulk modulus, 78, 79
   adiabatic, 79
   isothermal, 79
Burko, 159
Butikov, 341

Calogero, 362, 363
   model, 366
cart wheels
   motion on an inclined plane, 445–447
Cartesian coordinates, 1
Cauchy–Riemann equations, 123
causality, 59
Cavendish, 187
central force, 145
central isotropic force, 93, 146, 290
centre of gravity, 331

uniqueness of, 331
centre of mass, 5, 287, 288, 314, 326
   equation of motion for, 289, 314
   motion in a non-inertial frame, 546
   of a continuous system, 327
   of a rigid body, 400, 410
centre of momentum, 288
centre-of-mass frame, 290, 316, 337, 401, 421
centre-of-mass motion, 379
centre-of-mass trajectory, 315, 321
centrifugal force, 9, 218, 309, 316, 343, 544–546
chain falling, 393–397
Chambers, 156
Chandrasekhar, 5, 152, 332
chaos and chaotic systems, 341, 497–505
circular orbit, 260
   conditions for stability, 248
   in momentum space, 267
   stable, 221, 223, 227, 244
   unstable, 227
classical electron radius, 199
closed orbits, 251
   condition for, 281
coefficient of kinetic friction, 85, 175, 423, 429
coefficient of restitution, 130–142
coefficient of sliding friction, 390
coefficient of static friction, 85, 417, 427
collision
   elastic, 130
   inelastic, 130
   totally elastic, 130
   two-body, 137–142
compass needle
   in oscillatory **B** field, 500–503
Compton wavelength, 29
conic sections, 231, 266
conservation of angular momentum, 145, 146, 291, 328, 424, 583
   and Lagrangian formulation, 540
conservation of energy, 42, 95, 146, 390, 395, 421, 427, 583, 585
   and Lagrangian formulation, 398
conservation of momentum, 127–129, 138, 327, 585
   Lagrangian formulation, 525
conservative force, 95, 368
   conditions for, 96–98
conserved tensor
   for harmonic oscillator, 270–276
conserved vectors, 263–270
constants of the motion, 265, 272
constraint equations, 446
continuum array
   equilibrium of, 360
continuum solutions, 361
coordinate space, 267
Coriolis force, 9, 343, 347, 544–546, 549, 553–554
   effect on gyroscope, 441

Coulomb potential, 113, 222, 230, 263, 284
   energy diagram for, 222
   perturbed, 243, 253, 279
   time-dependent, 255–259
   trajectory, 255–259
   unique property of, 285
Coulomb problem, 263–270
   conserved vectors for, 263–270
   in Cartesian coordinates, 212–215
   relativistic, 580
   two-body, 293
Coulomb repulsion, 316
Coulomb's law, 14
coupled oscillators, 302–304
covariance, 528, 570
Crane, 555
critical damping, 70
critical velocity, 178, 547
cycloid, 68, 191, 445
   curtate, 193
   prolate, 193
cyclotron, 189, 198
cyclotron frequency, 189, 190, 202, 315, 580

damped oscillator, 90, 487
damping
   critical, 70
   negative, 87
   over, 69
   under, 69
   weak, 71, 72
Darboux, 91
dark matter/energy, 374, 536, 537
de Broglie wavelength, 355
deflection
   of falling object, 553
degeneracy, 275, 276
   accidental, 276
   lifting of, 276
degenerate minima, 310
density, 24, 26, 47, 53
   Planck, 28
   variable, 55
determinism, 11, 13, 31, 499
differential
   imperfect, 99
   perfect, 98, 99
dimensional analysis, 20, 21, 23, 25, 27
dimensionless arguments, 20
Dirac delta function, 59, 367
direct problems, 247
dislocation pile-ups, 357
dissipation, 52, 100
double pendulum
   normal modes of, 305
doubly periodic motion, 251
drag, 42, 43, 48, 53, 55, 56, 68, 168, 170, 382, 448
drag coefficient, 44, 380, 448–452
Dresden, 499

drift, 205
  in electromagnetic field, 192
driven pendulum
  chaos in, 497–500
duality transformation, 279–280
Duffing oscillator, 468
  damped driven, 481–487
dynamic pressure, 384

Earth
  angular velocity of, 441, 444
  atmosphere of, 377
  circumference of, 186
  g of, 552
  magnetic field of, 205
  motion about Sun, 322, 532–534
  motion near, 40, 54, 55
  rocket lift-off from, 382–384
  rocket motion near, 260
  tides of, 534
eccentric anomaly, 258, 280, 332
eccentricity, 231, 259
effective gravitational acceleration, 47, 535, 550–552
effective gravity, 531
effective mass, 308
effective potential, 218, 280, 308, 316, 342, 511, 513, 516
  for Coulomb problem, 222
  for inverse-cube force, 242
  for isotropic harmonic oscillator, 220
  for perturbed Coulomb potential, 243, 253
  for Yukawa, 224–228
Einstein, 10, 528
  on creating relativity, 532
Einstein's equivalence principle, 532
Einstein's relativity principle, 528, 557
elastic collision, 130
electric dipole moment
  origin dependence of, 522
electric field
  inside a shell, 371
electric quadrupole moment
  origin dependence of, 522
electronic charge, 49
electrostatic field, 368
energy, 63, 72, 76, 81, 91, 94, 108, 146, 155
  and the Hamiltonian, 398
  conservation of, 42, 95, 146, 390, 395, 421, 427, 583, 585
  critical, 104, 106
  heat, 99
  in polar coordinates, 218
  kinetic, 76
  loss for a raindrop, 380
  loss in, 393
  loss in a collision, 141–142
  of a rigid body, 404, 405
  potential, 76, 94, 102
  radiated, 196

  relativistic, 572, 579, 580, 583
  internal, 99
energy diagrams, 454–456
  for an inverse-square potential, 243
  for Coulomb potential, 222–223
  for isotropic harmonic oscillator, 221
  for one-dimensional motion, 103–108
  for power-law forces, 248
entropy, 99
epicycles, 318
epitrochoid, 318
equation of motion
  four-vector form, 570
  of centre of mass, 289, 314
  with variable mass, 378
equation of state, 355
equations
  constraint, 446
  two-body, 286
equations of motion, 3, 4, 8, 286
  for centre of mass, 326
  Galilean invariance of, 528
  in a non-inertial frame, 544
  in a rotating frame, 543–545
  in polar coordinates, 229
  multi-particle, 326
  third-order, 57
equilibrium
  and classical polynomials, 358
  distribution function, 361
  neutral, 311
  of a rigid body, 416
  of multi-particle systems, 355–361
  stable, 65, 66, 103, 104, 108, 309
  unstable, 65, 103, 104, 108, 309
equipotential curves, 123–124
equivalence principle, 22, 158, 534–536
  Einstein's, 532
  weak, 15, 532
Eratosthenes
  and Earth's circumference, 186–187
escape velocity, 260, 376, 377
  for two-body problem, 376
Euclidean space, 1, 527
Euler, 335, 348, 399
Euler's equations, 407
exhaust velocity, 379
expansion of universe, 536
experiment: 'most beautiful' in physics
  Eratosthenes', 187
  Foucault's pendulum, 555
  Galileo's rolling-ball, 31
  Millikan's oil-drop, 50
extensive variables, 17
external force, 286

falling chain, 393–397
falling rod, 426
Faraday's law, 99, 528
field

electrostatic, 368
gravitational, 367
irrotational, 99, 126
lines, 123
solenoidal, 99, 120, 125
uniform static, 121
vector, 120, 125
Fitzgerald, 527
'flat-Earth' approximation, 159
flux quantum, 29
forbidden region, 103
force, 2, 9, 44
  azimuthal, 9, 544–546
  central, 145
  central conservative, 100
  central interparticle, 328
  central isotropic, 93, 101, 113, 146, 229, 247
  central power-law, 352
  centre of, 233, 236
  centrifugal, 9, 218, 309, 316, 343, 544–546
  conservative, 6, 95, 98–101, 115, 116
  constant, 30, 108
  Coriolis, 9, 343, 347, 544–546, 549, 553–554
  Coulomb, 146
  dissipative, 100
  dual pairs of, 285
  effective, 9
  electromagnetic, 121
  electrostatic, 48, 99, 120
  exponentially decaying, 40
  external, 3, 286, 326
  fictitious, 9, 531
  friction, 68, 85
  frictional, 42–44, 47, 99
  gravitational, 13, 146, 532–533
  harmonic, 73
  impure, 573
  inertial, 9
  interparticle, 3, 286, 326, 352
  inverse-cube, 34, 242, 247
  inverse-square, 36, 212, 246
  linear, 60, 63, 68, 85, 102, 115, 208, 247
  linear time-dependent, 32
  Lorentz, 549, 570
  magnetic, 189
  Magnus, 448
  non-conservative, 95, 98–101
  non-inertial, 9
  oscillatory, 33, 107
  position-dependent, 42, 94
  pure, 573
  series expansion of, 66
  tidal, 533
  time-dependent, 34, 101
  translational, 9, 289, 507, 531–533, 544–546
  uniform gravity plus friction, 50, 53
  velocity dependent, 54
forces
  interparticle, 355

form invariance, 528, 561, 563–564
  of four-tensor equations, 567
Foucault, 555
Foucault pendulum, 554, 556
four-acceleration, 569
four-force, 570
four-momentum, 569
four-tensor, 567
four-tensor equation, 567
four-vector, 567
four-velocity, 569
Fourier approximations
  for non-linear oscillations, 462–480
frame of reference, 2
  centre of mass, 337
  geocentric, 337
  heliocentric, 337
  inertial, 2, 292
  non-inertial, 7, 289, 309, 506
  primary, 2
  rotation of, 537
free fall
  greater than g, 396
free motion
  on horizontal plane, 556
freely falling particle, 552–553
frequency, 63
  amplitude-dependent, 458–485
  angular, 63
  natural, 311
friction, 68, 85, 99, 487, 506

galactic halo, 372, 374
  effect on motion, 372–374
Galilean invariance
  of Newton's second law, 527
Galilean transformation, 526–530, 559, 561, 563, 570, 587
  and electromagnetism, 528
Galileo, 518
Galileo's experiment
  falling objects, 16, 54
  rolling balls, 31
  two-body correction, 378
Galileo's law of free fall, 15, 378
gauge function, 124
gauge invariance
  global, 121
  local, 121
Gauss's law, 120, 528
Gauss's theorem, 117, 118
gegenschein, 348
generalized coordinates, 6
generalized forces and momenta, 7
generalized velocities, 6
geocentric frame, 337
geometric degeneracy, 276
geometric orbit, 219
geometric symmetry, 276
golf ball and motion of, 448–451

Gonzaléz, 419
Gratton, 419
gravitational acceleration, 14, 18, 24, 26, 31, 45, 55, 68, 157, 168, 330
G varying, 322
gravitational billiards, 504
gravitational field, 367
   of a disc, 371
   of a shell, 370
   of a uniform sphere, 368–369
   of an annulus, 371
gravitational force, 532–534
gravitational mass, 14, 16, 158, 162
gravitational potential, 367
   of a disc, 371
   of a shell, 370
   of a uniform sphere, 368–369
   of an annulus, 371
gyration, radius of, 427
gyrocompass, 443
gyroscope, 438–443

Hall effect, 29, 322, 550
Hamilton, 269
Hamilton vector, 264–270
Hamilton's equations, 398
Hamiltonian, 91, 190, 398
Hamiltonian and energy, 398
harmonic and inverse-cube interactions, 364
harmonic approximation, 66, 106
harmonic oscillator, 60, 65, 66, 87, 188, 190
   anisotropic, 115, 279
   anisotropic two-dimensional, 210
   conserved tensor for, 270–276
   coupled, 302–304
   damped, 68–71, 78
   damped, driven, 73–76, 83
   isotropic, 115, 277
   isotropic and energy diagram for, 221
   isotropic three-dimensional, 209
   isotropic two-dimensional, 209
   relativistic, 576
harmonic resonance, 485
heliocentric frame, 337
Helmholtz's theorem, 125
Hermann
   solution for inverse-square force, 235
Hermann–Bernoulli–Laplace vector, 264
Hermite polynomial, 358, 360, 366
hidden symmetry, 215, 246, 251, 276
Hilbert transform, 361
Hipparchus, 318
hodograph, 267–269, 275
homogeneity of space
   and conservation of momentum, 526
homogeneity of time
   and conservation of energy, 398
horseshoe orbits, 352
Hubble's constant, 536
Hubble's law, 536

Huyghens, 68
   and the relativity principle, 529, 587
hydrogen atom
   classical, 198
   lifetime of, 198, 199
   quantum mechanical, 200, 269
   relativistic, 583
hysteresis, 483, 485

ideal gas, 355
improper length and time, 565
impure force, 573
indeterminism, 11
induced stability, 510, 511
   condition for, 513
inelastic collision, 130
inertia tensor
   origin dependence of, 521–522
inertial frame, 2, 5, 127, 292, 525, 539, 558
   special properties of, 9
inertial mass, 14, 16, 158, 162
initial conditions, 57, 60, 68
   and Coulomb problem, 215
   sensitivity to, 34, 64, 179, 189, 318, 341
initial velocity
   sensitivity to, 11, 12
intensive variables, 17
interacting charges
   in a **B** field, 314–322
internal energy, 355
interparticle force, 286, 352, 355
   central isotropic, 290
intrinsically non-linear oscillations, 456, 461
invariance, 10
   breaking of, 440, 441
   partial, 526, 540
   under rotations, 539
invariant interval, 566
inverse problems, 247
inverse-cube interparticle forces, 362
inverted pendulum, 510, 511, 513, 514
irrotational field, 99, 120, 126
isochronous oscillations, 68
isolated system, 525, 539
isothermal atmosphere, 55, 172

Jacobi constant, 348
Jauch–Hill–Fradkin tensor, 271
jerk, 56, 57, 83
jet, 379
Joule, 355
jumps in amplitude, 483

Kapitza, 511
Kepler problem, 215, 258, 263–270, 325
Kepler's
   equation, 258, 332
   first law, 233
   second law, 146
   third law, 27, 252

Kepler's laws
    for two-body problem, 302, 341
kinetic energy, 6, 52, 76, 92, 130, 134, 189, 193, 354
    loss in, 425
    loss per cycle, 196, 198
    of rigid body, 403–405
    of two-body system, 292
    relativistic, 196, 571
Kolmogorov, 23
Kronecker delta function, 362
Kuhn
    structure of scientific revolutions, 201

laboratory frame, 440, 443, 551, 552, 556
ladder, equilibrium of, 417–419
Lagrange, 5, 6, 335, 348
Lagrange formulation
    usefulness of, 396
Lagrange points, 342–348
    motion near, 346–352
Lagrange's equations, 6, 19, 151, 313, 394, 395, 397, 525, 530, 540, 545
Lagrangian, 6, 17, 151, 308, 309, 313, 393, 394, 397–398, 530
    and conservation of angular momentum, 539–540
    and conservation of energy, 398
    and conservation of momentum, 525–526
    for damped oscillator, 90
    generalized, 91
    in a rotating frame, 545
    in a translating frame, 530
    rotational invariance of, 539
    translational invariance of, 525
    two-particle, 540, 541
Laguerre polynomials, 357
Landau levels, 190
Landau theory, 312
Laplace, 377
Laplace vector, 264–270
    first, 264
    second, 264
Larmor effect, 549
Larmor frequency, 549
Larmor's theorem, 549
Lawrence, 198
Lax equation, matrices, 364
LC circuit, 66, 156
Legendre polynomial, 358
Legendre transformation, 398
length contraction, 564
length Planck, 28
Lennard-Jones potential, 114
Levi-Civita tensor, 271, 408
Lévy-Leblond, 561
lifetime
    of classical hydrogen atom, 198, 199
lift, 448
lift coefficient, 448–452

limit cycle, 90
linear 'gravity', 331
    Newton's method, 332
linearity
    of space-time transformations, 559
Lissajous figures, 211
Littlewood, 152
Livingstone, 198
logarithmic decrement, 80
Longuet-Higgins, 523
Lorentz, 527
    force, 549, 570
    function, 77
    transformation, 570
Lorentz's pendulum problem, 152
Lorenz, 499
LRC circuit, 75
Lyapunov exponents, 500

Mécanique Analytique, 6
magnetic mirror, 205
magnetostatic field, 126
    non-uniform, 201
    uniform, 187
Magnus effect, 450
Magnus force, 448
major axis, 233, 266
mass, 2, 9, 19
    active gravitational, 13
    additive property of, 17
    constant, 571–572
    dipole, 18
    effective, 308
    four-scalar nature, 569
    gravitational, 14, 16, 158, 162
    inertial, 14, 16, 158, 162
    invariance of, 527
    negative, 18
    passive gravitational, 13
    Planck, 28
    proper (or rest), 569
    reduced, 142, 289
    relative, 16
    standard, 21
    variable, 156, 378–390, 573
massless particles, 114, 562
Mathieu's equation, 508
Maxwell, 4, 23
Maxwell's notation, 21
Mehra, 200
Millikan's oil-drop experiment, 49
Minkowski space, 566
minor axis, 233, 266
mirror field, 205
mirror point, 204
mode
    anti-symmetric, 303, 304, 307
    symmetric, 303, 304, 307
modulation of gravity, 507
moment of inertia, 403–415

momenta
  canonical, 541
  Newtonian, 541
momentum, 2, 3, 127
  conservation of, 127–129, 138, 327, 525, 585
  of two-body system, 292
  origin independence of, 520
  total, 287, 326, 378
momentum space, 267
Moon, 285, 341, 348, 377, 534
  a headache for Newton, 336
  effect on Earth's tides, 534
motion
  bounded, 33, 63, 64, 105–107, 189, 223, 315, 455, 547
  unbounded, 33, 43, 64, 103, 106–107, 189, 223, 455, 547
multi-particle systems, 325

N-body problem
  reduction of, 352–353
neutral equilibrium, 311
Newton, 2, 4, 5, 159, 247, 325, 332, 336, 341, 518, 529
  and breaking of hidden symmetry, 285
  and duality, 285
  and Galilean invariance, 529
  and Kepler's equation, 258
  and origin independence, 524, 529
  and pendulum experiments, 16
  and rotational invariance, 529
  and tennis, 451
  and the relativity principle, 529, 587
  on freely falling objects, 553
Newton's law of impact, 129–142
Newton's law of universal gravitation, 13, 285
Newton's laws
  first law, 2, 5
  second law, 2, 5, 9, 16
  third law, 2, 13, 16, 287, 291, 326, 525, 526
Newton's second law
  Galilean invariance of, 527
  origin independence of, 523
  rotational invariance of, 538
Newtonian relativity, 519, 536, 543, 557
Nobel prize, 50
Noether's theorem, 128
non-inertial effect, 443
non-inertial frame, 7, 289, 309, 316, 444, 506, 531, 548
non-linear equations, 356
non-linear oscillators, 454–517
normal frequencies, 304, 307
normal modes, 304–306
nuclear explosion, energy of, 27
nutation, 434, 436, 447

oblate body, 432
orbit
  closure of, 252–254
orbit equation, 272
orbital angular momentum, 330
  origin dependence of, 520
orbits
  closed, 245, 316
  open, 245
order parameter, 312, 490
origin dependence, 518–523
origin independence, 145, 518–526
orthogonal transformation
  four-dimensional, 565
orthogonality relations, 537, 565
oscillations, 22, 33
  isochronous, 68
  linear, 60
oscillator
  critically damped, 70
  damped, 68–71, 78
  damped, driven, 73–76, 83
  Duffing, 468, 481–487
  harmonic, 60, 65, 66
  overdamped, 69
  relativistic, 576
  underdamped, 69, 78–80, 87
  weakly damped, 78
  with variable mass, 156, 385–390
oscillator potential, 284
oscillators
  coupled, 302–304
  non-linear, 454–517
overdamping, 69

parallel-axis theorem, 409
particle, 1
Pauli, 269
Peck, 447
pendulum, 15, 65, 105
  accelerated, 531
  cycloidal, 68
  double, 305–307
  driven, 497–500
  energy diagram for, 105
  equilibrium of, 106
  flip-flop, 87
  in laboratory frame, 554
  inverted, 510, 511, 513, 514
  Lorentz, 152
  Newton's work on, 15
  of varying length, 151–156
  period of, 15, 21, 23, 65, 111
  pivot-driven, 506–517
  potential energy for, 105
Penrose, 528
perihelion, 582
period, 15, 21, 26, 63, 65, 78, 81, 103, 108, 180, 228, 374, 462, 468, 474
  amplitude-independent, 66
  angular motion, 249–252
  integral formula for, 109

of a gyrocompass, 444
of a gyroscope, 441
of anharmonic oscillations, 110, 454–517
radial motion, 249–252
periods
   for Coulomb potential, 249–252
   for isotropic oscillator potential, 249–252
   for perturbed Coulomb potential, 253
Pesic, 285
phase, 63
phase space, 90
phase trajectory or phase portrait, 87, 493
Planck units, 28–29
Planck's constant, 27, 355
plane polar coordinates, 215, 216
Poincaré, 499, 527
Poincaré section, 87, 89, 500
Poisson's equation, 367, 368
polar aurora, 205
position vector, 1, 520
positronium decay, 201
potential, 116
   anharmonic, 110
   Coulomb, 113, 222, 230
   dual, 279, 280
   effective, 218
   for central isotropic forces, 113
   for pendulum, 105
   inverse-square, 237
   isotropic harmonic oscillator, 220
   Lennard-Jones, 114
   non-central, 279
   non-uniqueness of, 99
   perturbed Coulomb, 243, 253
   scalar, 94, 98, 120, 123, 125
   spherically symmetric, 94, 216
   two-body, 526, 528
   vector, 117, 120, 121, 123, 125
   wells, 107
   Yukawa, 113, 224
potential energy, 6, 76, 94, 95, 102, 193
   gravitational, 367
   of a rigid body, 404, 405
power, 93
power law, 21
   use in dimensional analysis, 22
power radiated, 196
preacceleration, 59
precession, 244, 252, 285, 433–443, 447
   of relativistic orbit, 582
pressure, 354
Price, 159, 171
principal axes, 407, 431
Principia Mathematica, 5, 336, 529
products of inertia, 408
projectile, 157–172
   closest approach, 165–167
   'flat-Earth' approximation for, 159
   from moving platform, 167–168
   maximum height, 158

maximum range, 158–161
moving target, 161–163
on inclined plane, 159
with drag, 168–172
prolate body, 432
proper
   acceleration, 576
   length, 565
   time, 565
proper time interval
   invariance of, 568
Ptolemy and Ptolemaic system, 291, 318
pure force, 573

quality factor, 79, 80, 497

race between two cylinders, 429
radius of gyration, 427
raindrop, 379
Rayleigh, 23
Rechenberg, 200
reciprocity, 559
reduced mass, 142, 289
   and translational force, 289
relative mass, 16
relativistic Coulomb problem, 580
relativistic motion
   constant force, 574–576
   in a uniform **B** field, 579
   of a one-dimensional oscillator, 576
relativistic rocket, 584–586
relativity
   universality of, 561
relativity principle, 10, 528
   Einstein, 557
relaxation time, 71, 78
resonance, 77
Reynolds number, 26, 44, 54, 56, 169
rigid body, 399–453
   angular momentum of, 403–408
   angular velocity of, 402
   centre of mass, 400, 410
   degrees of freedom, 399
   energy, 404, 405
   equations of motion of, 401
   equilibrium of, 416
   Euler's equations for, 407
   free motion of, 431–434
   kinetic energy of, 403–405
   moment of inertia of, 403–415
   potential energy, 404, 405
   precession of, 433–443
   products of inertia, 408
   velocity of, 402
Rindler, 557
rocket, 128, 260, 379
   in free space, 380
   relativistic, 584–586
   take-off from a planet, 381–384
rod, falling, 426

rolling motion, 423–431
Romano, 171
rope, sliding, 390–393
rotating frame, 7, 8, 309, 316, 440, 543–545, 549
rotating wire and bead's motion on, 546
rotation curves for stellar velocities, 374
rotation of frame of reference, 537
rotational field, 99
rotational invariance
    and central forces, 541
    and conservation of angular momentum, 539
Rüchardt's experiment, 78, 80
run-away solutions, 58
Runge–Lenz vector, 264
Rydberg constant, 200

scalar potential, 94, 98, 120, 123, 125
    for electrostatic field, 99
    magnetic, 126
Schrödinger equation, 398, 587
second-order phase transition, 312
sliding
    on a circular wire, 180–183
    on a curved wire, 183–186
    on a rough sphere, 175–179
    on a smooth sphere, 174–175
sliding rod, 419–422
sliding rope, 390–393
slipping motion, 423–431
solenoidal field, 99, 120, 125
solid of revolution, 411
Solvay Congress, 152
Sommerfeld Puzzle, 583
Soodak, 553
space, 1, 9
    absolute, 527
    homogeneity of, 9, 525, 559
    isotropy of, 9, 539, 558–559
    Minkowski, 566
spherical polar coordinates, 94, 146
    nabla in, 101
spherically symmetric potential, 216
spin, 448
    backspin, 425
    topspin, 424
spin angular momentum, 329, 424, 436
    origin independence of, 520
spontaneous symmetry breaking
    in a mechanical system, 312, 489, 490
spool
    rolling, 430–431
    slipping, 430–431
stability
    of atoms, 201
stable equilibrium, 65, 66, 103, 104, 106, 108, 221, 309, 456
standard configuration, 527, 557
steady flow in a pipe, 26

steady-state solutions, 74
Stephenson, 510
Stieltjes, 358
Stokes's law, 26, 49
Stokes's theorem, 96, 117, 118
Sun's effect on Earth's tides, 534
surface tension, 24
suspension of particles, 47
symmetry
    additional, 276
    and conservation laws, 526, 540
    broken, 312
    geometric, 276
    hidden, 276
    rotational, 275
symmetry axis, 408, 431
symmetry breaking, 276, 310
    explicit, 312
    spontaneous, 312, 489, 490
synchrotron, 198

tautochrone problem, 68
Taylor, 27
Temple, 327
tennis ball and motion of, 452
tensor
    Jauch–Hill–Fradkin, 271
    transformation of, 538
terminal velocity, 45–49, 55
Thomson
    and ions, 196
    and isotopes, 196
Thomson's experiment, 195
three-body problem
    restricted, 336–352
    triangle solution, 332–336
three-tensor, 538
tidal force, 533
Tiersten, 553
Tikochinsky, 281
time, 1, 9
    absolute, 527
    homogeneity of, 9, 398
    isotropy of, 9
    Planck, 28
time dilation, 564, 569
top, 434–438
topspin, 448
torque, 144, 400
    of translational force, 511, 512, 515
    total, 291, 328
trajectory, 1, 30, 32, 33, 36, 38
    closed, 12, 189, 210–212, 316
    Coulomb potential, 255–259
    crossing a stream, 173
    for two-body Coulomb problem, 293–301
    from the Laplace vector or Hamilton vector, 267
    in **E** and **B** fields, 549
    in a Coulomb potential, 230–235

in a non-uniform **B** field, 201–208
in a perturbed Coulomb potential, 243–247
in a uniform **B** field, 187–190, 548
in an inverse-square force, 212–215
in an inverse-square potential, 237–243
in an oscillator potential, 235–236
in momentum space, 267, 275
in uniform **E** and **B** fields, 190–195
inverse form of, 42, 108, 219, 255–259, 394
obtained from orbit equation, 273
of a beam, 195, 196
of anisotropic harmonic oscillator, 210
of centre of mass, 315, 321
of isotropic harmonic oscillator, 209
of relativistic Coulomb problem, 580–583
of the relative vector, 290
open, 12, 189, 210–212
polar, 254–259
projectile, 158, 169–172
relativistic in a constant force, 574–576
relativistic in a uniform **B** field, 579
time-dependent, 254–259
transformations
   between Coulomb and oscillator problems, 280
transformed orbits, 279
transient, 74, 77, 89
translation of the reference frame, 7, 8, 518
translational force, 9, 289, 507, 531–533, 544–546
   torque exerted by, 511, 512, 515
translational invariance
   and conservation of momentum, 525
trochoid, 193
Trojan asteroids, 348
tunnel through a planet
   motion along, 374–376
turning points, 103, 104, 106, 107, 221, 223
two-body potential, 526
   and Galilean invariance, 528
two-body problem, 286
   separation of, 290, 314
   with inverse square force, 293–301
two-body system
   angular momentum, 292
   kinetic energy, 292
   momentum, 292

unbound motion, 33, 43, 64, 103, 106, 107, 189, 223, 455, 547
uncertainty
   in initial conditions, 11, 12, 497–502
underdamping, 69
uniform gravity plus friction, 44
unit
   of force, 10
   of length, 21
   of mass, 10, 21
   of time, 21
units, 19, 21
   absolute (natural), 28, 29

arbitrariness of, 10, 17
Planck, 28
'relativity principle' for, 20
universal constant V^2, 559
universal constant of gravitation, 13
universal speed, 557, 561
universality of relativity, 561
unstable equilibrium, 65, 103, 104, 106, 108, 309

V^2-Lorentz transformation, 557–570
Van Allen belts, 205
Van Vleck, 520
variable mass, 378–390
variable-mass oscillator, 385–390
vector
   Hamilton, 264
   Hermann–Bernoulli–Laplace, 264
   Laplace, 264
   potential, 120
   Runge–Lenz, 264
   transformation of, 538
vector field, 125
   equivalences for, 120
vector potential, 117, 121, 123, 125
   condition for, 118
   formula for, 118
   non-uniqueness of, 119
velocity, 1
   critical, 547
   in plane polar coordinates, 217
   of a rigid body, 402
   origin independence of, 520
   root-mean-square, 354
   terminal, 45, 47
   transformation, 542, 544
velocity transformation
   relativistic, 561
viscosity, 25, 26, 47, 48

wave equation
   form invariance of, 563, 568
weak damping, 71, 72
weak equivalence principle, 158, 532
wedge
   ball bouncing on, 503–505
wheel
   rolling, 423–429
   slipping, 423–429
work, 92, 99, 147
work–energy theorem, 42, 92, 147, 149, 174–176, 180, 181, 189
   for a rigid body, 404

Yukawa potential, 113, 224
   energy diagrams for, 226–228
   range of, 113

Zeeman effect, 550
zero mass, 562